D0423186

Friday Harbor Symposia

MARINE

BORING

AND FOULING ORGANISMS

Edited by Dixy Lee Ray

SEATTLE • UNIVERSITY OF WASHINGTON PRESS • 1959

The Friday Harbor Symposia are contributions from the Friday Harbor Laboratories of the University of Washington.

Library of Congress Catalog Card No.: 59-14772
Printed in the United States of America

Photos of participants by Richard A. Cloney; photo of Friday Harbor Laboratories and part-title photos by Norman Jensen

PREFACE

Among the many questions of keen scientific interest that are encompassed by the broad field of marine biology, none has greater economic significance than the problem of marine biological deterioration. Even casual association with the sea serves to illustrate the extensive damage caused by marine animals that burrow into piles, floats, wooden drydocks, and boats, and the deleterious effects of other organisms such as barnacles that foul ships' hulls. These animals have always hampered maritime activities, yet many aspects of their biology remain largely unelucidated.

For the scientist, especially for the biologist and biochemist, there are questions of fundamental importance to be examined. Do marine wood-borers actually feed upon the wood into which they burrow and which indubitably passes through their digestive system? If so, how do they handle so unpromising a material as wood to extract from it the carbohydrates necessary for energy and for growth? From where do they get their nitrogen supply? How do they find the wood that they come eventually to inhabit? What ecological conditions govern or modify their distribution, and what determines where a barnacle will grow? For the harbor engineer, the wood preserver, and all those charged with maintaining water front and marine structures, the questions are no less fundamental and no less difficult. The preventative and control measures now in use require still further study to establish their effectiveness in a wide variety of environmental situations. It is our firm belief that only through a more thorough knowledge of the basic biology of boring and fouling organisms can we hope for ultimate understanding and control of this complex problem.

To this end the University of Washington, through its Department of Zoology and with the aid of the Biology Branch of the Office of Naval Research, sponsored this symposium. It was held at the Friday Harbor Laboratories during the first week of September, 1957. Joining with the university and the Office of Naval Research were twenty cosponsoring industries and maritime agencies. Scientists from this country and abroad were invited to meet together with representatives from interested industries to examine critically the basic research approaches to the problem of marine biological deterioration.

The present volume constitutes the proceedings of this symposium. Although many months have passed since these papers were given orally, the value of the results and of the ideas here recorded has not diminished during the interval. For those who attended the conference, further investigations have already been stimulated; for the wider audience that can now be reached, these pages may convey something of the freshness and enthusiasm that characterized the many thought-provoking discussions.

It is with pleasure and gratitude that we express our sincere appreciation to all those who contributed to the success of this meeting. For the enthusiastic and cooperative support of President Henry Schmitz and Dean Lloyd S. Woodburne of the University of Washington and Dr. Sidney R. Galler of the Office of Naval Research, we are deeply indebted. The generosity of our cosponsoring industries is gratefully acknowledged, and their continuing interest in marine biological problems is a source of pleasure and satisfaction. It has been a privilege to help inaugurate the Friday Harbor Symposia, but the real credit, and my heartfelt thanks, go to my deserving colleagues in the Department of Zoology and to our loyal and hardworking graduate students.

Dixy Lee Ray

Seattle, Washington
May, 1959

This symposium was sponsored by the University of Washington and the Office of Naval Research, with the support of the following agencies:

Alaska Packers Association
Bernuth, Lembcke Company, Incorporated
Canada Creosoting Company Limited
Crown Zellerbach Foundation
General Mills, Incorporated
General Petroleum Corporation
Humble Oil and Refining Company
Koppers Company, Incorporated
Newport News Shipbuilding Company Foundation
Northwest Pulp and Paper Association
Port of Bellingham
Port of Seattle
Puget Sound Foundation
Puget Sound Freight Lines
Reynolds Metals Company
Shell Oil Company
Skinner Corporation
Timber Preservers Limited
Todd Shipyards Corporation
West Coast Wood Preserving Company

Friday Harbor Laboratories of the University of Washington
San Juan Island, Washington

CONTENTS

CELLULASES

ECONOMIC IMPLICATIONS AND EVALUATION

Marine Boring and Fouling Organisms

WELCOMING ADDRESSES

On Behalf of the University of Washington

Henry Schmitz
President, University of Washington

I am delighted to be able, on behalf of the University of Washington, to extend to this distinguished group of scientists a most warm and cordial welcome. We of the University of Washington are very proud of the fact that the participants in this symposium represent not only every section of the United States but a number of foreign countries as well. I hope that your stay at the Friday Harbor Laboratories will be not only interesting and rewarding but pleasant as well.

As president of the University of Washington, I wish publicly to acknowledge the great debt the university owes Dr. Dixy Lee Ray for all the planning, work, and imagination she contributed to assure the success of this symposium. Dr. Ray's enthusiasm, as those of you who have had the good fortune to know her know, is contagious, her judgment flawless, and her energy boundless. I am sure that as this symposium proceeds all of you will share with me a feeling of indebtedness to Dr. Dixy Lee Ray for all she has done to make this symposium a notable scientific occasion.

I also want to take this opportunity to express the grateful thanks of the University of Washington to the Office of Naval Research and to a large number of private companies who have not only supported the symposium generously in a financial way but who have sent representatives to participate in it. I am certain that all of you will understand when I say that without the help and support of these agencies it would not have been possible to hold this symposium.

I am happy, too, that all of you are here so that you may see for yourselves what, at least we at the University of Washington believe to be, are the matchless opportunities for research in marine biology. We realize, of course, that our physical facilities do not yet match the research opportunities, but I am confident that it will be only a matter of a few years before our physical facilities will meet the needs of even the most discriminating researcher.

I have a personal reason, too, for being happy to be here. For many years I engaged in a research program dealing with the physiology of the wood-destroying fungi and with the preservative treatment of wood. I know of course that the process of wood destruction by Limnoria and by shipworms is quite unlike that of the destruction of wood by wood-destroying fungi, but the two processes do nevertheless have much in common. It will be a genuine pleasure to lay aside for a day or two routine administrative chores and to revel in the discussions on marine boring and fouling animals.

May I again extend, not only on behalf of the University of Washington, but also on behalf of Dr. Ray, a most cordial welcome to all of you, and may I express the hope that this may be only your first visit to the Friday Harbor Laboratories. In the years ahead I hope that many of you will return either for study or to participate in another symposium arranged by Dr. Ray.

Thank you all for helping to make this symposium an outstanding success.

On Behalf of the Office of Naval Research

Sidney R. Galler

Head, Biology Branch, Office of Naval Research

Historically the approach to the problem of controlling marine boring and fouling organisms has tended to be largely empirical, a trial and error evaluation of chemical agents, techniques, and structural materials to determine their efficacy in protecting marine structures against biological deterioration. This is not to suggest that this approach has been entirely unrewarding since the use of one of our most effective marine preservatives, creosote, evolved in this fashion. On the other hand, empiricism has served as a handicap to full utilization of the research and development resources existing in the United States and abroad for improving the protection of marine structures.

It is encouraging, therefore, to observe that within recent years increasing attention has been focused on another time-honored approach which has been followed successfully in such diverse fields as preventive and therapeutic medicine and insect control. This principle in its simplest form consists of a recognition that a thorough knowledge of the behavior and functions of "normal" organisms in relation to their environments is a requisite for developing methods and materials for controlling their activities. This principle serves as the basis for the present symposium. For the first time, a group of specialists representing a broad spectrum of the biological, chemical, physical, and engineering sciences has come together to assess the "state of the art" in the field of the protection of marine structures, with particular reference to the fundamental behavior of those animals and plants that cause marine deterioration.

It is particularly gratifying to note that the participants in this meeting include not only university scientists but also the technical representatives of the producers and consumers of marine protective agents. The second group will not only profit substantially from the information presented by the scientists engaged in basic research, but in turn will contribute materially to the objectives of this meeting by inviting the attention of the basic researchers to the applied problems involved in protecting marine structures against the ravages of boring and fouling pests. This symposium will lead to a fuller and more sympathetic understanding of the problems and interests of each group represented with the result that a common community of interest in this field will be firmly established.

On behalf of the Chief of Naval Research of the United States Department of the Navy, I join President Schmitz of our host institution, the University of Washington, and the industrial cosponsors who helped make this meeting possible, in wishing you the fullest success in these important deliberations.

On Behalf of the Friday Harbor Laboratories and the Department of Zoology of the University of Washington

Arthur W. Martin

Executive Officer, Department of Zoology of the University of Washington

Dr. Henry Schmitz has greeted this distinguished company for the state and the University of Washington. In his felicitous words he has made clear his own long-continued interest in one of the central problems before the symposium, the preservation of wood against biological deterioration, and his real concern for a basically oriented attack on the problem.

I have now the privilege of greeting you for the Friday Harbor Laboratories of the University of Washington. In the quiet and beautiful surroundings of this campus we expect you will accomplish much useful work. One of my own interests in this symposium is that we here attain a level of excellence that will lead to the establishment of a series of symposia, for we believe this location can hardly be surpassed for its combination of working comfort and opportunity for quiet contemplation and uninterrupted discussion.

It is also my pleasure to greet you for the Department of Zoology. This symposium has resulted from the dreams of the entire department faculty. The painstaking work of planning, of organization, and of fund-raising, however, has been the contribution of Dr. Dixy Lee Ray. The Department is proud of her work and believes that the success of this symposium and of the volume resulting from these presentations will be a lasting tribute to her devotion.

The symposium has been made possible by the support of the Office of Naval Research through Dr. Sidney Galler, and by generous contributions from industrial organizations and public agencies. We are happy that a number of our sponsors are in attendance, and we here express our gratitude for both the gracious manner and the generosity of their support.

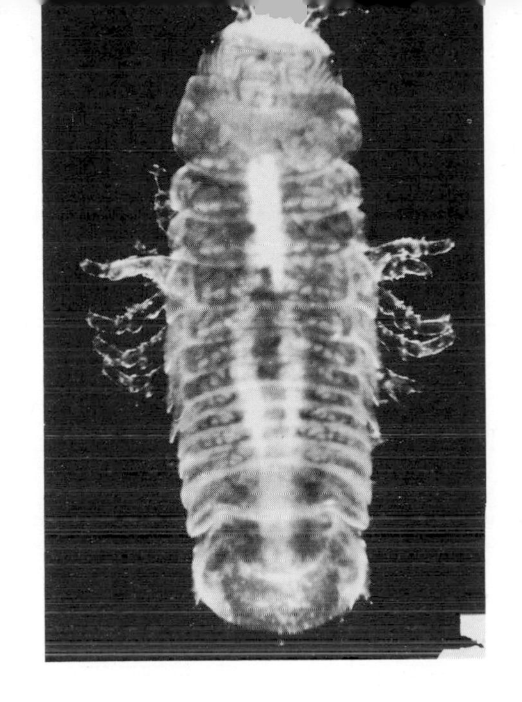

Limnoria

INTRODUCTION

Arthur W. Martin

The topic of marine boring and fouling organisms has many facets. It would not be possible to touch upon them all during the course of our few days here, but through the presentation of papers, informal discussions, and the ready interchange of ideas we hope to achieve a synthesis of present knowledge in this field and to evaluate the prospects for future progress.

The program has been organized to deal first with those animals that are prime agents of biological deterioration in the sea: Limnoria, the Teredinicns, and the barnacles. Their general biology will be our first concern for only on the basis of a thorough knowledge of their morphology, function, habits, distribution, and interrelationships can we build that framework of understanding upon which control of these important animals ultimately depends.

It is my privilege now to open the first session of scientific papers. We shall direct our attention to the biology of that very destructive marine wood-borer commonly known as the gribble, crustaceans of the Isopod genus Limnoria.

THE IDENTIFICATION AND DISTRIBUTION OF THE SPECIES OF LIMNORIA*

Robert J. Menzies

Introduction

The object of this paper is to provide a discussion of morphological structures utilized in the identification of Limnoria and Paralimnoria. World maps are provided in which the distribution of the various species is given. Taxonomy and detailed descriptions have been purposely avoided, and readers are referred to a monograph by Menzies (1957) covering these aspects.

Specimens should be adequately preserved (Menzies, 1957). A powerful stereoscopic microscope with a reflected light source as well as a standard microscope are needed for proper idendification.

Identification

The gross morphology of Limnoria is shown in figures 1-3. The genus Limnoria may be distinguished from the genus Paralimnoria through an examination of the uropods. Both rami of the uropods of Paralimnoria bear terminal, sharply pointed claws, whereas only the outer ramus (exopodite) of the uropods of Limnoria has a terminal claw.

---

*Contribution from the Lamont Geological Observatory (Columbia University) No. 313, Biology Program No. 34. This work was supported by a contract between the Biology Branch, Office of Naval Research, Department of the Navy, Nonr (266)-41, and the Lamont Geological Observatory of Columbia University, and a gift from the Rockefeller Foundation RF 54087 for research on marine biology.

The antennae

The first antennae (fig. 3, M) of all species of Limnoria are about the same. The second antennae (fig. 3, N) have a 5-jointed peduncle and a multiarticulate flagellum. The number of articles to the flagellum is useful in separating groups of species, e.g.:

With 3 or 4 Articles	With 5 Articles
1. *L. (L.) lignorum*	11. *L. (L.) quadripunctata*
2. *L. (L.) pfefferi*	12. *L. (L.) tripunctata*
3. *L. (L.) platycauda*	13. *L. (L.) sasboensis*
4. *L. (L.) simulata*	14. *L. (P.) algarum*
5. *L. (L.) japonica*	15. *L. (L?) septima*
6. *L. (L.) unicornis*	16. *L. (L.) foveolata*
7. *L. (L.) multipunctata*	17. *L. (L.) sublittorale*
8. *L. (L.) insulae*	18. *L. (P.) segnis*
9. *L. (P.) antarctica*	19. *L. (P.) nonsegnis*
10. *L. (P.) rugossima*	20. *L. (P.) stephenseni*

The mandibles

The mandibles (fig. 2, G-J) of the subgenera Limnoria and Phycolimnoria of the genus Limnoria (*sensu lato*) are distinctive. Those of Limnoria, the wood borers, have a rasp and file arrangement on the incisive parts of the mandibles; those of Phycolimnoria lack a rasp and file. The laciniate seta of the left mandible is distinctive for some species. Most species of Limnoria have a triarticulate mandibular palp; however, the species *L. (P.) segnoides* has none. The palp of *L. (L.) unicornis* has one article, and that of *L. (P.) segnis* has two articles.

The maxilliped (fig. 3, O)

The length and shape of the maxillipedal epipod provides a useful characteristic. It is apically pointed and short in most species but is straplike in others.

The fifth somite of the pleon

In each species of Limnoria the middorsal surface of the fifth somite of the pleon has a typical ornamentation. Different ones are shown in the key.

The pleotelson

The pleotelson shows the maximum number of unique characteristics:

1. *Tuberculations.* The arrangement of carinae and tuberculations has been used in the selection of a name for the species. Thus, *tripunctata* has three tubercles; *quadripunctata,* four; *platycauda,* none; *multipunctata,* many; *unicornis,* one; etc.

2. *Setae.* Each species appears to have its own arrangement of setae along the margin of the pleotelson. In *lignorum, nonsegnis, segnoides, segnis, antarctica, sublittorale, septima,* and *quadripunctata* there is a dorsally directed fringe of spike-like setae (Menzies, 1956). Several different marginal setal arrangements are shown in the key.

3. *Marginal tubercles.* These are present in several species and lacking in others.

Geographical Distribution

Paralimnoria

The distribution of the single known species of Paralimnoria is shown on map 1. It is evident that the species shows a pantropical (cosmopolitan) distribution.

Limnoria

Limnoria (Limnoria), wood borers, xylophagous. Fifteen species of the subgenus Limnoria are known. Their distribution is shown on maps 2 and 3. None is known from the Arctic or Antarctic. Six species are known from the Atlantic and Caribbean, whereas ten are known from the Pacific Ocean. Three are known from the Atlantic and Pacific; these are *L. (L.) lignorum* (Rathke), *L. (L.) quadripunctata* Holthuis, and *L. (L.) tripunctata* Menzies. The last is the only species known to inhabit all three oceans. Nine species have a tropical distribution; four have a temperate-subtropical distribution, and only one has a boreal distribution. The species *L. (L.) quadripunctata* appears to be stenothermic, occurring everywhere in the world where the sea water temperature averages between 11.4°C and 16.2°C seasonally for at least five successive months of the year (Menzies, 1957, p. 160). The species *L. (L.) tripunctata* Menzies appears to be eurythermic, inhabiting both temperate and tropical regions of the oceans.

Four species show a distribution in water below the littoral zone. These are *L. (L.) japonica* Richardson from 163 fathoms off Suduz Mizaki Light, Japan; *L. (L.) septima* Barnard from 185 to 250 fathoms off the Andaman Islands, Indian Ocean; *L. (L.) foveolata* Menzies from 25 fathoms off the Kai Islands in the Banda Sea; and *L. (L.) sublittorale* Menzies from 60 fathoms off New South Wales, Australia.

Phycolimnoria, kelp borers, phycophagous. Seven species are known to belong to this subgenus. Their distribution is shown on map 4. Only one is known from the Atlantic and Indian oceans. Six occur in the Pacific. None is known from the Arctic or Antarctic. Only one tropical species is known. Four occur in the temperate-subtropics and three in boreal regions. *L. (P.) stephenseni* Menzies is almost circumantarctic. None are known from the sublittoral zone.

LITERATURE CITED

Menzies, R. J. 1957. The Marine Borer Family Limnoridae (Crustacea, Isopoda). Bull. Mar. Sci. Gulf and Caribbean, 7: 101-200.

-------, and G. Becker. 1957. Holzzerstörende Limnoria-Arten (Crustacea, Isopoda) aus dem Mittelmeer mit Neubeschreibung von *L. carinata.* Z. angew. Zool., 44: 85-92.

1. Both rami of uropods similar, each with apical claw
Paralimnoria andrewsi (Calman)

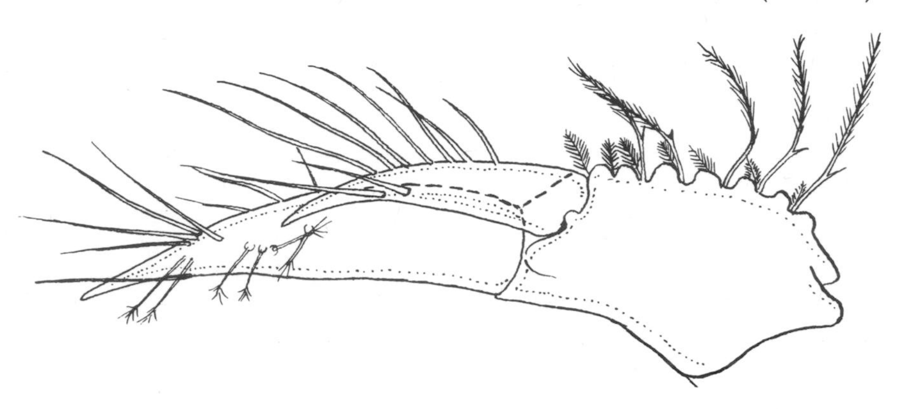

1. Uropodal exopod small and clawlike; endopod long and lacking apical claw *Limnoria* (sens. lat.)
2

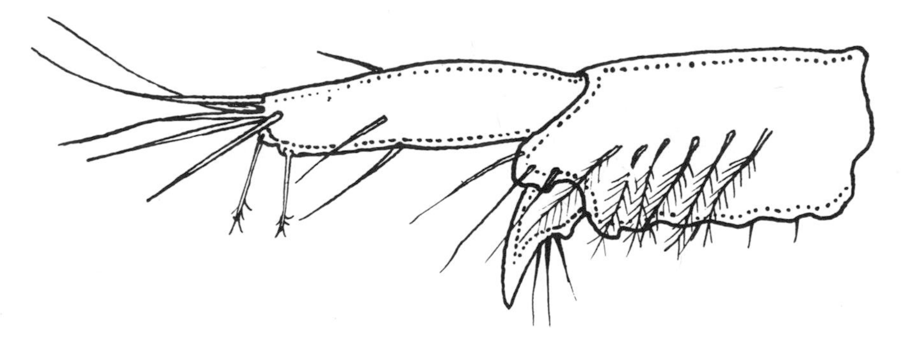

2. Mandibles with rasp and file incisive processes
Limnoria sub. gen.
3

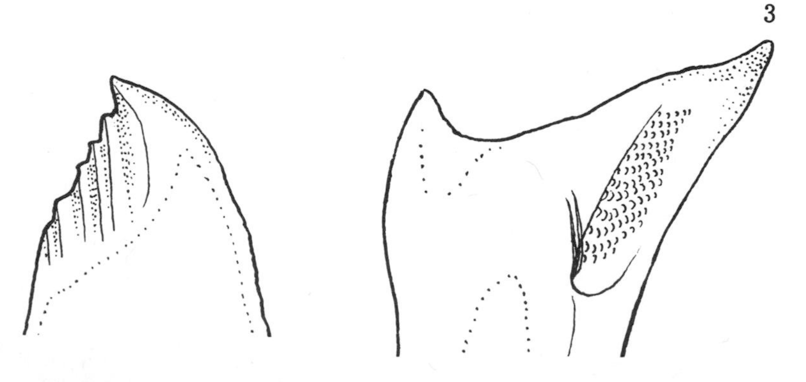

2. Mandibles without rasp and file incisive processes *Phycolimnoria* sub. gen. 17

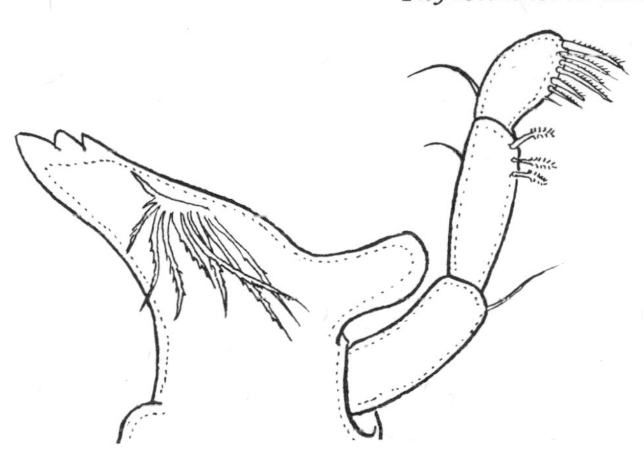

3. Flagellum of second antenna with 3 or 4 articles 4

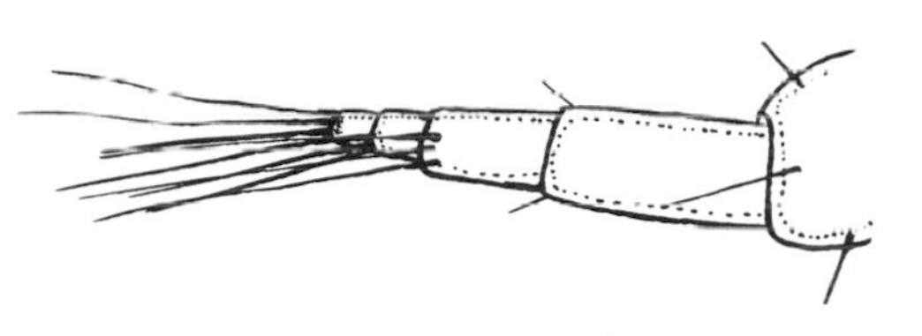

3. Flagellum of second antenna with 5 articles 12

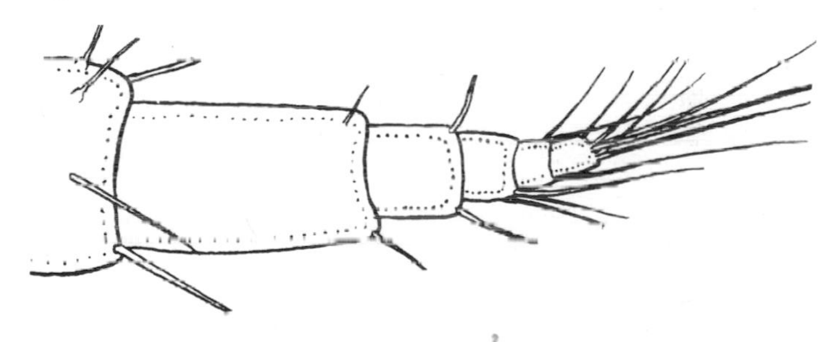

4. Mandibular palp with 1 article *L. (L.) unicornis* Menzies

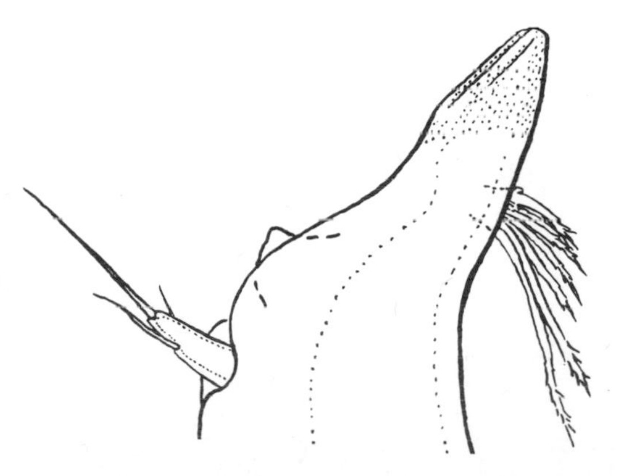

4. Mandibular palp with 3 articles 5

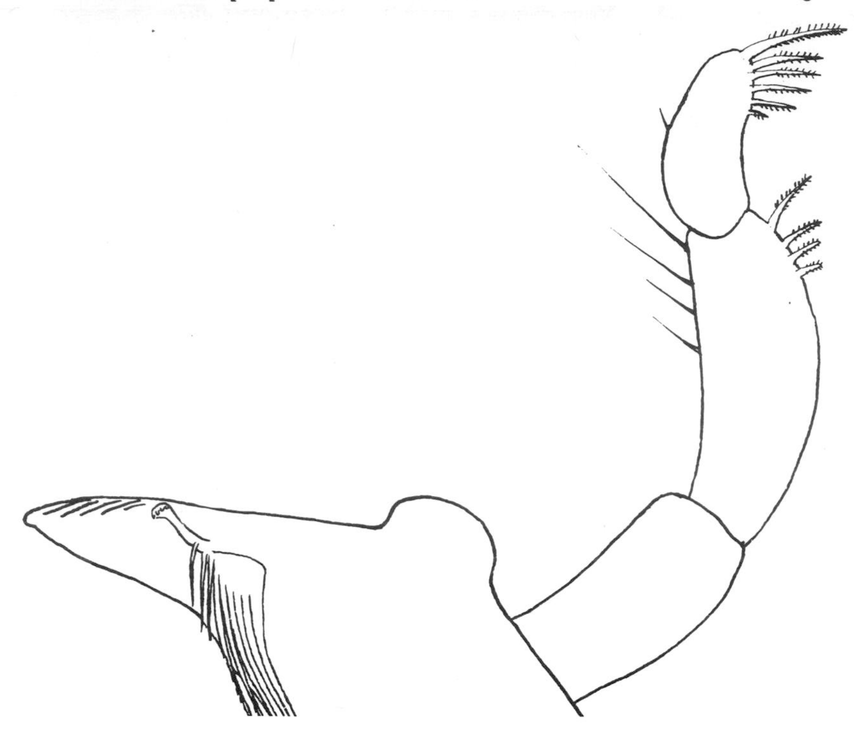

5. Maxillipedal epipod straplike and reaching beyond juncture of palp with endite 6

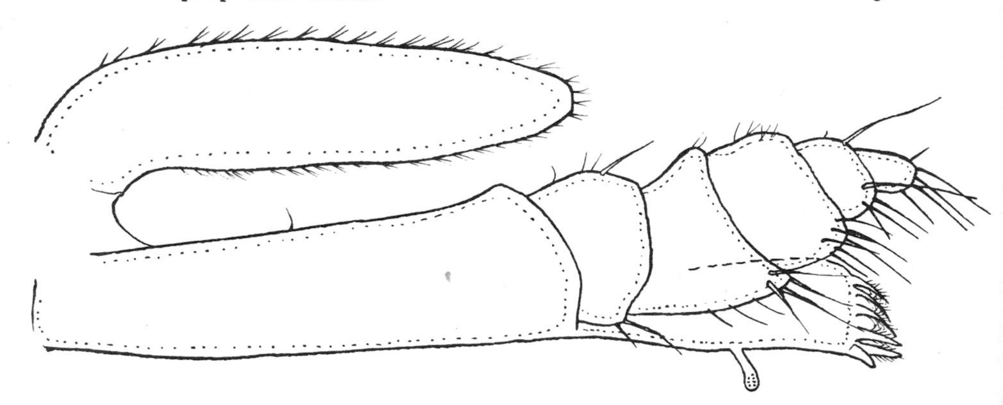

5. Maxillipedal epipod triangulate and not reaching to juncture of palp with endite 9

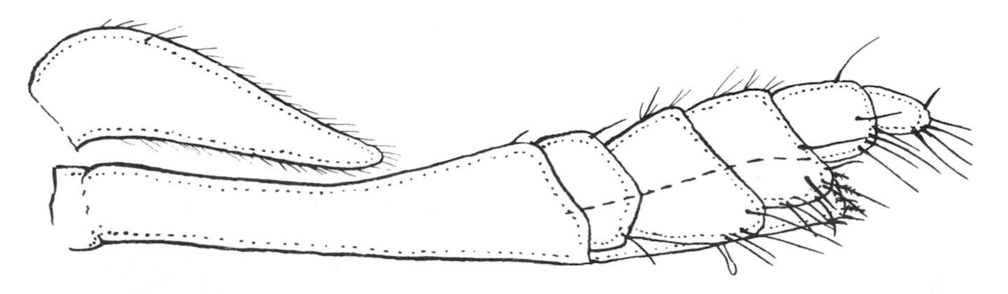

6. Margin of pleotelson with dorsally directed fringe of spike-like bristles *L. (L.) lignorum* (Rathke)

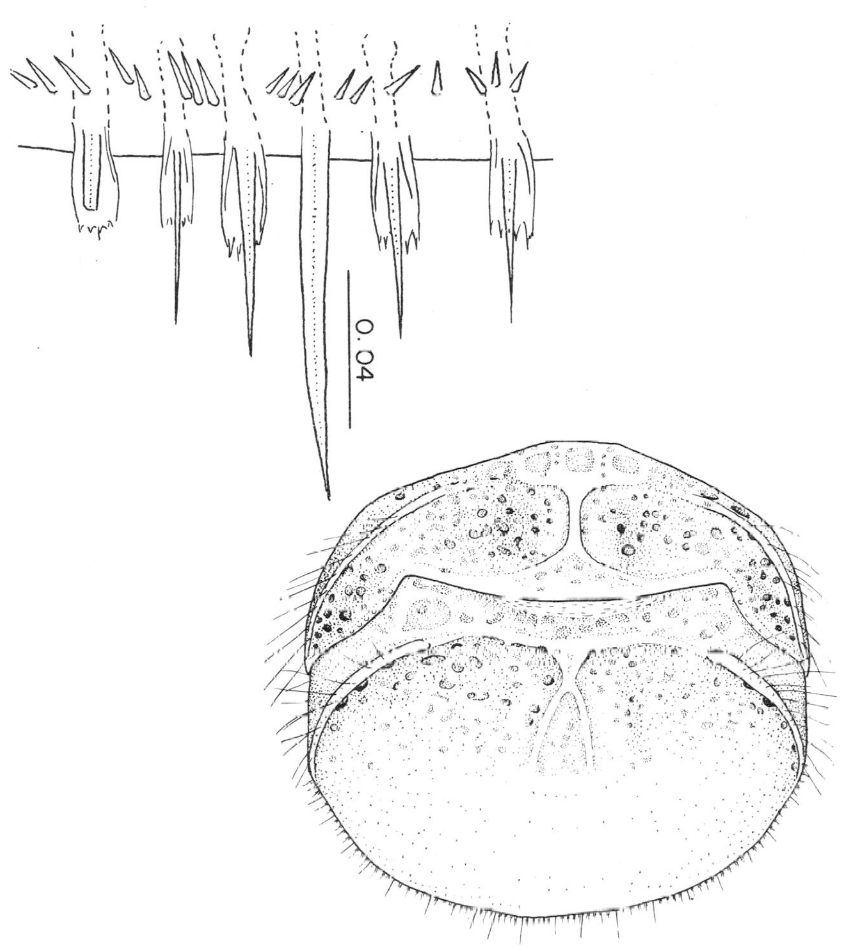

6. Margin of pleotelson without fringe of dorsally directed spikelike bristles 7

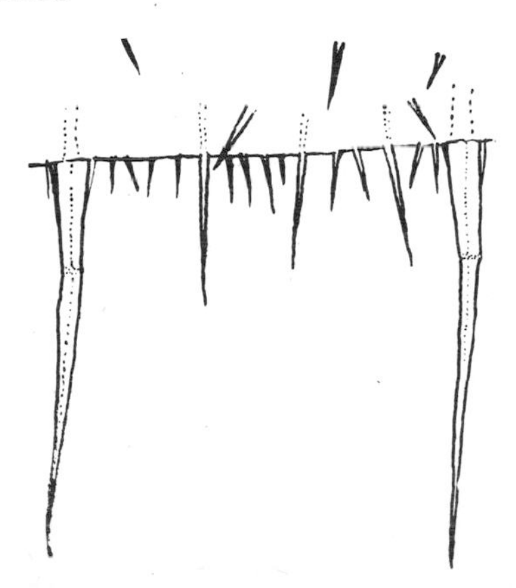

7. Dorsum of margin of pleotelson provided with tubercles
8

7. Dorsum of margin of pleotelson lacking tubercles
L. (L.) pfefferi Calman

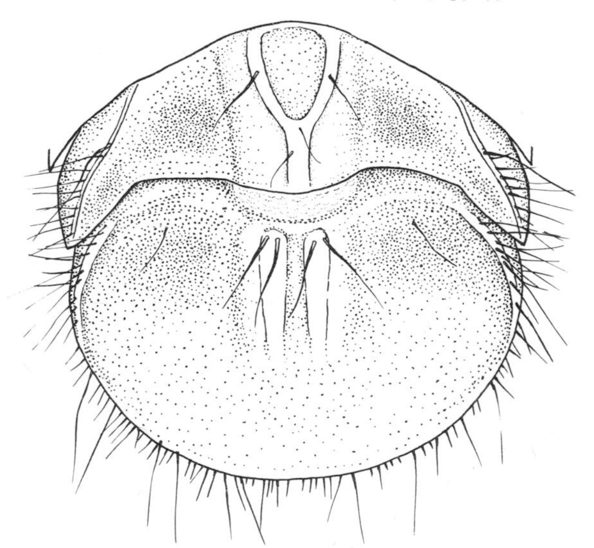

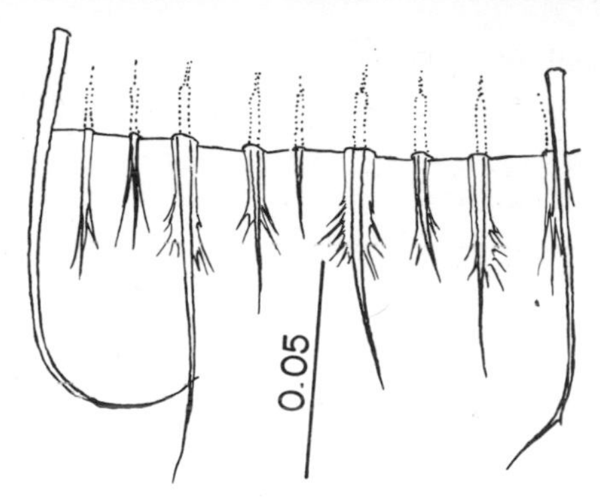

8. Dorsum of pleotelson lacking carinae or tubercles, except at margin *L. (L.) insulae* Menzies

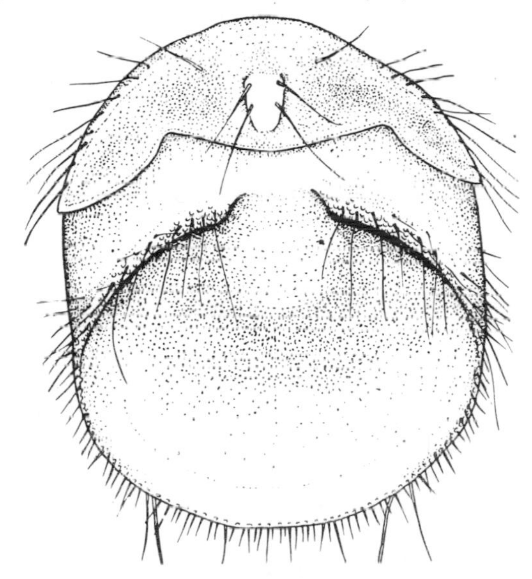

8. Dorsum of pleotelson with 8-10 tubercles and a single carina *L. (L.) multipunctata* Menzies

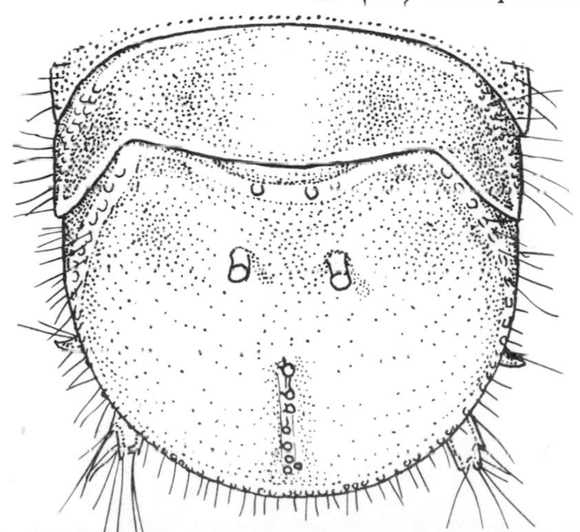

9. Dorsum of pleotelson without tubercles
L. (L.) platycauda Menzies

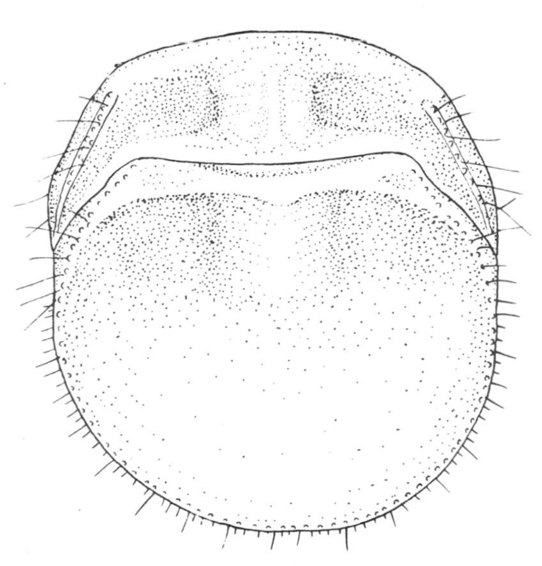

9. Dorsum of pleotelson with tubercles 10

10. Dorsum of pleotelson with 2 tubercles; carinae lacking on fifth somite *L. (L.) simulata* Menzies

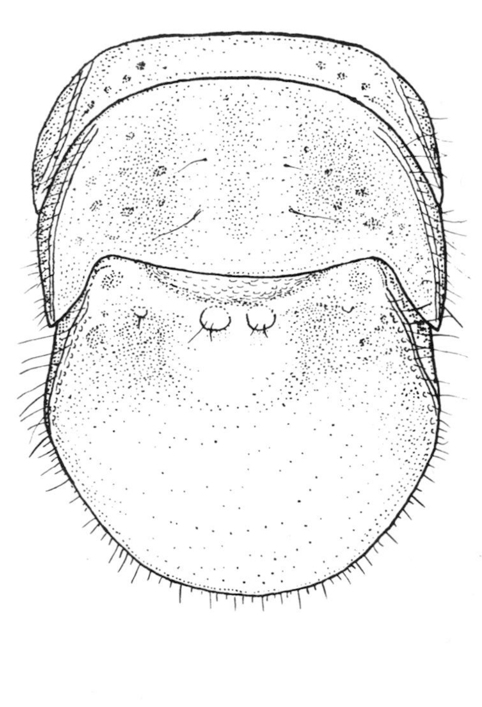

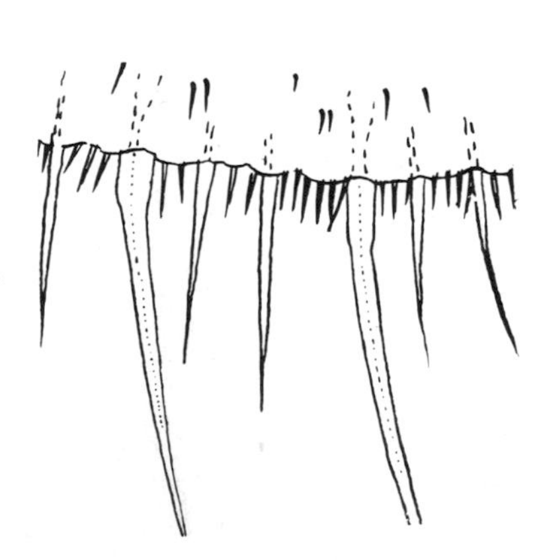

10. Dorsum of pleotelson with 3 or more tubercles; fifth somite of pleon carinate 11

11. Dorsum with 3 tubercles; fifth somite of pleon with 2 tubercles *L. (L.) japonica* Richardson

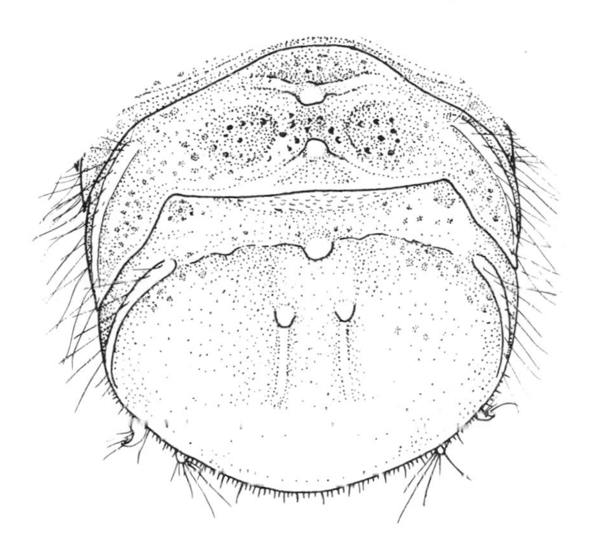

11. Dorsum with 6 tubercles; fifth somite pleon with yolk-shaped carina *L. (L.) carinata* Menzies and Becker

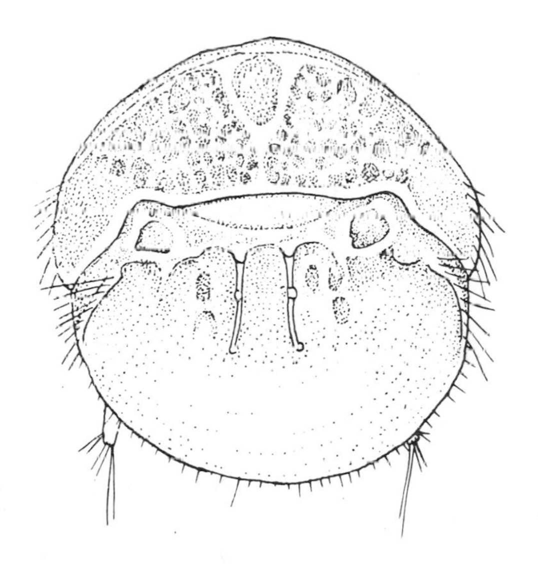

12. Posterior margin pleotelson tuberculate 13

12. Posterior margin pleotelson not tuberculate 14

13. Pleotelson with 3 tubercles and carinae besides marginal tubercles *L. (L.) tripunctata* Menzies

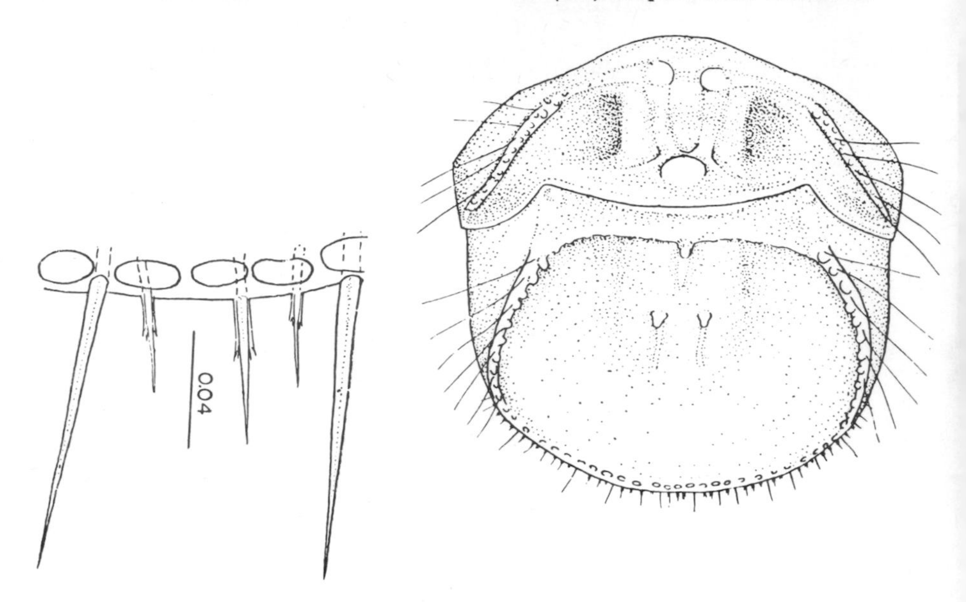

13. Pleotelson with 2-4 tubercles besides marginal tubercles *L. (L.) saseboensis* Menzies

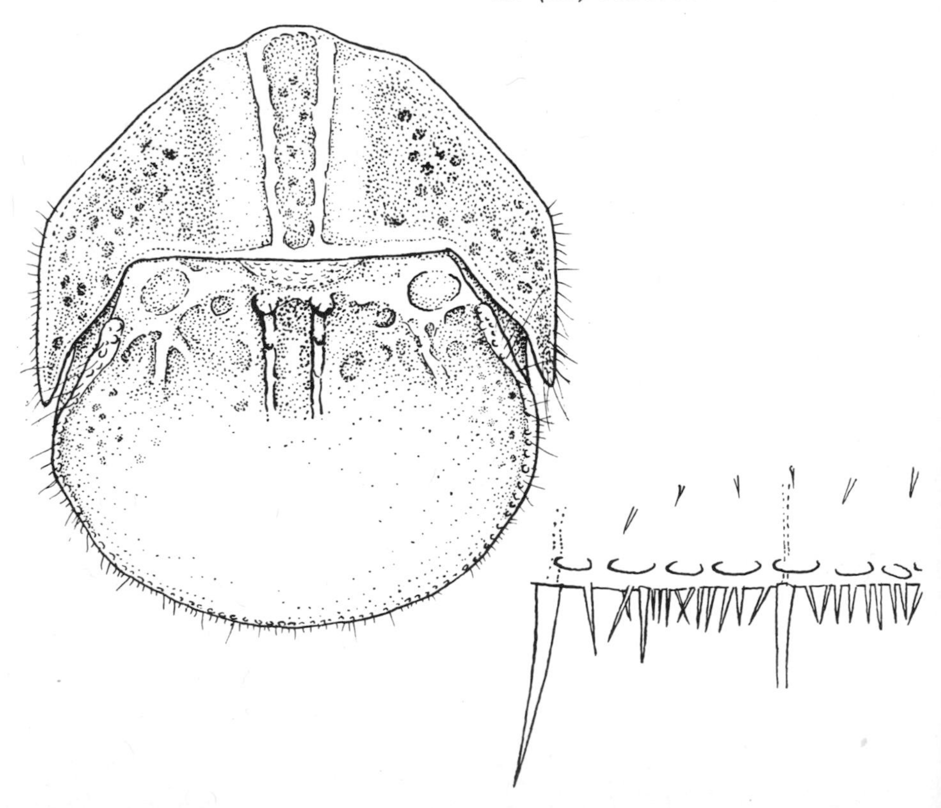

14. Posterior margin pleotelson without dorsally directed fringe of spikelike setae *L. (L.) foveolata* Menzies

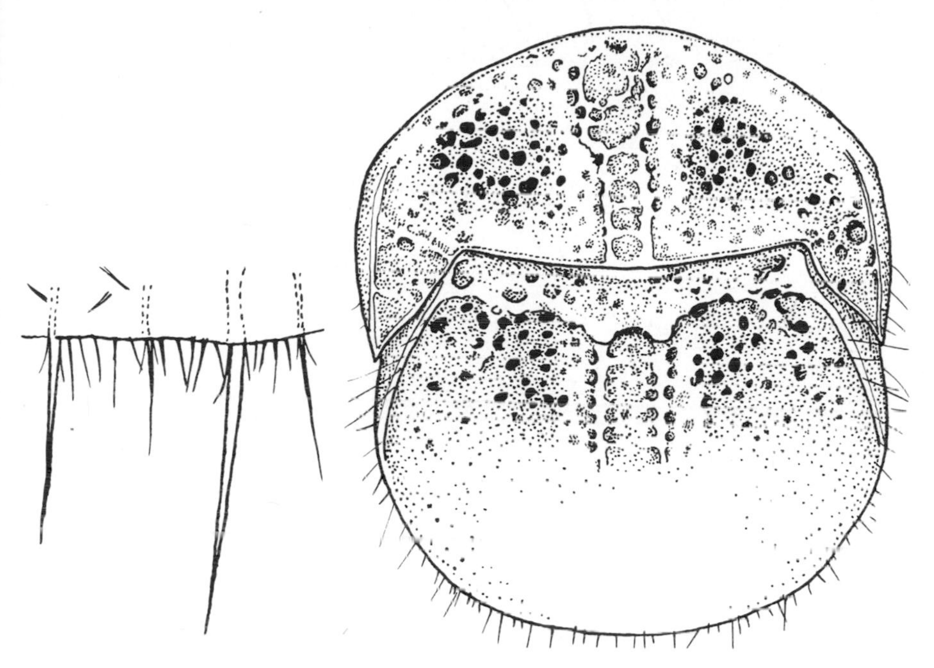

14. Posterior margin of pleotelson with dorsally directed fringe of spikelike setae 15

15. Carinae of fifth somite of pleon consisting of subparallel ridges *L. (L.) sublittorale* Menzies

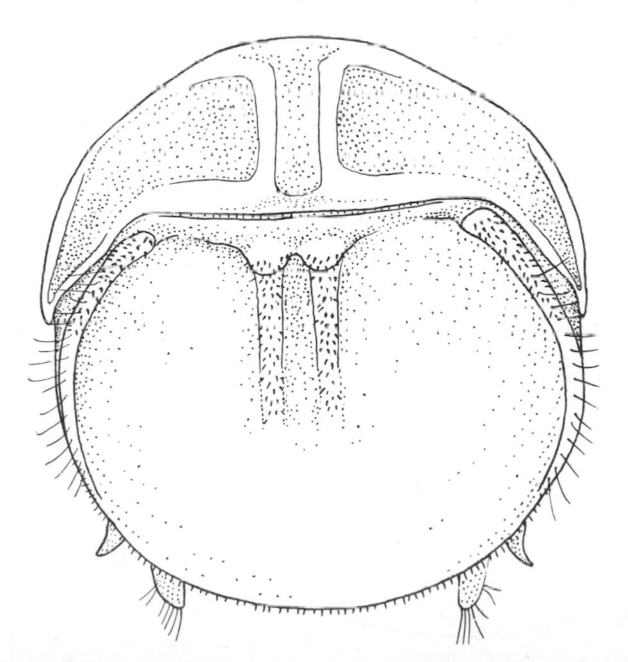

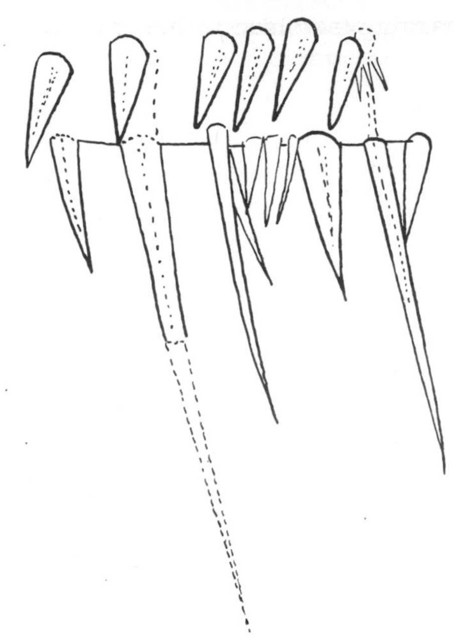

15. Carinae of fifth somite of pleon not subparallel 16

16. Carinae not joined at midline *L. (L.) septima* Barnard

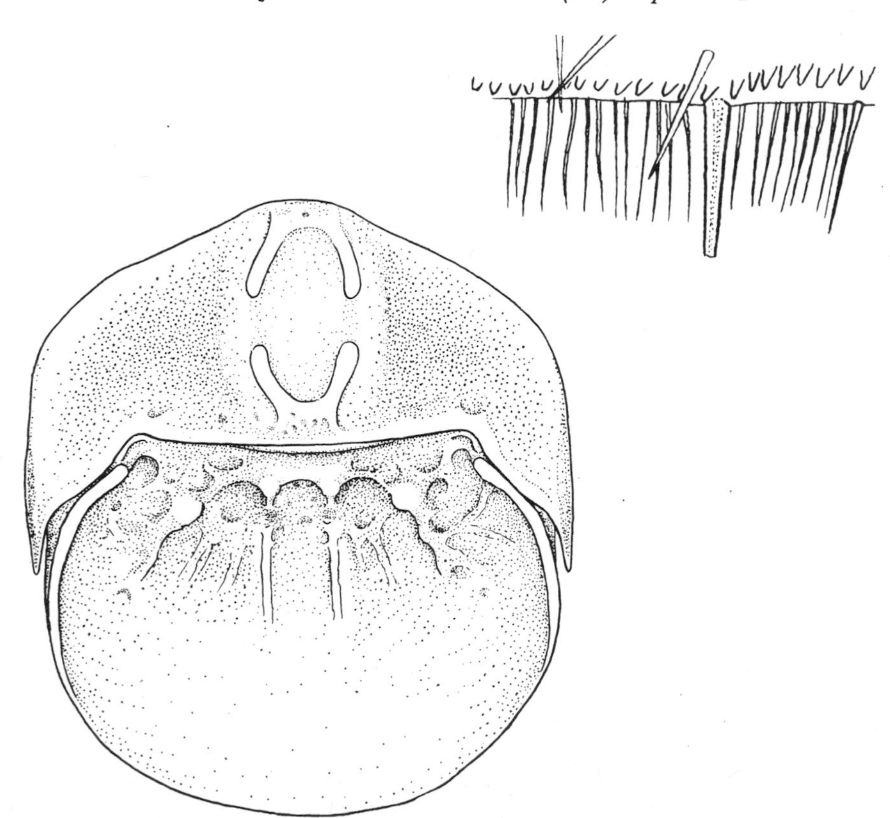

16. Carinae X-shaped *L. (L.) quadripunctata* Holthuis

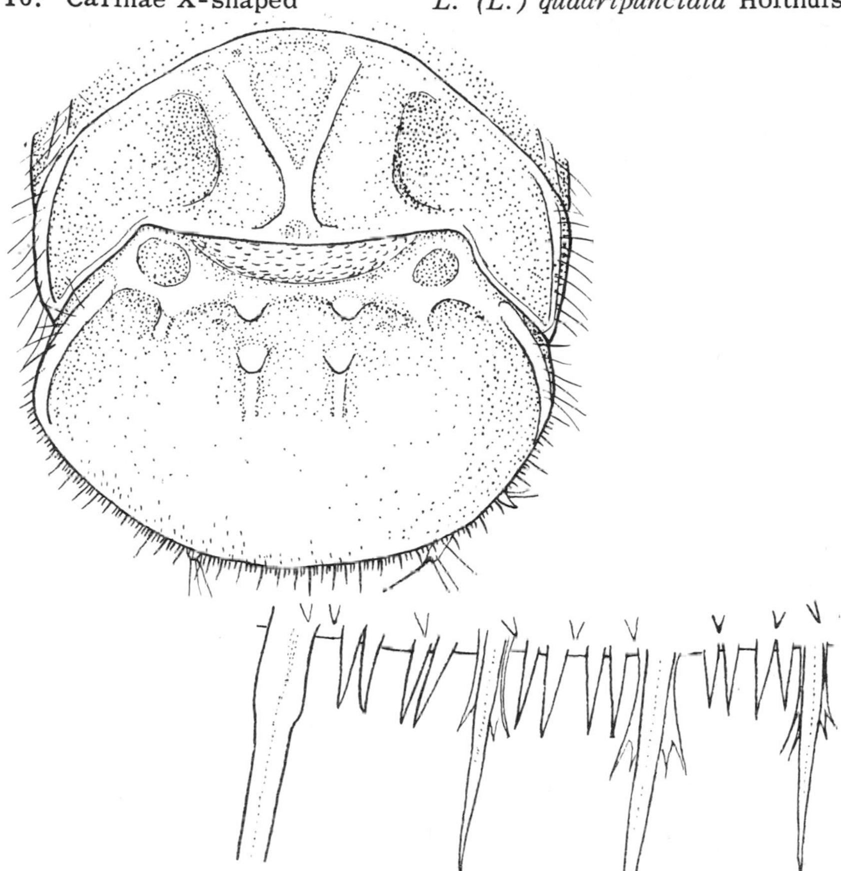

17. Mandibular palp with fewer than 3 articles 18
17. Mandibular palp with 3 articles 19
18. Mandibular palp lacking *L. (P.) segnoides* Menzies

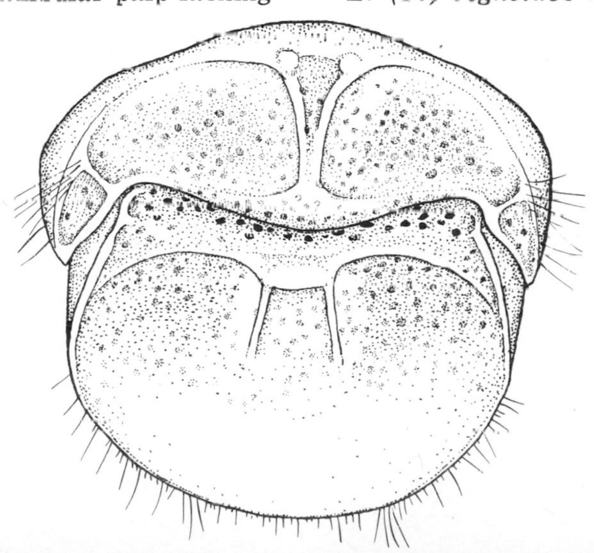

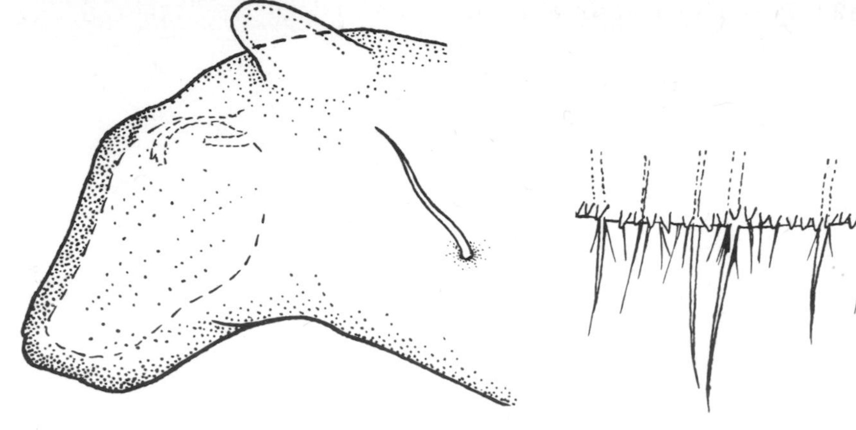

18. Mandibular palp with 2 articles

L. (P.) segnis (Chilton)

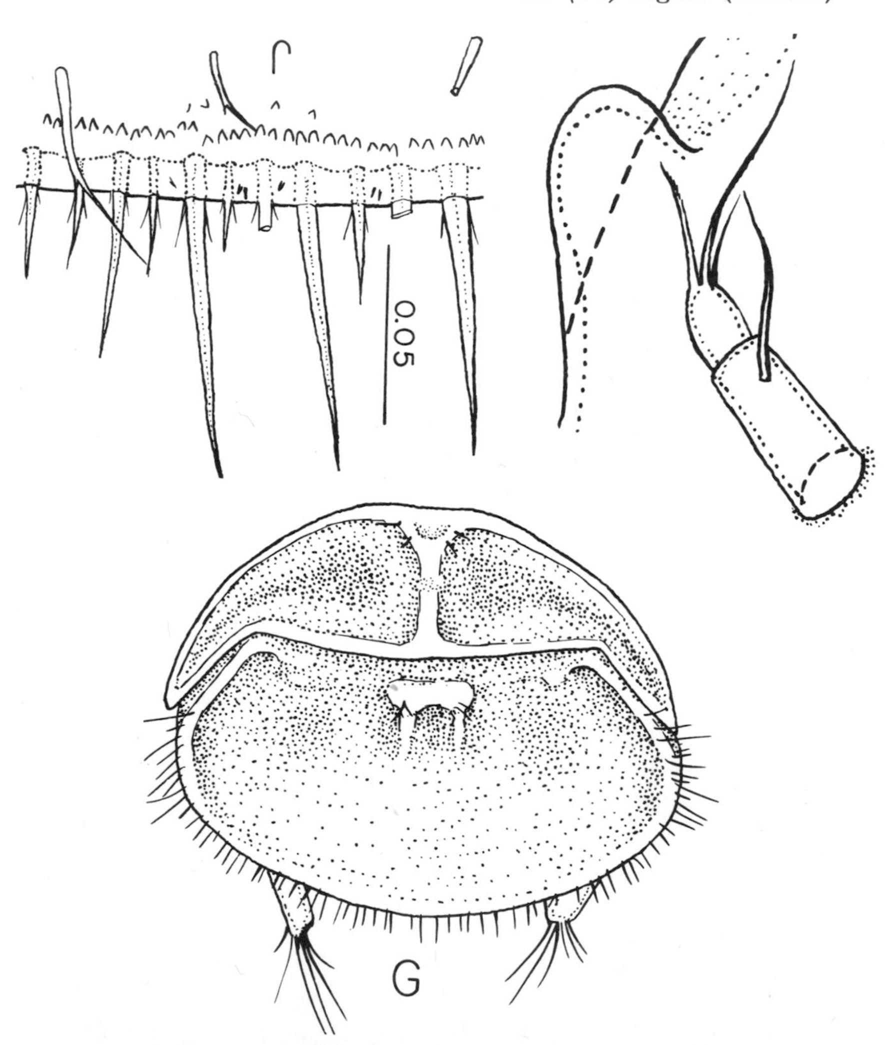

19. Posterior margin of pleotelson with dorsally directed fringe of spikelike bristles; dorsum of pleotelson carinate 20

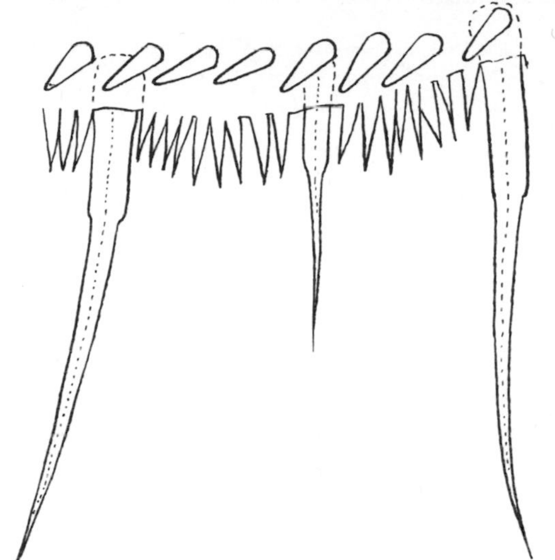

19. Posterior margin of pleotelson without dorsally directed fringe of spikelike bristles; pleotelson not carinate, smooth *L. (P.) stephenseni* Menzies

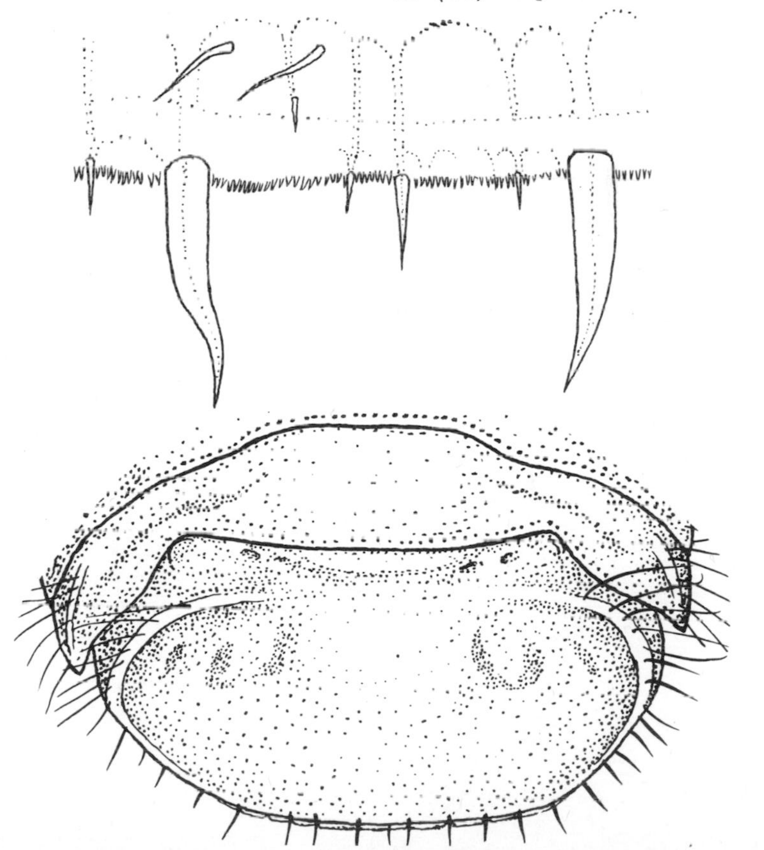

20. Epipod of maxilliped straplike and reaching articulation of palp with endite 21

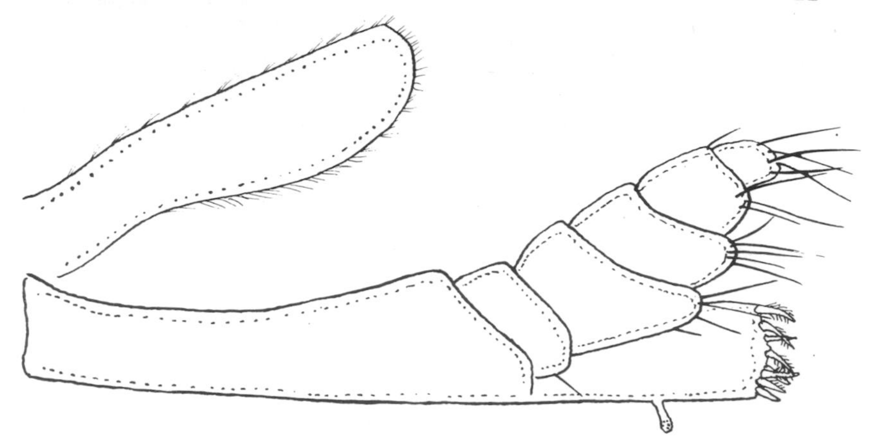

20. Epipod of maxilliped not straplike and not reaching articulation of palp with endite 22

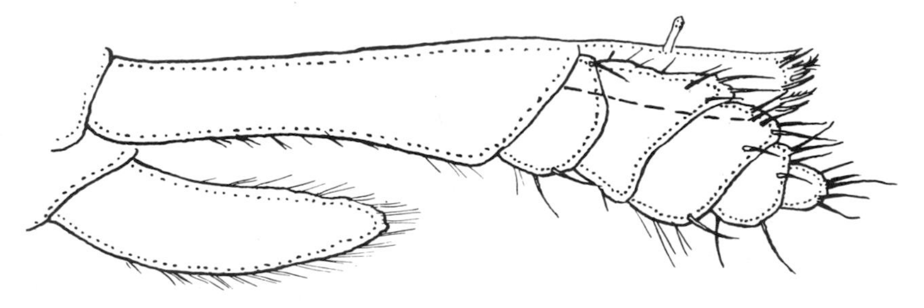

21. Fifth somite of pleon with midlongitudinal carina
L. (P.) nonsegnis Menzies

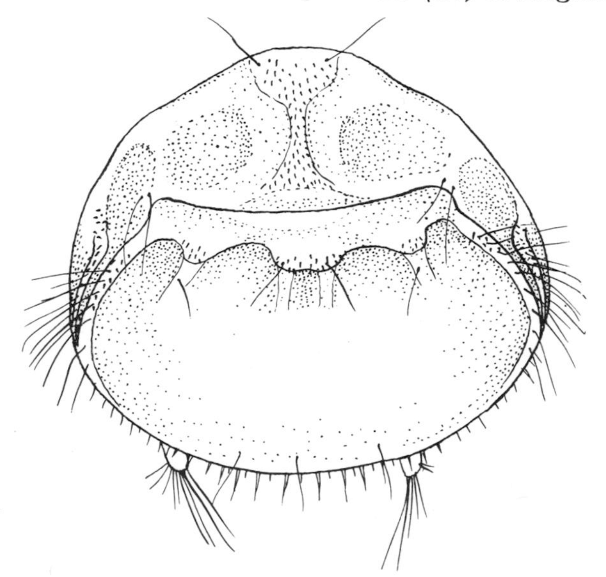

21. Fifth somite of pleon with a pair of midlongitudinal carinae
L. (P.) antarctica (Pfeffer)

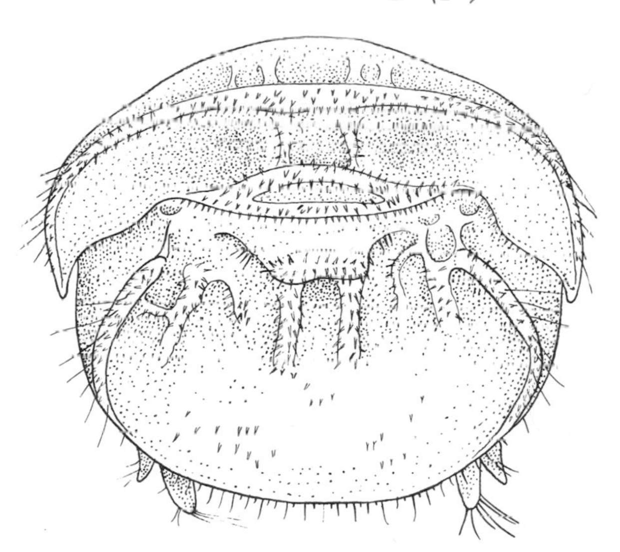

22. Major carinae of pleotelson subparallel

L. (P.) algarum Menzies

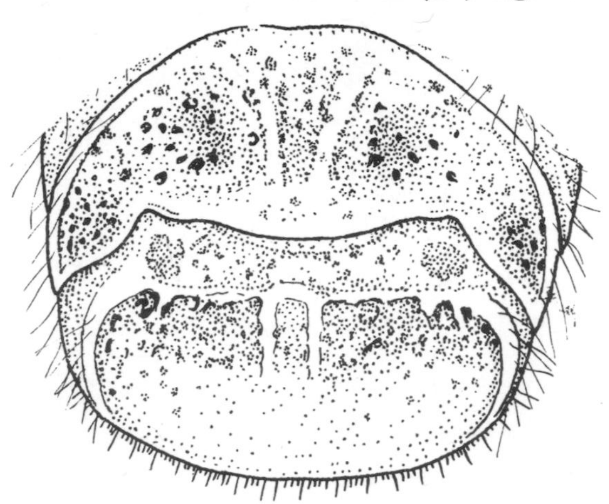

22. Major carinae of pleotelson arcuate

L. (P.) rugosissima Menzies

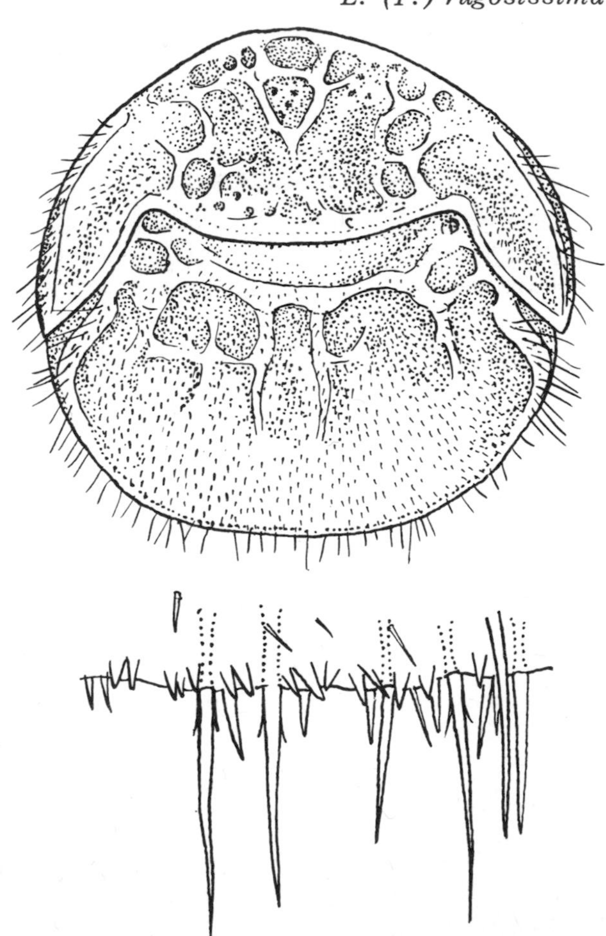

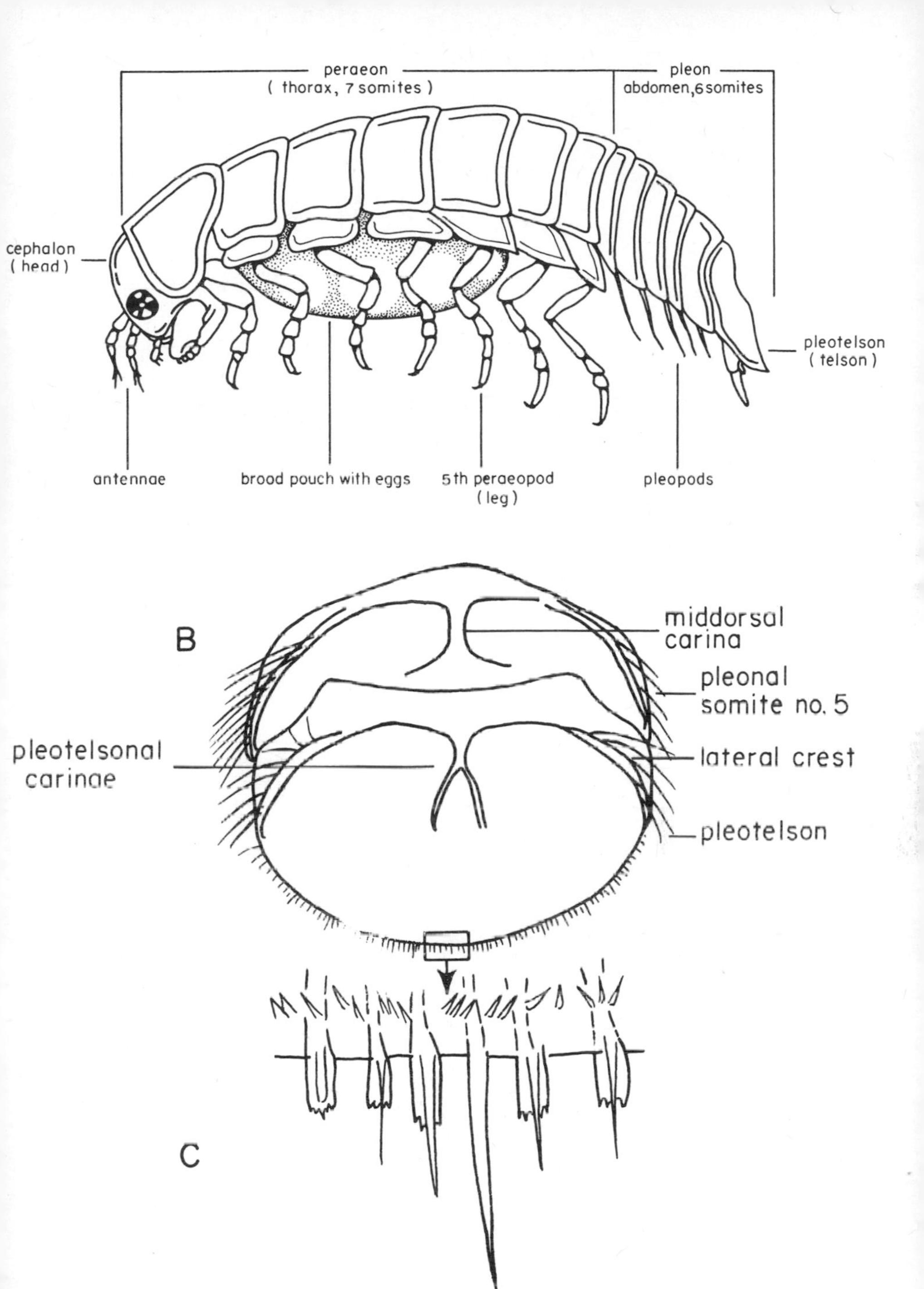

Fig. 1. Morphology of Limnoria. A, whole animal, lateral view; B, fifth pleonal somite and pleotelson; C, setae at pleotelsonal margin

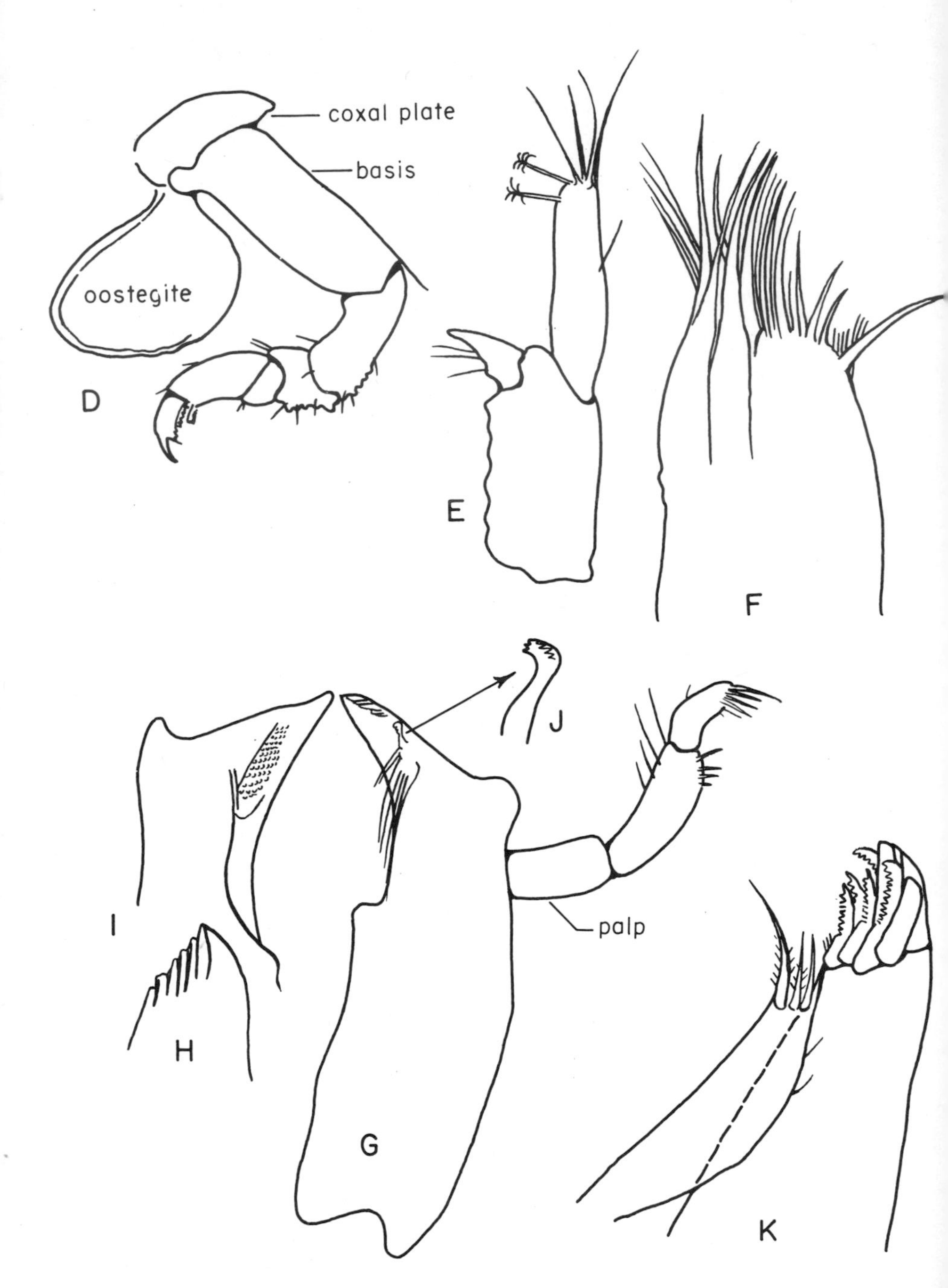

Fig. 2. Morphology of Limnoria (cont.). D, peraeopod; E, uropod; F, first maxilla; G-J, mandible; K, second maxilla

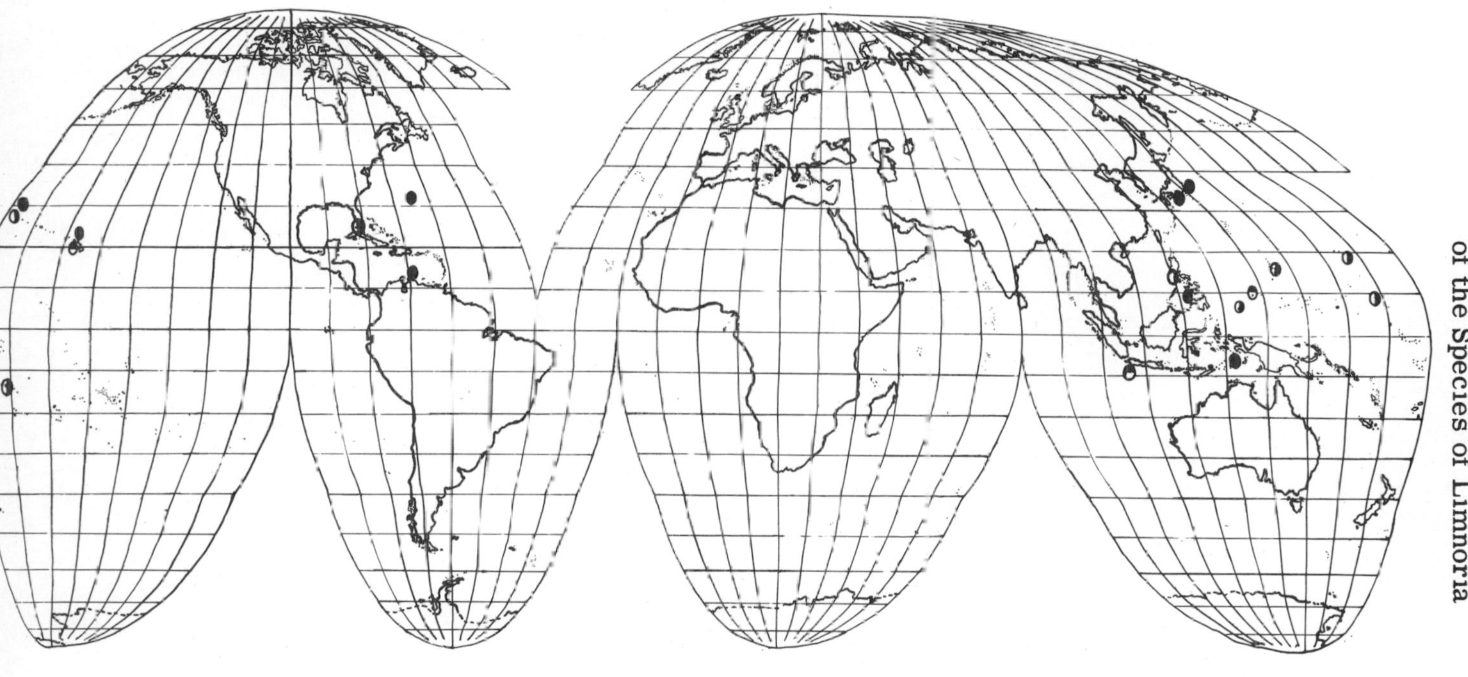

MAP I WORLD-WIDE DISTRIBUTION OF *PARALIMNORIA ANDREWSI* (CALMAN)

Symbols: ◐ Forma typica, ● Forma A, ○ Forma B

MAP 2 DISTRIBUTION OF LIGNOPHAGOUS OR WOOD-BORING LIMNORIA

DISTRIBUTION OF THE SUBGENUS LIMNORIA IN THE ATLANTIC OCEAN

Symbols: ● *LIGNORUM*, ■ *QUADRIPUNCTATA*, ▲ *TRIPUNCTATA*, ⊘ *SASEBOENSIS*

⊠ *PFEFFERI*, ⊞ *SIMULATATA*, ○ *PLATYCAUDA*, ⊖ *MULTIPUNCTATA*

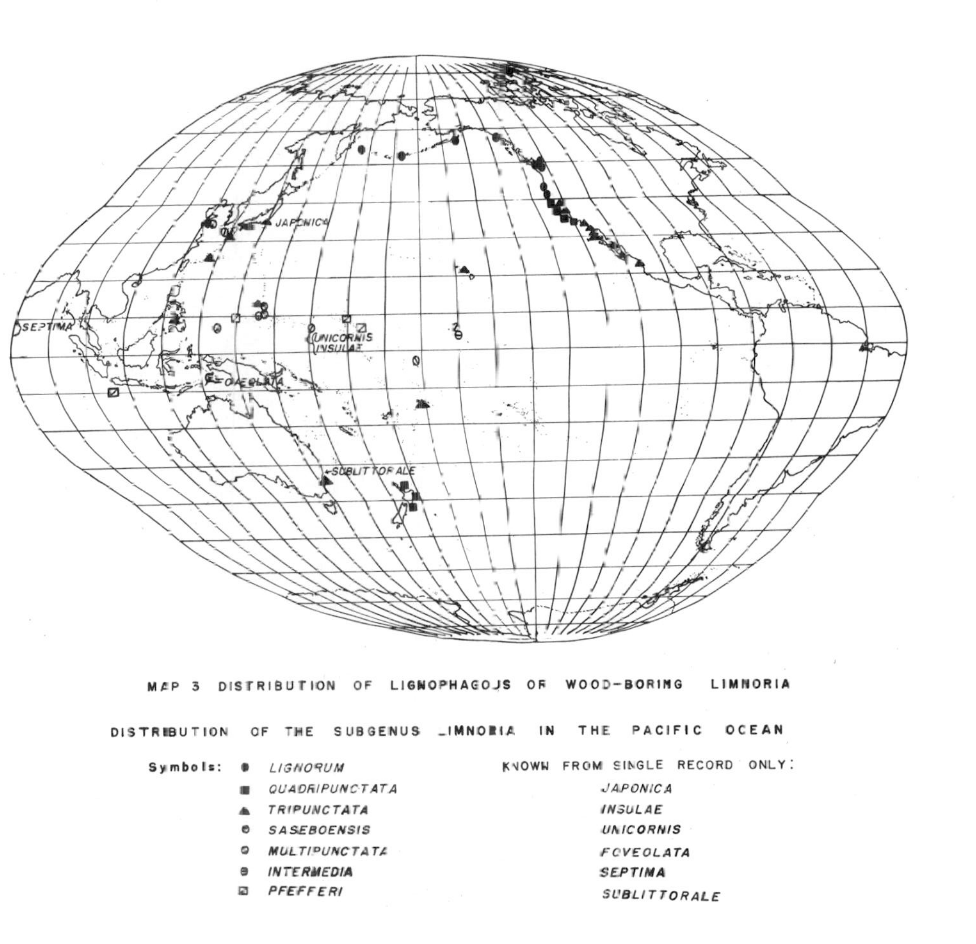

MAP 3 DISTRIBUTION OF LIGNOPHAGOUS OR WOOD-BORING LIMNORIA

DISTRIBUTION OF THE SUBGENUS LIMNORIA IN THE PACIFIC OCEAN

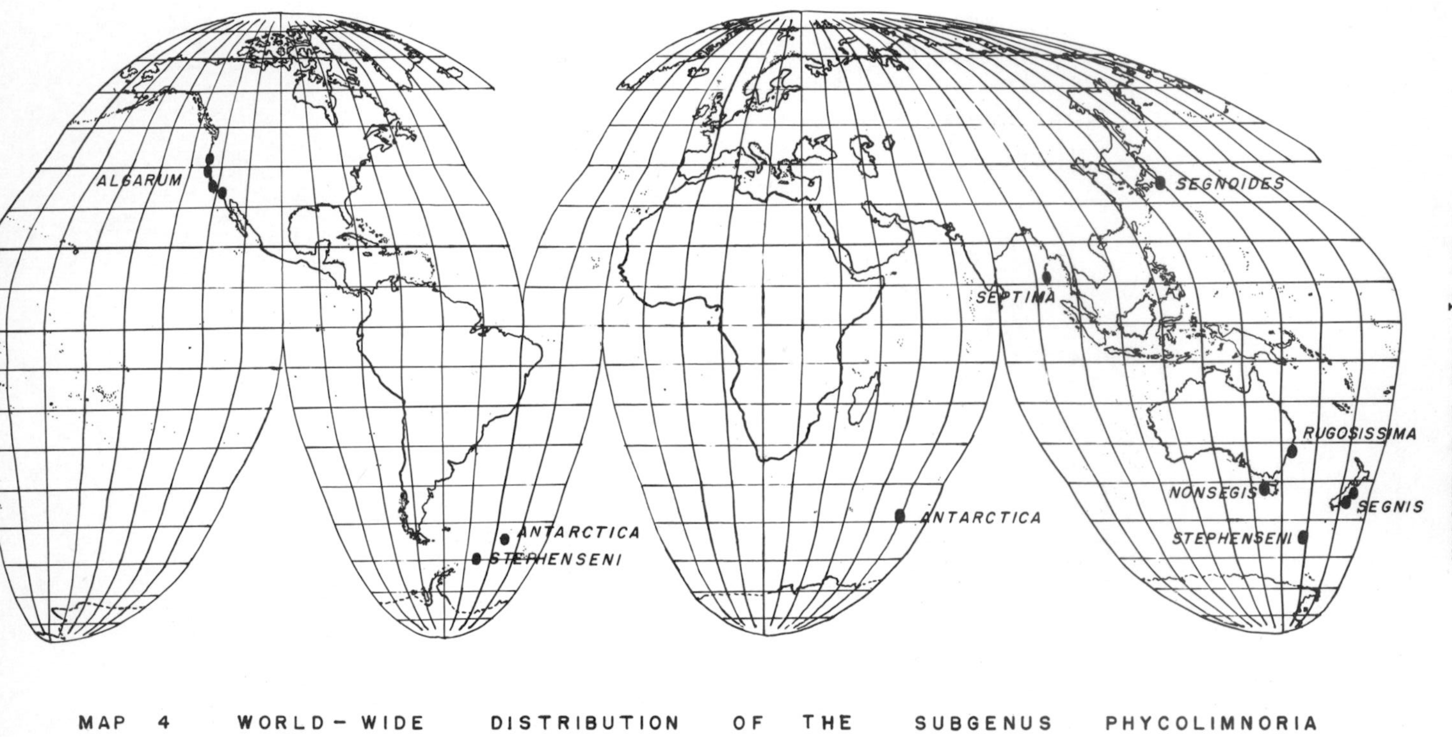

MAP 4 WORLD-WIDE DISTRIBUTION OF THE SUBGENUS PHYCOLIMNORIA

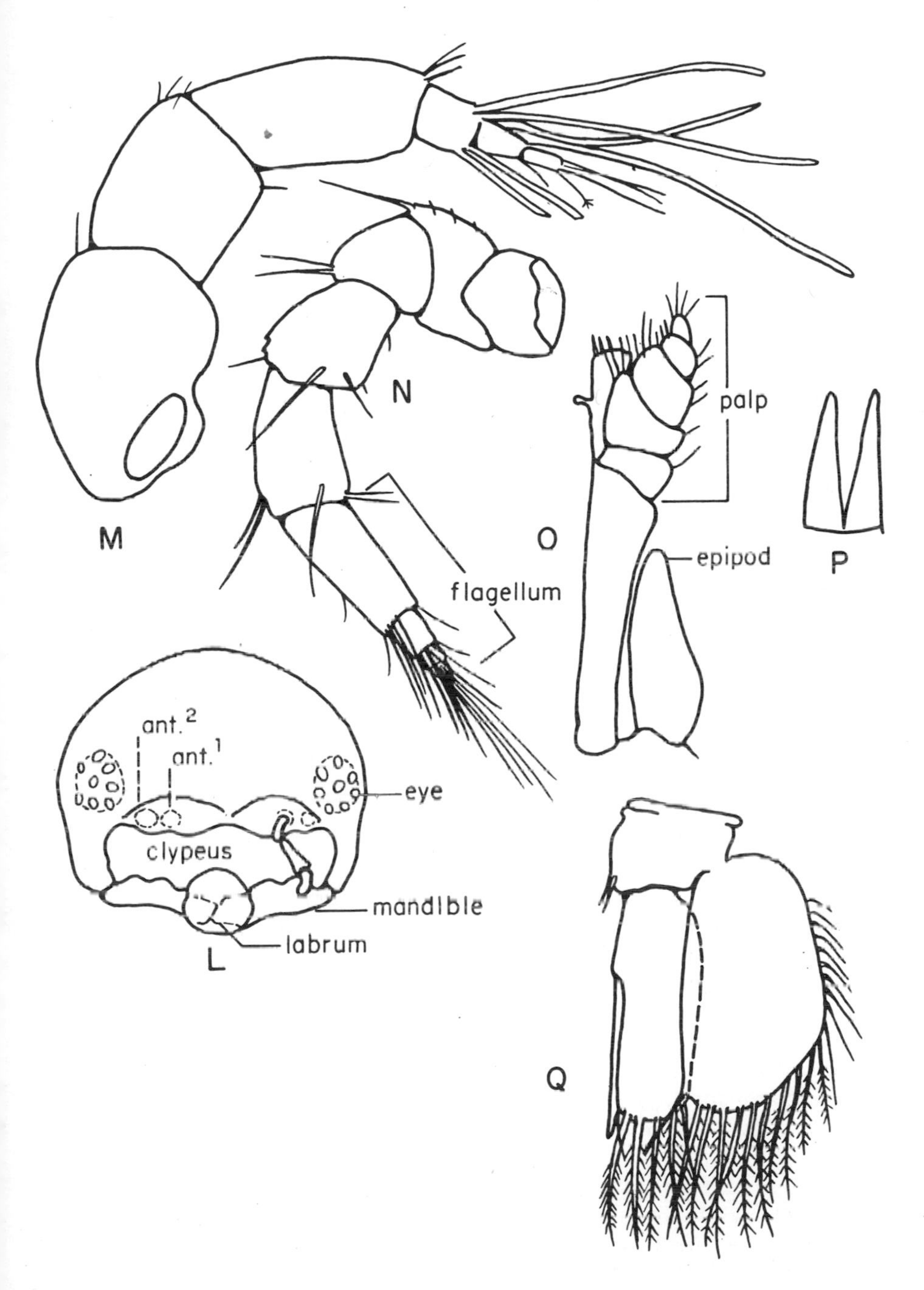

Fig. 3. Morphology of Limnoria (cont.). L, front of cephalon; M, first antenna; N, second antenna; O, maxilliped; P, penes; Q, second male pleopod (after Menzies, 1957)

THE GENERAL HISTOLOGY AND NUTRITION OF LIMNORIA*†

Charles E. Lane

The genus Limnoria of the family Limnoriidae (Crustacea, Isopoda) includes many species. Most of these may destroy wood that is exposed to the environment of the seas. The genus is apparently world-wide in its distribution, having been reported from all seas and continents. The total annual damage wrought by members of this genus has been variously reported, but all estimates agree that it amounts to millions of dollars. This is no mean accomplishment for a group of animals, the individuals of which rarely exceed 2 mm in total length.

Since mine is the introductory paper of the group that will be concerned with the general subject of the biology of Limnoria, it seems appropriate to begin my consideration with a brief examination of the morphology of this animal. The sagittal section (fig. 1) shows Limnoria to be a fairly generalized and typical isopod. The body consists of head, thorax, and abdomen. The head bears specialized and sturdy mouth parts with which the burrow is is chiefly excavated. The complex of mouth parts is moved by powerful muscles that fill much of the volume of the head. The thorax contains 7 free segments, each of which bears a pair of walking legs. The abdomen consists of 6 somites, the first 5 of which are free, the last being fused with the telson. Each of the first 5 abdominal segments bears a pair of thin, platelike appendages, the pleopods, which are employed both for swimming and

---

*These studies were supported by contract Nonr 840(03) between the Office of Naval Research and the University of Miami Marine Laboratory.

†Contribution number 199 from the Marine Laboratory, University of Miami.

as gills in respiration. The body terminates posteriorly in the telson, the superficial sculpturing of which is of considerable taxonomic importance. The work of Hoek (1893) should be consulted for further details of the external anatomy and sexual dimorphism of Limnoria.

The sagittal section (fig. 1) and cross section (fig. 4) show the gut as a relatively simple straight tubular structure that extends from the mouth to the anus with no convolution or extensive folding. Near the juncture of the foregut with the hindgut there is a pair of diverticula each of which is bifurcated to form 4 fingerlike projections that occupy the body cavity and extend posteriorly parallel with the gut. These are the midgut caeca, digestive glands, or "livers."

In figure 1 the well-developed ganglionated nervous system ventral to the gut is a prominent anatomical feature. In general, the nervous system of Limnoria adheres closely to the usual arthropod plan. There is an anterior supraesophageal ganglion or "brain." This connects with the rest of the ganglionated nervous system, which lies ventral to the gut, by a pair of broad esophageal connectives. The prominent optic lobes (fig. 5) extend dorsolaterally from the "brain." These receive nerve fibers from the compound eyes. The ventral ganglionated chain contains 4 large ganglia in the head, 7 in the thorax, and 5 in the abdomen.

Dorsal to the gut and extending through most of the abdominal segments is the tubular heart. This organ provides the motive power for the circulation of blood through the sinusoidal circulatory system. Figure 6 shows that the muscle fibers of the tenuous wall of the heart are oriented obliquely to the long axis of the structure. Contraction of these elements decreases both the length and the diameter of the heart. A vessel may be seen in figure 1 which appears to drain the heart and to conduct blood to the vicinity of the origin of the midgut diverticula. Further details of the circulation have not been worked out.

The gonads lie between the heart and the gut. The ovary is generally insignificant except just prior to ovulation, when the oocytes grow enormously with the accumulation of large quantities of yolk. Then the ovary may fill most of the space in the body cavity (fig. 2), and so compress the gut as to interfere with normal feeding. The testes (fig. 7) are generally quite small and relatively insignificant among the other organs of the body cavity. They occupy the last thoracic segment. The testicular ducts, or *vasa deferentia,* open on the ventral surface of the last thoracic segment, in the median plane of the body.

The detailed structure of the midgut diverticula has not been described. Kofoid and Miller (1927) mention the occurrence of crystal inclusions in the cells of these diverticula. They were described as being diamond-shaped and yellowish in color. The suggestion was made that these crystals may be uric acid or some other metabolic waste product.

In routine histological preparations of the midgut glands of *Limnoria tripunctata* from the local population, large, Feulgen-positive, crystalline inclusions were frequently seen in the cytoplasm of certain cells of the midgut. The crystals rarely, if ever, occur in the lumen of the diverticulum. The crystals are bipyramidal in form and range in size from 1 to 20 μ in their long axis. These crystals have been observed in material prepared by lyophilizing living tissue and after standard histological and cytological fixation. They have also been seen in surviving living material examined with phase optics. It therefore appears that there can no longer be any question of the reality of these crystalline bodies in the living system. They have been observed in animals fixed *in situ*, as well as in animals that have been denied access to wood in the laboratory for up to 16 days. When the animals are starved the crystals appear to increase in number. After histological fixation the crystals are insoluble in water and in dilute alcohols. There is some evidence to suggest that crystalline inclusions are more frequent in the midgut glands of *L. tripunctata* than of other species. These apparent variations in interspecific distribution should certainly be studied more extensively. It is of some interest that crystals of the same general description are also found in the digestive diverticula of Idotea (H. B. Moore, personal communication).

These inclusions achieve considerable interest and significance because of their possible physiological role. The only function so far attributed to the midgut of Limnoria has been that of elaborating the complex of enzymes responsible for the hydrolysis of cellulose. These crystals may represent presecretion cellulase enzyme. Since this work on cellulases in Limnoria was initiated in Dr. Ray's laboratory and has been pursued here with considerable diligence, it is most probable that we shall be hearing a great deal more about it during the course of this symposium. For that reason I shall not expand further upon our somewhat preliminary observations.

The cells of the epithelial lining of the midgut diverticula (figs. 8, 9, and 10) are characteristically large, pyramidal elements with a pronounced tendency toward binuclearity. In contact with the basement membrane and wedged between the larger cells are smaller cells that typically contain only a single nucleus. The nuclei of both types of cells are large, quite vesicular, and with well-developed chromatic masses. The cytoplasm, aside from the granular or crystalline inclusions previously described, also contains numerous mitochondria and a well-developed Golgi apparatus. There is a tenuous muscle layer encompassing the entire diverticulum, in which the component fibers are arranged somewhat obliquely and are extremely loosely organized. It is doubtful, from structural appearance at least, whether the muscle layer contributes significantly to the movement of materials contained in the lumen of the diverticulum toward the lumen of the main digestive canal. Gut contents rarely appear to enter the lumina of the midgut diverticula.

In addition to its function of trituration of food materials, the foregut gastric mill (fig. 3) appears to be a sorting and classifying system which protects the orifices of the midgut caeca and prevents the entrance of intestinal contents. For details of the structure of the gastric mill see Hoek (1893).

The processes of feeding and nutrition in Limnoria differ from those in Teredo (Lane, 1959) in several important ways, among which is the absence of any machinery for collecting suspended particulate materials from the ambient water and for transporting them into the digestive tract. For this reason it is clear that Limnoria can make no direct use of planktonic animals for food but must rely entirely upon the wood within which it lives to supply all of its nutritional requirements. One should, therefore, expect to find a rather close resemblance between the amino acid spectrum of the proteins of the wood and of the animals living therein, making necessary allowances for interconversion of amino acids and for their biological synthesis from simpler nitrogenous compounds.

To test this hypothesis, acid hydrolyzates of *Limnoria tripunctata* and of newly milled southern pine heartwood were compared by paper chromatography (Schafer and Lane, 1957). The amino acids in Limnoria were conspicuously different from those of the wood within which the animals may be found. Some of the qualitative differences are shown in table 1.

It was recalled that Meyers and Reynolds (1957) have emphasized that wood exposed to the sea is rapidly and universally attacked by various species of marine fungi. Within a very few days, in tropical waters, the wood becomes interpenetrated by

TABLE 1
AMINO ACID CONTENT OF *LIMNORIA TRIPUNCTATA*, STERILE SOUTHERN PINE, AND THE MARINE FUNGUS *LULWORTHIA FLORIDANA*

Amino Acid	Limnoria	Southern Pine	*Lulworthia floridana*
Alanine	X	X	X
Aspartic acid	X	X	X
Cystine	X	0	X
Cysteine	X	0	0
Glutamic Acid	X	X	X
Glycine	X	X	X
Histidine	X	X	X
Isoleucine	X	X	X
Leucine	X	X	X
Lysine	X	0	X
Methionine	X	X	X
Serine	X	0	X
Threonine	X	0	X

a lacework of fungal mycelia. These authors present morphological evidence of structural thinning and weakening of the wood fibers penetrated by mycelia. It has been suggested (Isham and Tierney, 1953) that preinfection with marine microorganisms is a necessary preliminary to the establishment of a molluscan marine borer infection.

To evaluate the effect of an established marine fungus infection on the amino acid composition of wood, a further experiment was performed in which Limnoria were removed from an infected southern pine heartwood timber and hydrolyzed as before. The remnants of the same timber, carefully freed of all borers and of surface fouling, were also hydrolyzed. When the resulting hydrolyzates of animal and of wood were compared by paper chromatography, a good qualitative and quantitative agreement in amino acid composition was observed.

The nutritive role of marine fungi was evaluated by a series of feeding experiments in which Limnoria were allowed access only to sterile wood in sterile sea water. These experiments were repeated with animals of different ages. In no single instance did the experimental animals survive longer in the presence of sterile wood than they were able to survive in the complete absence of wood. The absence of fecal pellets under the conditions of these experiments suggests that no attempt was made even to feed upon the wood.

Through the courtesy of Dr. E. S. Reynolds and Dr. S. P. Meyers of our laboratory, cultures of some of the more common marine fungi of Biscayne Bay were made available to us. These had been grown on artificial media of defined composition and with a single known nitrogen source. When fragments of living mycelial mat were made available to Limnoria, under the same conditions as were employed for the presentation of sterile wood as a dietary substrate, the animals fed vigorously and survived normally. Indeed, it was observed that the animals frequently converted the entire myccelial fragment to fecal pellets. This is in marked contrast with the usual behavior in wood, which which is never consumed entirely. The avidity with which Limnoria gorged themselves on fungal mycelium suggests that this material may constitute an important component of the natural diet.

When the amino acid spectrum of acid hydrolyzates of Limnoria and of the fungus *Lulworthia floridana*, grown on artificial sea-water medium with asparagine as the sole nitrogen source, were compared (see table 1), there was exact correspondence in qualitative composition. There was even a very close quantitative resemblance in the two preparations. Since the gut of the experimental animals was crammed with mycelial material, this resemblance was not surprising. However, in another series of animals the digestive tracts were separated from the rest of the body, and the gut-free residue was hydrolyzed. In these animals, too, the resemblance was exact. It thus appears that amino acid derived from fungal proteins had been absorbed from the gut and had been incorporated in nondigestive tract tissue proteins.

These observations help to explain the resemblance between the amino acid content of hydrolyzates of borers and of the wood in which they live, despite significant discrepancies between the composition of Limnoria and of sterile wood of the same kind. A varying degree of exposure to the marine environment may also help to explain certain inconsistencies and discrepancies that have appeared in published descriptions of the amino acid content of wood, (Drisko and Hochman, 1957). It is apparent that fungal infection varies with duration of exposure and with the season during which the wood has been exposed to fungal attack. Differences in degree of infection will be reflected in differences in amino acid composition.

Finally, these observations suggest that Limnoria, while undeniably capable of digesting wood and of absorbing the digested products of some components of natural wood, does so primarily to provide itself with a suitable concentration of the fungal proteins which exposed wood always contains.

LITERATURE CITED

Drisko, R. W., and H. Hochman. 1957. Amino Acid Content of Marine Borers. Biol. Bull., 112:325-29

Hoek, P. P. C. 1893. Rapport der Commissie uit de Koninklijke Akademie van Wetenschappen . . . betreffende de levenswijze en de werking van *Limnoria lignorum*. Verhandel. Koninkl. Akad. Wetenschappen, (tweede sectie) 1, No. 6:1-103.

Kofoid, C. A., and R. C. Miller. 1927. Biological Section, pp. 188-343, in Marine Borers and Their Relation to Marine Construction on the Pacific Coast. Final Report, San Francisco Bay Marine Piling Committee.

Lane, C. E., 1959. Some Aspects of the General Biology of Teredo. In this volume.

Meyers, S. P., and E. S. Reynolds. 1957. Incidence of Marine Fungi in Relation to Wood Borer Attack. Science, 126: 969.

Schafer, R. D., and C. E. Lane. 1957. Some Preliminary Observations Bearing on the Nutrition of *Limnoria*. Bull. Mar. Sci. Gulf and Caribbean, 7:289-96.

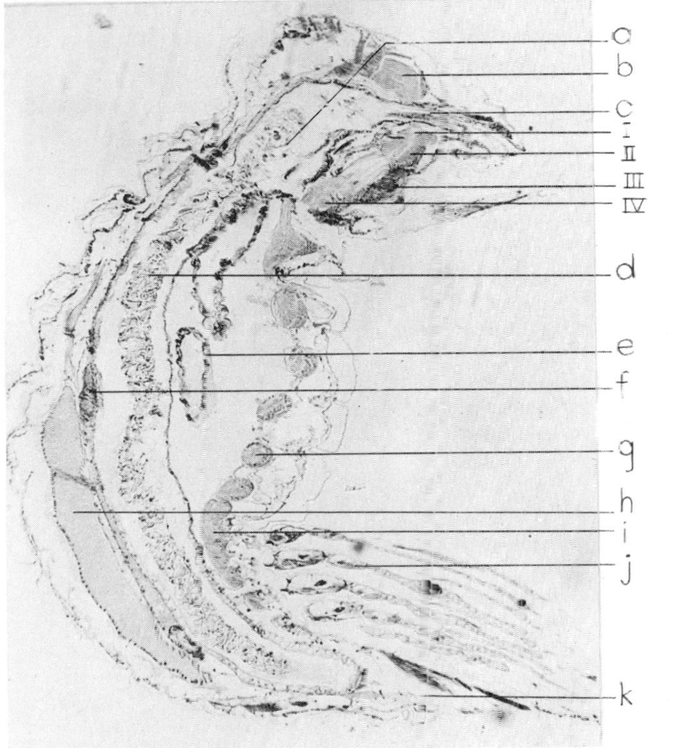

Fig. 1. Midsagittal section of female *Limnoria tripuncta-ta,* fixed in boiling Bouin's fluid, stained with hematoxylin and orange G. a, stomach; b, brain; c, esophagus; d, intestine; e, digestive gland; f, ovary; g, thoracic ganglion #5; h, heart; i, abdominal ganglion #1; j, pleopod; k, telson. I, II, III, IV are cephalic ganglia. x 106

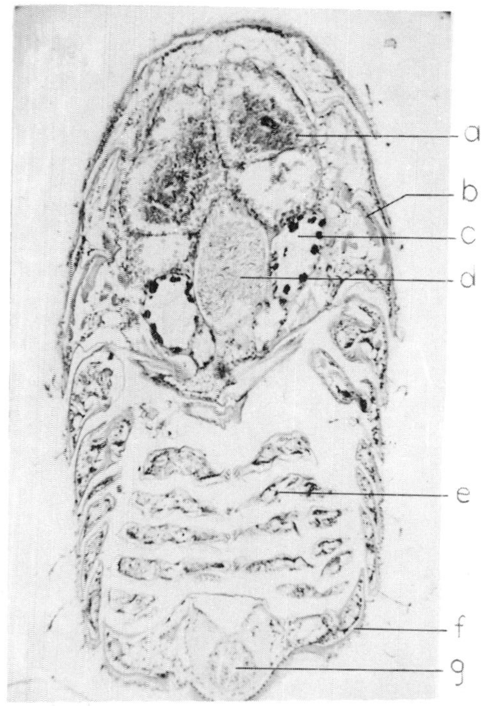

Fig. 2. Cross section of mature female Limnoria. In this animal the maturing oocytes occupy all the space in the body cavity without obvious occlusion of the gut or the digestive glands. In more mature specimens these organs tend to be more compressed by the ovary. a, oocyte; b, muscle; c, digestive gland; d, intestine; e, pleopods; f, telson; g, intestine. x 65

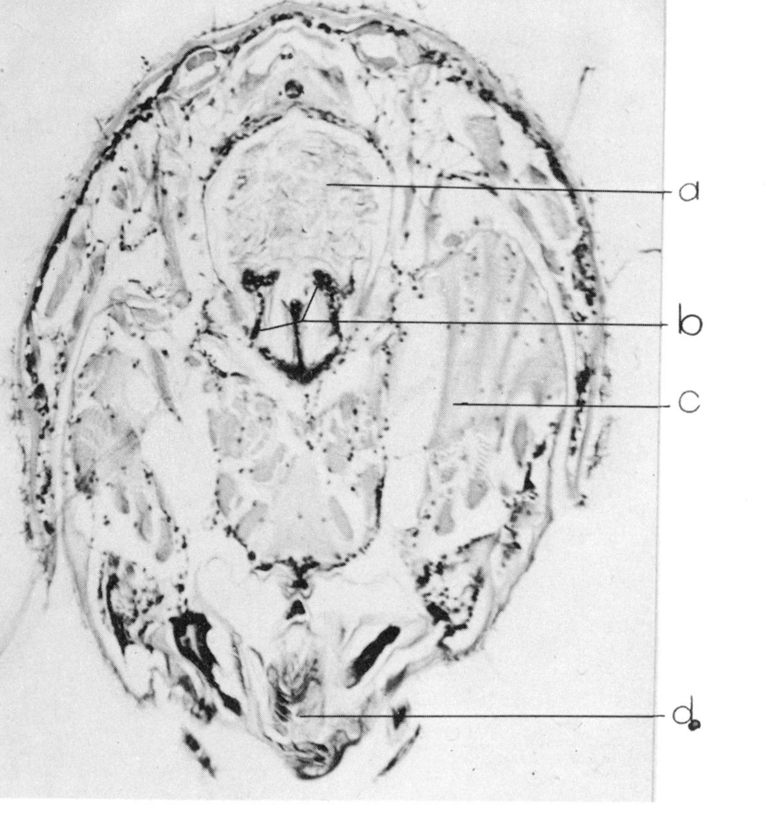

Fig. 3. Cross section of Limnoria. The median and lateral processes of the gastric mill. The powerful musculature of the mouth parts is also shown. a, stomach; b, processes of gastric mill; c, muscle; d, mouth. x 148

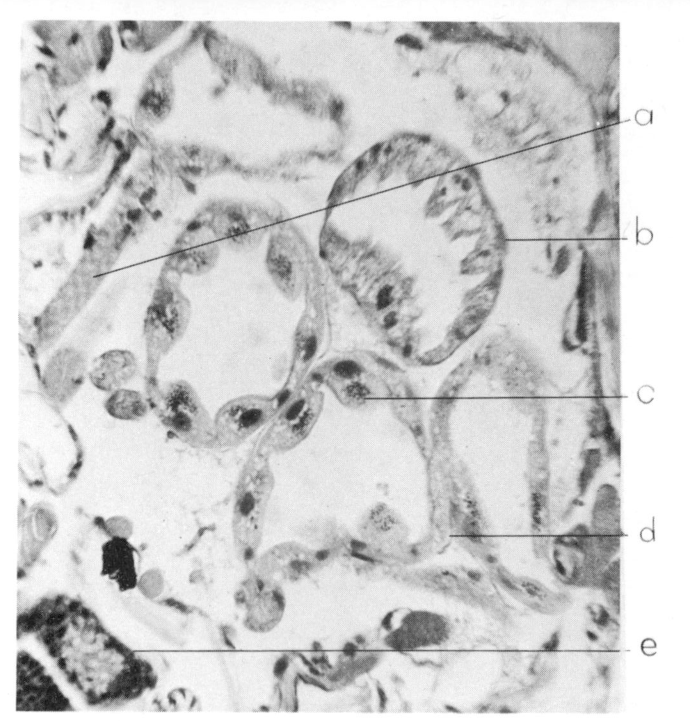

Fig. 4. Cross section of Limnoria. This section shows the single intestine and 4 sections through the digestive gland. The perinuclear concentration of granules in the cells of the midgut diverticula is noteworthy. Fixation in Flemming's solution, stained with iron hematoxylin and orange G. a, muscle fiber; b, intestine; c, precrystalline granules; d, digestive gland; e, ganglion. x 430

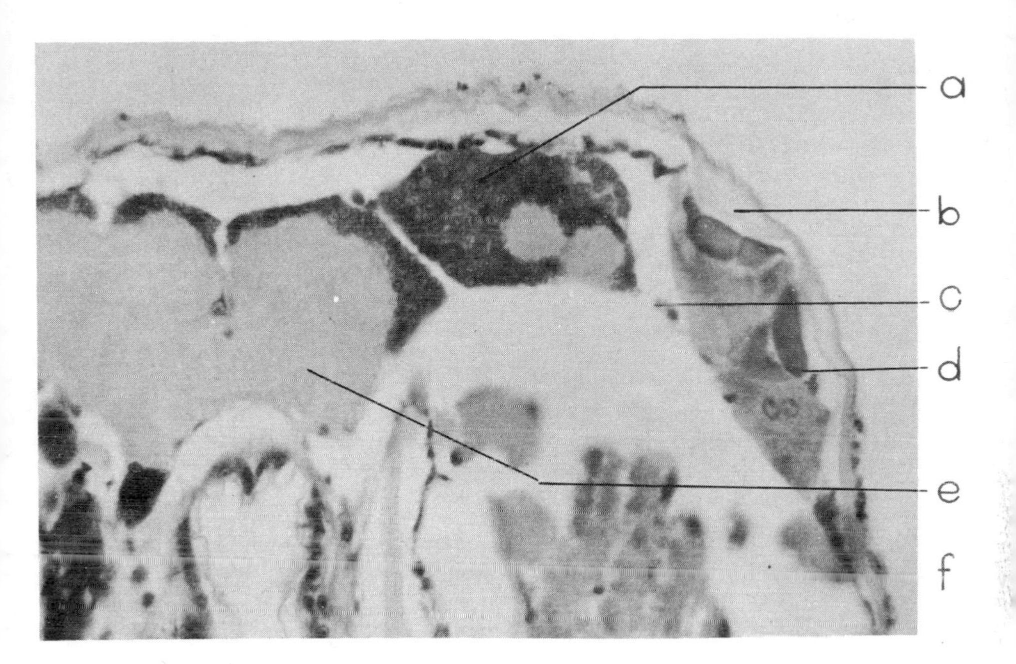

Fig. 5. Frontal section to show the relation between the "brain," the compound eye, and the optic lobes. a, optic lobe; b, cornea; c, optic nerve; d, lens; e, supraesophageal ganglion; f, esophagus. x 495

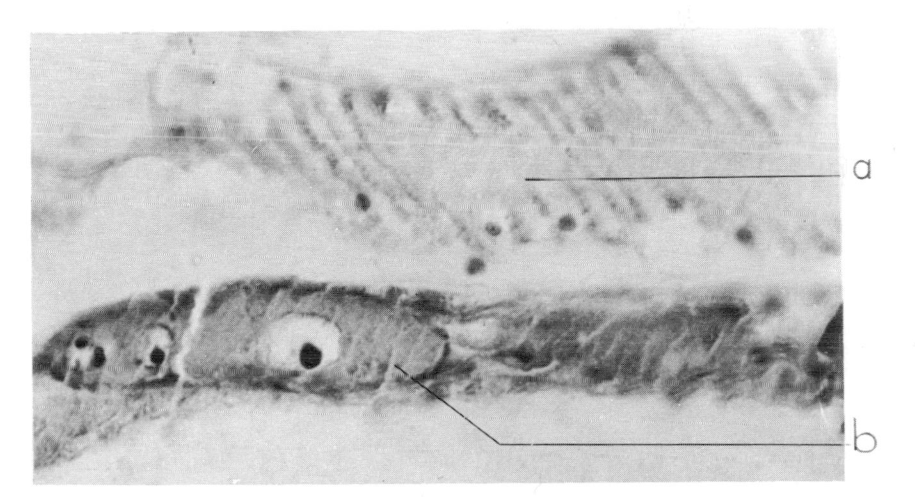

Fig. 6. Sagittal section. The oblique direction of the muscle elements in the wall of the heart are well shown. The ovary contains only immature oocytes. a, heart; b, oocyte. x 640

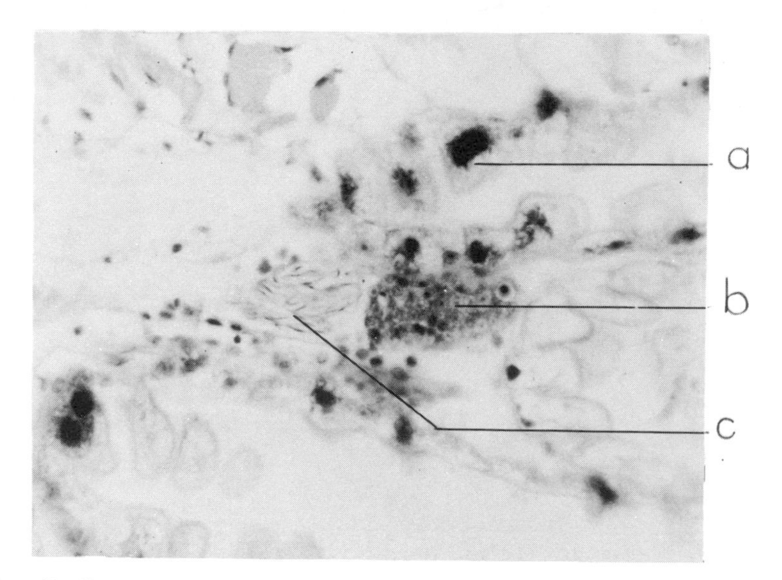

Fig. 7. Sagittal section. The posterior portion of the testis and proximal portion of the *vas deferens* are shown. a, digestive gland cell; b, spermatogenetic region of testis; c, spermatozoa in efferent testicular duct. x 435

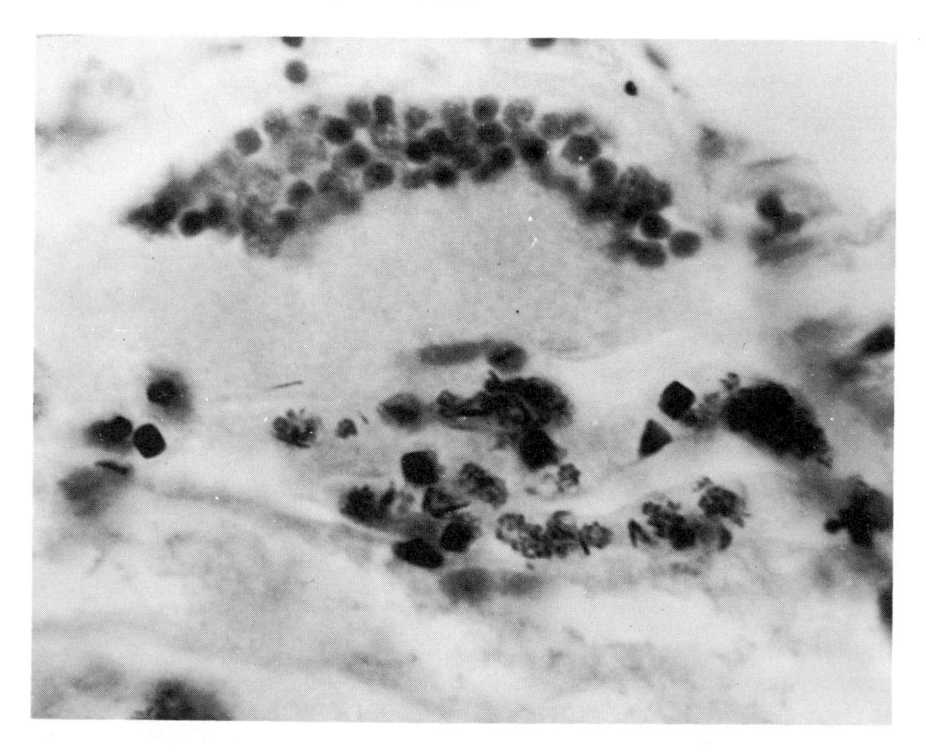

Fig. 8. Sagittal section. This is a longitudinal section through the posterior end of one of the digestive glands. A ventral ganglion is also shown. x 1320

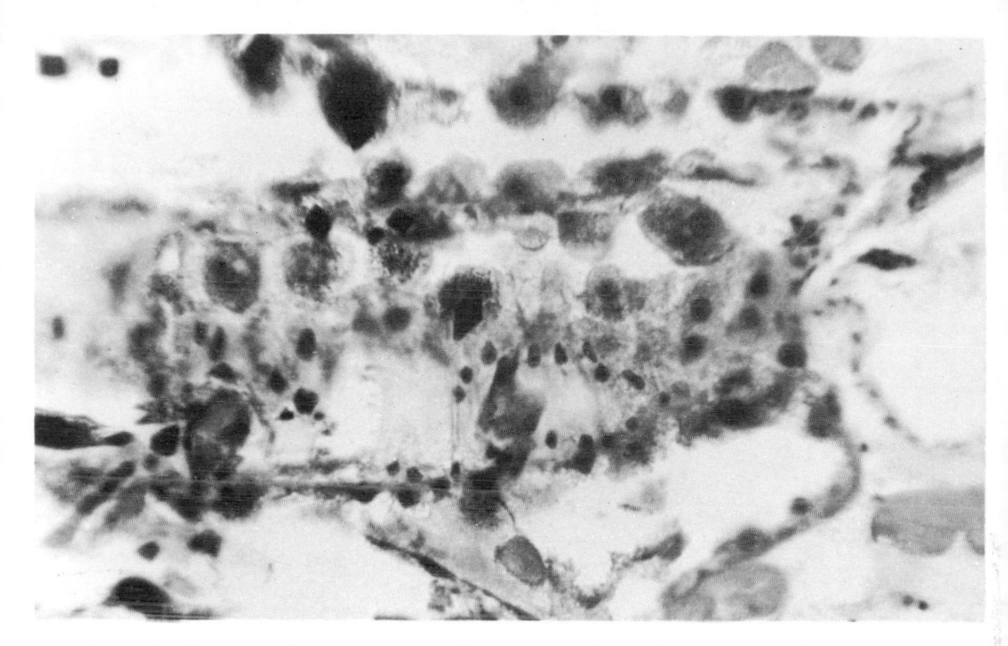

Fig. 9. Longitudinal section through a digestive gland. Various stages of crystal formation are shown. x 1050

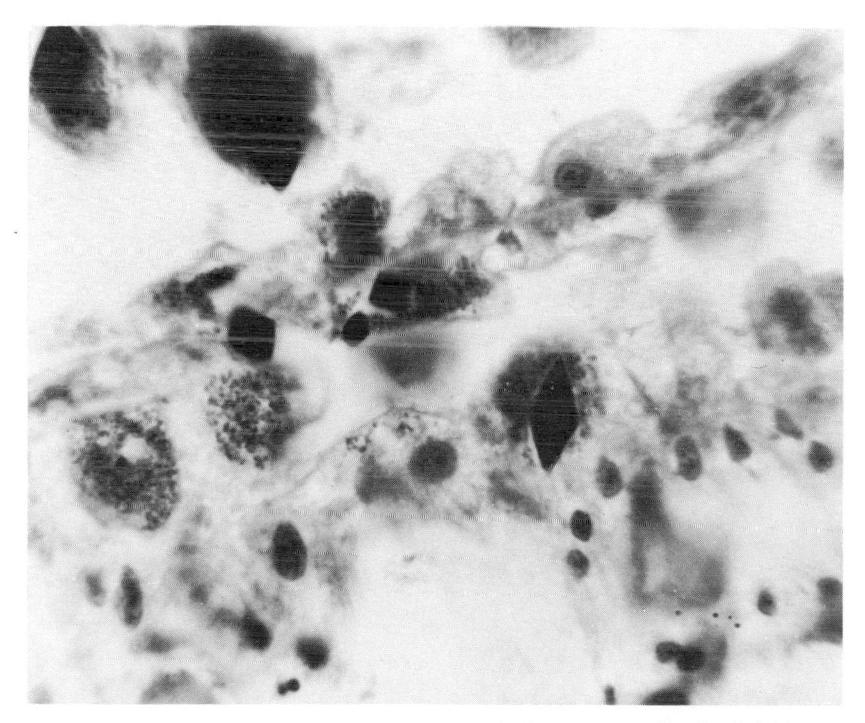

Fig. 10. Longitudinal section through digestive gland of Limnoria. Fixation in Flemming's solution, sectioned in paraffin at 5 μ, stained in iron hematoxylin and orange G. x 780

NUTRITIONAL PHYSIOLOGY OF LIMNORIA*

Dixy Lee Ray

That Limnoria burrows into wood and often causes severe damage to marine structures has long been recognized and acknowledged. But, until recently, the question of the role that burrowing plays in the life of the animals has remained unclarified. Contemporary workers have generally favored the explanation offered by Yonge (1927, 1950, 1951) that Limnoria burrows solely for protection, but the older opinion that the animals may feed upon the wood into which they tunnel has been supported by the investigations of Larkin (1951), Ray (1951, 1953), and Ray and Julian (1952). Whether the wood does or does not provide some or all of the nutriment for Limnoria is an important question to answer, not alone for the insight that might be gained into the biological utilization of cellulose by animals, but also because such knowledge might provide a basis for more effective control or protective procedures. Considering the enormous economic importance of Limnoria and the far-reaching biological significance of the questions that are implicit in the activities of an animal that chews its way into wood and remains there all the days of its life, our knowledge of the nutritional physiology of Limnoria has been scanty indeed. It is to this problem that we have directed our main attention.

Early investigators agreed with the suggestion of Rathke who, according to Hoek (1893), reported to a meeting of the Naturhistorie-Selskabet in Copenhagen in 1797 that wood is the only

---

*These studies were aided by a contract between the Office of Naval Research, Department of the Navy, and the University of Washington, NR 104-142.

food of Limnoria. Smeaton (1812), Coldstream (1834), Dalyell (1851), Bate and Westwood (1868), and many others also assumed that Limnoria feeds on the wood into which it burrows. But it was Hoek (1893) who first examined this question critically and wrote as follows (p. 29): "Investigating the food [brei] which fills the intestinal tract of Limnoria, sometimes over its entire length, we always found that wood fibers, minute chips of wood, etc. formed the main component. We always succeeded in finding at least *some* elements, among these fibers, which by the presence of bordered pits betrayed their origin in an unmistakable manner. Whether wood fibers are the only food which Limnoria ingests, we dare not say"; and on page 45: "The main reason why they [Limnoria] burrow in wood probably is that they feed on the wood fibers. Even if we admit that it is not necessary to assume that they live *exclusively* on these fibers, the latter in all probability do form at least the main course on their menu"; and, finally: "This Isopod understands the art of concentrating from the wood that which she needs for her nutrition. Whether or not this is her only food, is hard to decide."* It is obvious that these statements are cautiously made, and that they rest upon keen observation and thoughtful reflection, but there is no evidence that the tentative conclusions reached were ever put to experimental test. Clearly, Hoek believed that Limnoria eats wood, but reserved judgment whether this is its *only* food. The situation remained unchanged for several decades; most authors either implied or stated flatly that Limnoria feeds on wood, and as recently as 1924 Henderson writes, "Certain it is that at least part of the wood swallowed is digested, and forms the chief food source." No evidence is cited in support of this statement.

Finally in 1927 Kofoid and Miller, in the only comprehensive report on the biology of Limnoria that has been published in this century, recorded the results of preliminary experiments that revealed a difference in the appearance of the cellulose fibers of wood before and after passage through the gut. This they determined by applying the iodine-zinc chloride color reaction as a test for modified cellulose. They also conducted survival tests in which one group of animals was supplied with fil-

---

*I am indebted to Dr. B. J. D. Meeuse, Department of Botany, University of Washington, for translation from the original Dutch of the pertinent portions of Hoek's important paper.

ter paper, and a control group was kept in sea water alone. The control animals died within seven weeks, whereas some of those provided with filter paper produced fecal pellets from it and remained alive for the nearly ten weeks' duration of the experiment. These results suggest that Limnoria does eat wood, but by this time a widespread reluctance to believe that an animal could derive its nourishment directly from wood had grown up, and this reluctance persists to the present.

Such an attitude of skepticism is justified, for there is little readily hydrolyzable material in wood, and the burden of proof rests upon those who claim its biological utilization. Further, there is now a far greater understanding of the difficult problems inherent in enzymatic attack on an insoluble substrate. Cleveland's (1924, 1925a, 1925b) pioneering studies clearly implicated microbial symbionts rather than the animals themselves in the digestion of wood by termites, and these investigations, followed by the elegant experiments of Trager (1932, 1934) and Hungate (1936, 1938), firmly established that at least most species of wood-eating termites are entirely dependent upon the activities of symbiotic protozoa. In view of this, the 1927 report by Yonge, that he tested homogenized Limnoria for cellulolytic enzymes and found amylase but no evidence for the presence of cellulase, gave adequate grounds for skepticism.

And so the idea that Limnoria could feed on wood has not been widely accepted in recent years, and Yonge's findings seemed to obviate this interpretation. But two fundamental and related questions remained unanswered: (1) For what purpose does Limnoria burrow? and (2) On what does the animal feed? In 1950 Yonge wrote: "There is no evidence that wood itself is digested, although the gribbles may obtain some of their food from the microscopic algae and bacteria which multiply on rotting wood. They bore primarily for protection"; and again, in 1951: "There is no evidence that these animals [Limnoria] can digest any constituent of the wood and they presumably burrow for protection." A number of considerations, some of which had also been noted by Kofoid and Miller (1927), led us to doubt that Limnoria burrows only for protection, and to undertake a re-examination of this whole question. The pertinent observations underlying this decision may be briefly summarized as follows:

1. *In habitat Limnoria is restricted, almost exclusively, to wood.* The only other natural substratum used by some of the animals (Phycolimnoria) is the stipe and holdfast of certain brown algae (Laminariaceae). The primary constituents of both

wood and algae are complex polysaccharides, and the constant association of Limnoria with these carbohydrate-containing structures is suggestive. If the animals burrow for protection alone, then conceivably other substances having similar mechanical properties but composed of materials other than polysaccharide ought to serve Limnoria equally well, but this is not the case.

2. *Limnoria burrows by chewing its way into the wood.* Tiny particles are rasped and bitten off, and the great majority of these are swallowed. They pass through the gut and are eliminated as compact, cylindrical, fecal pellets of uniform diameter but slightly varying length. It may be that the gut represents only the most direct route to the exterior of the burrow, but the passage of wood fragments through it provides an opportunity for digestion to take place.

3. *No materials other than wood* (or bits of algae in the case of the algal dwelling forms) *have ever, consistently, been found in the gut of Limnoria.* No microorganisms occur in the gut.

4. *Limnoria continues to burrow throughout life.* Hatched within the protective confines of the parental burrow, the juveniles begin to bore side chambers within a few hours of their release from the brood pouch. If protection is the primary consideration in constructing a burrow, then, once the body of the animal is buried, continued boring serves no useful purpose, and indeed it results only in the destruction of that very protection that the animal is presumed to be seeking.

5. *The normal position of the animal within the burrow is head down in the deepest part of the blind tunnel* (fig. 1). Lim noria has rarely been observed to leave its burrow except under conditions of severe environmental stress or population pressure. This habit is hardly adapted to browsing on the microorganisms of the wood surface. There is no evidence that Limnoria is a predator, and no conceivable prey.

6. *Limnoria lacks the necessary anatomical structures to feed by any method other than rasping and chewing.* To strain planktonic or particulate matter from sea water requires morphological adaptations that Limnoria does not possess. No alternative supply of food other than wood is discernible.

It seemed, therefore, that it would be instructive to look once again, as carefully as possible, into the problem whether Limnoria is capable of utilizing wood for food. We began by reinvestigating the animal's digestive enzymes. A preliminary report of this aspect of the work has been made (Ray, 1951; Ray and Julian, 1952). We found that cellulolytic enzymes are indeed present, produced by the cells of the midgut diverticula, and that these are able to attack several constituents of wood as well as to degrade cellulose to glucose. Considerable data on the cellulase of Limnoria have been accumulated; these are reviewed elsewhere in this volume (Ray, 1959).

But the fact that cellulolytic enzymes can be extracted from body tissues leaves two questions still unanswered: (1) Does the animal (in this case Limnoria) actually produce its own cellulase? and (2) Does wood really serve as the major or only food?

Does Limnoria Produce Its Own Cellulase?

In the case of most animals, exclusive of Protozoa, from which the extraction of celluloytic enzymes has been demonstrated, it is no easy task to determine whether the cellulase has been produced by cells of the animal's body or by microorganisms present within the gut. With Limnoria, on the other hand, the situation is perfectly clear, for these animals harbor no microorganisms anywhere in the digestive system. Indeed, it is one of the remarkable features of Limnoria, for which at present no explanation is forthcoming, that, within the entire alimentary tract, no bacteria, protozoa, or other microorganisms can be found. This conclusion is based upon the results of direct microscopical examination of the gut and its contents in many hundreds of healthy individuals. Both freshly dissected whole mount and smear preparations, and fixed, stained serial sections have been thoroughly studied. Further, the digestive system of Limnoria has none of the anatomical specializations that are requisite for harboring microbial symbionts where these play an essential role in the nutritional economy of the host (see Fahrenbach, 1959, fig. 1). In the absence of any evidence whatsoever that microorganisms occur in the gut of Limnoria, the conclusion that the animal elaborates its own cellulolytic enzymes is warranted.

Does Wood Serve as the Major or Only Food For Limnoria?

Efforts to answer this question have approached the problem in three different ways. (1) Survival tests have been carried out using groups of animals kept under the same laboratory conditions, with and without wood as the only food source. (2) Dry weights have been determined on a series of test samples of wood before and after attack by a known number of specimens of Limnoria; dry weights of the fecal pellets produced during attack have been determined, and, from these data, calculations have been made of the amount of wood probably used by the animals. (3) Analyses have been made of the chemical composition of the wood used in the experimental setups, with particular reference to the following constituents: α-cellulose, polyuronide hemicelluloses, noncellulosic polysaccharide cell wall components (xylans, galactans, etc.), and lignin. Similar analyses have been made of the fecal pellets.

The results in every case point to the likelihood of utilization of wood. In the survival tests, more than 500 individual animals have been removed from their burrows and separated, alone or in groups of 20 or 30, into dishes of clean sea water. The water was changed in some cases daily, in others weekly, and in all cases maintained at a temperature of 10° to 13° C. Similar numbers of Limnoria have been segregated into finger bowls or stender dishes of sea water under identical conditions except that a small block of wood (generally Douglas fir) was added. In the absence of wood, 80 per cent of the animals were dead within 60 days. The remainder died within the third month and apparently had survived for the longer time through cannibalism. In the culture dishes that lacked wood no fecal pellets appeared after the first 72 hours. Whether diatoms or other microorganisms developed in the experimental cultures had no effect upon the outcome of these tests. Serial sections of the bodies of these test animals showed ample evidence of death from starvation; most of the internal organs had been resorbed. In those containers that included wood, however, fecal pellets continued to accumulate at a rate of 10 to 100 per animal per day, the animals molted about every 60 to 90 days, young were hatched and grew, and the population survived for more than a year and a half, after which these tests were terminated.

To determine whether there is any loss of weight in the wood that passes through the gut of Limnoria, small pieces of Douglas fir (all cut from the same block) were soaked in fresh water until no further chloride could be extracted, and then dried to constant weight. The weighed samples were then reimmersed in sea water and placed in individual labeled stender dishes, and 10 to 20 specimens of Limnoria were introduced into each dish. The animals were allowed to burrow for 2 months, during which the sea water was changed, and all material chewed off the wood and all fecal pellets produced were collected, every 48 hours. When Limnoria burrows, a certain amount of wood in the form of exceedingly fine particulate material is chewed off but not swallowed. This is referred to simply as "fluff" in the tabulated data to distinguish it from the fecal pellets, which represent wood that has passed through the digestive tract. The appearance of fluff in the experimental dishes is particularly evident during the early stages of the establishment of a burrow, and becomes noticeably less in amount as the animals move deeper into the wood. The fluff and pellets were collected by pouring each 48-hour accumulation, together with the water from the experimental dish, onto a sintered glass filter of fine pore size. The fluff and pellets were washed with distilled water until no further chloride was detectable in the filtrate, and were then transferred to a small beaker of distilled water, separated, and dried to constant weight. Although there are inevitable losses of material when such small amounts are handled through a series of washing, drying, and weighing, experimental checking of the procedures with comparable amounts of finely ground and ball-milled wood showed that the handling error amounted to only 0.5 to 2 per cent. Weight determinations of a total of 76 samples of wood, and of collections of pellets representing the activities of 1,320 animals, have been made in this manner. Since the results in each of the 76 experimental samples are comparable, only enough of these data have been tabulated to illustrate the outcome of the experiment. Reference to table 1 will show that, of the wood actually passed through the gut, there is an approximate 45 per cent loss in weight (experimental range, 36 to 53 per cent). This weight loss represents wood presumably utilized by the animals and compares favorably with the results from a similar experiment with termites (Hungate, 1944) where it was found that 40 to 60 per cent of the weight of the wood eaten is represented by the weight of the pellets, and the rest is presumably utilized.

That the wood is actually eaten by Limnoria is also indicated by chemical analyses of Douglas fir before and after passage through the gut. 500 mg wood samples were finely divided in a Waring blendor and analyzed for the major constituents. Whole Douglas fir as well as separated sapwood and heartwood portions were used in these analyses, and the results were compared with similar analyses of 500 mg samples of fecal pellets. The matter of experimental technique and choice of procedure presented some problems, for it has been emphasized repeatedly and supported by a multitude of experimental results (Wise, 1944; Hägglund, 1951; Bailey, 1952; Ott, Spurlin, and Grafflin, 1954) that the constituents of wood may vary in their percentage composition within rather wide limits, not only between different species of land plants, but also between different individuals of the same species and even within different parts of the same tree. Furthermore, particularly with cellulose, the analytical

TABLE 1*
DRY WEIGHT DETERMINATIONS OF DOUGLAS FIR BEFORE AND AFTER ATTACK BY LIMNORIA AND WEIGHT OF FECAL PELLETS PRODUCED

Wood Sample Number	Original Weight†	Final Weight+ Fluff	Weight Loss	Pellet Weight	Unaccountable Weight Loss	% Wood Utilized
1	202	210.5	51.5	30	21.5	42
2	212	164	48	27	21	44
3	293	257.3	37.5	19.2	16.5	46
4	363	309.5	53.5	29.5	24	45
5	307	263.5	43.5	20.5	23	53
6	346	285	61	30	31	51
7	319	271	48	26	22	46
8	295	246	49	26.5	22.5	46
9	364	328	36	18	18	50
10	530	473.5	56.5	30	26.5	47
..	..	..	..	..	..	..
25	427	368.5	58.5	30	28.5	49
26	347	295.2	51.0	29.3	22.5	43.5
27	346	286.5	59.5	36.5	23	39
28	316	249	67	43	24	36
29	212	161.5	50.5	28.5	22	43.5
30	227	174.5	52.5	28.5	24	46
31	313	264.5	48.5	27.5	21	43.5
32	309	263.5	45.5	25.5	20	44
33	402	330	72	40	32	44.5
34	240	184.5	55.5	29.5	26	47

*This table includes results from only 20 of the 76 samples included in the experiment. Results from the remaining 56 are within the same range of variation as these illustrative examples.

†All weights expressed in mg after drying to constant weight at 105° C.

procedure followed may itself affect the results. As Hägglund (1951) points out after a careful consideration and comparison of the various techniques for determining the amount of cellulose in wood, "It is evident that the yield of cellulose depends very much upon the method of isolation." Wise (1944) had earlier reached the same conclusion with regard to isolation of α-cellulose, and remarked, "Hence, the only criterion of accuracy is that of reproducibility." With these difficulties in mind, every effort was made to select the best analytical methods that could be adapted to small quantities of material and to standardize all procedures. The data in tables 2 and 3 illustrate the degree of reproducibility that was obtained from aliquots of the same wood sample.

For direct determination of α-cellulose, the Hägglund (1951) process, modified in the last step by substituting the Flanders (1952) extractions for removing xylan and other noncellulosic cell wall materials, was used. This procedure gave an average figure of around 36 per cent for α-cellulose. For sapwood alone, the figure was 34.64 per cent (range 34.2 to 35 per cent), for heartwood, 37 per cent (range 36.4 to 37.6 per cent). Additional samples of whole Douglas fir, analyzed according to the monoethanolamine-ethanol (1:1; v/v) process (Vincent, 1946; Ott, Spurlin, and Grafflin, 1954), gave similar results but a wider range of variation, from an α-cellulose low of 31 per cent to a high of 40 per cent.

Polyuronide hemicelluloses, when calculated by difference as is necessary in the Hägglund process, accounted for only 8.73 per cent of the original dry weight of the wood. When determined directly by the probably more accurate method of extracting with cold 4 per cent NaOH following mild chlorination and extraction with ethanolamine, this figure averaged 15.5 per cent (experimental range 13.5 to 18 per cent).

The noncellulosic polysaccharides, determined by the Flanders process, account for 12.3 per cent and 11.84 per cent of the original weight of heartwood and sapwood, respectively. Bonner (1950) comments that the woods of temperate tree species are characterized by "variable" amounts of the noncellulosic xylans, mannans, etc. By using a different procedure (extraction with 17.5 per cent NaOH after removal of lignin and hemicellulose), we found amounts varying from 9 to 12.5 per cent in the whole Douglas fir samples.

The Klason process for direct determination of lignin, repeated a number of times with several modifications, gave val-

TABLE 2

WEIGHTS OF DOUGLAS FIR HEARTWOOD AND SAPWOOD SAMPLES AT DIFFERENT STAGES DURING THE PROCEDURE FOR DETERMINATION OF α-CELLULOSE

	Heartwood Sample 1	Heartwood Sample 2	Heartwood Sample 3	Heartwood Sample 4	Heartwood Average of 4 Samples	Sapwood Sample 1	Sapwood Sample 2	Sapwood Sample 3	Sapwood Sample 4	Sapwood Average of 4 Samples
Original weight	500	500	500	500	500	500	500	500	500	500
Weight after first cooking; Sulfite process	448	440	445	440	443.2	419	427	403	408	414
Weight after delignification and bleaching	243	243	248	252	246.5	231	228	236	234	232.4
Weight after Flanders extraction = α-cellulose content	185	182	188	187	185	174	175	173	171	173.2

All samples dried to constant weight at 105° C.

All weights expressed in mg.

TABLE 3

PER CENT OF α-CELLULOSE IN DOUGLAS FIR HEARTWOOD AND SAPWOOD SAMPLES, DETERMINED BY HÄGGLUND-FLANDERS PROCESS; PER CENT COMPOSITION OF OTHER COMPONENTS BY CALCULATION. ALL FIGURES REPRESENT PER CENT OF ORIGINAL WEIGHTS

	Heartwood Sample 1	Heartwood Sample 2	Heartwood Sample 3	Heartwood Sample 4	Heartwood Average of 4 Samples	Sapwood Sample 1	Sapwood Sample 2	Sapwood Sample 3	Sapwood Sample 4	Sapwood Average of 4 Samples
% loss after first cooking; Sulfite process	10.4	12	11	12	11.36	16.2	14.6	19.4	18.4	17.2
% loss after delignification and bleaching; equivalent to lignin plus polyuronide hemicelluloses	41	39.4	39.4	37.6	39.34	37.6	39.8	33.4	34.8	36.32
% loss after Flanders extraction; equivalent to noncellulosic polysaccharides	11.6	12.2	12	13	12.3	11.4	10.6	12.6	12.6	11.84
% α-cellulose; by direct determination	37	36.4	37.6	37.4	37	34.8	35	34.6	34.2	34.64

ues of around 29 per cent (range 27.4 to 32 per cent), whereas lignin analyses by the ethanol-monoethanolamine extraction procedure resulted in figures of 31 to 37 per cent. We suspect that the latter figures are higher than the true value, for the 29 per cent lignin concentration determined by the Klason process is closer to that reported for Douglas fir by the Institute of Paper Chemistry (Ott, Spurlin, and Grafflin, 1954).

The average value for each major constituent of Douglas fir is given below and compared with the results from similar analyses of dried fecal pellets.

	Average Composition of:	
	Douglas fir	Pellets
α-cellulose	36.0%	33.6%
polyuronide hemicelluloses	15.5	0.2
noncellulosic polysaccharides	12.0	0.8
lignin	29.0	63.2

On the basis of these figures it is apparent that only a barely detectable amount of the polyuronide hemicelluloses and noncellulosic cell wall components remained in the pellets, and that the cellulose to lignin ratio had been drastically changed. If it is assumed that Limnoria is incapable of attacking lignin, then only about half of the expected amount of α-cellulose is present in the pellets; the rest has presumably been hydrolyzed by the animals. Since it is known that the digestive enzymes of Limnoria can degrade cellulose in vitro, this conclusion seems to be fully warranted. In summary, the chemical analyses of the wood that passes through the gut of Limnoria indicate that the animal is capable of removing from it nearly all of the polyuronide hemicelluloses and noncellulosic carbohydrates and about half of the cellulose.

We conclude, as a result of these investigations, that Limnoria does indeed feed directly upon the wood into which it burrows. Like Hoek and the other early workers, we believe that wood serves as the animal's chief nutriment, but, and here again we must agree with Hoek, whether wood fulfills all of Limnoria's dietary requirements we cannot say with certainty. There are still unanswered questions regarding the nutritional physiology of Limnoria, and paramount among these is the problem of nitrogen metabolism. It can safely be assumed that these animals require some nitrogen, but how much they need and where it comes from is not at present known. The fact that Limnoria has no recognizable excretory organ and that no nitrogenous excretory products have ever been reliably identified might be interpreted as indicating that these animals conserve nitrogen in a highly efficient manner and may require only a small outside supply. But there are no quantitative data as yet available that could help to answer this question.

LITERATURE CITED

Bailey, I. W. 1952. Biological Processes in the Formation of Wood. Science, 115:255-59.

Bate, C. Spence, and J. O. Westwood. 1868. A History of the British Sessile-eyed Crustacea. Vol. II, Isopoda. J. Van Voorst, London.

Bonner, James. 1950. Plant Biochemistry. Academic Press Inc., New York.

Cleveland, L. R. 1924. The Physiological and Symbiotic Relationships between the Intestinal Protozoa of Termites and Their Host, with Special Reference to *Reticulitermes flavipes* Kollar. Biol. Bull., 46:178-227.

-------. 1925a. The Method by Which *Trichonympha campanula,* a Protozoan in the Intestine of Termites, Ingests Solid Particles of Wood for Food. Biol. Bull., 48:282-88.

-------. 1925b. The Feeding Habit of Termite Castes and Its Relation to Their Intestinal Flagellates. Biol. Bull., 48:295-308.

Coldstream, J. 1834. On the Structure and Habits of *Limnoria terebrans*. Edinburgh New Philos. J., 16:316-34.

Dalyell, John Graham. 1851. The Powers of the Creator Displayed in the Creation. Vol. I. London.

Fahrenbach, W. H. 1959. Studies on the Histology and Cytology of Midgut Diverticula of *Limnoria lignorum*. In this volume.

Flanders, C. A. 1952. Hemicellulose Fractions from Some Hays and Straws. Arch. Biochem. Biophys., 36:425-29.

Hägglund, Erik. 1951. Chemistry of Wood. Academic Press Inc., New York.

Henderson, Jean T. 1924. The Gribble: A Study of the Distribution Factors and Life History of *Limnoria lignorum* at St. Andrews, N.B. Contrib. Can. Biol., N.S., 2:307-26..

Hoek, P. P. C. 1893. Rapport der Commissie uit de Koninklijke Akademie van Wetenschappen. . . . betreffende de levenswijze en de werking van *Limnoria lignorum*. Verhandel.

Koninkl. Akad. Wetenschappen, (tweede sectie) 1, No. 6:1-103.

Hungate, R. E. 1936. Studies on the Nutrition of *Zootermopsis*. I, The Role of Bacteria and Molds in Cellulose Decomposition. Zentralbl. Bakter., Ser. 2, 94:240-49.

-------. 1938. Studies on thé Nutrition of Zootermopsis. II, The Relative Importance of the Termite and the Protozoa in Wood Digestion. Ecology, 19:1-25.

-------. 1944. Termite Growth and Nitrogen Utilization in Laboratory Cultures. Proc. and Trans. Texas. Acad. Sci., 27: 91-98.

Kofoid, Charles A., and Robert C. Miller. 1927. Biological Section, pp. 188-343, in Marine Borers and Their Relation to Marine Construction on the Pacific Coast. Final Report, San Francisco Bay Marine Piling Committee.

Larkin, Francis T. 1951. Cellulose Utilization by Limnoria. Rept. Marine Borer Conf., U.S. Naval Civ. Eng. Res. and Eval. Lab., Port Hueneme, Calif. Pp. L-1 to L-4.

Ott, Emil, Harold M. Spurlin, and Mildred W. Grafflin (eds.). 1954. Cellulose and Cellulose Derivatives. Part I. Interscience Publ. Inc., New York.

Ray, D. L. 1951. The Occurrence of Cellulase in *Limnoria lignorum*. Rept. Marine Borer Conf., U.S. Naval Civ. Eng. Res. and Eval. Lab., Port Hueneme, Calif. Pp. K-1 to K-5.

-------. 1953. Digestion of Wood by *Limnoria lignorum* (Rathke). Proc. XIV Internat. Cong. Zool. Copenhagen (abstract), p. 279.

-------. 1959. Some Properties of Cellulase from Limnoria. In this volume.

-------, and Jean R. Julian. 1952. Occurrence of Cellulase in *Limnoria*. Nature, 169:32.

Smeaton, John. 1812. The Report of John Smeaton, Engineer, upon the State of the Bridlington Piers, with the Most Probable Means of Preserving the Same from the Destruction of the Worm. [Dated 15th May 1778.] In Reports of the late John Smeaton . . . 3:187-93. London.

Trager, W. 1932. A Cellulase from the Symbiotic Intestinal Flagellates of Termites and of the Roach, *Cryptocercus punctulatus*. Biochem. J., 26:1763-71.

-------. 1934. The Cultivation of a Cellulose-Digesting Flagellate, *Trichomonas termopsidis*, and of Certain Other Termite Protozoa. Biol. Bull., 66:182-90.

Vincent, Rex. 1946. The Importance of Hemicellulose in Wood Pulp. Paper Trade J. 122:53-57.

Wise, Louis E. 1944. Wood Chemistry. Reinhold Publ. Co., New York.

Yonge, C. M. 1927. The Absence of Cellulase in *Limnoria*. Nature, 119:855.

-------. 1950. The Sea Shore. Collins, 14 St. James's Place, London.

-------. 1951. Marine Boring Organisms. Research, 4:162-67.

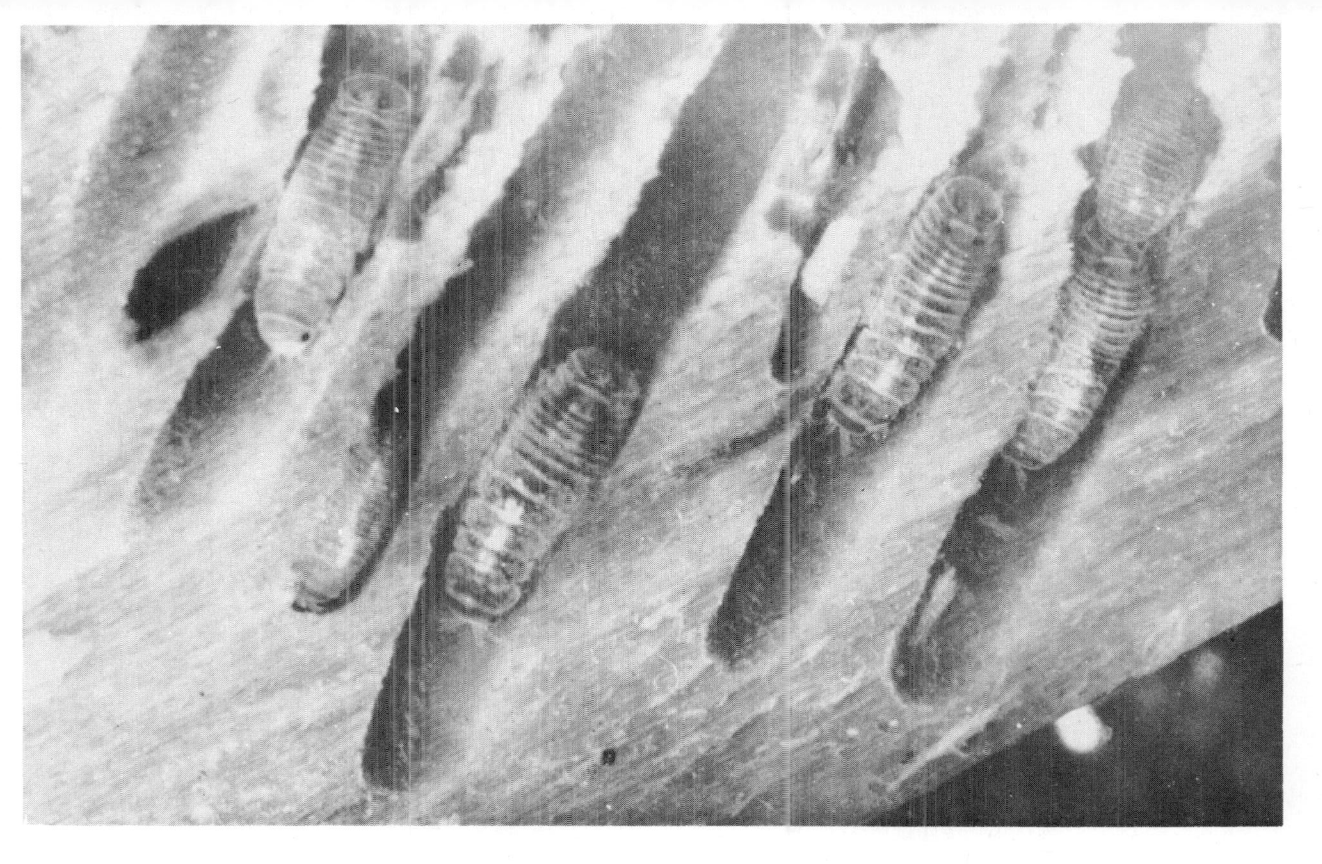

Fig. 1. Limnoria photographed alive immediately after wood was split to show normal position of animals within their burrows

BIOLOGICAL INVESTIGATIONS ON MARINE BORERS IN BERLIN-DAHLEM

Günther Becker

Marine borer investigations were started in Berlin-Dahlem in August, 1937. They were interrupted in 1944 and were resumed in 1951. Since 1954 Dr. W.-D. Kampf has been engaged in zoological research work on Limnoria and Teredo, and since 1956 Dr. J. Kohlmeyer has been dealing with marine fungi. This report summarizes the results so far obtained.

Geographical Distribution of Borer Species

Teredo navalis L. is a common pest in the North Sea and the western Baltic Sea. According to earlier records, until 1929 its limit of distribution was the harbor of Warnemünde. During the hot summers of 1933 to 1936 this borer extended its range up to the Darsz peninsula (Becker, 1938). Here the salinity is about 8‰; it decreases rapidly in an easterly direction. *Teredo navalis* is still an important pest at the western shores of the Darsz.

Limnoria (fig. 1) was collected at several places in the Mediterranean. Contrary to many reports, *Limnoria lignorum* Rathke, the classical Limnoria species, does not occur in the Mediterranean. The most common species seems to be *Limnoria tripunctata* Menzies, distributed from the western to the eastern parts of the Mediterranean and from the northern coasts to Africa. A second species, *Limnoria carinata* Menzies and Becker, has been found up to now only along the western coast of Italy (Menzies and Becker, 1957).

The genus Limnoria was collected by the author for the first time on Indian continental coasts in the southern parts of the east coast near Ceylon and in Madras harbor. Most specimens belong to a new species, *Limnoria indica* Kampf and Becker, which is a Limnoria species with distinct sexual dimorphism. In Madras *Limnoria tripunctata* was also detected.

Breeding of Borers in the Laboratory

It was an idea of the late Dr. William F. Clapp to use marine borers in the laboratory which was adopted by B. Schulze when he appointed the author in 1937 to develop laboratory methods of breeding and testing. The advantage of such methods is so well known that a discussion is unnecessary. All results mentioned below are based on laboratory investigations conducted far from the coast in artificial sea water.

Limnoria species and the rhythm of development of L. tripunctata

Limnoria lignorum from the North Sea survived for only a few weeks when cultivated at a temperature of 20°C in normal air-conditioned rooms of the institute. No reproduction took place. The conclusion that a constant temperature of 20°C is too high for this species from northern regions (Becker, 1944) was later verified by observations of the natural distribution (Kühl, 1957). Since it is too expensive to cool down a number of aquaria or a whole room for months, cultures were arranged with Limnoria from the Mediterranean. They proved to be very successful in artificial sea water at a salinity of 35‰. Individuals remain alive for a long time, and many generations have been bred at a constant temperature of 20°C as well as at 27° to 28°C. New attack of immersed timber takes place in the aquaria by active, swarming gribbles and after passive transfer from infested wood. But, as under natural conditions in the sea, a precondition for boring new tunnels and for development is that the timber blocks or panels must have been exposed to the sea water for a sufficiently long time, usually a few weeks.

For breeding purposes it seems to be advisable to use aquaria that are not too small. A capacity of not less than about 5 liters is recommended. Continuous or at least frequent aeration is necessary. Recurrent filtration is advantageous if there is a large number of gribbles in one aquarium. In the case of smaller cultures, change of water once or twice a week or even less frequently will do. The water lost by evaporation must be replenished by adding distilled water, of course, and a water-level mark must therefore be fixed in advance. The pH should approximate the neutral point. With continual filtration and aeration, it was possible to breed Limnoria in glass tanks holding about 30 liters of artificial sea water unchanged for several years and many new generations were obtained. The two gribble species from the Mediterranean have now been cultivated in the laboratory without interruption for more than six years.

The swarming behavior, the periods of development and reproduction, and the total life span have been determined by Kampf (1957). *Limnoria tripunctata* and *L. carinata* swarm throughout the whole year when kept at a constant temperature of 20°C in the aquarium. With test samples replaced monthly, no seasonal rhythm of swarming activity could be detected, as has otherwise been observed in nature. Thus the rhythm of swarming seems to be dependent on temperature. In the laboratory females with eggs and young participated in swarming, while according to Johnson and Menzies (1956) this is not the case in the sea.

The periodicity of reproduction in Limnoria species that was detected by several observers in nature has repeatedly been observed in connection with a change of temperature. But, when the temperature is maintained constantly at 20°C, as it has been under our experimental conditions for a total of six years, there seems to be an endogenous rhythm of propagation with the bulk of the eggs and embryos appearing in the marsupium of the female during April and May.

Teredo cultures and lunar periodicity of T. pedicellata

Teredo navalis L. from the North Sea has been maintained in the laboratory for a long time. Some individuals have survived for about three years (Becker, 1944); this is contrary to the field experience and to the widespread belief that *T. navalis* has a normal life span of only one or two years. It may be that unfavorable nutritional conditions slowed the rate of development in aquaria. But never, in the course of years, was a new generation of shipworms observed. Probably the swarming period of the larvae of this species, which is said to be 3 or 4 weeks in the sea and might be somewhat shorter at 20°C, is too long for normal development under artificial conditions in the laboratory. Imai, Hatanaka, and Sato (1950) described the laboratory breeding of a Teredo species which they called *T. navalis*. They mention that the larvae of their species swarmed for only a few days. Therefore this Teredo of Japan seems to be a different species or at least a physiological variety.

In 1940 Roch published his monograph on the Teredinidae of

the Mediterranean. According to his investigations *Teredo pedicellata* Quatref. and *T. utriculus* Gmelin are larviparous, and the larvae settle on the timber after a swarming period of 1 or 2 days. He was able to breed them in the laboratory for several generations in running sea water at Rovigno d'Istria. During the war these species were also cultivated in artificial sea water in Berlin-Dahlem until the investigations were interrupted in 1944. However, probably because of the lack of plankton, reproduction in the aquarium was irregular. Since 1955 we have *Teredo pedicellata* again in Berlin-Dahlem, and we have now successfully reared this species through four generations. This seems to be the first case of successful cultivation of Teredo far from the sea in artificial sea water in a laboratory.

Culture methods for Teredo and Limnoria have already been described by Becker and Schulze (1950). We are now adding phytoplankton to Teredo cultures, but it seems that they are able to develop in aquaria without special enrichment of plankton. When maintained continuously at 20°C, young *T. pedicellata* first became reproductive after 8 weeks. This is in line with recorded observations on the time lapse between the settling of young animals and the appearance of fertile individuals in subtropical seas.

A curious feature of *Teredo pedicellata* is a marked lunar periodicity. This has been observed and described by Roch (1940). Larvae are discharged only during the time span extending from 24 to a few hours before astronomical full moon, the maximum rate of elimination occurring at about 10 hours before full moon. This periodicity is independent of tides, light, and thick brick or concrete walls. Observations during the last two years have shown that each new generation swarms only during the above-mentioned interval, and that the first young larvae of the next generation appear two full moons later. The most mysterious fact, however, is the behavior of the animals during total lunar eclipse, which was observed once by Roch in Venice in 1931 and again by us in Berlin-Dahlem in 1957 with the same result. The discharge of the larvae is interrupted before the normal maximum of swarming, and the retained larvae are eliminated later, more or less coinciding with the first quarter of moon. Remarkably enough, the animals react to the eclipse several hours before the earth's shadow begins to fall on the moon. We are trying to find out whether the normal lunar periodicity is based on gravity, on radiation, or on long electrical waves.

The Teredo species which has been used for laboratory cultures in Miami since 1950 is also called *T. pedicellata.* But Isham and Tierney (1953) mention that it seems to be doubtful whether this species is indeed identical with the species that has been described from Europe. It is evident from their publication the the shipworms from Miami do not show a lunar periodicity at all. Therefore the Miami specimens probably represent at least another physiological variety. We should try to determine whether specimens of *T. pedicellata* from the Caribbean and the Mediterranean can be crossed.

Ecological Investigations on Limnoria

The effects of environmental conditions on *L. tripunctata*, discussed below, have been studied by Kampf (1957). The experiments were possible because we had been successful in growing the organisms in the laboratory.

Temperature

Contrary to *L. lignorum*, with its above-mentioned low temperature optimum and maximum, *L. tripunctata* develops well at high temperatures; for many years it has been reared satisfactorily at 27° to 28°C. But mortality of eggs, embryos, and adults is high, and it seems that only a small number of individuals can develop normally under these conditions. Kampf has found 30°C to be the upper limit for this species; the lowest temperature for full development is 7°C.

As expected, the generation time decreases with an increase in temperature (fig. 2), but this also entails a progressive decrease in the number of eggs in the marsupium (table 1). It must be stated, however, that in these experiments the fertile females were immediately transferred from aquaria at 20°C to those at the higher temperatures. When transferred to aquaria at 6° to 7°C, only those embryos that already showed visible eye formation developed further. Earlier stages were resorbed.

The longest life span is about 4 months at 30°C and up to approximately one year at 20°C (fig. 3). If the water temperature remains below 20°C for a certain length of time, as is the case in the Mediterranean and other moderate zones, Limnoria is capable of living significantly longer than one year.

In cultures maintained at a constant temperature of 26°C the feeding activity, measured by the amount of excrement, was two or three times, the number of swarming gribbles about

TABLE 1
DEPENDENCE OF THE DEVELOPMENT OF
LIMNORIA TRIPUNCTATA ON TEMPERATURE

Temp. °C	Embryonic Development		Total Development
	Time in Days	Percentage of Hatched Young Animals	Maximum Length of Life in Weeks
20	17	85	50
22	15	65	44
26	13	51	28
30	11	42	17

twice, and the share of females participating in swarming about three times as great as in cultures kept constantly at 20°C.

The most favorable temperature range for permanent breeding seems to lie between 22° and 26°C, but at 20°C and below the animals still develop well. The upper limit of 30°C should explain why on the continental coast of India, where the water temperature reaches and exceeds 30°C, Limnoria does not develop plentifully and become a pest; because of its scarcity, it was found there only last year (Becker and Kampf, 1957). Only where more heat-resistant Limnoria species or races occur can we expect this genus to be important in tropical oceans.

Salinity

If specimens of *Limnoria tripunctata*, after having been kept at a constant salinity of 35‰, are suddenly brought into water of lower salinity (30 to 25‰), the animals continue to develop well. But if they are exposed to 20‰ most of them die. Some individuals were capable of adjusting to this decrease in salinity and survived, but they were no longer able to propagate. When specimens were suddenly transferred from 35 to 15‰ salinity, all the animals died after 11 days at the latest; in water with a still lower salinity death resulted after a few hours to one day. Exposure for less than 30 minutes to slightly saline or even distilled water apparently did not hurt the animals, provided they were afterward transferred to more concentrated salt solution. Whether the animals can gradually be acclimatized to salinities below 25‰, which latter they tolerate well, has not yet been tested.

If specimens of *L. tripunctata* are transferred from an environment with a salinity of 35 to one of 25 to 30‰, the activity of the gribbles is significantly stimulated. During the first few weeks the animals produce 4 to 5 times as much excrement as those maintained at 35‰ salinity. After some weeks, however, they begin to feed more slowly; at 30‰ the decline is eventually overcome, but not at 25‰ (fig. 4). Kampf concludes that the most favorable breeding conditions involve an intermittent decrease in salinity by about 5‰. These laboratory results agree well with Sømme's (1941) observations on *L. lignorum* in the open seas around Norway, where fluctuations in the fresh water inflow from rivers affected the gribbles favorably. Observations in India with Sphaeroma species likewise indicate that recurrent changes in salinity stimulate activity and provoke swarming (Pillai and John, personal communication).

Hydrogen-ion concentration

Some tests on the effect of hydrogen-ion concentrations on *Limnoria tripunctata* showed a pH range of 6 to 9 to be suitable. At a pH down to 5 and up to 11 the gribbles still remained alive for about 3 weeks. According to these observations this Limnoria species is more sensitive to an acid than to an alkaline environment. Walker *et al.* (1926) and White (1929) obtained somewhat different results. The question is not yet answered whether the various species exhibit differences in their response to hydrogen-ion concentration.

Nutrition of Limnoria and Its Symbiosis with Fungi

At present our knowledge about the nutrition of the genus Limnoria is not sufficiently clear. The wood that is rasped during the boring of the tunnels is eaten, and according to Kofoid (1921) and others it is said to be digested. Yonge (1927) could not find cellulase within Limnoria and therefore assumed that the gribbles bore their tunnels only for protection and feed on the microfauna and -flora living on the wood surface. Ray and Julian (1952), however, demonstrated that there is a cellulase in the midgut diverticula of Limnoria. Thus they concluded that wood is utilized as nutriment, but they did not regard cellulose as necessarily the only source of carbohydrates.

It is known that wood freshly immersed in sea water is attacked neither by larvae of wood-destroying shipworms nor by wood-inhabiting crustaceans during the first weeks. Even in tropical waters at least 10 days pass until attack by the borers takes place. As already mentioned, Limnoria develops in the laboratory only if the wood has been submerged in sea water for several weeks (Becker and Schulze, 1950). According to Zo-Bell (1946) and Deschamps (1952), cellulose- and lignin-decomposing bacteria and fungi causing a chemical and mechanical change of the wood surface may create the preconditions for an attack by wood-eating animals. Isham and Tierney (1953) suggested that the effects of either the surface microflora, the sea water itself, or a combination of both is responsible for the softening and partial removal of intracellular constituents of the wood.

The observation that wood-inhabiting fungi regularly occur in laboratory cultures of *Limnoria tripunctata* and *L. carinata* induced the author to suggest that Kohlmeyer undertake a careful investigation of the wood inhabited by Limnoria. According to his examinations all samples of soft- and hardwood species contained mycelium of fungi after having been submerged in sea water for several weeks. This was especially evident in the secondary walls and conspicuous in the thick-walled summer-wood tracheids (figs. 5 and 6). The fungi were found to be localized in the outer parts of the wood samples and--especially interesting here--in the vicinity of the Limnoria tunnels (fig. 7). The deterioration pattern of these fungi corresponds fully with that of the soft rot fungi occurring in wet wood on land (Findlay and Savory, 1950, 1954). These fungi, as well as those already known from sea water (Barghoorn and Linder, 1944; Meyers, 1953; and others), are Ascomycetes and Fungi Imperfecti. A fungus belonging to the genus Helicoma was regularly and most commonly found in the aquarium cultures in Berlin-Dahlem where Limnoria had settled on the wood. Wood samples from the North Sea, the Mediterranean, and the Indian Ocean with and without Limnoria burrows also regularly contained fungi. These are eaten by Limnoria, as is obvious from the ever present conidia in the excrement.

The detection of this regular association of Limnoria with fungi in wood induced us to carry out a large number of systematic tests on the nutrition and habits of Limnoria; the results have recently been reviewed (Becker, Kampf, and Kohlmeyer, 1957). Comparable populations of *L. tripunctata*, placed in Petri dishes (5 cm) each containing 10 ml of artificial sea water, which was renewed at weekly intervals, were incubated at 20° and 28°C, respectively. The cultures were supplied with various ingredients and maintained until the animals were dead. Where absence of microorganisms was aimed at, the Petri dishes with sea water and food were autoclaved once or twice weekly so as to minimize contamination of the foodstuffs by the viable conidia voided by the gribbles with the excreta. The results discussed below are statistically significant to the extent of 95 to 99 per cent.

In the absence of a food supply the animals remained alive for about 9 weeks at 20°C, or 4 weeks at 28°C. Animals fed fungus-free pine wood died earlier than starved ones. Hence, pine wood unaltered by fungi or other microbes is unsuitable as food for Limnoria. That the animals die more rapidly in dishes with sterile wood than in the control dishes may be a consequence of energy consumption for the obvious rasping, not compensated by an energy supply (fig. 8).

Another test, using cellulosehydrate-foil, showed the decisive importance of the fungi for behavior and nutrition of Limnoria.* If sterile, the material was hardly attacked by the gribbles, the survival time of which was again somewhat shorter than that of starved animals. But after attack by fungi the foil was severely destroyed (fig. 9), and the animals feeding on it lived significantly longer than the controls (fig. 10).

It does not seem probable that Limnoria can eat only wood that has been mechanically softened by microbes--as assumed by ZoBell and Deschamps--because the gribbles obviously rasp at fresh wood and also attack tropical woods which, even after fungal attack, are harder than unattacked indigenous wood. The fungus-free cellulosehydrate-foil and the wood treated with hypochlorite were not eaten, although both are mechanically softer than wood invaded by fungi.

Now it seemed necessary to answer the question whether the gribbles feed exclusively on fungi and wood serves only for housing purposes--as Yonge believed--and as substrate for the fungi. Therefore some of the cultures were supplied with mycelium of an imperfect fungus--probably *Helicoma sp.*--from an aquarium, alive as well as autoclaved, and also with live mycelium of a

---

*The entrance of the hyphae of Ascomycetes and Fungi Imperfecti into such foils as well as the decomposition of this cellulose were described by Kohlmeyer (1956).

number of brown-rot basidiomycetes of the genera Coniophora, Polyporus, and Lenzites, and of the white-rot fungus *Pleurotus ostreatus*. As shown by the accumulation of the thick-walled conidia of the imperfect mold in the excreta--in contrast to the mycelium, the conidia are not digested--the mycelium was actually eaten, as was also that of the basidiomycetes. Nevertheless, the gribbles in these tests did not on the average outlive those in the controls. Only in the case of Pleurotus were the animals able to utilize the mycelium better; here they survived markedly longer than in the controls. Thus fungi alone do not suffice as food for Limnoria, which seems to need certain components of the wood that are produced under the influence of the molds.

In the presence of fungus-free starch paste the life span was similar to that of the controls, even though Limnoria contains amylase (Yonge, 1927). Neither was a vegetable fat ("Palmin") by itself suitable as food. But endosperm from date-palm kernels (*Phoenix dactylifera* L.), which consists mainly of hemicelluloses, as well as the pulp of the common sunflower (*Helianthus annuus* L.), supplied as mold-free substrates, could be utilized by the animals. If sunflower pulp was given exclusively, the animals' life span even reached the limit of those maintained on wood with fungal attack. But with this exclusive nutrition they remained strikingly small and apparently did not reproduce.

Hence the gribbles can utilize hemicelluloses and other carbohydrates which are more easily decomposed enzymatically than the native cell wall substance of wood, but--just like other wood-destroying animals--they cannot get along with carbohydrates alone. In case hemicelluloses or other readily decomposable carbohydrates are not available, they obviously need for their natural nutrition certain decomposition products of the wood. These products are formed by the action of the fungi. Besides, they need the mycelium of these fungi, probably largely for their protein, vitamin, and fat metabolism.

Whenever Helicoma or some other fungus reached a copious development in the wood in our laboratory cultures, Limnoria was obviously inhibited or even damaged. Repeatedly this caused extinction of the gribbles in the wood pieces, darkly colored by the spores and hyphae of the fungi. This unfavorable influence of an excessive mold growth on Limnoria cannot yet be explained. It could be caused by a depletion of certain carbohydrates that are important for the gribbles, or by a toxic chemical effect of the high concentration of the fungi. It may be expected that this effect varies with the species as has been found in the case of interrelations between wood-destroying beetles and fungi (Becker, 1942, 1943).

In spite of the just-mentioned observations, the results obtained so far seem to prove that Limnoria is obligatorily dependent on a symbiosis with wood-inhabiting fungi. The advantage to the fungi may well be restricted to the effective distribution of their spores through the excrement of the gribbles; perhaps metabolic products of the latter also contribute to the fungus growth. This relation of a distinctly one-sided benefit presents an entirely new aspect of ectosymbiosis between animals and plants among marine organisms, although comparable cases have long been known in various manifestations among terrestrial wood-destroying animals.

The knowledge of this compulsory dependence of the gribbles on wood-decomposing fungi creates a new basis for protection of wood against these animals. If the timber can be protected against the fungi, it should be possible also to prevent a marked attack by Limnoria.

The Relation of Larval Settling and Development of Teredo to Marine Fungi

As has long been known, the larvae of the shipworms also settle on wood and enter it after their metamorphosis provided the wood has been in sea water long enough. In connection with the nutritional physiology of Limnoria it has already been mentioned that some authors believe that the microorganisms and their wood-decomposing effect create the necessary conditions for attack. Pertinent data based on experimental investigations with Teredo are not known to us. It seems to be noteworthy that the formation of a primary film, which according to Daniel (1955) and others is said to be an indispensable requirement for the settling of fouling organisms, proceeds in a far shorter time (only a few days) than the development of conditions for attack by marine borers. It is known that for the fouling organisms a surface cover of algae, diatoms, and bacteria is sufficient. The longer lag in the case of attack by marine borers indicates that microbial processes, penetrating deeper into the wood and affecting it more intensively, must have been taking place. This at once stresses the decisive participation of the fungi in particular.

Some time ago the author was struck by the fact that in wood samples from the Mediterranean the entrance holes of *Teredo pedicellata* and *T. utriculus* were mostly situated in the summerwood rings. Moreover, pine wood samples in the Berlin-Dahlem aquaria, into which *Teredo pedicellata* had bored, showed lime tubes predominantly in the summerwood rings (fig. 11). Nevertheless, the literature contains several references to the effect that settling larvae display a particular preference for springwood. In accordance with these statements, examination of test pieces from aquaria cultures showed that settling larvae do not prefer the summerwood at all; but the mortality of young shipworms in the layers of springwood is significantly higher than in the summerwood. Under natural conditions the shipworms normally cross summerwood as well as springwood when boring into poles. Therefore such observations are to be expected more with test panels in the aquaria. As the fungi mostly decompose the thick secondary cell walls of the summerwood, there could also exist a relation between shipworms and fungi. Therefore this question was examined.

In Petri dishes, U tubes, and other containers the influence of several materials on the settlement of *Teredo pedicellata* was followed up.* It was difficult to eliminate the influence of unilateral dim light. This was possible only by reversal tests. Materials whose influence was to be tested were exposed to the larvae partly directly and partly behind a partition of nylon gauze in order to give or to prevent the possibility of boring. Details of methods and analysis of these tests are omitted.

The results establish that wood infested by fungi shows a marked attracting effect if the larvae can come into contact with it. Neither organic acids, which in earlier tests (Harrington, 1921) were said to be attracting, nor timber free from fungi, nor fungi alone showed the same effect. These observations indicate that there is no tactic response of the animals to substances released by the wood and dissolved in the water. Whether only a softening or other changes of the wood surface caused by fungi makes the larvae remain on the wood has not yet been cleared up by enough different experiments. It is equally possible that Teredo is attracted and induced to settle by compounds formed by the activity of fungi in the wood, or by a specific combination of products of breakdown of wood constituents

---

*I wish to thank Drs. W.-D. Kampf and J. Kohlmeyer for carrying out these experiments.

by fungal enzymes and substances contained in or produced by fungi. These investigations are not yet finished, and therefore the questions cannot be answered definitively.

Besides this attraction effect there may be expected a physiological influence on nutrition: the fungi may be essential for the animal's protein, vitamin, and fat metabolism. Lasker and Lane (1953) showed in nice tests that Teredo also needs plankton as well as wood for its growth. Some of the animo acids contained in the shipworm body are not present in wood but do occur in plankton. For certain wood-boring insects the protein content of the wood is of paramount importance (Becker, 1943b); and Koch (1956) has shown that wood is deficient in many vitamins that are needed for the development of animals. Hence the marine wood-inhabiting fungi may be important to the shipworms as suppliers of protein and vitamins, so that their role may be compared with that of the plankton. This will be greater where the plankton content in the sea water is lower. It could be expected that the fungi have more influence in the Mediterranean, which is known to be poor in plankton, or in our aquaria, than in eutrophic seas. This aspect of the nutritional physiology must be followed in the future.

Whether the role of the fungi in the life of shipworms and gribbles is similar is thus not yet clear; that the fungi do influence the settling behavior of shipworm larvae, and are nutritionally important after metamorphosis, seems, however, reasonably certain at this stage.

LITERATURE CITED

Barghoorn, E. S., and D. H. Linder. 1944. Marine Fungi: Their Taxonomy and Biology. Farlowia, 3:395-467.

Becker, G. 1938. Die Bohrmuschel Teredo, der gefährlichste Holzzerstörer an deutschen Küsten. Holz als Roh- u. Werkstoff, 1:249.

-------. 1942. Untersuchungen über die Ernährungsphysiologie der Hausbockkäferlarven. Z. vergl. Physiol., 29:315-88.

-------. 1943a. Beobachtungen und experimentelle Untersuchungen zur Kenntnis des Mulmbockkäfers (*Ergates faber* L.) II, Die Larvenentwicklung. Z. angew. Ent., 30:263-96.

-------. 1943b. Ernährung und Klimaabhängigkeit holzzerstörender Insekten. Forsch. u. Fortschr., 19:179-80.

-------. 1944. Holzschutzaufgaben gegen Meerwasser-Schädlinge. Z. hyg. Zool. u. Schädlingsbek., 36:51-67.

-------. 1950. Zerstörung des Holzes durch Tiere. In Mahlke-Troschel-Liese, Handbuch der Holzkonservierung, 3. Aufl., pp. 111-65. Berlin, J. Springer.

-------. 1955. Ueber die Giftwirkung von anorganischen Salzen, -Chlornaphthalin und Kontaktinsektiziden auf die Holzbohrassel Limnoria. Holz als Roh- u. Werkstoff, 13:457-61.

-------, and B. Schulze. 1950. Laboratoriumsprüfung von Holzschutzmitteln gegen Meerwasserschädlinge. Wiss. Abh. Dt. Materialprüfungsanst., II, 7:76-83.

-------, and W. -D. Kampf. 1955. Die Holzbohrasseln der Gattung Limnoria (Isopoda) und ihre Lebensweise, Entwicklung und Umweltabhängigkeit. Z. angew. Zool., 42:477-517.

-------, W. -D. Kampf, and J. Kohlmeyer. 1957. Zur Ernährung der Holzbohrasseln der Gattung Limnoria. Naturwiss., 44:473-74.

-------, and W. -D. Kampf. 1957. Funde der holzzerstörenden Isopodengattung Limnoria und der Festlandküste Indiens und Neubeschreibung von *Limnoria indica*. Z. angew. Zool., 44.

Daniel, A. 1955. The Primary Film as a Factor in Settlement of Marine Foulers. J. Madras Univ., B. 25:189-200.

Deschamps, P. 1952. Xylophages marins. Protection des bois immergés contre les animaux perforants. Peintures, Pigments, Vernis, 28:607-10.

Findlay, W. P. K., and J. G. Savory. 1950. Breakdown of Timber in Water-cooling Towers. Proc. Int. Bot. Congr., 7:315-16.

-------, and J. G. Savory. 1954. Moderfäule. Die Zersetzung von Holz durch niedere Pilze. Holz als Roh- u. Werkstoff, 12:293-96.

Harrington, C. R. 1921. A Note on the Physiology of the Shipworm (*Teredo norwegica*). Biochem. J., 15:736-41.

Imai, T., M. Hatanaka, and R. Sato. 1950. Breeding of the Marine Timber Borer *Teredo navalis* in Tanks and Its Use for Antiboring Test. Tohoku J. Agric. Res., 1:199-209.

Isham, L. B., and J. Q. Tierney. 1953. Some Aspects of the Larval Development and Metamorphosis of *Teredo (lyrodus) pedicellata* de Quatrefages. Bull. Mar. Sci. Gulf and Caribbean, 2:574-89.

Johnson, M. W., and R. J. Menzies. 1956. The Migratory Habits of the Marine Gribble *Limnoria tripunctata* Menzies in San Diego Harbor, California. Biol. Bull., 110:54-68.

Kampf, W. -D. 1957. Ueber die Wirkung von Umweltfaktoren auf die Holzbohrassel *Limnoria tripunctata* Menzies (Isopoda). Z. angew. Zool., 44:359-75.

Koch, A. 1956. Das Verhältnis zwischen Symbionten und Wirt. Verhandl. d. Zool. Ges. Erlangen 1955, pp. 328-48.

Kofoid, C. A. 1921. The Marine Borers of the San Francisco Bay Region. Rep. San Francisco Bay Marine Piling Survey, 1:23-61.

Kohlmeyer, J. 1956. Ueber den Cellulose-Abbau durch einige phytopathogene Pilze. Phytopath. Z., 27:147-82.

Kühl, H. 1957. Der Befall durch Bohrmuscheln und Bohrkrebse in Norderney, Wilhelmshaven, List auf Sylt und Kiel in den Jahren 1953-55. Z. angew. Zool., 44:258-79.

Lasker, R., and C. E. Lane. 1953. The Origin and Distribution of Nitrogen in *Teredo bartschi* Clapp. Biol. Bull., 105:316-19.

Menzies, R. J. 1954. The Comparative Biology of Reproduction in the Wood-boring Isopod Crustacean Limnoria. Bull. Mus. Compar. Zool. Harvard College, 112:363-88.

-------, and G. Becker. 1957. Holzzerstörende Limnoria-Arten (Crustacea, Isopoda) aus dem Mittlemeer mit Neubeschreibung von *Limnoria carinata*. Z. angew. Zool., 44:85-92.

Meyers, S. P. 1953. Marine Fungi in Biscayne Bay, Florida. Bull. Mar. Sci. Gulf and Caribbean, 2:590-601.

Ray, D. L., and J. R. Julian. 1952. Occurrence of Cellulase in Limnoria. Nature, 169:32-33.

Roch, F. 1940. Die Terediniden des Mittelmeeres. Thalassia, 4:5-147.

Sømme, O. M. 1941. A Study of the Life-History of the Gribble *Limnoria lignorum* (Rathke) in Norway. Nytt Mag. Naturvidenskapene, 81:145-205.

Walker, H. W., H. S. McQuaid, M. S. Allen, R. H. Carter, and T. F. McCabe. 1926. Marine Piling Investigation. Bull. Am. Railway Eng. Assoc., 28:1-102.

White, F. D. 1929. Studies of Marine Wood-borers. I, The Toxicity of Various Substances on *Limnoria lignorum*. II, The Effect of Experimental Variations of Salinity and Hydrogen-Ion Concentration upon the Wood-borers of the Pacific Coast of Canada. Contrib. Can. Biol. and Fish., 4:3-18. B

Yonge, C. M. 1927. The Absence of Cellulase in Limnoria. Nature, 119:855.

ZoBell, C. E. 1946. Marine Microbiology: A Monograph on Hydrobacteriology. Waltham, Mass., Chronica Botanica., pp. 1-240.

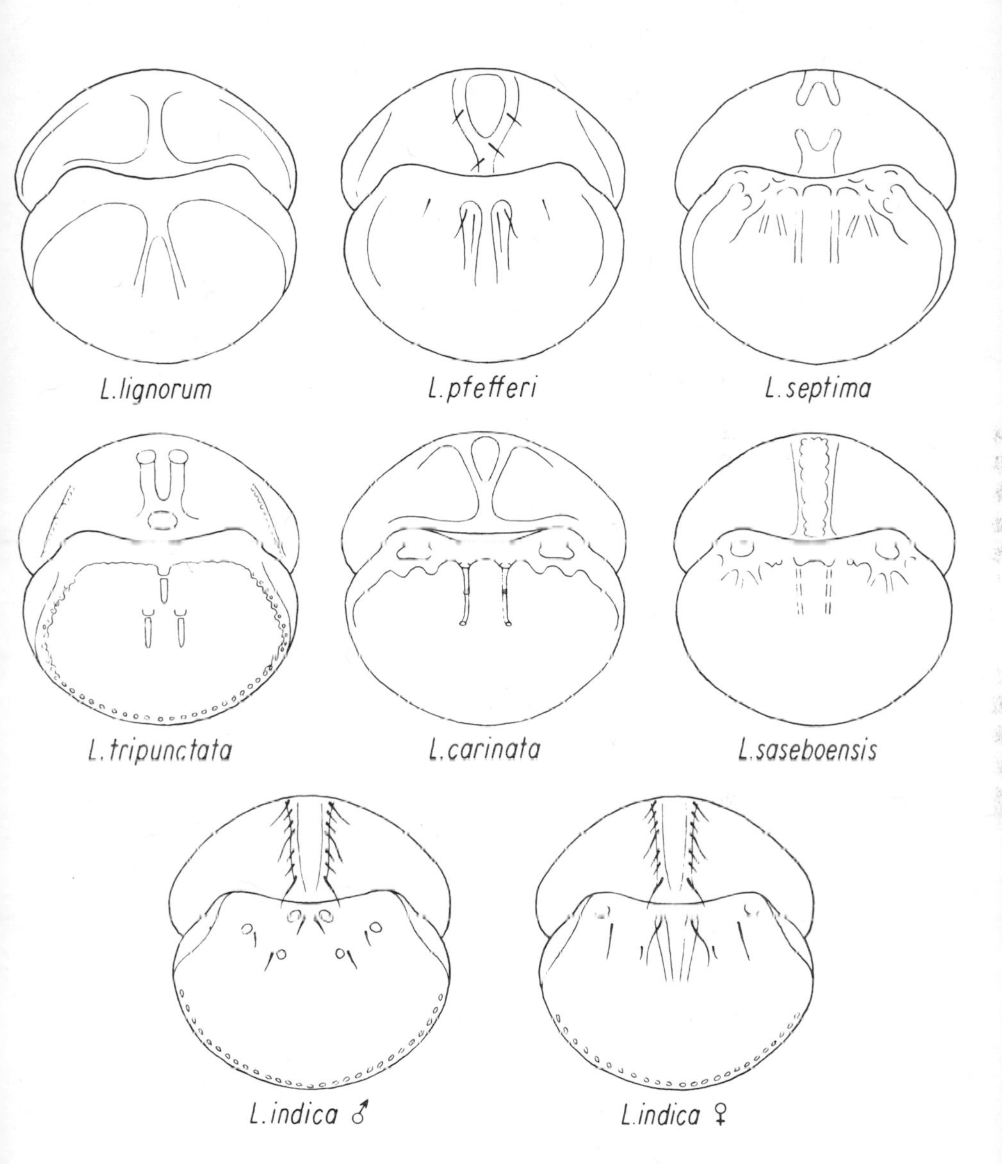

Fig. 1. Profile pattern of the telsons of different Limnoria species (according to G. Becker and W.-D. Kampf, 1957; and R. J. Menzies, 1957)

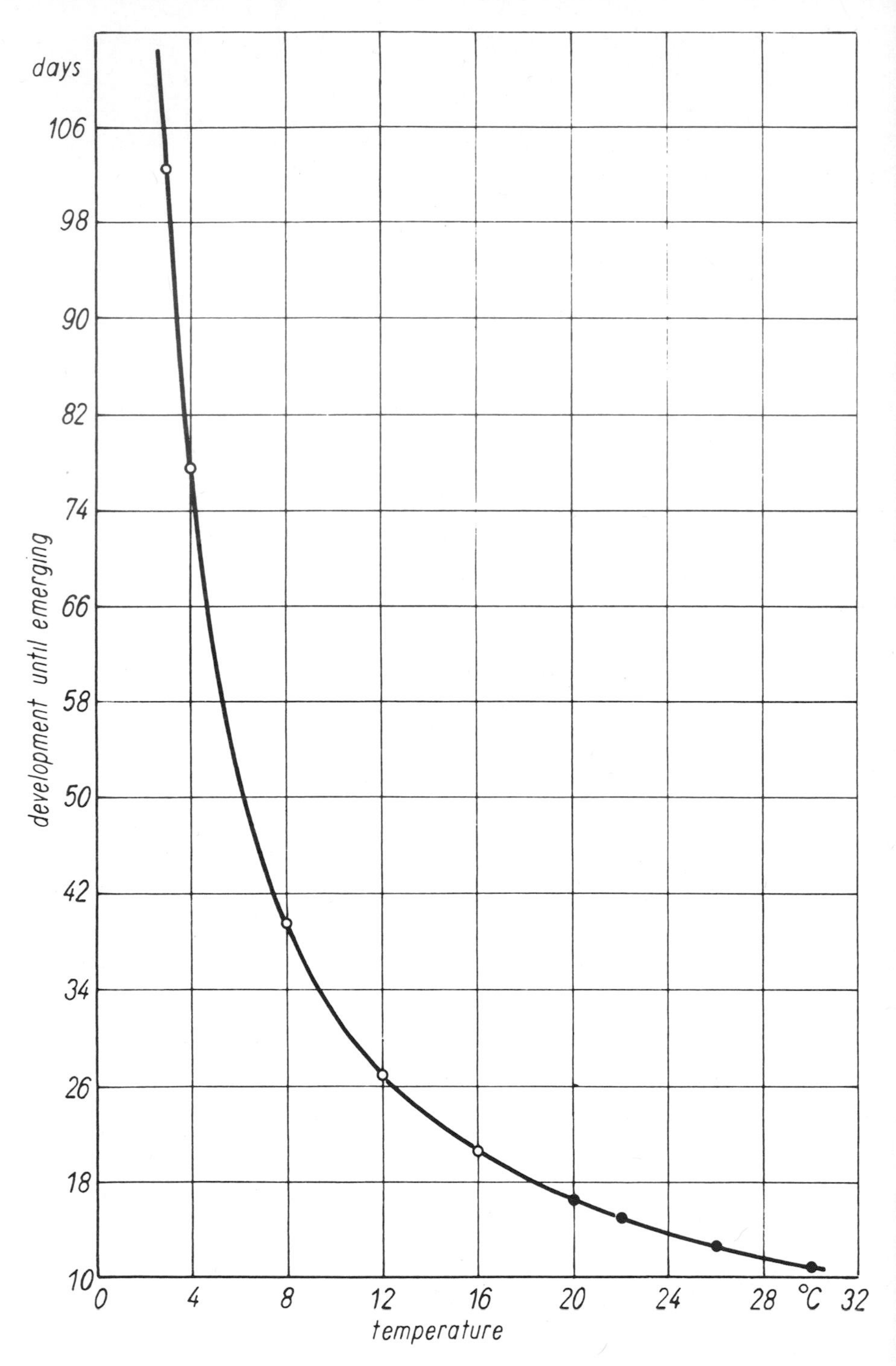

Fig. 2. Temperature dependence of the embryonic development of *Limnoria tripunctata* Menzies (according to W.-D. Kampf, 1957)

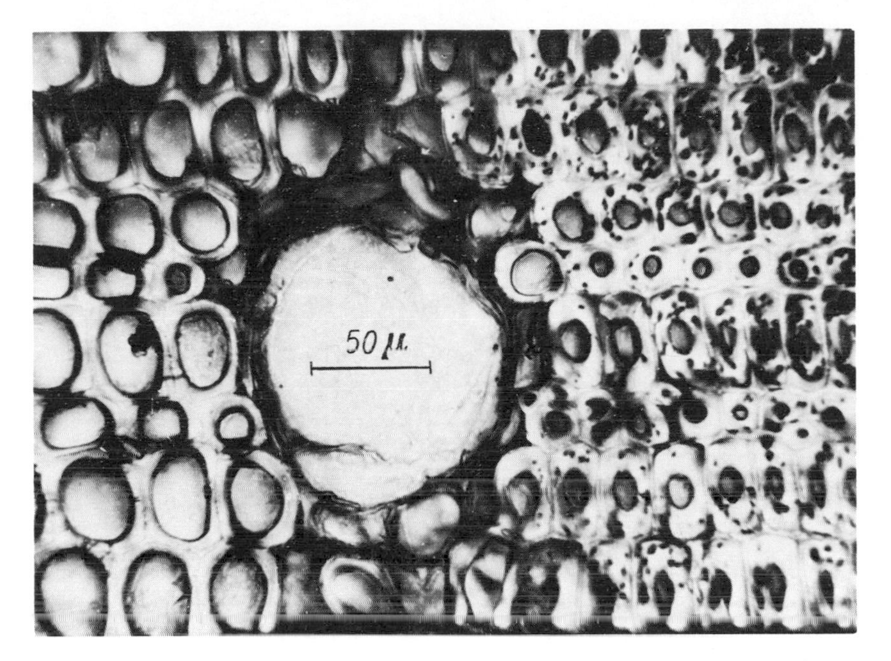

Fig. 5. Mycelium of marine fungi in the secondary layer of the cell wall of summer wood. The spring wood cells are free of fungi (photograph by J. Kohlmeyer, Berlin-Dahlem)

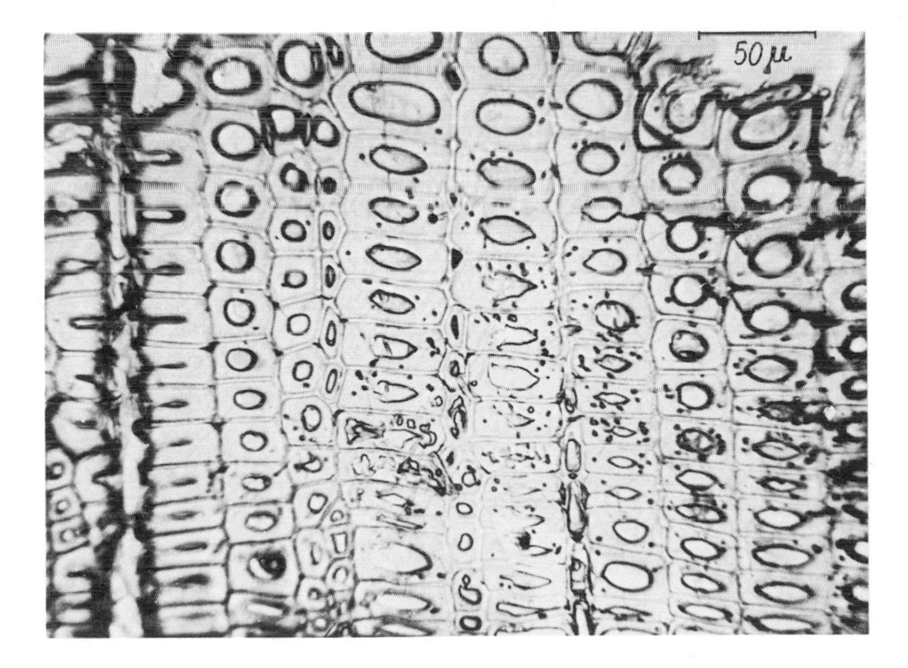

Fig. 6. Deterioration of cell walls by marine fungi (photograph by J. Kohlmeyer, Berlin-Dahlem)

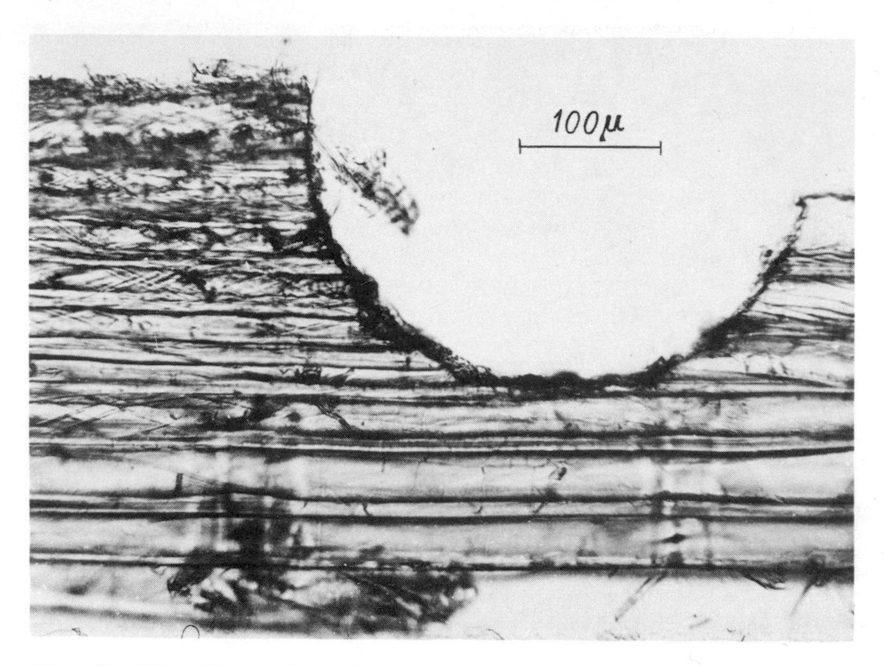

Fig. 7. Mycelium of marine fungi and cell walls decomposed by them in the vicinity of a boring hole of Limnoria (photograph by J. Kohlmeyer, Berlin-Dahlem)

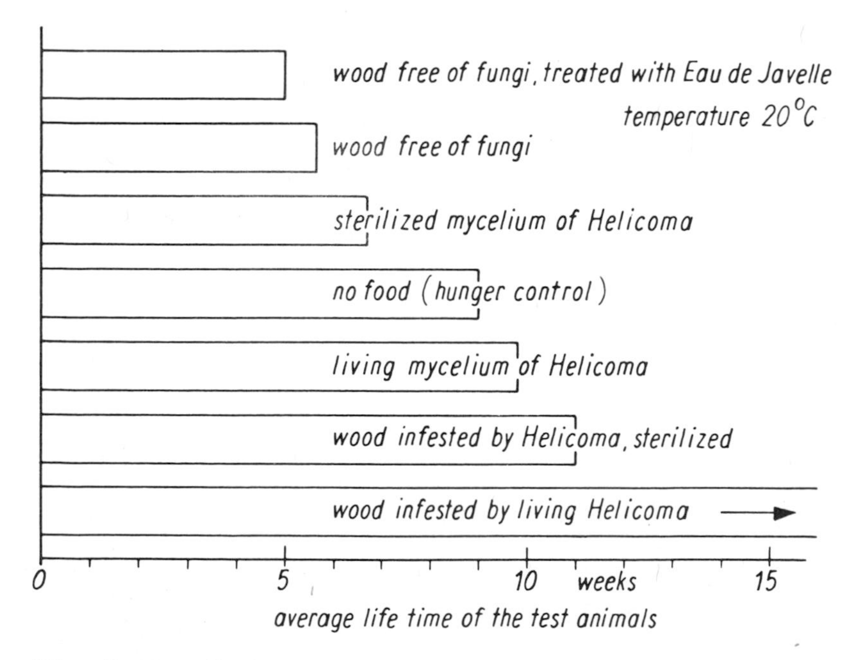

Fig. 8. Result of nutrition tests with *Limnoria tripunctata* in sapwood of *Pinus silvestris* L. Test fungus· *Helicoma sp.* (according to G. Becker, W.-D. Kampf, and J. Kohlmeyer 1957)

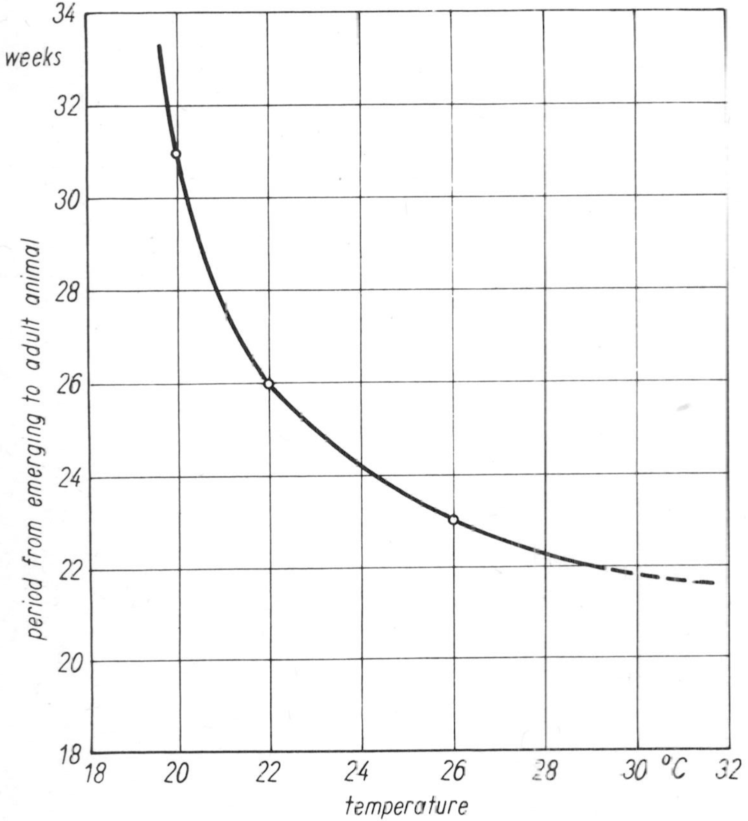

Fig. 3. Temperature dependence of the growth of *Limnoria tripunctata* Menzies (according to W. -D Kampf, 1957)

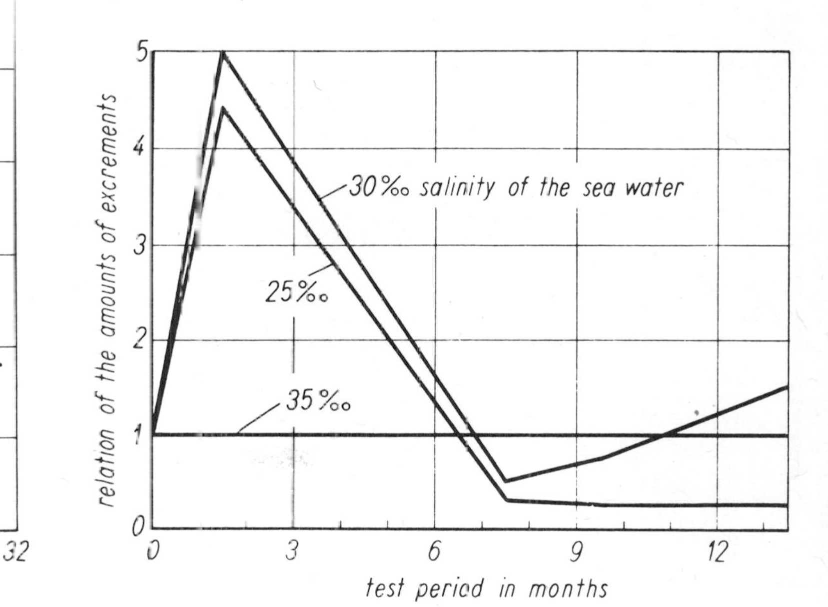

Fig. 4. Influence of changed salinity on the boring activity of *Limnoria tripunctata* Menzies (according to W. -D. Kampf, 1957)

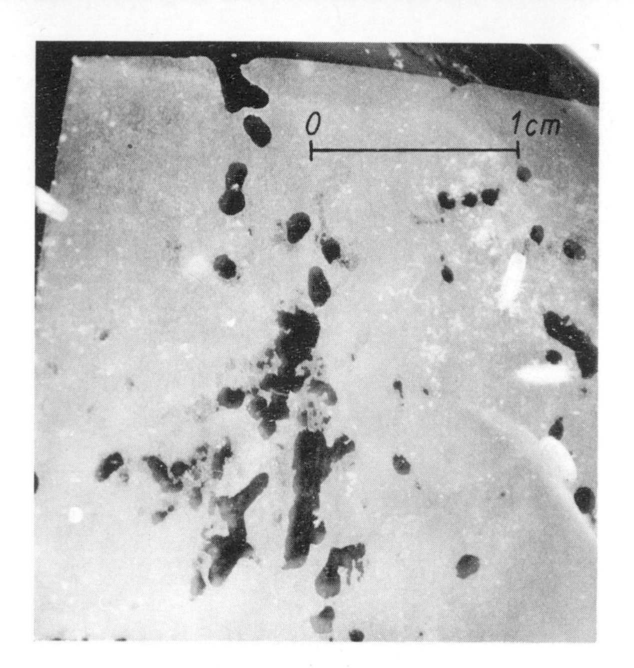

Fig. 10. Result of nutrition tests with *Limnoria tripunctata* in cellulose hydrate-foil. Test fungus: *Helicoma sp.* (according to G. Becker, W.-D. Kampf, and J. Kohlmeyer, 1957)

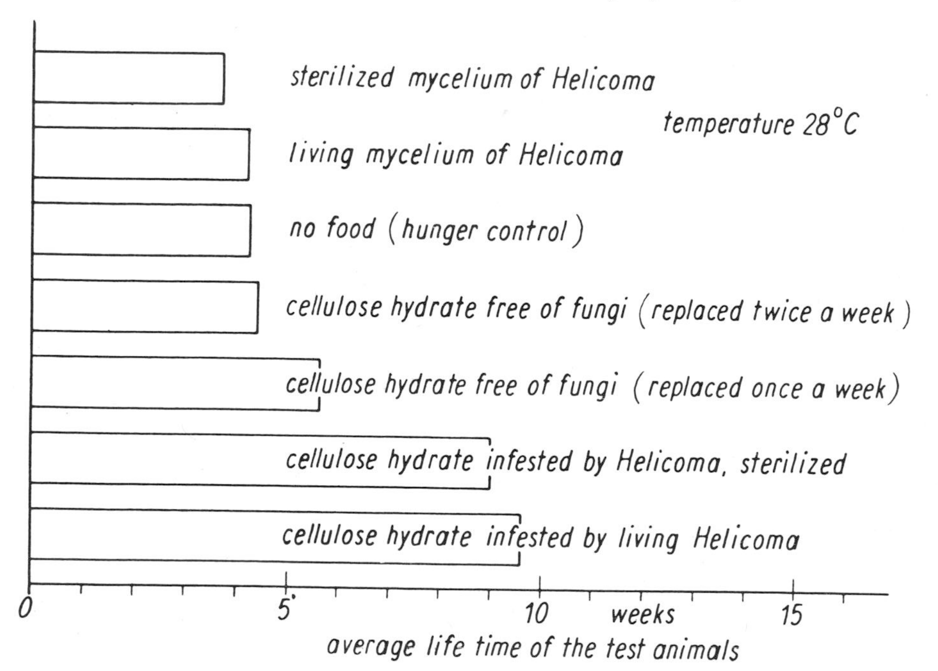

Fig. 9. Deterioration of cellulose hydrate-foil by *Limnoria tripunctata* after infestation by *Helicoma sp.* (photograph by J. Kohlmeyer, Berlin-Dahlem)

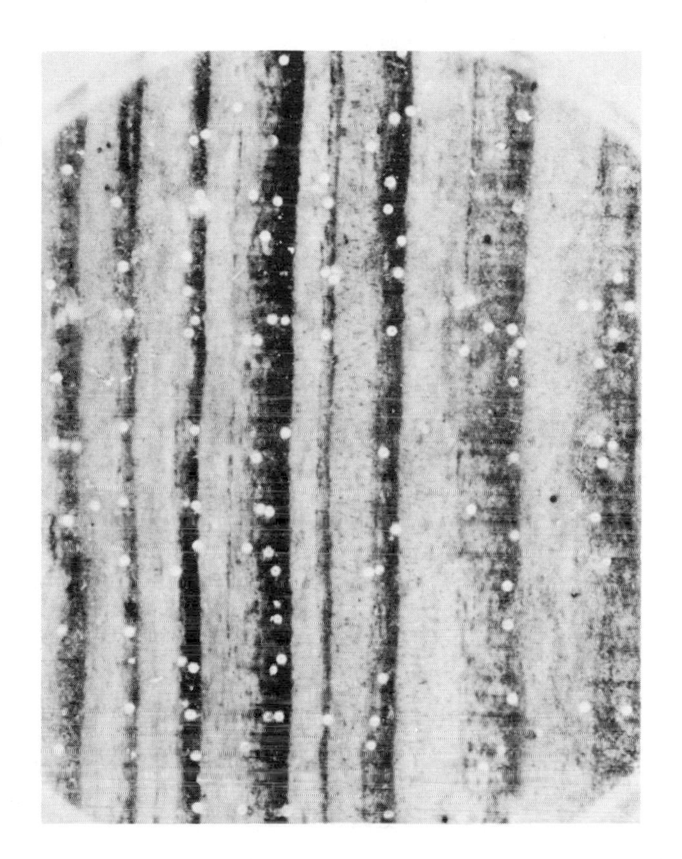

Fig. 11. Favored development of *Teredo pedicellata* Quatref. in the summer wood rings of *Pinus silvestris* L. in aquaria cultures in Berlin-Dahlem (photograph by M. Gersonde, Berlin-Dahlem)

ON THE PROTOZOAN ASSOCIATES OF LIMNORIA

John L. Mohr

As a rule, gregarious animals, because they provide recurring contacts, are more likely to be parasitized than solitary organisms. Those occupying a restricted area (or those fixed) are more exposed to attack than animals moving about freely. Animals that burrow, because they provide shelter for more than themselves, very frequently have "boarders" as Fisher and MacGinitie (1928) and others have shown for Chaetopterus, Urechis, and other tubemakers. Some years ago the writer and Mr. J. A. LeVeque decided that Limnoria, because it is gregarious, occupies a restricted area, and also burrows, i. e., it belongs to all three of these categories, must be sufficiently parasitized to be interesting on that account. Limnoria has still another characteristic especially favorable to epizoans that attach externally; the female holds its young in a ventral brood basket until they are active subadults, in this manner providing prolonged protection from some dangers but also prolonging exposure to infection by those organisms that parasitize the mother and her neighbors.

We have studied the protozoan associates (generally ignoring the nematodes and various crustaceans that frequently occur with gribbles) and have directed our attention to the external rather than the internal parasites. None of the organisms so far observed destroys or even effectively limits the increase of gribble populations. The fact is, one does not find evidence of epizootics in the field, and animals in culture are free from pathogens. Such parasitic groups as microsporidians, which annihilate populations of some other invertebrates, are unknown in Limnoria. Internal parasites appear to be rare. I have seen a small holotrichous ciliate that works inside the chitinous skeleton destroy a very few Limnoria at Friday Harbor, but I have observed none in southern Californian waters. This ciliate appears to be the organism that sometimes destroys single Epinebalia. Invasion is facultative, nonspecific, and limited probably to single old or injured individuals. Workers in Dr. Ray's group have seen, very rarely, an internal, parasitic dinoflagellate. Except for this occasional dinoflagellate, all the protozoan associates of the gribbles are ciliates. Some of these (here termed casuals) have been found elsewhere or appear to lack obligate relationship with Limnoria; the remainder (obligate associates) appear to occur only with Limnoria.

The casual associates are members of at least three groups: the sessile peritrichidans, suctoridans, and conidophryids, of which the first are much the more numerous. Among them Vorticellas of the "small-mouthed" group (*V. microstoma* and closely related forms) are most common. They may attach on the head, dorsum, and especially about the proximal joints of the appendages. Somewhat less common are colonial peritrichs. On occasional Limnoria, a large Vaginicola occurs on the dorsolateral surface; these individuals appear not to agree with described forms. Their assignment to the "casual" group is made with some misgiving.

The Suctorida have so far been represented by one species (fig. 3), possibly undescribed. Occasionally it occurs in such numbers on the dorsal surface as to make a "felt." The protozoan is stalked, roughly obtriangulate in profile, and has tentacles in two prominent bundles.

Conidophrys pilisuctor (fig. 4), which for most of its life appears like a tiny teardrop impaled at one end upon a marginal bristle of its host, occurs on the small, cylindrical amphipod, *Corophium acherusicum*, which occasionally invades gribble galleries, particularly old ones. At Newport Beach, California, we have found Corophium and Limnoria associated and Conidophrys on both crustaceans, but the occurrence on Limnoria is rare.

It may be noted that the position on the host that is occupied by these casual associates is not rigidly circumscribed. One does not associate their occurrence with a particular site unless it is one as broad as "the bristles of the dorsum." This lack of site specificity appears to fit the general pattern of facultative association.

The ciliates found only on or with Limnoria are also confined to three taxa: the Folliculinidae (Heterotrichida), the Peritrichida, and the Chonotrichida. The folliculinids (from Latin, "little windbag") are flask-secreting heterotrichidan ciliates somewhat like Stentors. The most common of them, *Mirofolliculina limnoriae* Dons (fig. 1), occurs commonly on the dorsal surface of the pleotelson and adjacent segments of *Limnoria lignorum* and may well be the dorsal "spots" illustrated in the papers of Dicquemare (1783) and of Rathke (1799); it is clearly represented in a number of more recent papers. Typically *M. limnoriae* has a large test or flask, expanded at the base and produced into a number of small lateral pouches. Workers at the United States Navy Civil Engineering Research and Evaluation Laboratory, Port Hueneme, California, have noted considerable color variation in the flasks which may represent more than subspecific variation. Kahl (1932) points out that animals of strikingly different nuclear composition have been assigned to *M. limnoriae*. We have observed further that the development of side pouches in the test is much less marked in examples (fig. 2) from lower (that is, temperate zone) latitudes from both northern and southern hemispheres (cf. specimens from Port Etienne, French Morocco, figured by Th. Monod, 1925, pl. LII). We suspect that we are dealing with several species of folliculinids with multilobed tests, all of them confined to the posterior dorsum of gribbles. However, study of this genus has been slight and limited too much to examination of the tests rather than of the organisms themselves.

A second kind of folliculinid, *Folliculinopsis gunneri,* less common than Mirofolliculina, attaches to the undersurface of the thorax. It has not been found on any other part of Limnoria. Like Mirofolliculina it is large for a ciliate, but, unlike that genus, *F. gunneri* has a brown, wide-mouthed, short-necked test without outpocketings. Possibly because of its size *F. gunneri* is often solitary, but Limnoria with two or three or, more rarely, up to eight individuals may be found. It appears to occur only in colder waters (Norway, Washington, Oregon, northern California, New Zealand).

Folliculina lignicola, described by Fauré-Fremiet (1936) from Woods Hole and Wimereux, and reported by the writer from Friday Harbor, is an inhabitant of the tracheids of the wood attacked by gribbles. While it cannot be stated that this slender, blue folliculinid does not occur elsewhere, it does appear likely that it has taken advantage of the many tracheids opened by the gribble burrows and that it may have evolved with the ready availability of a new habitat.

The presence of three folliculinids with *Limnoria lignorum* at

Friday Harbor, all three occurring associated with a single gribble population, raises the question whether they arise from a single species. In both body and test these three kinds differ greatly; so much so that the writer inclines to the view that three different free-living folliculinids have become adapted to life in a Limnoria association. One of these, succeeding in attachment and maintenance on the dorsum of the gribble, gave rise to several kinds of Mirofolliculina.

At least two of the peritrichidan ciliates are obligate associates of Limnoria. These are *Cothurnia limnoriae* Dons (fig. 5) and *Lagenophrys sp.* (fig. 6). Both appear to be widespread and common on the host. Cothurnia attaches to the joints of the legs, characteristically in small numbers. Lagenophrys apparently has been overlooked because its low-domed tests adhere to the flat surfaces of the pleopods (abdominal appendages, swimmerets, or "gills"). Here they are not easily seen unless the swimmeret is dissected from the gribble. They are common and may be present in some numbers (e. g., 15 to 25) on a single host. Site specificity (location on host) is rigid for both the peritrichidans. The Cothurnia occur only on the legs, particularly at the joints, and the Lagenophrys attach only on the broad surfaces of the pleopods.

The last of the protozoan orders occurring on Limnoria is the Chonotrichida or collar ciliates. These are slender, vaselike ciliates with the anterior produced into a flaring collar or funnel within which lies the cytostome. On the inner surface of the funnel are always two fields of cilia, a horizontally elongate field with cilia beating toward the second field, and a smaller vertical patch with cilia beating in an oblique pattern toward the cytostome. The cilia in these two peristomial fields are the only ones these ciliates possess.

Attached to the marginal bristles of the pleopods of Limnoria are collar ciliates of a sort not found elsewhere. These attach to the shaft of the pleopodal bristles by an adhesive disc that appears to be peculiar to this chonotrichidan genus. The most obvious characteristic of the gribble collar ciliates is a moderately flared collar simple except for border lappets produced on either side of the peristomial ciliary fields. This pair of lappets differs in prominence in different populations or species.

The collar ciliates may be considered as the best adapted of any of the Limnoria associates, however confined they are in habit. So far they have occurred in every colony or collection of wood gribble we have examined. These include collections from Norway (from Tromsø southward), from England; Spain; the Adriatic; Thasos in the eastern Mediterranean; Capetown; the Atlantic, Gulf, and Pacific coasts of the United States; northwest Vancouver Island; Hawaii; New Zealand; Australia (including Tasmania); and the Philippines. They may also be found on the gribbles burrowing in algae, but possibly not in every colony.

When we began work on the gribble protozoans, we assumed that we were dealing with a single host species, *Limnoria lignorum*. Subsequently Dr. Robert J. Menzies pointed out that there are three common species on the Pacific Coast and a fourth common in Hawaii. These occupy overlapping thermal areas, *Limnoria lignorum* occurring in colder, *L. quadripunctata* in cold, *L. tripunctata* in warm, and *L. (Paralimnoria) andrewsi* in warmer waters. There is a region of overlapping between *L. lignorum* and *L. quadripunctata* in northern California (Eureka, Crescent City), between *L. quadripunctata* and *L. tripunctata* in central and southern California (San Francisco, Los Angeles harbors), and between *L. tripunctata* and *L. (Paralimnoria) andrewsi* in Honolulu harbor. In some of the areas of overlapping for which we have thermal data (particularly Los Angeles and San Francisco harbors) there is evidence of thermal separation of species in microhabitat; however, this does *not* appear to be the case between *L. lignorum* and *L. quadripunctata* at Eureka, California, where Mr. Price Peterson and I have collected both species at one level of the same wharf.

The question arises whether the gribble chonotrichidans are the same on the several Limnorias. We have studied these protozoans closely on only three of the hosts--*Limnoria lignorum* from Friday Harbor, Washington, and Coos Bay, Oregon; *L. tripunctata* from Los Angeles (San Pedro) and Newport Bay, California; and *L. (Paralimnoria) andrewsi* from Honolulu, Oahu, Hawaii. The collar ciliates on these three, though obviously congeneric, are clearly separable by differences in proportions and especially by the relative prominence of the marginal lappets of the collar. The lappets are long on the collar of the *lignorum*-chonotrich (fig. 7a), intermediate in that of *L. tripunctata* (fig. 7b), and very small in that of *L. (P.) andrewsi* (fig. 7c).

Although judgment must be provisional, the differences among these ciliates do not seem to be so subtle as to require specially fixed materials for them to be recognized. The characteristic "*lignorum*"-form is easily recognized in alcohol-preserved materials from northern, western, and southern Norway provided by Mrs. Olaug Sømme; while *"tripunctata"*-form is clear in variously fixed examples from the Mediterranean (Thasos, Greece); the Adriatic; Douro estuary, Portugal (from Dr. Braga); and Manila harbor, Philippines (from Dr. E. Y. Dawson).

Whether clear-cut differences will remain when the ciliates of *Limnoria quadripunctata* have been examined more closely, and particularly when adequate materials from areas of overlapping have been studied, remains to be seen. Further, such differences as have been observed may be explainable as results of different growth at different temperatures or as population variants within the range of a single species. However, for the present Mr. LeVeque and I choose to consider them as characteristic of three separate species.

The evolution of separate lines of collar ciliates as ectocommensals on various crustacean groups seems to be a most important manifestation of the order Chonotrichida. One or two genera are confined to each of the following: the mouth parts of true crabs; the thoracic or abdominal appendages of Nebaliaceans; the thoracic or abdominal appendages of several lines of gammaridean amphipods; the appendages or body surface of cyamids (whale lice); and the abdominal appendages of several lines of isopods.

Such parallel evolution of protozoan parasites with hosts is well known in flagellates of termites (from the works of Katzin and Kirby, 1939, some aspects of which have been ably summarized by Professor McBee in this symposium) and in opalinids of amphibians and some other animals (from the work of Metcalf, 1923). Metcalf was so impressed with the firmness and significance of the relationships between opalinids and their hosts that he undertook a drastic revision of the classification of the anurans on the basis of these (although not without protests from herpetologists).

In the case of the Chonotrichida and the host Crustacea, unlike the situation involving flagellates in termites and opalinids in anurans, the data so far are all consistent except for a single minor and explainable anomaly; i.e., *Stylochona sivertseni* has been described as occurring on a marine filamentous alga on Tristan da Cunha Island in the south Atlantic Ocean. Each of the genera or genus complexes (e.g., *Stylochona-Kentrochona*) is confined to one group of closely related crustacean species. Further, site-specificity is characteristically rigid. So far, however, a pattern of thermal limitation is discernible only in the group of gribble chonotrichidans.

A final word may be said on the gribbles of algae. These are limited in the extent of the galleries they can excavate because the algal stipes or other parts they bore into are slender. Further, those we have studied are intertidal, exposed to surf. Accordingly, the algae will break loose whenever the Limnoria galleries weaken the stems too much. Although this may serve to release floating portions of the algae, thus disseminating the gribbles long distances, it places a low upper limit on the gribble population in any one plant. The advantage of contacts spoken of above is thus greatly reduced, but the factors of small territory, protective burrows, and retention of young in brood pouches remain. We have studied the algal-dwelling Limnorias only a little but have found that both folliculinids and chonotrichidans do occur on them.

In conclusion it may be stated that the gregarious, wood-tunneling gribbles do have the range of associates that their mode of life would lead one to expect; that these organisms include casual and obligate associates; and that among the latter the Chonotrichida appear to exhibit an unusually high level of evolution parallel with hosts.

LITERATURE CITED

Dicquemare. 1783. Ueber ein holzzernagendes See-Insect. Lichtenbergs Magazin, 11:49-53.

Fauré-Fremiet, E. 1936. La famille des Folliculinidae. Mem. Mus. Roy. Hist. Nat. Belg., 2nd Ser., 3:1129-77.

Fisher, W. K., and G. E. MacGinitie. 1928. The Natural History of an Echiuroid Worm. Ann. Mag. Nat. Hist., Ser. 10, 1:204-13.

Kahl, A. 1932. Urtiere oder Protozoa. 1, Wimpertiere oder Ciliata Spirotricha. p. 474 in Dahl, F., Die Tierwelt Deutschlands, 25:399-650.

Katzin, L. I., and H. Kirby, Jr. 1939. The Relative Weights of Termites and Their Protozoa. J. Parasitol., 25:444-45.

Metcalf, M. M. 1923. The Origin and Distribution of the Anura. American Naturalist, 57:385-411.

Monod, Th. 1925. Tanaidaies et Isopodes aquatiques de l'Afrique Occidentale et Septentrionale. Bull. Soc. Sci. Nat. Maroc. Rabat, No. 6, pl. LII.

Rathke, J. 1799. Jagttagelser henhorende til Indvoldeormemes og Bloddyrenes Naturhistorie. Skrivter af Naturhistoric Selskabet, 1:61-153.

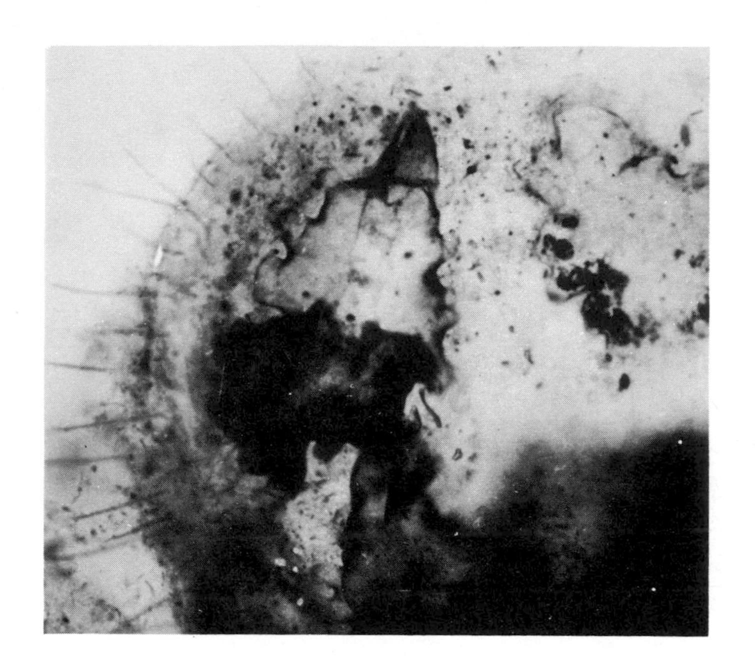

Fig. 1. *Mirofolliculina limnoriae*

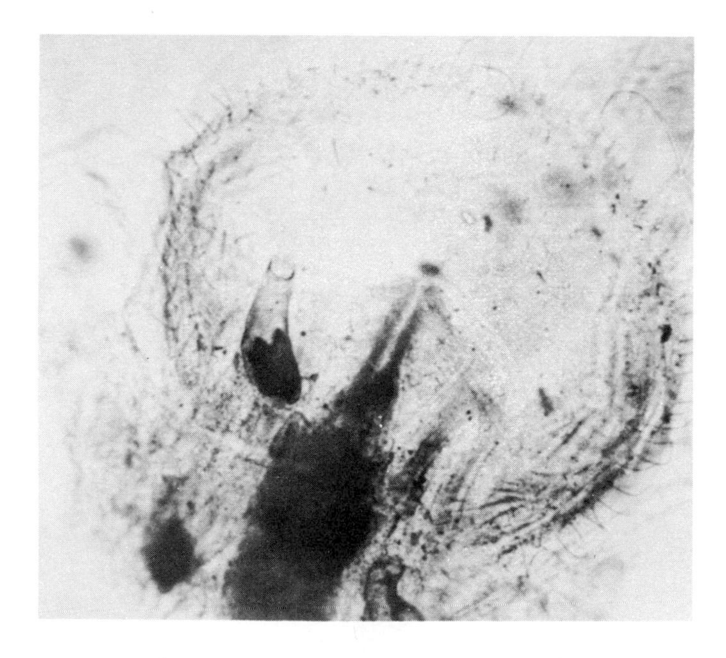

Fig. 2. *M. limnoriae* from temperate zone

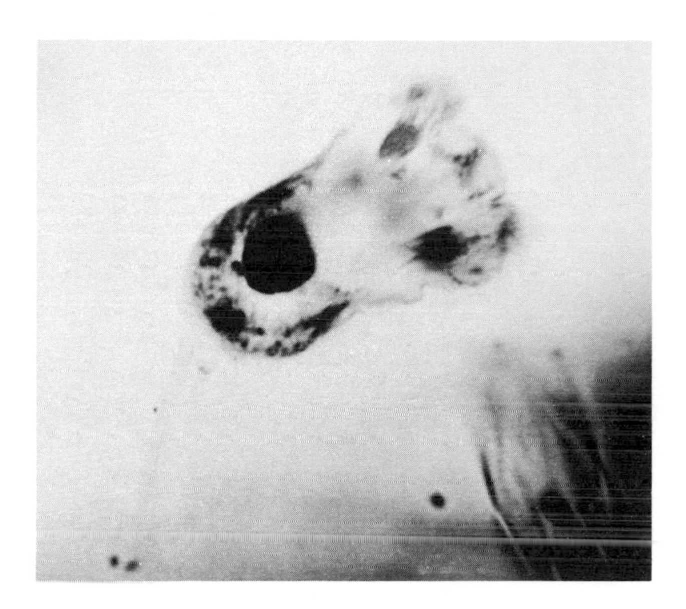

Fig. 3. Species of Suctorida

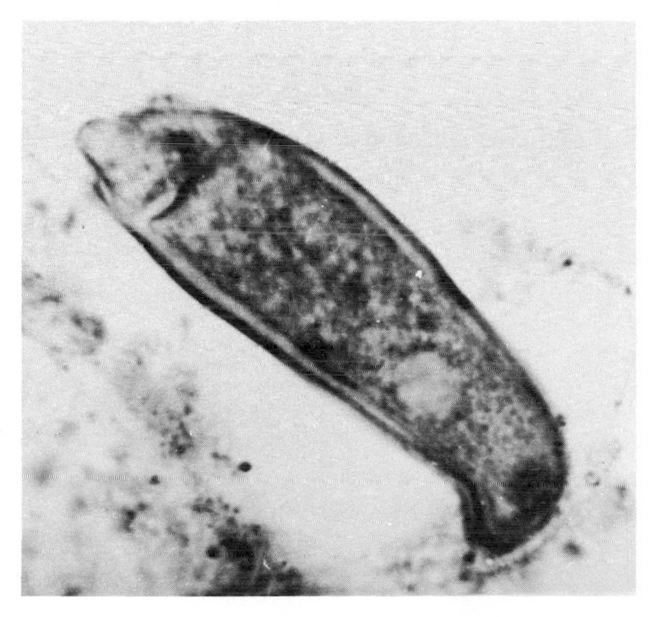

Fig. 4. *Conidophrys pilisuctor*

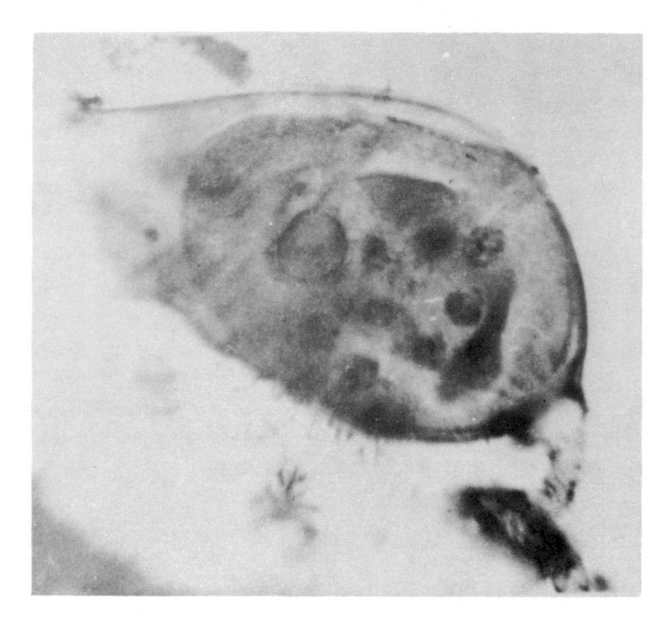

Fig. 5. *Cothurnia limnoriae*

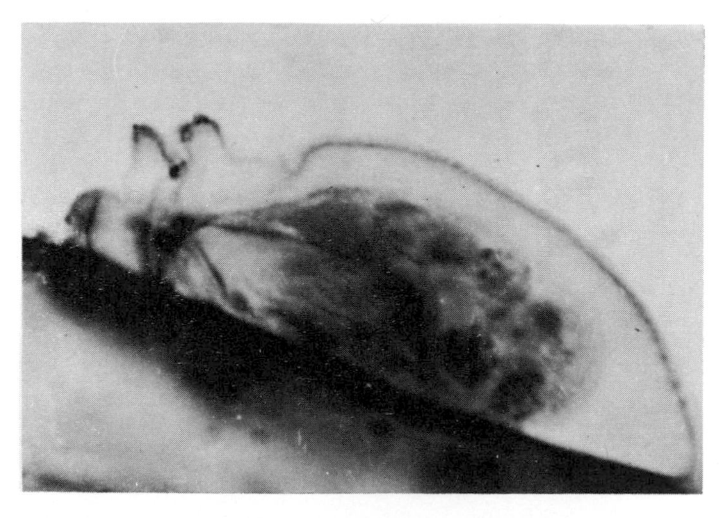

Fig. 6. *Lagenophrys sp.*

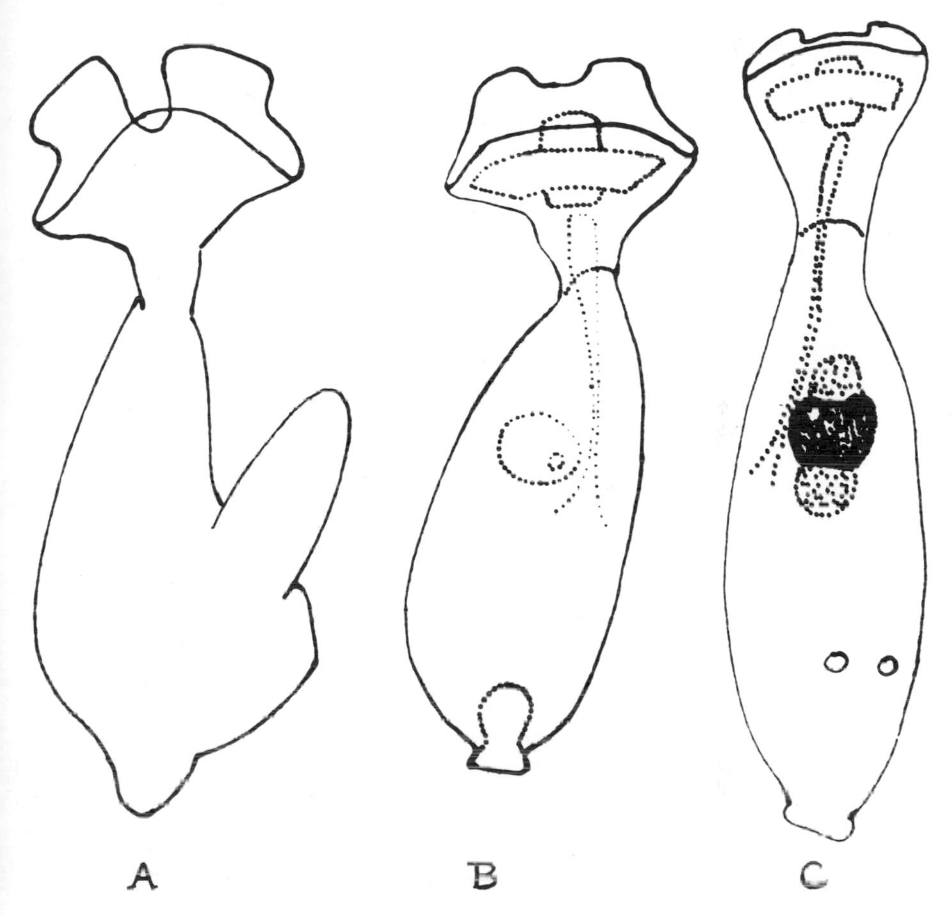

Fig. 7. Chonotrichidans from three different species of Limnoria. a, from *L. lignorum;* b, from *L. tripunctata;* c, from *L. (P.) andrewsi*

STUDIES ON THE HISTOLOGY AND CYTOLOGY OF MIDGUT DIVERTICULA OF *LIMNORIA LIGNORUM*[*]

Wolf H. Fahrenbach

In previous investigations little attention has been paid to the histology of the midgut diverticula of Limnoria. Hill and Kofoid (1927) described the gross morphology of the diverticula and noted the presence of crystals in them. In view of the extensive studies on the wood-boring habits of Limnoria and the knowledge that the diverticula secrete cellulolytic enzymes (Ray and Julian, 1952; Ray, 1958), it seemed appropriate to study this part of the digestive tract in more detail.

The two pairs of diverticula constitute the entire midgut of this animal's digestive system (fig. 1). They arise at the posterior extremity of the foregut, just behind the small but complex gastric mill, and lie in the body cavity lateral and parallel to the hindgut. The diverticula are blind, finger-shaped, tapering pouches that have often been called salivary glands, digestive caeca, hepatic caeca, livers, etc., but, since each of these terms carries implications of functions not known with certainty to be carried out by this tissue, it is preferable to use the simple and descriptive term, midgut diverticula.

A diverticulum can be divided into several distinct histological components. Outermost is a thin envelope incorporating epithelium and some muscle cells, and internal to this are the cells lining the lumen. The latter fall into two groups: α-cells, large binucleate cells not in contact with each other, and β-cells, smaller, granulated cells that form a latticelike network between the α-cells (fig. 2). Both α- and β-cells appear to be

---

*These studies were aided by a contract between the Office of Naval Research, Department of the Navy, and the University of Washington, NR 104-142.

secretory in function, the cell surface adjacent to the lumen being a brush border (fig. 7). The lumen is filled with a viscous, clear liquid and is found to be consistently free of any ingested wood particles or other formed matter.

The envelope (fig. 4) is transparent and relatively resistant to mechanical injury. Regularly arranged muscle bands separated by very thin, roughly rectangular epithelial areas are visible on the surface of the envelope. These bands show no histological details under conditions of ordinary light microscopy, but there is good evidence for their muscular nature. In diverticula of Armadillidium, which are very similar in structure to those of Limnoria, the muscle bands of the envelope bear marked chevron striations. Furthermore, both Limnoria and Armadillidium diverticula show the same active contractions, change of shape, and the same type of nodelike constrictions, produced after the diverticulum is removed from the body. Also, in Limnoria, the bands are birefringent (fig. 3) and, further, they can be demonstrated by intense silver impregnation.

The α-cells, each with two prominent vesicular nuclei ranging in diameter from 14 to 22 μ, project deeply into the lumen. The cytoplasm usually contains a certain number of oil droplets (fig. 2), granules, and, to a varying degree, small and large crystals (figs. 5, 6, and 8). These crystals constitute one of the most interesting aspects of the diverticula and merit considerable attention.

They are conspicuous, yellowish in color, tetragonal and bipyramidal in shape with a ratio of the axes of 1:1:2.13, and range in length from 0.1 to 28 μ (lower measurement as determined by electron microscopy). As many as 27 crystals have been counted in a single α-cell. The crystals are found most concentrated in the circumnuclear and basal parts of α-cells. Crystal concentration or other cytoplasmic inclusions or reactions to be subsequently mentioned are always more pronounced in the proximal portion of the diverticula, the distal portion usually showing less clear histological differentiation.

The crystals are insoluble in both fresh and sea water, are fixed by common fixatives, and do not dissolve in organic solvents, e.g., methyl, ethyl, propyl, butyl, and amyl alcohols; carbontetrachloride; acetone; and toluene. They remain unaffected by acids or bases between pH 2.1 and pH 11.8. At pH values over 1/10 of a unit below and above these limits, respectively, the crystals decompose slowly, losing their color and leaving a ghost, while a further increase in acidity or alkalinity produces immediate dissolution. The crystals stain with mercuric bromphenol blue, an indication of their proteinaceous nature. The crystals are birefringent (fig. 3). Careful measurements, for which I am indebted to Dr. Shinya Inoué of the Department of Biology, University of Rochester, show that the birefringence is positive with respect to the long axis, and that the strength of birefringence (retardation/thickness) is somewhat weak for ordinary dry proteins, but within the range reasonable for a fairly hydrated protein. The axes are parallel throughout; therefore the crystals are not liquid crystals, nor is it likely that lipids are incorporated into their structure. Dr. Wayne Thornberg, Department of Anatomy, University of Washington, has measured the absorption spectrum of single crystals and finds a strong absorption peak at 4100 Å, a fact that suggest that the crystal protein is a porphyrin.

The most easily demonstrated chemical characteristic is that the crystals contain large amounts of iron, as shown by the Turnbull Blue reaction and by exposing the diverticula to hydrogen sulfide or ammonium sulfide. The latter reaction was found to be a simple test for the presence of iron; it consists of adding a drop of full strength $(NH_4)_2S$ to several diverticula in a depression slide and noting the reaction after about 10 seconds. The presence of intracellular iron is shown by the development of the typical green color.

While carrying out these qualitative tests on the crystals, it was discovered that iron is also accumulated in granular inclusions in the cytoplasm of α-cells, even in the absence of microscopically detectable crystals. The degree of iron concentration was recorded in a semiquantitative fashion as 1+ to 4+, * and the presence of crystals was noted separately. In this connection it was observed that a population of animals showing presence of crystals in some individuals will have low iron "titres" (0-2+), mostly in the very small animals, whereas high concentrations of iron and crystals occur in the older animals. It was further noted that in the 3+ condition very tiny crystals may be present, their size (1-2 μ) only slightly exceeding that of the surrounding granules of the α-cell. In these larger, therefore older animals with a high iron content, the crystals are large

---

*1+ = slight greenish tinge; 2+ = strong green color, but α-cells still transparent; 3+ = some cells dark olive green, but most lighter; 4+ = most cells dark olive green to black.

and more numerous, a situation that apparently leads to the conditions of unusual dimensions and concentrations mentioned previously. It seems, considering the degree of correlation between presence of crystals and a high cytoplasmic iron content, and the relative distribution of high and low iron concentrations in different age groups, that the presence of iron in the α-cells is the result of a slow accumulative process. This process probably extends over many months and culminates in the formation of crystals.

The structure of the β-cells (figs. 5 and 6) is generally obscured by oil droplets and densely packed, opaque granules, distinctly different from the crystals in color, size, and shape. Using a variety of techniques, particularly supravital staining, it could be demonstrated that most, if not all, β-cells are binucleate, with nuclei ranging from 9 to 14 μ, and that the cell membranes between adjacent β-cells are quite thin and fragile. The latticelike network formed by these cells, not coinciding with the muscle bands of the envelope, is usually more distinct both in older animals and near the proximal end of any diverticulum. In cross section (fig. 7) the β-cells are sometimes found to be quite flat and occupy a peripheral position between the larger α-cells. The thickness of the β-cells seems to vary in different individuals.

Microchemical tests showed that the β-cells frequently give an intense sulfhydryl reaction with sodium nitroprusside, but sometimes show no trace of the bright red color indicative of a positive reaction. Further investigation revealed that the presence of sulfhydryl groups in β-cells is directly and proportionally related to the iron content of the α-cells. Uranyl acetate-silver nitrate impregnation of the diverticula resulted in blackened β-granules, which can be explained by the presence of reducing sulfhydryl groups (fig. 9).

In order to arrive at some understanding regarding the conditions governing iron storage and possible reasons for it, iron distribution in different populations was investigated. A random sample of 50 animals from a given locality was tested for iron, and the percentage distribution of various degrees of iron and crystal presence was noted. It was found that great variations occurred between populations. The bar diagrams illustrate the difference in iron distribution in two extreme populations (fig. 10). Each graph represents a sample of 50 animals. These tests could be repeated with a high degree of consistency, and the distribution remained identical within statistical limits for periods of several months.

To determine what possible factor may be responsible for these differences, groups of animals from populations with known iron distributions were subjected to varying conditions. Groups with high or low average iron content showed no statistically significant change when subjected to extended exposure to air in a moist chamber, sublethal starvation, change to different types of wood, and high concentrations of iron in both water and wood. At the same time, control groups maintained the same distribution when kept on untreated pine for 2 months. One positive and simultaneously quite interesting result was obtained. A test group, with 68 per cent of the animals having no iron at all in the α-cells, after one month on soaked creosoted wood, showed a shift in the mode of concentration from negative to 2+, with no animals giving a negative reaction. This increase was found to be highly significant when subjected to statistical analysis by the Chi square test.

Nothing has so far been said regarding the function of the crystals. The various ideas regarding this subject should be discussed here.

1. A frequently mentioned suggestion is that the crystals are connected with cellulase secretion. In my opinion there is at present no positive evidence in support of this hypothesis.

In the first place, the crystals occur in highly varying proportions in different populations. They are consistently absent in young animals and give all indications of being produced and accumulated over a period of months.

Second, cellulase is not inhibited by cyanide (Ray, personal communication), indicating that no iron is incorporated into the cellulase molecule, which, in fact, is the case for all known hydrolytic enzymes. The crystals contain iron and are intimately associated with cytoplasmic iron accumulation as described previously.

Last, a cell secreting an enzyme should show some evidence of cyclical secretory phenomena, in this case correlated with the nutritional status of the animal. This has not been observed to occur in the α-cells. On the other hand, it was repeatedly noticed that the granules of the β-cells vary in amount in different animals irrespective of age. In cross sections of the diverticula the β-cells are found to be of varying thickness, sometimes extending as far toward the lumen as the α-cells, sometimes occupying only a small part of that space. On no occasion has there been found any iron or crystals in the β-cells. This problem obviously needs further investigation.

2. The idea that the crystals are formed in response to a virus infection has met objections in the form of electron microscopi-

cal evidence and simple distribution patterns. Frequently, Limnoria populations from adjacent logs on the same beach will show widely different average iron concentration, although within each log the distribution is constant.

3. Another hypothesis is that the crystals represent the final result of a digestive detoxification process. The results of the experiment with creosoted wood seem to support this idea. But, considering the complexity of creosote with its carcinogenic coaltar derivatives and numerous unknown impurities, it is impossible to say whether the increased iron accumulation is due to an actual detoxification mechanism, to a tissue reaction caused by irritant ingested substances, to excess iron ingestion, or to the presence of some substance that depresses the oxidative metabolism and leads to excessive porphyrin synthesis.

4. A last suggestion, and perhaps the simplest, is that Limnoria is incapable of excreting iron and stores it bound to protein in the form of crystals when the cytoplasmic iron content rises to a critical level. It is known that the iron is bound tightly in the crystals since it cannot be removed by prolonged action of chelating agents. The coincident increase in sulfhydryl groups in the β-cells possibly serving to maintain the oxidation-reduction equilibrium may be cited in support of this hypothesis. Furthermore, the differing iron contents of different woods, treated and untreated, could account for the variations in iron content found in Limnoria populations. On the other hand, animals feeding on wood impregnated with iron salts showed no significantly higher iron content in comparison to the control group.

The presence of sulfhydryl groups in the β-cells should also not be neglected when the possible secretory function of these cells is considered.

To determine causative agents for crystal formation, it would be essential to establish whether they are formed in response to any environmental stimulus or to a specific irritant, or as a result of excessive iron intake. It would also be worthwhile to try to correlate the condition of the β-cells with the nutritional status of the animal. *

---

*I wish to thank Dr. D. L. Ray for her stimulating suggestions and advice. I am indebted to Dr. Wayne Thornberg for the determination of the crystal absorption spectrum, to Dr. Shinya Inoué for the polarization microscopy (fig. 3), and to Mr. F. L. Clogston for the remaining photomicrographs.

LITERATURE CITED

Hill, C. L., C. A. Kofoid, *et. al.* 1927. Marine Borers and Their Relation to Marine Construction on the Pacific Coast. Final Report, San Francisco Bay Marine Piling Committee.

Ray, D. L. 1959. Some Properties and Activities of Cellulase from Limnoria. In this volume.

-------, and J. R. Julian. 1952. Occurrence of Cellulase in Limnoria. Nature, 169:32.

Fig. 1. Showing the complete digestive tract of Limnoria, with the foregut and gastric mill at the top of the picture, the long straight hindgut, and the two lateral pairs of midgut diverticula. x 44

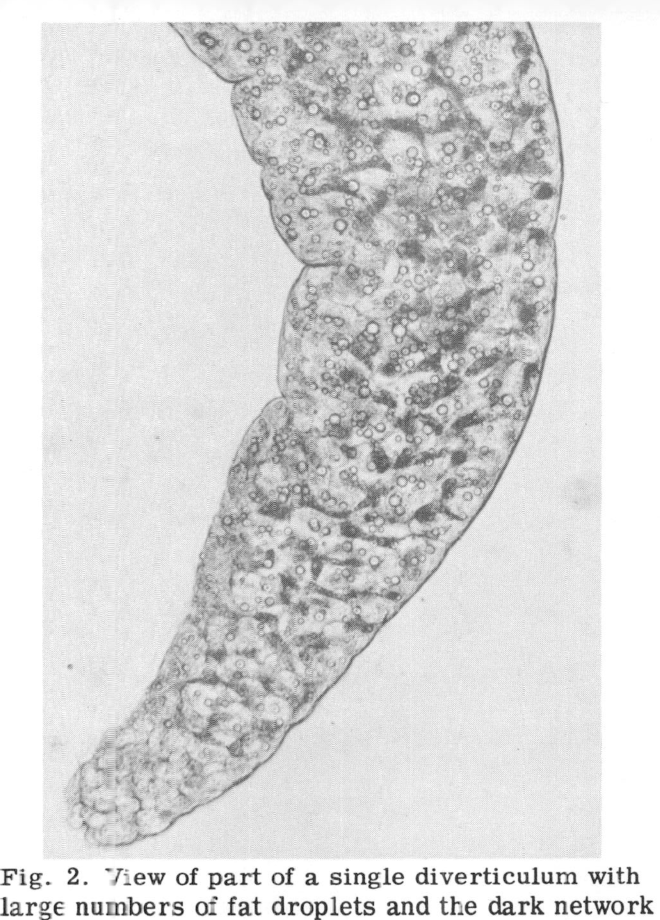

Fig. 2. View of part of a single diverticulum with large numbers of fat droplets and the dark network of β-cells. Note the faint outline of the envelope on the concave border of the diverticulum. x 152

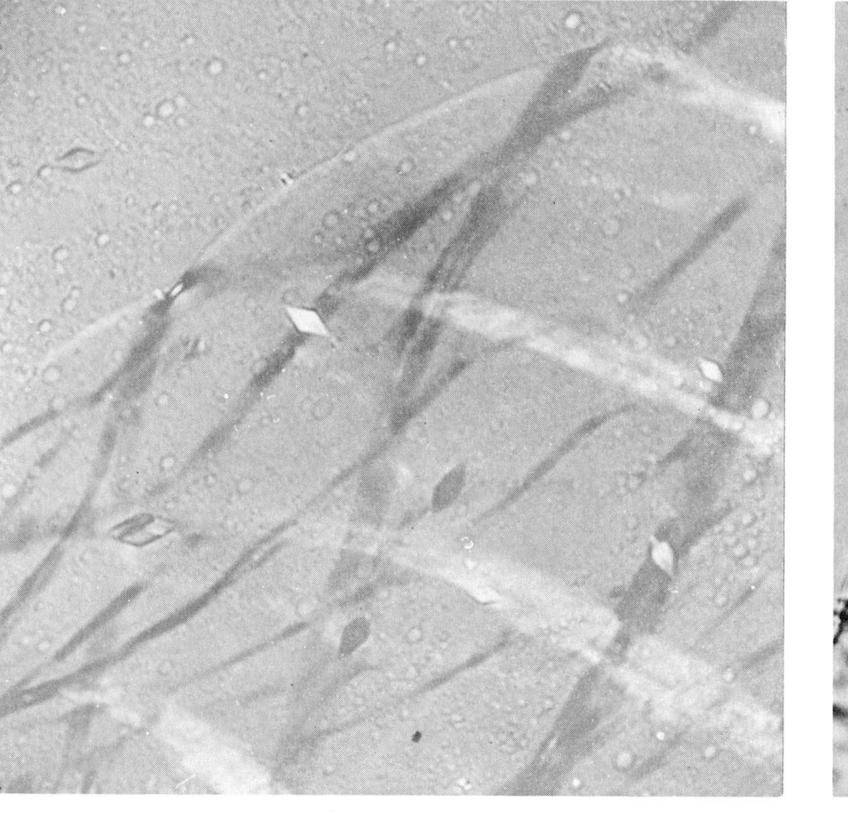

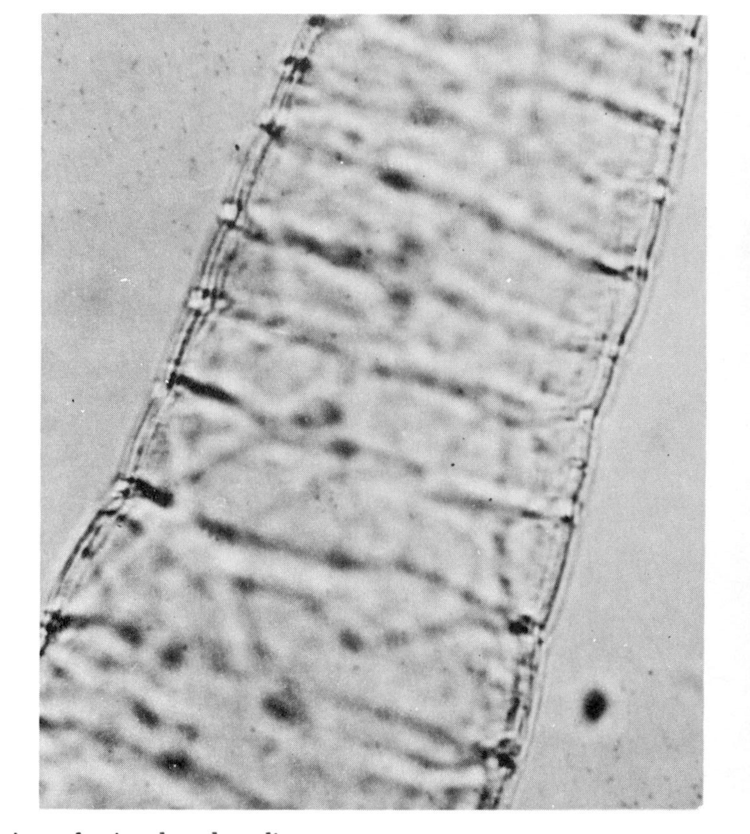

Figs. 3 and 4. Showing the envelope in polarized and ordinary light, respectively. Both photographs illustrate the regular pattern of transverse muscle bands. Figure 3 also illustrates the birefringence of both crystals and muscle bands. x 950, x 170

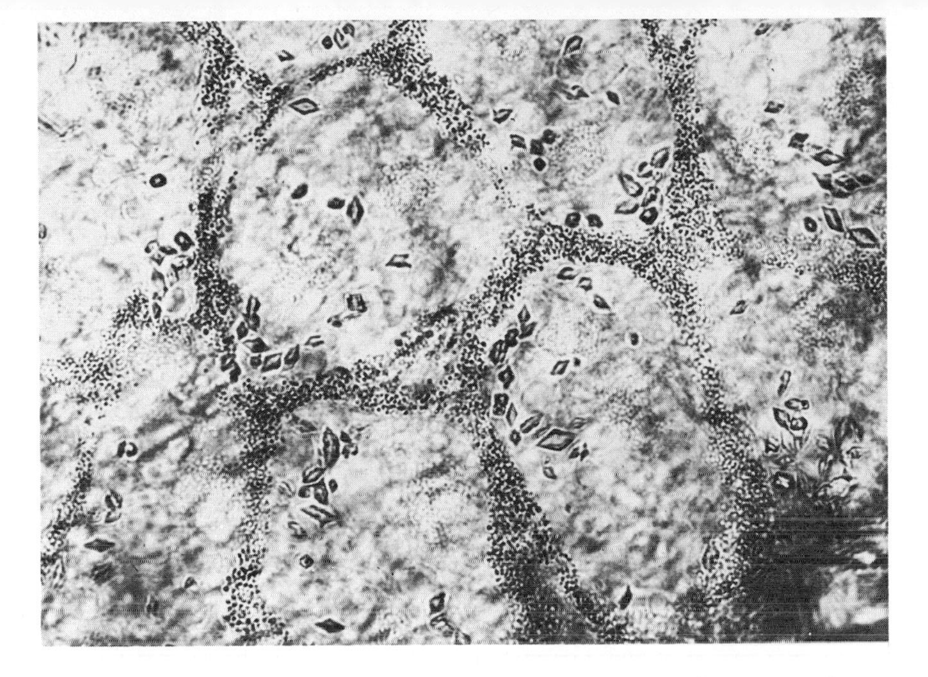

Fig. 5. View of α- and β-cells in a fresh, unfixed diverticulum. The α-cells are seen to contain numerous crystals, faintly visible oil droplets, and two nuclei. The β-cells and their granules form the network between the α-cells. x 790

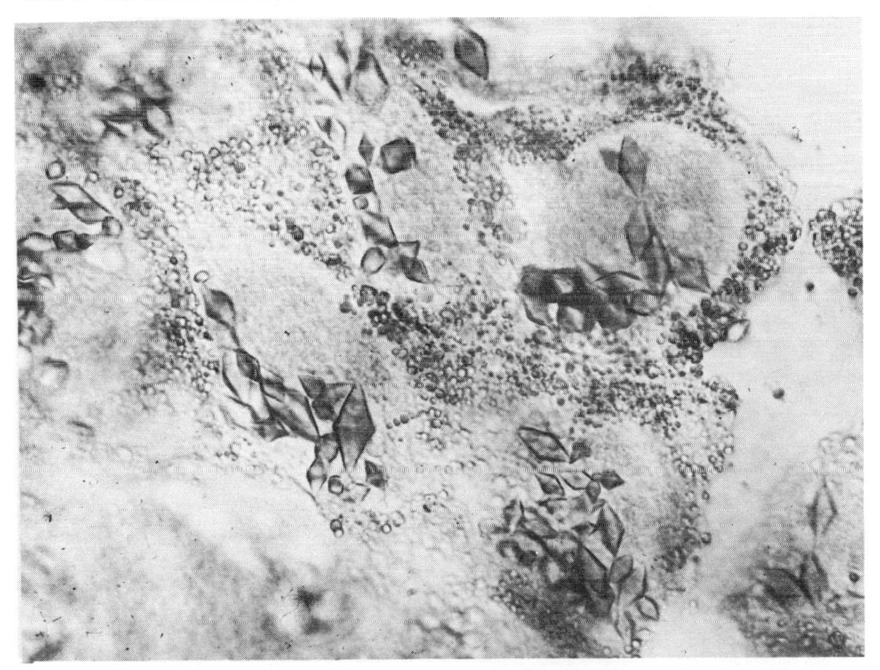

Fig. 6. A higher magnification view of α- and β-cells, illustrating the shape of the crystals as viewed from different angles. x 910

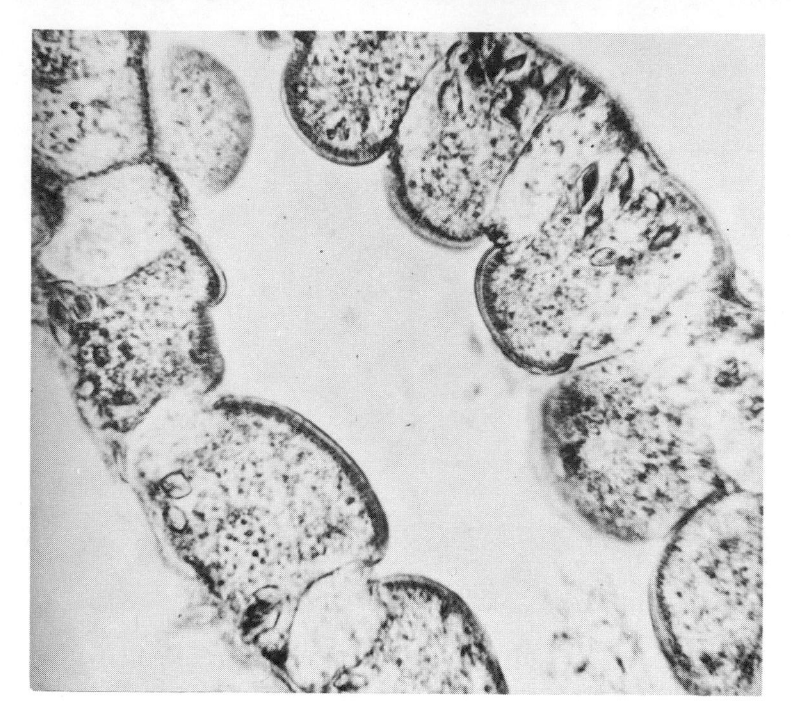

Fig. 7. View of a longitudinal section of a diverticulum, showing large, dark α-cells with peripherally located crystals and lighter, generally smaller β-cells. x 820

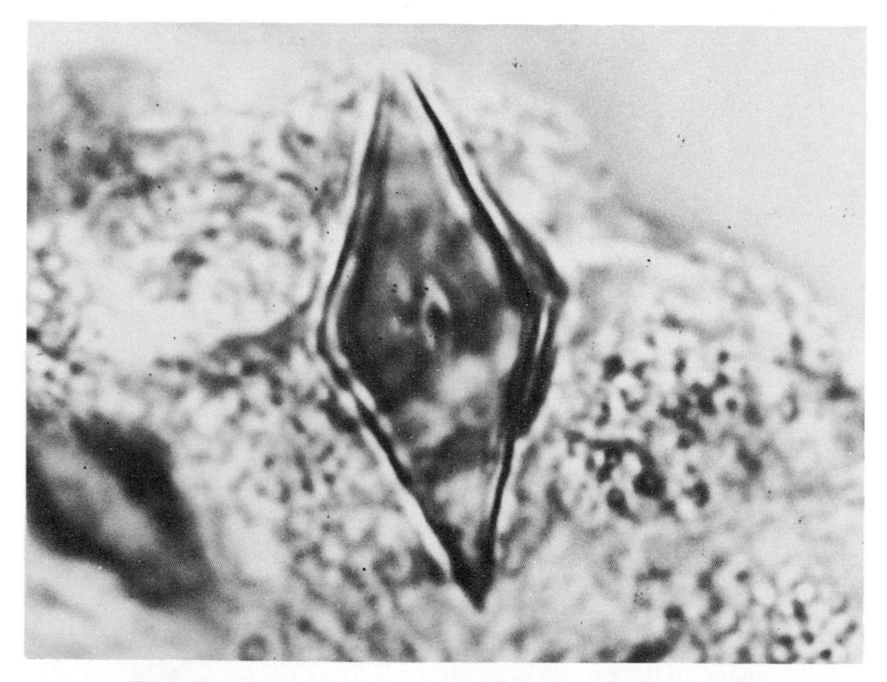

Fig. 8. Single large crystal in an α-cell. x 3,600

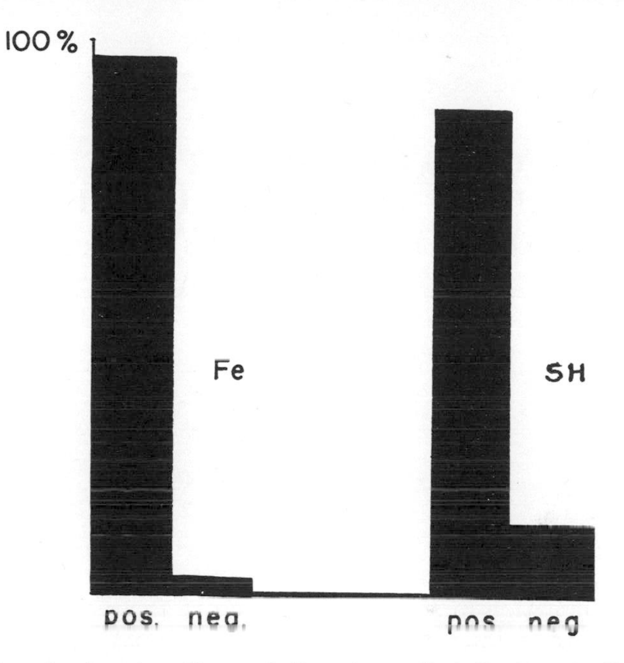

Fig. 9. Graph showing the relative iron distribution in 2 populations. The diverticula of 50 animals in each group were tested

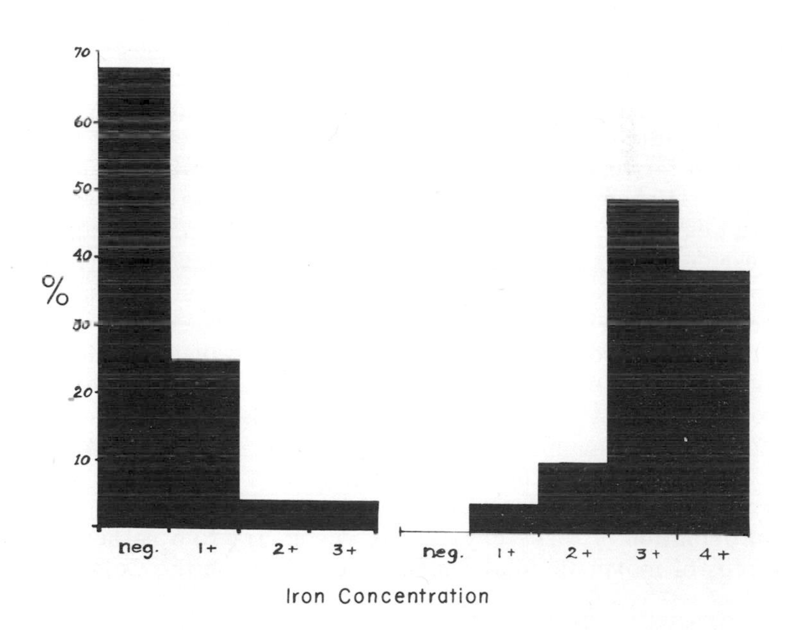

Fig. 10. Graph showing the coincidence between iron and sulfhydryl reactions. The experiment consisted of using 2 diverticula of each animal for the iron test and the remaining 2 for the sulfhydryl reaction. The graph represents the results obtained from 62 animals

THE FORMATION OF INTRACELLULAR CRYSTALS IN THE MIDGUT DIVERTICULA OF *LIMNORIA LIGNORUM**

Stanley W. Strunk

The presence of intracellular crystals within the midgut gland cells of Limnoria has been recognized for many years, but very little information is available relevant to their composition, function, or method of formation. Kofoid (1927) suggested that these crystals, by virtue of their color and shape, resembled uric acid and concluded that they were secreted by the midgut cells.

The present report consists of certain observations, made with the electron microscope, on the structure of these crystals and their method of formation. In order to study this material successfully with the electron microscope, the midgut diverticula were rapidly removed from the animals and fixed in a 1 per cent osmium tetroxide solution buffered to pH 7.4 with Michaelis-Acetate-Veronal buffer; the entire mixture was made isotonic with sea water by the addition of appropriate amounts of sodium chloride. After fixation and rapid dehydration with ethyl alcohol, the glands were embedded in a monomeric plastic mixture of 15 per cent methyl and 85 per cent n-butyl methacrylate, which was subsequently polymerized with heat and catalyst. Sections ranging in thickness from approximately 200 Å to 700 Å were cut from the blocks of tissue and observed with the electron microscope. Initial magnifications ranged from approximately 15,000x to 20,000x, and further magnification was obtained by photographic enlargement.

Formation of these crystals is secondary to a rather complex series of morphological changes occurring in the Golgi material

---

*This study is supported by a Medical Student Training Fellowship in the Department of Anatomy.

of the midgut gland cells. During the course of this study numerous Golgi membrane configurations were encountered within these midgut cells. They were analyzed and compared with each other and were finally arranged in an order thought to represent successive stages in time as they might appear in a single cell during the production and elaboration of the crystals. The description and illustrations presented here are based on this premise.

The initial stages of this process are characterized by membrane systems such as illustrated in figure 1.* These systems are typical examples of the Golgi complex (g) as described by many investigators, both in vertebrate and invertebrate forms (see, for example, Beams, Tahmisan, Devine, and Anderson, 1956; Clermont, 1956; Datton and Felix, 1956; Lacy, 1957; Sjöstrand and Hanzon, 1954; Sjöstrand, 1956). They are composed of units, 1 to 2 μ long, of closely packed, flattened cysternae with cysternal cavities 70 Å or more in width.

These cysternal cavities (c) subsequently dilate by the intracysternal accumulation of materials, thus forming systems such as seen in figures 2 and 3. Some of these distended cysternae are arranged in nearly parallel fashion (fig. 2), whereas others appear to curve about or to encircle some central focus (fig. 3). Within the cavities, fine filamentous strands of dense material (f) are noted. These strands are the first representatives of larger accumulations of this material to be seen in later stages.

As development progresses, the entire cystern-membrane complex becomes enclosed by a single large vesicle which is bounded externally by limiting membranes (m) in part of single thickness and in part three or four ply (fig. 4). Larger intracysternal accumulations of dense material are noted at this stage. Very occasionally, tiny crystals (x) are observed within the systems of this type (fig. 5). More often, however, these tiny crystals are encountered in more differentiated systems which are characterized by a decrease in the number of large distended vacuoles and an increase in the number and complexity of the in-

---

*All figures are photographic reproductions of electron micrographs taken from midgut glands of Limnoria fixed with buffered osmium tetroxide and embedded in methacrylate. They illustrate, in sequence, various stages in the development of the Golgi complex as it is associated with the production of intracellular crystals.

ternal membrane components (fig. 6). The dense material, so obvious in preceding stages, is still present, both in large dense clumps and scattered as finely divided particles between the membranes.

Crystals are formed on certain members of the internal membranes. These membranes (m) are arranged in a nearly parallel fashion, and the material (p) from which the crystals are built is deposited in double rows on each membrane (fig. 7). This figure also shows a dense body (f) from which the above membranes radiate. This body has a similar density to the dense accumulations of material noted earlier and to the subunits of the crystals themselves. In a speculative vein, one could postulate that this material is a reservoir from which small amounts of crystal precursor are withdrawn and transported along the membranes to the sites where crystal formation occurs. On arriving at these sites, this material is then possibly converted, through a process of molecular rearrangement, into a form suitable for crystallization. The concept of membrane flow and intracellular transport of materials has been postulated by Bennett (1956), and evidence in support of this mechanism has been found in many types of cells.

After a crystal has been formed, the membranes on which synthesis occurred disappear, leaving even numbers of double parallel rows of particles (p) without supporting membranes (fig. 8). These particles subsequently orient themselves in a hexagonal, close-packed pattern (fig. 9), thus giving rise to a fully formformed crystal. Each of the particles composing a crystal is grossly spherical with a diameter of 60 Å and a center-to-center distance of 90 Å.

Crystals of this type have a rather large size range, extending from below the limit of resolution of the light microscope to 1 to 2 μ. In cells that are actively forming crystals they are very numerous and occur throughout the cytoplasm but primarily in the apical portion of the cell. They appear to be the building blocks from which the large dipyramidal crystals, so easily visible with the light microscope, are formed. This is accomplished presumably by the accretion of large numbers of these small crystals into a large aggregate. Figure 10 shows two of these small crystals united with a cleavage plane (z) between them.

The large dipyramidal crystals have the same fine structure as their minute counterparts. There are many small defects within the interstices of the larger crystals which are taken to be crystal dislocations remaining from the original accretion process. Figure 11 depicts a typical large crystal.

In conclusion, it is clear that this study merely scratches the surface of understanding the process of crystal formation and the understanding, ultimately, of membrane function in the midgut cells of Limnoria. It has, however, elucidated the fine structure of these crystals and has given some insight into the role of intracellular membranes in the process of crystal formation. From a more general point of view, the study has shown that it is possible to observe the changes occurring in the Golgi complex during the elaboration of a cellular product and to observe how this organelle contributes, from a morphological standpoint, to the process of synthesis. Golgi function has been implicated in cellular product synthesis for many years, and elegant investigations have been carried out both with the light and the electron microscope which suggest that a close relationship between the Golgi material and secretion products exist. In this instance, a close correlation between Golgi membranes and product is evident and a definite close topographical relationship between them has been demonstrated. *

LITERATURE CITED

Beams, H. W., T. W. Tahmisan, R. L. Devine, and E. Anderson. 1956. Electron Microscope Studies on the Dictysomes and Acroblasts in the Male Germ Cells of the Cricket. J. Biophys. and Biochem. Cytol. Suppl., 4:123.

Bennett, H. S. 1956. The Concepts of Membrane Flow and Membrane Vesiculation as Mechanisms for Active Transport and Ion Pumping. J. Biophys. and Biochem. Cytol. Suppl., 99.

Clermont, Y. 1956. The Golgi Zone of the Rat Spermatid and Its Role in the Formation of Cytoplasmic Vesicles. J. Biophys. and Biochem. Cytol. Suppl., 119.

Dalton, A. J., and M. D. Felix. 1956. A Comparative Study of

*I wish to express my deep gratitude to Dr. Dixy Lee Ray for introducing me to this problem and for her encouragement during its development, and to Dr. H. Stanley Bennett for his guidance and suggestions and for making it possible for me to use the exceptional facilities of the Department of Anatomy of the University of Washington.

the Golgi Complex. J. Biophys. and Biochem. Cytol. Suppl., 79.

Kofoid, C. A., and R. C. Miller. 1927. Biological Section, pp. 188-343, in Marine Borers and Their Relation to Marine Construction on the Pacific Coast. Final Report, San Francisco Bay Marine Piling Committee.

Lacy, D. 1957. The Golgi Apparatus in Neurones and Epithelial Cells of the Common Limpet *Patella vulgata*. J. Biophys. and Biochem. Cytol., 3:779.

Sjöstrand, F. S., and V. Hanzon. 1954. Membrane Structures of Cytoplasm and Mitochondria in Exocrine Cells of Mouse Pancreas as Revealed by High Resolution Electron Microscopy. Exp. Cell Res., 7:415.

-------. 1956. The Ultra Structure of Cells as Revealed by the Electron Microscope. Int. Rev. Cytol., 5:455.

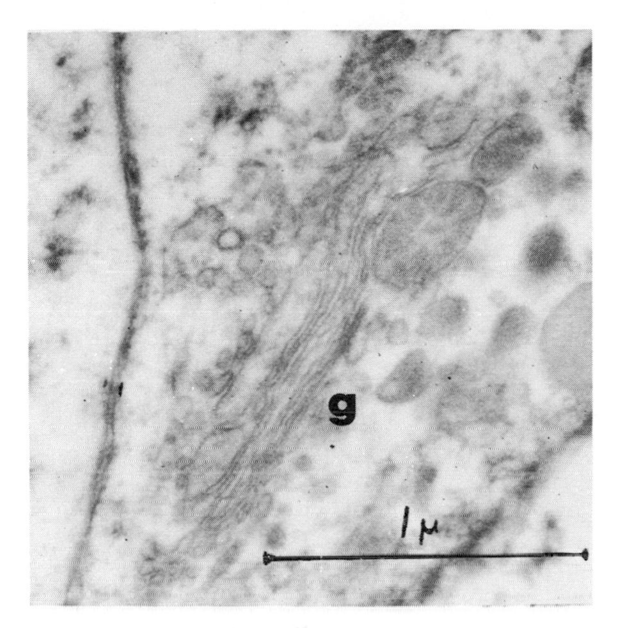

Fig. 1. A typical, unmodified Golgi membrane system showing flattened parallel cysternae associated with numerous vesicles. x 43, 000

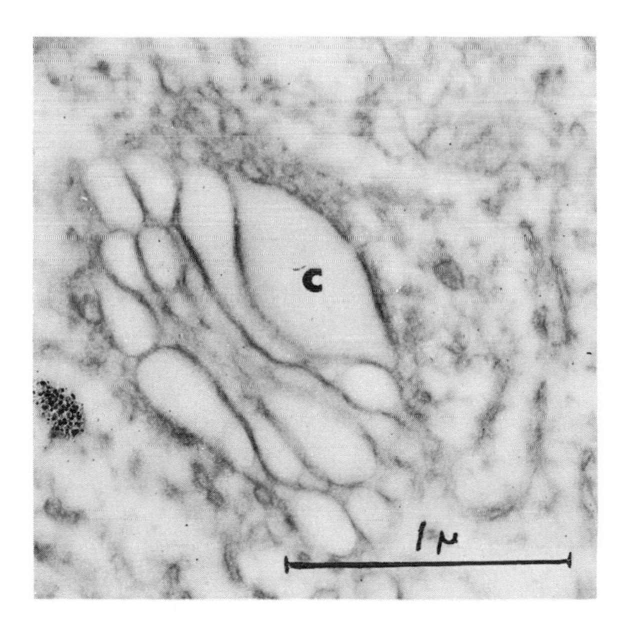

Fig. 2. A Golgi membrane system which is in the early stages of development in becoming suitable for crystal formation. The beginnings of cysternal dilatation (c) and intracysternal accumulations of dense materials are evident. x 38, 000

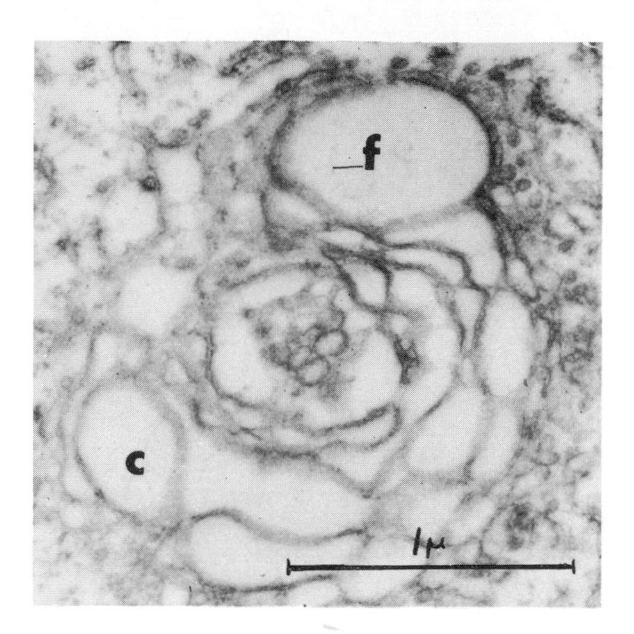

Fig. 3. A somewhat later stage in Golgi membrane development showing increased cysternal distention and larger accumulations of intracysternal densities. x 38, 000

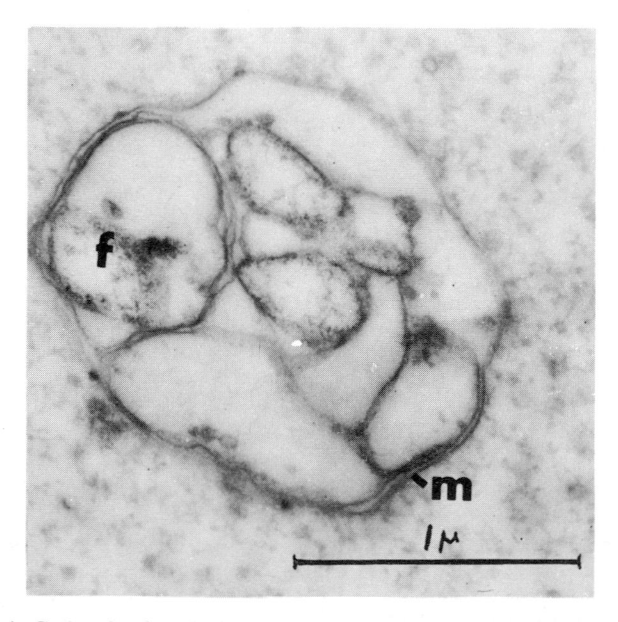

Fig. 4. A Golgi body showing further changes than those seen in figure 3. The entire complex is bounded externally by a limiting membrane partly of single thickness, and partly 3 or 4 ply. x 38, 000

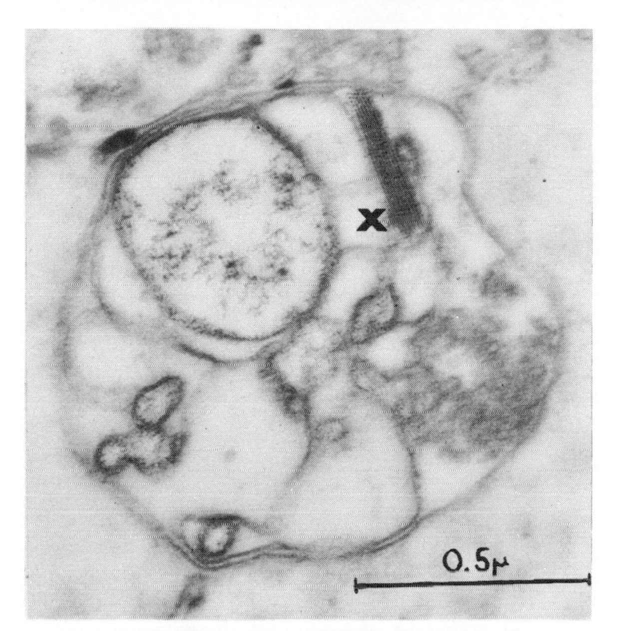

Fig. 5. A Golgi complex very similar to the one in figure 4; large quantities of dense material are present within the cysternae. A crystal (x) is situated within the membrane system, thereby coding this complex as one in which crystal formation occurs. x 63, 000

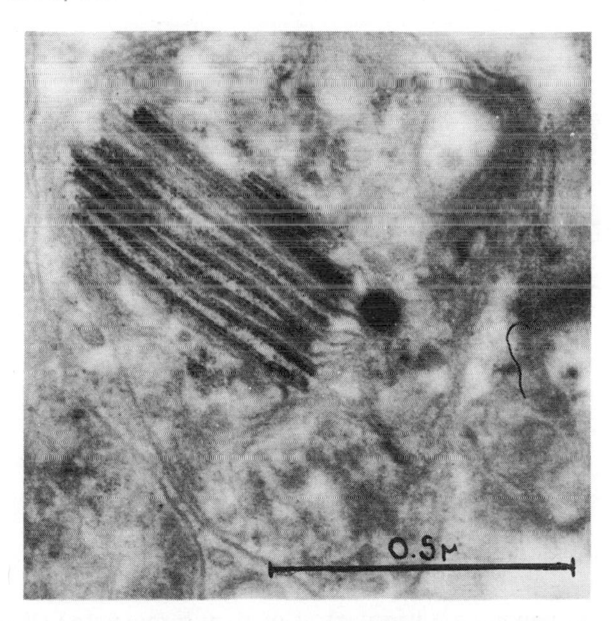

Fig. 6. A typical crystal-containing Golgi complex showing more extensive internal membrane elaboration, a crystal in the process of formation (an enlargement of this crystal is seen in figure 7), and small amounts of the dense material seen in previous figures. x 80, 000

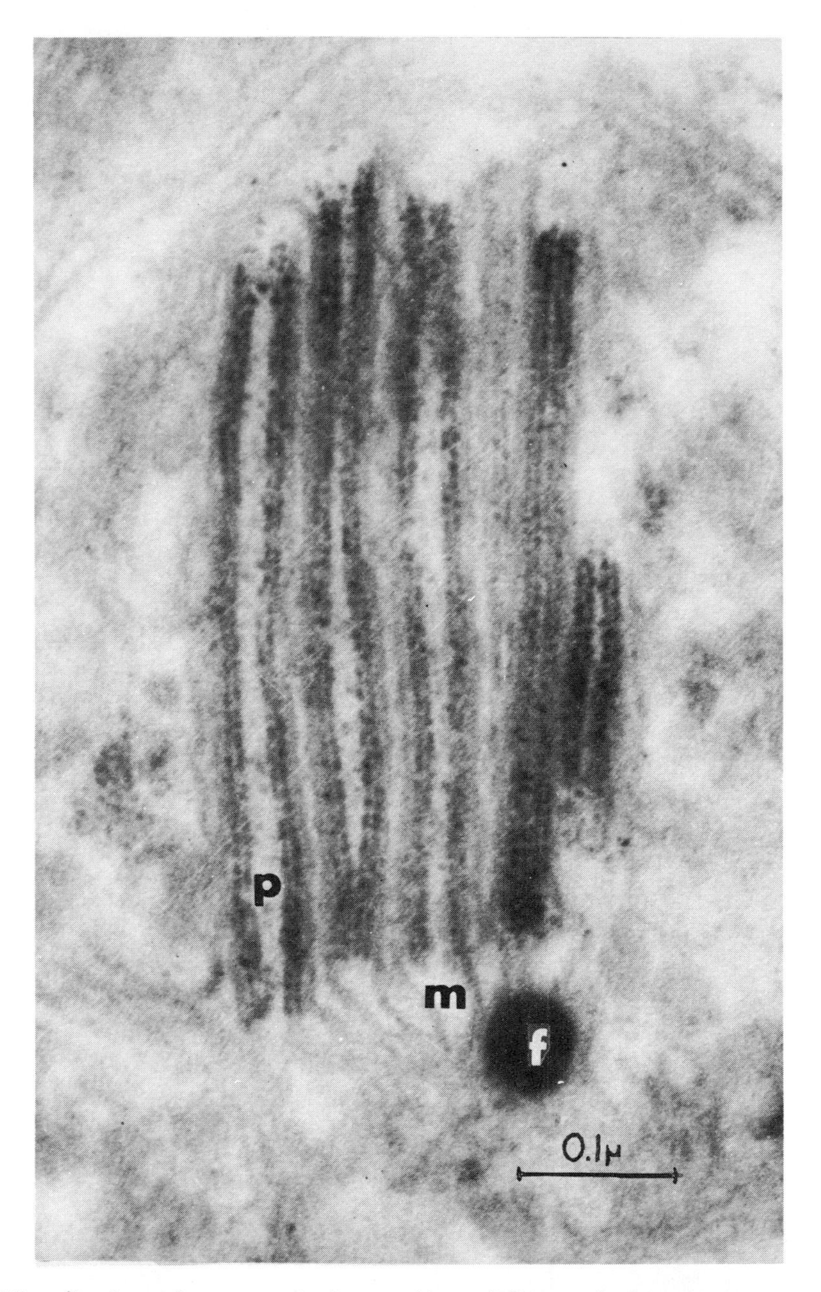

Fig. 7. An enlargement of a portion of figure 6 showing membrane and particle relations at an early stage of crystal formation. An amorphous accumulation of dense material (F) is surrounded by a membrane which is connected to flat parallel sheets of membrane material (m), each of which is adjacent to two layers of particles. x 210, 000

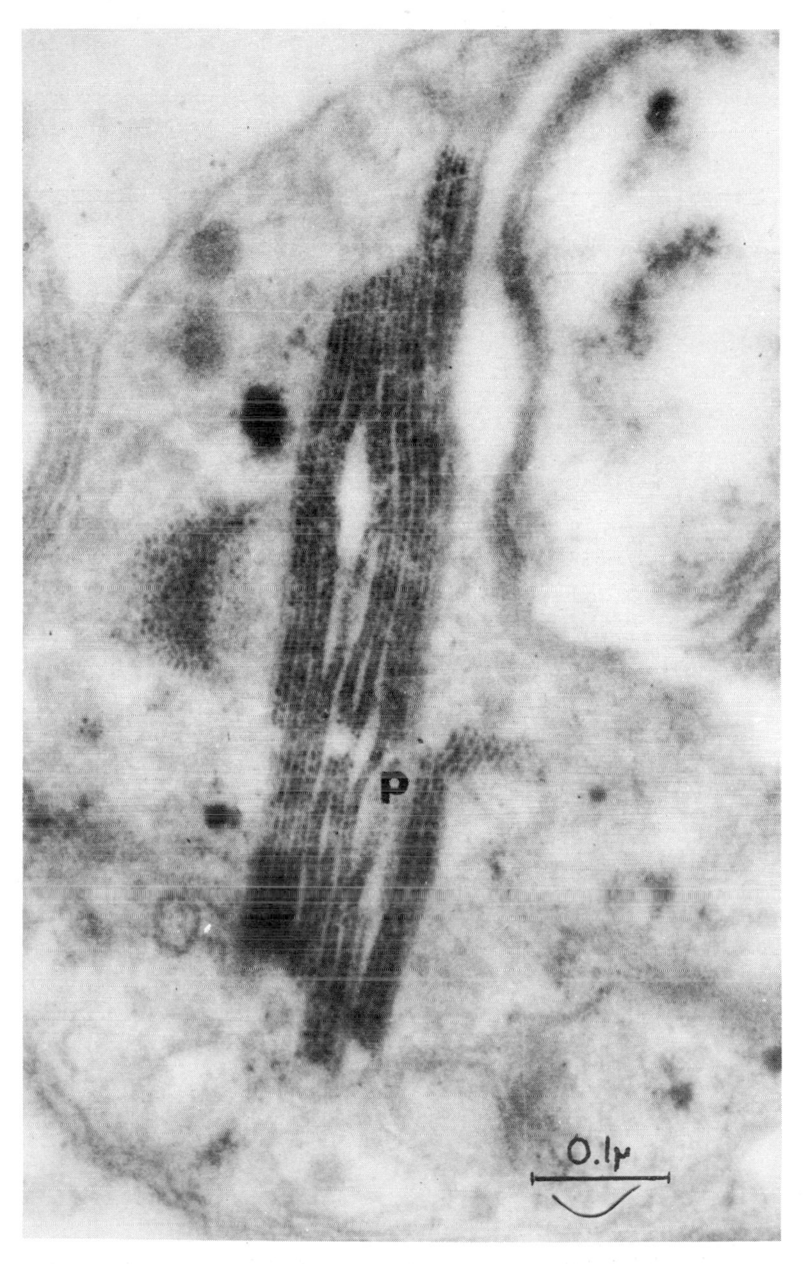

Fig. 8. A further stage in crystal formation. The membranes associated with the particles are disappearing, leaving the particles in parallel layers of double rows. x 180, 000

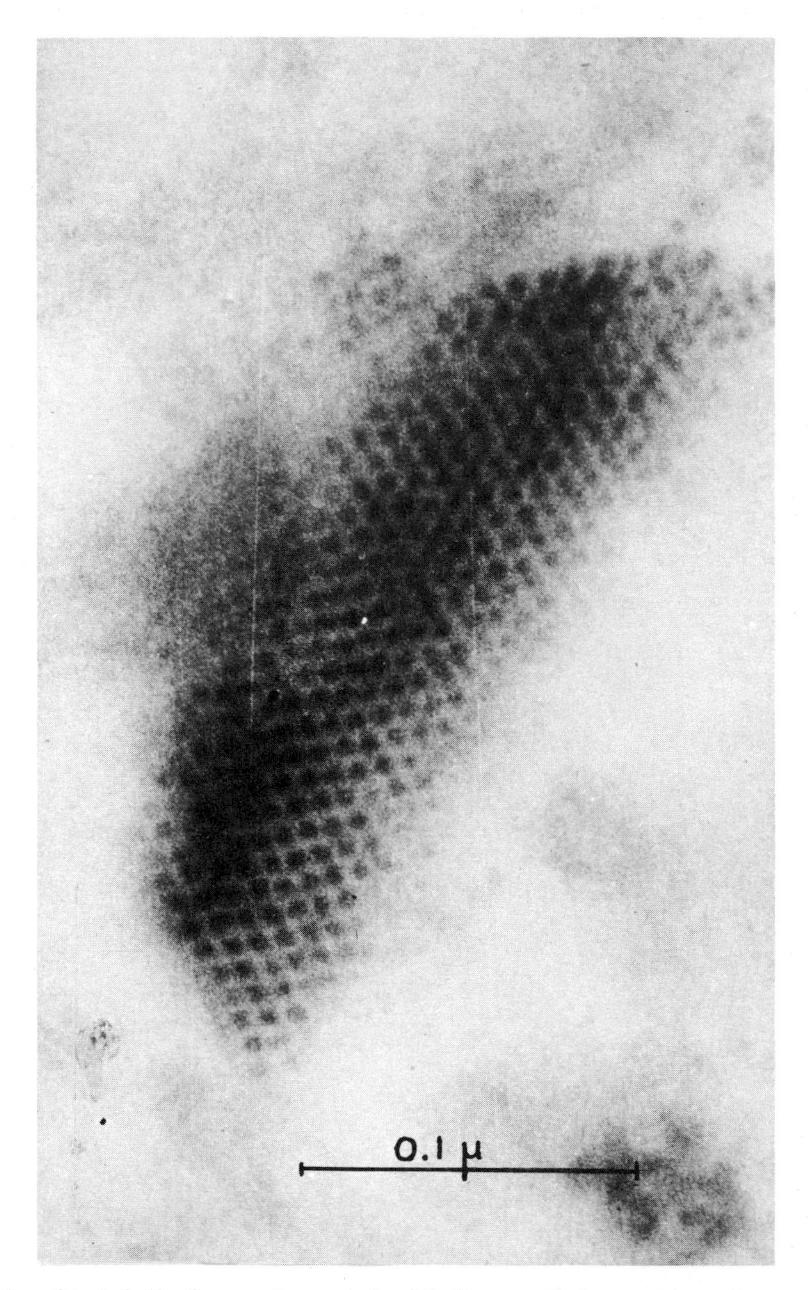

Fig. 9. A fully formed crystal with the particles arranged in a hexagonal close-packed pattern. The particles are about 60 Å in diameter and are about 90 Å apart (center-to-center measurement). x 450, 000

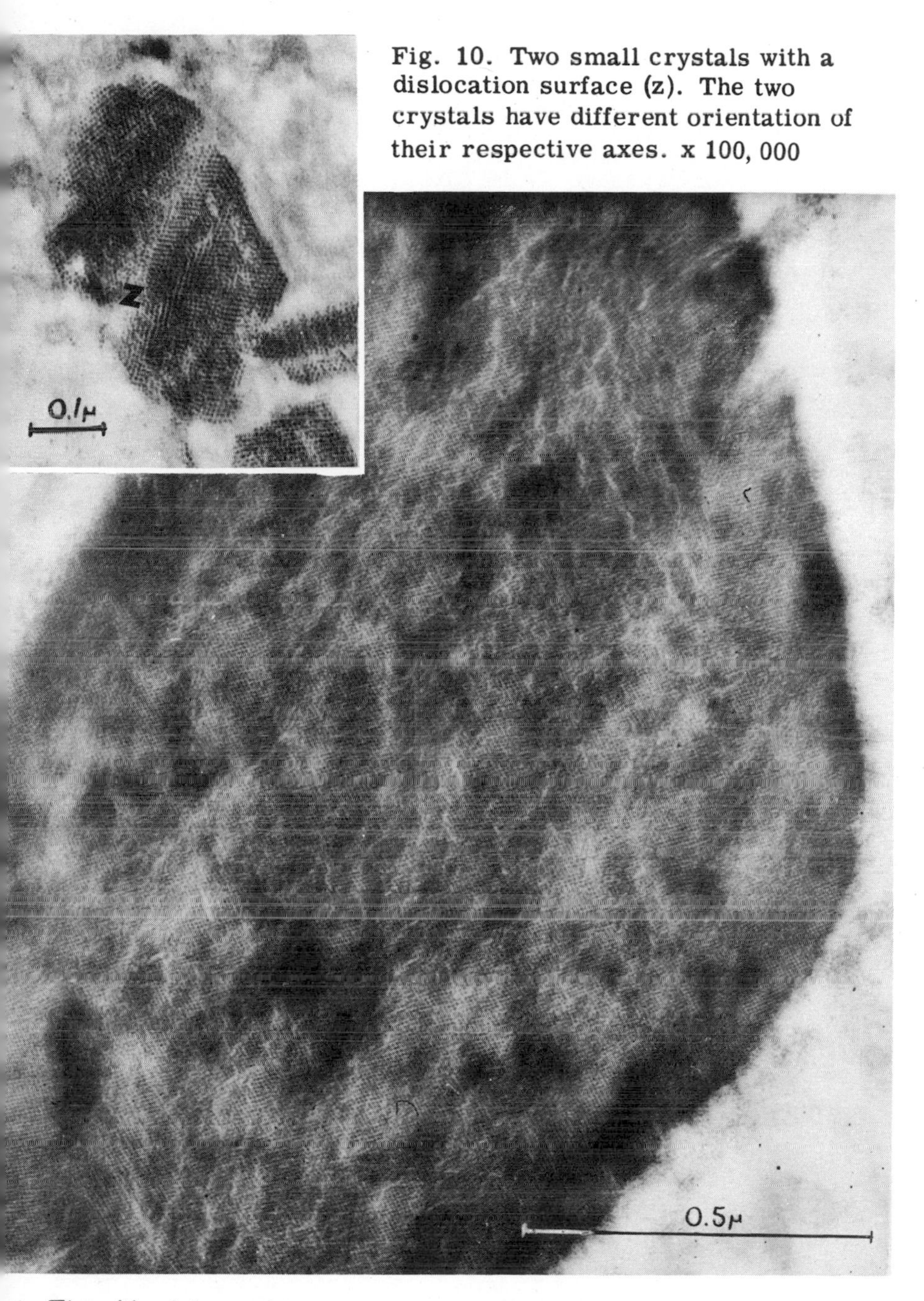

Fig. 10. Two small crystals with a dislocation surface (z). The two crystals have different orientation of their respective axes. x 100, 000

Fig. 11. A large bipyramidal crystal showing a number of dislocations. x 96, 000

Terediniens

INTRODUCTION

C. H. Edmondson

It was with a great deal of personal interest and pleasure that I accepted Dr. Ray's invitation to serve as chairman for this session of the symposium. I am very happy indeed to be at Friday Harbor and to preside at today's program.

Our topic is the biology of the Terediniens. Like the speakers who follow, I have devoted many years to the study of shipworms. These ubiquitous wood-borers occupy a position of preeminence among the agents of wood destruction in the sea. Though much is known about their depredations, we are far from understanding them completely and still farther from our goal of adequate and reliable control in all seas and under all conditions. Necessary to the achievement of this aim is the kind of information to be considered today: the systematics, morphology, physiology, and life histories of the terediene mollusks.

We are indeed fortunate to have as our participants scientists of the caliber of Dr. Turner, Dr. Lane, and Dr. Quayle. Their researches, to be reported now, help to measure the progress of our knowledge. So also, I am convinced, do meetings like this one play an important and valuable role in furthering understanding and in shedding new light on the many perplexing questions associated with problems like marine biological deterioration. It is therefore with grateful thanks to the University of Washington for the generous hospitality of these laboratories that I now present our speakers for today's session of scientific papers.

THE STATUS OF SYSTEMATIC WORK IN THE TEREDINIDAE

Ruth D. Turner

Man's fight against the shipworms probably goes back to the time when primitive man first learned to use rafts and small boats in sea water. His struggle against shipworms is well documented in the literature going as far back as Theophrastus in 350 B.C. The impact of these borers on the history of man is difficult to estimate, but it has certainly been tremendous. Reading the accounts of the voyages of Dampier, Cooke, Drake, and others it is evident that these early navigators dreaded the shipworm as much as they did scurvy and other deficiency diseases which they did not understand or know how to control. Columbus, on his fourth voyage, lost all of his vessels because of shipworms and had to be rescued from Jamaica where the *Capitana* and *Santiago* had been beached at what is now called St. Ann's Bay on the north coast. It appears certain that many of the early explorers who set out on voyages of discovery never to be heard from again probably were wrecked, not because of poor navigation, but because their vessels were so weakened by shipworms that they could not withstand the buffeting of heavy storms. Nor were those who were actually sailing the seven seas the only ones to suffer from the activities of the teredo for in 1730 the very existence of the Netherlands was threatened by the invasion of shipworms in the then wooden-faced dykes of that country. Out of this disaster came one of the finest pieces of research ever published on the Teredinidae, particularly when we consider the time at which it was done--the monograph by Godfrey Sellius, published in 1733. In his report he not only definitely proved that teredos belong to the phylum Mollusca but he recommended creosote as a means of protecting timber against them.

It is interesting that, even though man has been fighting against the ravages of shipworms since early historic times, no positive control has been developed. Scientists have been able to work out the very complicated life histories of many parasitic diseases of man and domesticated animals and to effect control in most cases. However, unlike the simple life history of the shipworms, there is always a "weak link" in the complicated cycle of these parasites, and it is at this point that control can be effected. There is no "weak link" in the life history of a shipworm, and the medium in which they live makes such procedures as poisoning almost impossible. It is certainly true that the Teredinidae are one of the most difficult groups of animals to control, and the classification of the family is so complex that no attempt has ever been made to monograph it on a world-wide basis. Papers usually cover a single collection or a small geographic area. Titles listed in a bibliography on the systematics of the Teredinidae would read as follows: The Teredinidae of the British Museum, the Natural History Museums at Glasgow and Manchester and the Jeffreys Collection, or Die Teredien im Koeniglichen Museum fuer Naturkunde zu Bruessel, or Die Terediniden der Zoologischen Museen zu Berlin und Hamburg, or Revision des Teredinidae Vivants du Museum National d'Histoire Naturelle de Paris, to give just a few examples of these types of papers which may be world-wide in scope but concern only the specimens contained in a given museum. Other papers, of which there are many, are geographic in scope, usually covering a rather small area; generally in such papers there has been little attempt to relate these areas to other parts of the world. Such titles would include Die Teredinidae des Mittelmeeres, A Systematic Account of the Teredinid Molluscs of Port Jackson, Queensland Cobra or Shipworms, Teredinidae of Hawaii, or Shipworms of the Philippine Islands. In addition there are, of course, innumerable small papers describing new species of teredo from this, that, or the other locality. The Monograph of American Shipworms by Bartsch is one of the largest single monographic works attempted in this group. This was published in 1922 and concerns species of both coasts covering the entire family for North America. A number of new species have been published and a great deal of new information has been brought together since that time, and so there is a real need to have this work revised and brought up to date. To my knowledge there has been only one detailed monographic study of a genus in this family covering a wide area, that of the genus Bankia in the western Atlantic. Two studies on a single species which are outstanding are those of R. C. Miller on the variations in the shell and pallets of *Teredo navalis* Linné in San Francisco Bay and the work of Th. Monod and M. Nicklès on the variation in *Teredo senegalensis* Blainville on the West African coast. Both studies show how extremely variable a single species can be even in a restricted locality. Knowing this we should examine all very closely related species from various parts of the world to determine whether they are good species or only variations of a single, widely distributed one.

It is not the purpose of this paper to give a detailed account of the classification of the Teredinidae. In the first place it would be impossible at this time, and, besides, a recitation on the relationships of genera and species would be most difficult to follow. Such an account must be published so that it can be used when the need arises. Rather, I want briefly to review the work that has been done, to present some of the problems inherent in the study of the family, and to outline the plans I have for further study.

The question is often asked, "Why study the systematics of the Teredinidae?" or, "They are all destructive so what difference does the name make?" Granted that today the procedures in treating wood to prevent their ravages are the same for all species, this is not all there is to the picture. In many other economic groups the studies of the systematist have been very important in working out control measures. At present there is little connection between the systematics and the control of Teredo, but it is too early to say that such will always be true. There are many reasons why a complete understanding of the taxonomy of the group is essential. (1) It is necessary to have a name in order to talk or write about a given animal. (2) Working from the opposite angle, if a particular animal is an economic problem, the name of the species is a key to all of the literature that has been published concerning it. (3) The various species of shipworms differ greatly in the extent to which they are distributed by floating wood and wooden vessels. Some are tolerant of a wide range of temperature and salinities and so are distributed widely and, on arriving at a new locality, survive. Others, for reasons which we do not yet understand, have a restricted distribution. However, at present the taxonomy of the group is so confused we cannot prove the validity of these distribution patterns though there is good evidence that they do exist. (4) It is a well-known fact in other groups that closely related species may have different life history patterns, different anatomical features, and different reactions even though superficially they appear quite similar. There are many cases in the literature of disagreement between two scientists who thought they were working with the same animal but produced very different results in their research. When the animals were submitted to a specialist in the group they proved to be two different species. However, in the Teredinidae it is often impossible even for the systematist positively to identify a species without a great deal of work for there are no good monographic studies to which one can refer. Certainly the experimental zoologist working on "teredo" is greatly hindered by this lack of up to date systematic work.

As already mentioned, the early Greeks and Romans were aware of the shipworms, and Ovid in 20 B.C. referred to them in his writings as *teredin navis*. It was from this that Linné undoubtedly took the scientific name *Teredo navalis*.

Sellius in his monograph of the group in 1733 gave a rather complete account of all that was known up to that time. Our classification, of course, dates from Linné and the tenth edition of his Systema naturae which was published in 1758. Here Linné introduced the generic name "Teredo" and the species "navalis," based largely on the work of Sellius. However, he did not follow Sellius in placing the shipworms in the Mollusca but classified them with the Vermes. The early history of the systematics of this group is complicated and is of relatively little concern to us here for it was largely on the generic and family level. Adanson in 1757 had come to the same conclusions as Sellius, but even so it was some time before the shipworms were generally classified as mollusks; O. F. Müller and Fabricius were among the early workers to do so. It is quite amazing to find that Lamarck was apparently unaware of the work of Adanson. Gray's monograph of the group in 1827 was little more than a list of the species then known. The family was poorly covered by Sowerby and Reeve in their illustrated works. In 1862 Tyron monographed the family, giving a list of all the papers on the group up to that time. He recognized 28 species included in 4 genera and placed in synonymy a large number of names. The work unfortunately was not illustrated. In 1865 Jeffreys gave a very fine account of the British species in his British Conchology. Following this, very little was done in the systematics of the group until about 1920 when the teredo problem really began to be felt in this country. Dr. Bartsch published a New Classification of the Shipworms in 1921 which was followed by his monograph of American species in 1922. From that time to the present numerous systematic studies, mostly by-products of various marine borer investigations, have appeared. Lamy in 1926 wrote a revision of the Teredinidae based upon the collections of the Paris Museum. This study gives rather complete synonymies and is most useful to anyone working on this group. Unfortunately, the paper was not illustrated and the descriptions are reduced to a few comparative remarks so that the paper loses much of its value so far as ecologists, physiologists, or other nonmalacologists are concerned. Both William F. Clapp and R. C. Miller have published important taxonomic papers, largely in conjunction with the National Research Council Investigation and the San Francisco Bay Marine Piling Investigation; the principal reports of these two investigations were published in 1924 and 1927. In 1936 the borers had become a problem in Australia and several reports were published there, mainly on Port Jackson and the Queensland coast. In the systematic portions of these reports several new species were described. A report on the Hawaiian Teredinidae was published by C. B. Edmondson in 1942. In later years the major specialists in the systematics of the Teredinidae have been Felix Roch and Friedrich Moll, two German workers who collaborated in many of their reports. Together since 1928 they have published some 20 papers and described at least 62 new species. Their papers have generally been limited in scope either to a geographic area or to the specimens contained in a given collection such as the British Museum or the Hamburg Museum. Other workers interested in the systematics of this group are P. Rancurel, who is with the Institut Français d'Afrique Noire, and R. Nagabhushanam of Andhra University, India.

For some years I have been working on a catalogue of the family Teredinidae, and my file at present lists about 400 specific and 35 generic names. Needless to say there are far more names available than there are good species. I would estimate that there are probably only about 50 to 60 species and at the most 8 genera. Many of the names in this catalogue have already been placed in synonymy, but there is still a great deal to be done and it will be a tremendous task for it is always much easier to describe a new species than to prove that one already described is a synonym.

Problems inherent in the systematic study of the Teredinidae

In reading through the various monographs on this group one continually comes upon statements to the effect that the Teredinidae are among the most difficult of all mollusks to study. Such statements go back to the time of Tryon and Jeffreys in the 1860's, and the situation has become if anything worse rather than better with the passage of time; the tremendous number of names that have been introduced into the literature is a good indication of this fact. There are several factors that have contributed to this. First, the animals themselves are difficult to obtain in proper condition for study. The specimens should be alive or only recently dead at the time they are preserved. The best preservative seems to be a mixture of 5 parts of 85 per cent alcohol and 1 part of glycerine. Preferably the entire animal--shell, soft parts, and pallets--should be preserved. However, if the soft parts have decomposed and the pallets are still in good shape, these should be preserved in glycerine alcohol and the shells and pallets of each specimen kept together in a small vial. It is a very time-consuming task to extricate the animals from their burrows, and one cannot expect an engineer, physiologist, or wood-preserving specialist to get out the large series of animals necessary for the systematist. The amateur shell collectors, who are a tremendous help in collecting other groups, are rarely interested in such specialized groups as teredos and pholads. For the systematist interested in this group to do all of his own collecting from all parts of the world would, of course, be ideal, but this is a virtual impossibility. Large collections are usually brought together only at times of stress of one kind or another such as the invasion of an area by shipworms or a war.

During World War II and continuing to the present the United States Navy has conducted a testing program with installations scattered all over the world. As I was working at the W. F. Clapp Laboratories at that time I have had access to this material and it, along with material that has come in since from this and other sources, forms the basis for the reports I am now preparing. However, as wonderful as this material is--and it is without question the best that has ever been brought together--there are many drawbacks to this type of collecting from a systematist's point of view. Although we have exact locality data on all of the material, including the exact time of submergence and removal of the test boards, temperature and salinity data, and the exact depth at which the board is submerged, the material is never preserved before shipment. Test boards arrive at the laboratory days and often weeks or months after removal from the water, and by this time the animals are often so badly decomposed that even the pallets have disintegrated or are in such poor shape that identification is difficult or impossible. This does not detract from the major purpose of the test boards, but it certainly disheartens one who is interested in the animals, their variation and geographic distribution, and not just their destructive ability. So much effort goes into the placing of these test boards that it would seem worth while to add just a little more to it and at intervals to have the boards preserved in alcohol before shipping.

Another drawback is that test boards are only submerged where there is an economic interest of one kind or another. Consequently there are many areas in the world from which we do not have a single specimen. This brings us to the second factor that makes the study of the Teredinidae difficult. Shipworms are perhaps the most readily distributed of all groups of mollusks. They are easily carried by wooden vessels, floating wood, and other plant materials such as nuts. Many West Indian shipworms were originally described from England; they were probably carried there in wooden ships that had been plying the West Indian waters or in floating wood carried across by the Gulf Stream. If they arrived early in the summer they might even reproduce, only to be killed off by the cold winter. As late as 1950 Dr. Tera Van Benthem Jutting of the Zoologisch Museum, Amsterdam, reported the occurrence of *Bankia fimbriatula*, a common West Indian species, in Holland. We know that shipworms can be carried great distances in this way, and if they find proper conditions of temperature and salinity when they arrive they will survive at least for a while. We also know that shipworms can endure long periods of adverse conditions merely by retracting their siphons and putting the pallets tightly into place and awaiting the return of suitable conditions. Consequently today there is no reason to doubt that a species occurring in Brazil might also be found in India or tropical Africa. In fact it now looks as though the species Dr. Clench and I described as *Bankia katherinae* from Brazil may be the same as *Bankia campanellata* from India. However, the original description and figures of the latter are so poor and the type specimens in such poor condition that, until we can examine the type, which is in the Paris Museum, and obtain material from India, we cannot be sure. In the past 25 years there have been 9 species described from Singapore, 3 from Ceylon, 6 from Togo (West Africa), 15 from Australia, and so forth. The descriptions and figures have been studied, and many of them appear very close to if not identical to species already described. However since we do not have material from these areas or the opportunity to study the type specimens, it is impossible to make a decision one way or the other.

A third factor making this a difficult group is the variability within the species. When a large series of individuals is taken from a single heavily infested board, the shape and size of the shells and pallets may vary tremendously depending upon the age of the specimen, the amount of crowding, the rate of growth, and the type of wood in which they are boring. A young set in a piece of wood already infested with borers will become stunted or stenomorphic with the pallets often completely different in proportion to the "normal." The shells, too, are affected, the auricle usually becoming greatly reduced in relative size, the number of denticulated ridges increased, and the shell generally much thickened. In the genus Bankia, particularly, it is easy to see these differences for in this group they are often very striking. In rapidly growing individuals the cones of the pallets are widely spaced, the stalk is thin and often quite flexible, the periostracal margin of the cones is wide and usually beautifully serrated, while the calcareous portion of the cone is much reduced. A slow-growing or stenomorphic specimen, on the other hand, has a short and proportionately broad pallet, the cones are close set, the stalk is thick and solid, the calcareous portion of the cones is larger than the chitinous portion, and the serrated edges are often nearly smooth. Similar variations are to be seen in *Teredo navalis* Linné as shown by Miller, whose paper I mentioned earlier. In some genera, particularly Psiloteredo the young and adult pallets are often quite different in form, and until a connecting series of all stages is obtained it would be easy to consider the two extremes as two different species.

Considering the normal variability of the species, the tremendous effect the environment has upon them, and the poor condition in which most of them are received for study, it is little wonder that early workers described as different species material that we now believe to be only of one species.

The fourth and one of the chief difficulties in working with this group is the fact that we must observe the law of priority in deciding what name should be applied to each species. Actually, if we could forget all previous work and start out anew, the task would be relatively simple. However, in this field, unlike other fields of zoology, a published systematic report, regardless of the quality of the work, cannot be overlooked. It is easy to imagine the confusion that would ensue if this were possible, as there would probably be a new set of names for every generation of workers. Consequently it is the task of the reviser of any group to consider each name that has been published and to determine whether or not it applies to a distinct species or to a variation of a previously described species. In a family such as the Teredinidae some species may be extremely variable and specimens living in different types of wood or in different localities may look quite different from one another, or there may be little resemblance between the young and adult specimens. In addition, shipworms deteriorate rapidly once they are removed from the water if they are not properly preserved. The appearance of the pallets--the organ most used in classification--differs greatly in the various stages of deterioration, and consequently one must consider the normal variation of the species, the age of the specimens, and the stages of decomposition for all of the "normal" variants. This results in virtually a three-dimensional problem in variation. The foundation on which any species rests is the holotype specimen, and, before placing a species in the synonymy of one described at an earlier date, it is best to examine the type specimen of both and to understand all types of variations in both in order to fit the type specimens into the complete picture of the species. This leads to other problems, for the location of the type specimens of many of the species is unknown, and many of those that are known are in very poor condition. The descriptions of many species that had been based on fresh specimens bear little resemblance to the dried up types as they exist today. In such cases, especially if the description was inadequate, it will probably be necessary to allow good specimens of what is believed to be the same species to dry out in order to produce something that looks like the type as it exists today. So it can be seen that it is often a long and very tedious task to determine which is the earliest name that should be applied to a species and which of those subsequently described should be placed in the synonymy of it.

Outline for future work

Now, having discussed the work that has been done, and the difficulties involved in working with the group, I should like to outline my program for a systematic study of the group and to point out the great need for cooperation of all workers in this group.

The first and most basic piece of work that must be done is to make a complete card file of every name that has ever been published in this group. As stated earlier, this is almost completed. The cards are arranged alphabetically by species and include references to the original description, the type locality, and the location of the type specimens if this is known. Two other files should also be made: one with the species arranged geographically so that it is possible to determine quickly what species have been described from a given country; the other arranged systematically so that when working on a given species one can readily refer to all other closely related species in the world.

A catalogue of all the literature on the Teredinidae is also being compiled. It is nearly complete for the systematic literature, reasonably so for papers on physiology, embryology, and life history studies, but weak on papers concerning control measures and the more purely economic side. However, these subjects have been well covered by the bibliographies of Clapp and Kenk. Any monographic study should include a discussion of, or at least references to, papers on the biology and economic status of the species, and such a file, when properly arranged, will give ready access to the material as it is needed. However, when referring to papers on the biology of shipworms one runs into difficulties. Many of the observations and experiments that have been published are of top quality, but often there is some doubt as to the identification of the species concerned. Consequently, it is difficult to know to just what species the work refers, particularly if there is no illustration of an adult specimen. It would, I believe, be a wise procedure for all physiologists, embryologists, anatomists, ecologists, and others to deposit, in a large museum specimens of the species on which they were working and record this fact in their report. In this way future workers could always check back on the determination of the species. There has been much disagreement among workers as to the names to be employed for the various species in the Teredinidae. Until these problems are resolved, an illustration of the species studied is absolutely essential when biological papers are published. Moreover, regardless of the status of the taxonomy of the group this is always a wise procedure, for systematics is a dynamic science and many changes in classification will undoubtedly be made in the century to come.

Bringing together a large collection preserved in glycerine-alcohol is, of course, essential. This collection should have as large a series from each locality as possible. Preferably the entire animal should be preserved, though in the past only the shell and pallets were considered important. However, the siphons often show very good specific characters, and these should, I believe, be given more attention. This was certainly true in the Pholadidae, particularly the genus Xylophaga. An effort should be made to get material from as many areas throughout the world as possible. To do this, it will be necessary to get government cooperation, for no private institution would ever be able to afford such an undertaking. We have received some specimens from marine laboratories in Brazil, Africa, India, and Japan, but while such material is extremely helpful it is relatively insignificant when we consider the large series from many localities that are needed really to understand this group.

An absolute essential for a complete study of the Teredinidae is an examination of all holotype specimens, so far as this is possible. In many cases we do not know where the types are located as no mention was made at the time the species was described. However, we do know that there are many types in the British Museum, the Paris Museum, the Museum at Hamburg, the United States National Museum, the Australian Museum in Sydney, the Harvard Museum of Comparative Zoology, and a few others. A visit should be made to these and to other museums to study, compare, and illustrate properly as many type specimens as possible. This should come after a careful study has been made of all the material available in this country so that the greatest possible benefit can be obtained from an examination of the types.

Plans for publishing on the Teredinidae include first the Catalogue of the Family Teredinidae, including all names, generic, subgeneric, specific, and subspecific, with a reference to the original description, the type locality, and the location of the type specimens. No attempt will be made to bring the nomenclature up to date or to place any species into synonymy in this paper. The monographic studies should, I believe, be published a genus at a time and, if possible, on a world-wide basis. This, of course, will depend upon the amount of material available for study. Illustrations of all type specimens should be included as well as a series of drawings or photographs to show the variation within the species. This will be expensive both in time and money, but without these illustrations the work will, I feel, lose 75 per cent of its value.

The need for cooperation among all workers in this group is, I think, quite evident from the preceding remarks. Any specimens contributed to the collection of the Museum of Comparative Zoology will be most welcome and will be put to good use in the studies now in progress. Determinations will be sent, just as rapidly as possible, to those requesting names.

Collecting shipworms for systematic study

Since it is almost impossible to remove entire specimens from a pile, it is best to set traps for them. There are several types of collecting boards or test boards in use depending upon the type of information that is desired. The simplest type can be made from any knot-free, straight-grained piece of wood, the softer woods being better, cut to about 12" x 6" x 1" though the exact size is not important. However, it should not be too small or the specimens attacking the wood will not have room to grow normally. The board can be hung from a wharf, buoy, or other structure and weighted to hold it down, or it can be secured directly to the wharf. When in place, the board should be about 2 feet off the bottom, and for rapid infestation it should be in the vicinity of infested wood.

The board should be examined at intervals to determine the extent of the attack. It should take about 2 to 6 months to get a heavy infestation in most tropical and many temperate localities. The rate of attack will vary with the locality, the time of year, and the proximity of infested wood so that it is impossible to give a definite time.

The second type of board is the laminated collecting board, made up of 6 or more layers of soft, straight-grained wood about 12" x 6" x $\frac{1}{2}$" with thin washers separating the layers to produce cracks large enough so that the shipworms will not cross from one layer to the next. In this type of board the borers make long straight tubes, and since the wood is thin they are easily extracted.

A third type, the monthly collecting panel, is far more elaborate. It is made up of a series of boards, one for each month, hung on a long rack, with a control panel in the center. The boards are numbered from 1 to 6, 8, 10, or 12, depending upon the number used, and one is removed each month in order from 1 to 12. The control panel is removed every month. In this way it is possible to determine the months in which the shipworms are breeding and the number that are settling per square inch as well as the rate at which they are growing. In northern waters a collecting panel having 12 boards may be successful, but in many tropical areas the boards may be completely destroyed by the end of 6 or 8 months. The bottom board on the rack should be about 2 feet off the bottom.

Regardless of the type of board that is used, the treatment once the board is removed from the water is the same for all. The best specimens are obtained if the animals are dissected out as soon as the board is removed from the water and are preserved in glycerine-alcohol. If this is impossible the board should be submerged in 80 per cent alcohol for about 2 weeks, then removed, wrapped in several layers of paper or cloth that has been saturated with alcohol, and placed in a plastic bag, which should be securely tied to prevent leakage. The bag can then be wrapped and shipped by ordinary parcel post. Packages sent from outside the United States should be labeled, "Scientific specimens. No commercial value." If there is no container available large enough to use for soaking the board, the plastic bag can be used for this purpose if the open end is tied securely and held up to prevent leakage.

SOME ASPECTS OF THE GENERAL BIOLOGY OF TEREDO*

Charles E. Lane

The genus Teredo is the best known of the related forms that together comprise the family Teredinidae. This group, which includes all the important molluscan wood borers, is world-wide in distribution; it has been recorded from all coasts of all seas and from many oceanic islands. The family comprises soft-bodied, boring mollusks. The body plan is modified by extensive elongation of the visceral mass that results in a wormlike appearance. Close examination, however, reveals that the visceral mass occupies only about 1/4 of the total length of the animal and that the rest is made up of gills and mantle. As in other bivalves the visceral mass is enclosed by the mantle. The gills occupy a specialized portion of the space enclosed by the two lamellae of the mantle; this space is known as the branchial cavity.

The body plan of the adult is shown in the composite photomicrograph (fig. 1). This is a series of overlapping photographs of a single sagittal section through the anterior third of a young animal. The body is that of a specialized pelycypod mollusk. The position of the mouth and of the anus is fixed, structurally, in relation to the shell and the shell musculature, but between these two fixed extremities the visceral mass has elongated. The gut is extended into a capacious caecum. This is continuous with the straight and relatively uncomplicated hindgut.

The foregut gives origin to numerous, extensive, glandular evaginations. Greenfield and Lane (1953) have suggested that

---

*These studies were supported by contract Nonr 840(03) between the Office of Naval Research and the University of Miami Marine Laboratory.

these foregut glands secrete the cellulolytic enzyme found in extracts of this portion of the animal. It is thought that this secretion is mixed with the abraded wood during its passage through the narrow foregut. Hydrolysis of cellulose is presumed to occur during the relatively long period that the food mixture must remain in the caecum.

Dependent from the foregut in this same general region is another characteristic molluscan digestive device, the crystalline style. This is incorporated into the substance of the reduced foot of the adult. The style is enclosed in the style sac. The walls of this organ are lined with an epithelium consisting of cuboidal to columnar cells that vary dramatically both in basic structure and in dimension. The cilia of the lining cells are large, robust organelles that are thought to rotate the style against the gastric shield. This latter is a specialized portion of the wall of the foregut opposite the orifice of the style sac.

The hindgut returns in a relatively straight course through the substance of the visceral mass to terminate in the anus, dorsal to the main adductor muscle.

The adductor muscle is a prominent feature of this section. In life its action causes the unceasing, rasping motion of the shells by means of which the burrow is extended. Microscopic teeth (fig. 2) make the shells resemble small circular wood rasps. Examination of the inner surface of the burrow of an adult borer reinforces this resemblance, for the surface is seen to be smooth and well polished.

The mantle, both lamellae of which are well shown in figure 3, secretes the calcareous inner lining of the burrow. The chief locus of shell secretion is in a specially modified collar of mantle cells near the anterior end of the animal. Secretory areas show considerable morphological diversity, and certain cell groups associate to form an aggregate that bears a striking histological similarity to a taste bud of a vertebrate (figs. 3 and 4). The calcified layer is chiefly responsible for the characteristic X-ray shadows cast by the animals (fig. 5).

The inner lamella of the mantle encloses the mantle cavity. This cavity houses the gills and serves to collect the reproductive and excretory products and the material from the hindgut, and to conduct these out of the body in the efferent siphonal stream. The most important structures in the mantle cavity, from the standpoint of the maintenance of the respiratory stream of water pumped over the gills, are the angular ciliated grooves and patches of cilia dispersed widely over the gills and general visceral surface of the body. In conjunction with a semilunar valvular mechanism in the efferent siphon (fig. 6) and with the general musculature of the mantle itself, these structures maintain a constant stream of water coursing through the mantle cavity so long as environmental conditions remain propitious.

The renal organ extends from the posterior end of the pericardial cavity, with which it is continuous (fig. 7), through the length of the body to the posterior surface of the adductor muscle. There the tubular structure turns through 180° and returns, parallel with the ascending limb, to the orifice into the exhalant mantle cavity in the neighborhood of the visceral ganglion (fig. 8). Near the adductor muscle the kidney tubule is extensively lobulated, with saccules extending into the surrounding tissues. Circulation of fluids in the tubular renal system is insured by the large and active ciliated cells at the point where the system connects with the pericardial cavity.

The heart (fig. 9) is tubular, and lies within the pericardium. Blood reaches the heart through vessels that drain the gills. The efferent flow from the heart is by way of a central artery. This enters the substance of the visceral mass and breaks up at once into a series of smaller vessels. The further ramifications of the arterial system have not been studied.

The gonad lies dorsal to the caecum and extends somewhat laterally from it. Teredo is protandrous, i.e., the testis develops first and the animal functions as a male. Later the gonad transforms into an ovary and the animal functions as a female. It is rare to find both male and female elements in the same gonad. Ripe ova are liberated into the branchial chamber by way of the short, heavily ciliated oviduct (fig. 10).

Development begins with internal fertilization. Spermatozoa suspended in the water enter the afferent siphon. Clapp and others (1946-57) have described what appears to be copulatory behavior in certain species of Bankia. The excurrent siphon of one animal was observed to enter the incurrent siphon of a neighboring borer. There appears to be a transfer of material that could include gametes between the animals. This behavior has not been described for Teredo, nor has it ever been observed in several years of close observation of captive populations in our laboratory. Early cleavage is presumed to be rapid process, which would be expected in a lightly yolked egg like that of Teredo (fig. 11). It has been suggested (Lane, 1955) that fertilization and early cleavage in Teredo are probably accompanied by some reduction in the rate of ventilation of the mantle cavity. Whatever may ultimately prove to be the mechanism responsible, the fertilized ova are retained in the mantle cavity of the maternal organism. Here they become implanted in the tissues of the gill. This process is associated with a "placentation reaction" which appears to be analogous to the same reaction in the endometrium of the mammalian uterus. This reaction in Teredo is associated with cellular proliferation, the development of local concentrations of glycogen, imbedding of the embryonic borers, and their retention during the balance of the embryonic period. They are liberated from the maternal gill as free-living, independent, larval organisms.

At liberation larvae of the different species of the local Teredo population are morphologically indistinguishable. Their average diameter is 250 microns. They are encased in a chitinous embryonic shell that is not calcified. There are no teeth on the shell at this stage of development. The chief activity of the animal during the first 24 hours is swimming by means of its velar cilia. The velum has been likened to the rotor of a helicopter because changes either in speed or direction are produced by tilting the velar axis. The most general fuction of the velum is to move the animal vertically. The swimming rate has been closely observed and measured (Isham and Tierney, 1953) and has been found to be totally inadequate to account for the horizontal distribution of the population from a center of dispersal. Final reliance for distribution must be placed upon tidal currents. This, together with a brief period of survival in the larval state, serves to explain the relative scarcity of teredine borers in test boards suspended from offshore lightships and lighthouses (Clapp *et al.*, 1946-57).

The larva is provided with a considerable store of glycogen as an energy reserve as well as with a significant concentration of cellulase (Lane, 1955). General metabolic considerations make it unlikely, however, that the animal goes through the entire free-living period without feeding. The oxygen uptake has been measured (Lane, Tierney, and Hennacy, 1954). Through the entire free-living larval life a reasonable average value, at 25° C, would appear to be 25 μ l oxygen per hour. If the free-living period is 100 hours duration then the total oxygen uptake would suggest that nearly 4 mg of glucose or glycogen were oxidized. Actually the total dry weight of a single larva does not significantly exceed 0.010 mg. Food organisms have not yet been identified, but it appears reasonable that they should be sought in the nannoplankton.

The oxygen consumption of the larva, denied access to wood, has been measured in suitable microrespirometers during the entire free-swimming period. This is shown graphically in figure 12. During the first 24 hours, while the animal is chiefly engaged in swimming with velar cilia, the oxygen consumption increases significantly. Thereafter, as though to reflect the decreased swimming activity and the increased tendency to crawl, the oxygen consumption begins a steady decline that is continuous until death. Larvae that are prevented from penetrating wood begin an involutional process around 96 hours. One result of this "aging" process is loss in ability later to penetrate wood.

There are several things difficult to understand in the penetration of wood by larval Teredo. In the first place the animal appears to seek and to select the site of initial penetration. The sensory mechanisms and, indeed, the structural machinery that would explain such a selection process have not yet been described. One needs only to observe such a larva, however, to be impressed by the apparently purposeful nature of much of its activity. The prehensile foot is used as an exploratory organ, probing irregularities and restlessly sweeping over the surface. Statistically, the animal is more likely to make its initial penetration into spring wood. There seems to be little preference in favor of entering at right angles to the grain of the wood in contrast to going into end grain. By the time the animal has penetrated to a depth of about 1 body diameter, the course of the burrow is parallel with the grain of the wood whatever may have been the direction of initial penetration.

It should be recalled that the larva is totally uncalcified at the time of its initial penetration and that the chitinous shells are still devoid of teeth. The foot is unarmed. There is no single organ with the strength and the rigidity one would think necessary to abrade the fibers of the wood to permit penetration by the larva. Photomicrohraphs of the larva *in situ* suggest that this initial penetration is, nevertheless, a physical process in which there is an actual abrasion of the wood fibers (fig. 13). At this stage of its development the larva possesses considerable concentrations of cellulase (Greenfield and Lane, 1953). It is not inconceivable that the initial penetration makes use of all these methods of erosion of the wood, physical as well as chemical.

Whatever may prove to be the manner of its initial entry into the wood, the larval gut is gorged with wood debris by the time the animal has penetrated to a depth of 1/4 of its diameter into the wood (fig. 14). By the time penetration has progressed to a depth of 1 body diameter, the larva begins its metamorphosis. Then, within a very few hours, it calcifies the shell, develops shell teeth, resorbs much of the substance of the foot, and begins the elongation that is to lead to the definitive body shape of the teredine borers.

Much of the recent literature relating to the physiology of the adult borer is reviewed by Lane (1955) and will not be repeated here. Some more recent observations which have not yet been published should, however, be made.

It has been stated (Kofoid and Miller, 1927) that the respiratory stream is interrupted when the oxygen tension of the surrounding water drops below a minimal level. In experiments conducted in our laboratory, in a different connection, it was shown that the siphons remained extended and apparently continued to conduct water through the mantle cavity in the complete absence of dissolved oxygen in the ambient water. For this work the animals were placed in water that had been previously degassed, and through which a mixing stream of oxygen-free nitrogen was passed. Oxygen concentration was always less than 0.01 ml/l. This condition was maintained for 23 days. During this time the animals were seen to continue normal siphonal activity, to continue to produce fecal pellets, and, by X-ray examination, to continue to extend their burrows. This apparent, relative insensitivity to oxygen deficiency probably reflects the absence of suitable chemoreceptors. It is doubtful if the animal in its natural environment ever experiences complete oxygen deficiency.

In the presence of irritant substances dissolved in the water, or if the water level should fall sufficiently low so as to expose the siphons to the air, they are promptly withdrawn into the burrow and the siphonal orifice is tightly closed by the pallets. Since these consist almost entirely of calcium carbonate they have not been preserved in most of the histological sections. One measure of their effectiveness, however, is seen in the survival of adult Teredo for at least 10 days when the block of wood within which it lives is removed from the water and placed on the laboratory shelf in air. The pallet closure is sufficiently good to insure the retention of enough water within the mantle cavity to support intramural circulation and to allow normal metabolism to persist even though the animal is entirely out of the water that is its normal environment.

Teredo under various degress of anaerobic stress have been studied biochemically in our laboratory (Lane, Sims, and Clancey, 1955). Glycogen levels decreased from 34.0 per cent to

14.3 per cent in 23 days of anaerobiosis. During the same period the lactic acid concentration rose from 0.02 per cent to 0.320 per cent and pyruvic acid values increased from the normal level of 0.01 per cent to 0.25 per cent. These data are all consistent with the hypothesis that survival of Teredo under conditions of severe oxygen deficiency is made possible by anaerobic glycolysis. To the buffer substances normally present in tissues and in sea water must be added the calcareous lining of the burrow itself; all these serve to prevent excessive accumulation of hydrogen ions during anaerobiosis. It appears, indeed, that this group of animals has evolved extremely efficient biochemical machinery for the metabolism of a diet that is predominantly carbohydrate in nature. A further suggestion of specialization is the very high normal concentration of glycogen in the tissues. Values of 50 per cent of the total dry weight have been found. This may be compared to 28 to 30 per cent glycogen for "very fat" oysters (Albritton, 1953).

The magnitude of the respiratory stream--that is, the volume of water that is circulated through the mantle cavity in unit time --has been studied (Lane and Gifford, 1954). Under normal conditions the average rate of ventilation was found to be 4.1 liters of water per hour per gram of dry weight of animal. This is nearly four times the average rate of circulation reported for the oyster by Galtsoff (1926) and others. This circulation is driven by cilia located on the gills and the visceral mass. A peristaltic wave frequently passes over the incurrent siphon from its distal toward its proximal end. This undoubtedly contributes to the general mantle circulation. Given adequate motive power, the only other requirement for a directed flow is the presence of suitable valves to insure circulation in only one direction. The exhalant siphon contains just such a semilunar valvular arrangement (fig. 6). Some suggestion of the complexity of the mechanisms that control the mantle circulation of this form is offered by the observation (Lane and Tierney, 1950) that the rate of oxygen uptake, and, therefore, the rate of circulation of water through the mantle and over the gills, varies cyclically. It is assumed that the "circulation center" is sensitive to the accumulation of carbon dioxide rather than to oxygen deficiency. This is consistent with the later observation that completely anaerobic conditions produce no overt signs of distress.

No studies of growth of individuals of the genus Teredo have been published. An unpublished series of observations made in this laboratory by spaced X-ray examination of infected panels has revealed that the average length of life of the adult, from the time of first penetration to death, is only 10 weeks. During this time the animal has changed from a globular, clamlike larva of 0.25 mm diameter to an elongated, wormlike form 5 mm in diameter by 100 to 125 mm in length. It has also functioned both as an adult male and an adult female and has destroyed a column of wood of the same dimension as its largest size.

LITERATURE CITED

Albritton, C. E. 1953. Handbook of Biological Data. Committee on the Handbook of Biological Data, AIBS, NRC.

Clapp, W. F., *et al.* 1946-57. Annual Progress Reports on Marine Borer Activity in Test Boards. W. F. Clapp Laboratories, Duxbury, Mass.

Galtsoff, P. S. 1926. New Methods to Measure the Pumping Rate Produced by Gills of Oysters and Other Molluscs. Science, 63:233-34.

Greenfield, L. J., and C. E. Lane. 1953. Cellulose Digestion in *Teredo*. J. Biol. Chem., 204:669-72.

Isham, L. B., and J. Q. Tierney. 1953. Some Aspects of Larval Development and Metamorphosis of *Teredo (lyrodus) pedicellata* de Quatrefages. Bull. Mar. Sci. Gulf and Caribbean, 2:574-89.

Kofoid, C. A., and R. C. Miller. 1927. Biological Section, pp. 188-343, in Marine Borers and Their Relation to Marine Construction on the Pacific Coast. Final Report, San Francisco Bay Marine Piling Committee.

Lane, C. E. 1955. Recent Biological Studies on *Teredo*--a Marine Wood-boring Mollusc. Sci. Monthly, 80:286-92.

-------, and J. Q. Tierney. 1950. Hydrodynamics and Respiration in *Teredo*. Bull. Mar. Sci. Gulf and Caribbean, 1:104-11.

-------, and C. A. Gifford. 1954. Ventilation in *Teredo*. Fed. Proc., 13:1, 84.

-------, J. Q. Tierney, and R. E. Hennacy. 1954. The Respiration of Normal Larvae of *Teredo bartschi* Clapp. Biol. Bull., 106:323-27.

-------, R. W. Sims, and E. J. Clancey. 1955. Anaerobiosis in *Teredo*. Am. J. Physiol., 183:3.

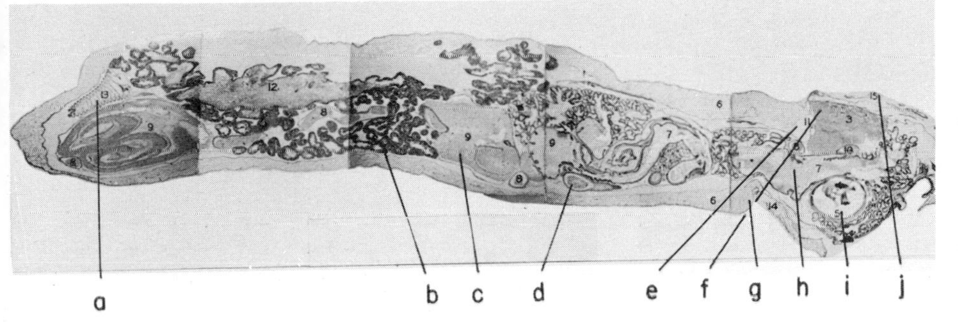

Fig. 1. Sagittal section through the anterior quarter of a small male Teredo. a, gill; b, testis; c, caeoum; d, hindgut; e, kidney; f, adductor muscle; g, mantle; h, precaecal gut; i, style sac; j, cephalic hood . x 7.0

Fig. 2. Single shell of Bankia . x 11.0

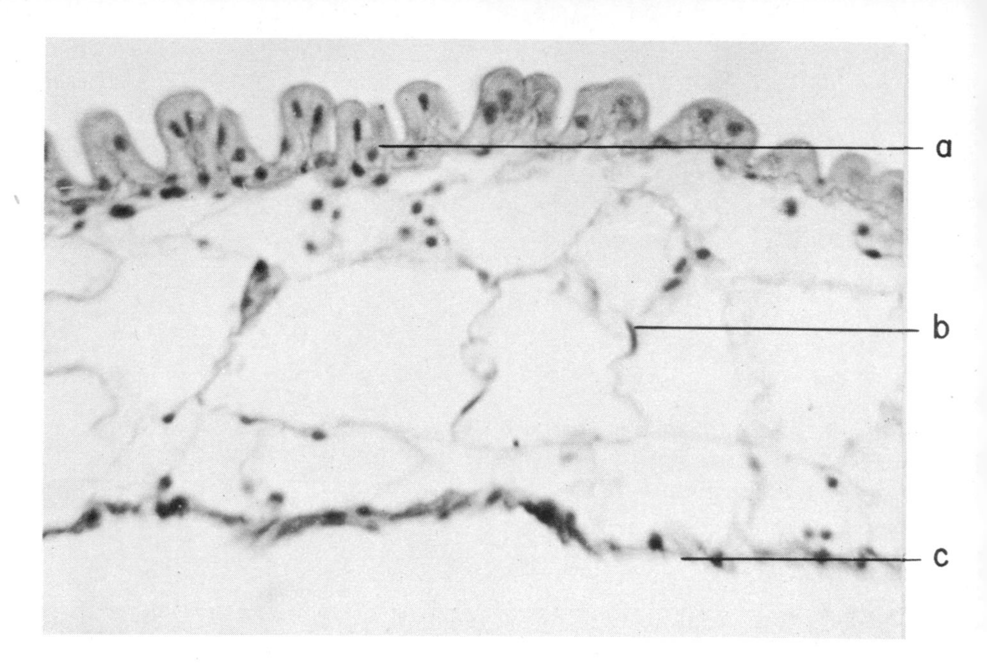

Fig. 3. Cross section of Teredo mantle. This section is made through the shell-secreting collar. a, secretory mantle epithelium; b, mantle connective tissue; c, mantle epithelium . x 720

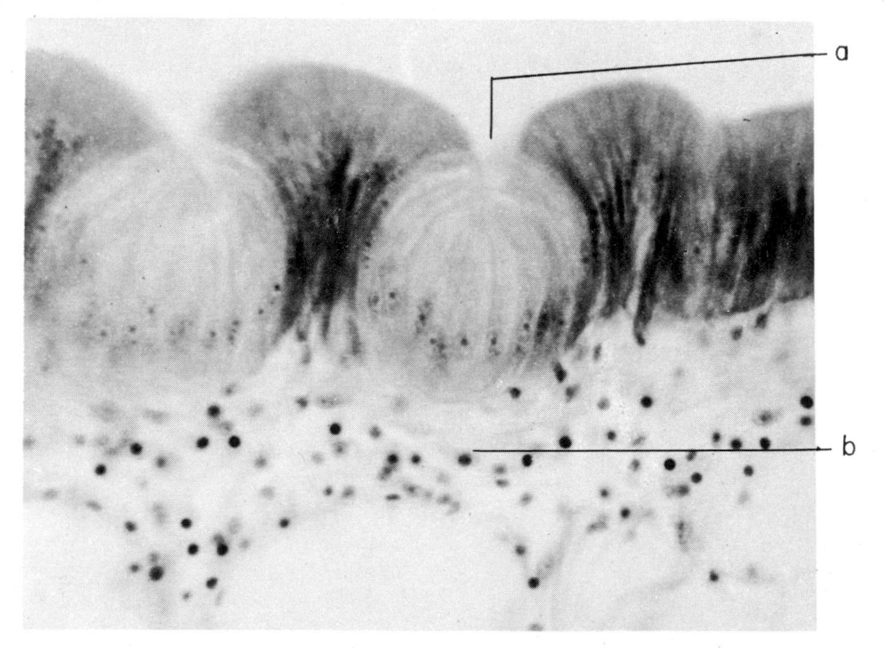

Fig. 4. Cross section of Teredo mantle. This section is made through the shell-secreting collar region and illustrates structural modifications that may develop in this region. a, mantle gland; b, mantle tissue . x 620

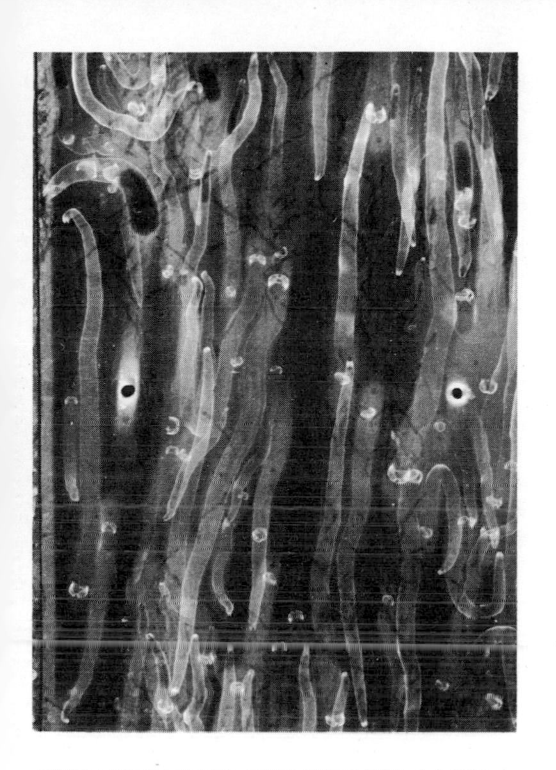

Fig. 5. X-ray photograph of infected panel. This population contains both Bankia and Teredo

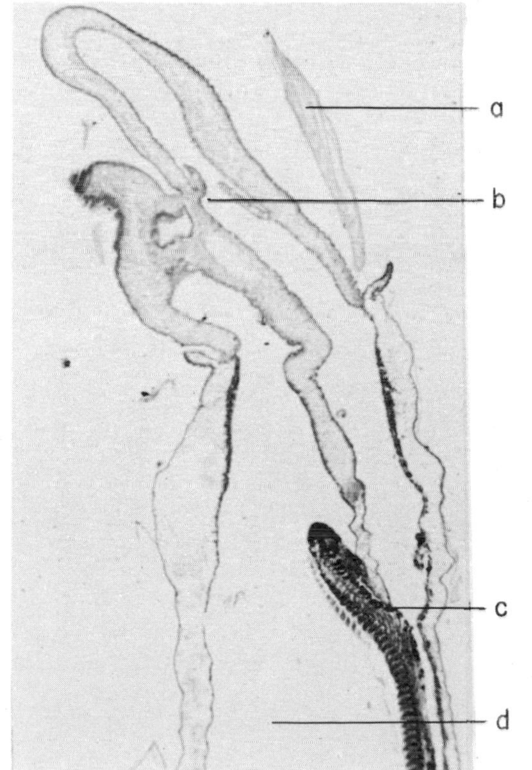

Fig. 6. Sagittal section through posterior extremity of small Teredo that had been anesthetized before fixation. In this way contraction was reduced to a minimum. a, blade of pallet; b, semilunar valves in excurrent siphon; c, gill; d, inhalant mantle cavity. x 40.5

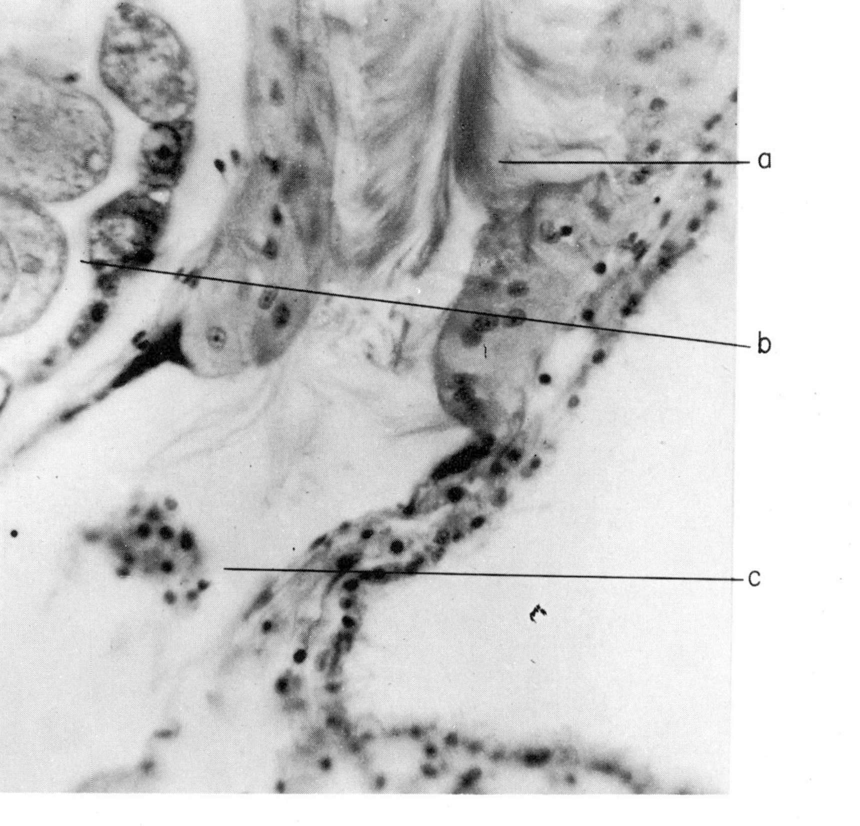

Fig. 7. Cross section through the connection between the pericardial cavity and the kidney tubule. a, kidney tubule; b, ovary; c, pericardial cavity . x 670

Fig. 8. Cross section through adult female Teredo to show the orifices of the kidney ducts. a, gill; b, larvae in inhalant mantle cavity; c, orifice of kidney duct; d, visceral ganglion; e, excurrent mantle cavity . x 97

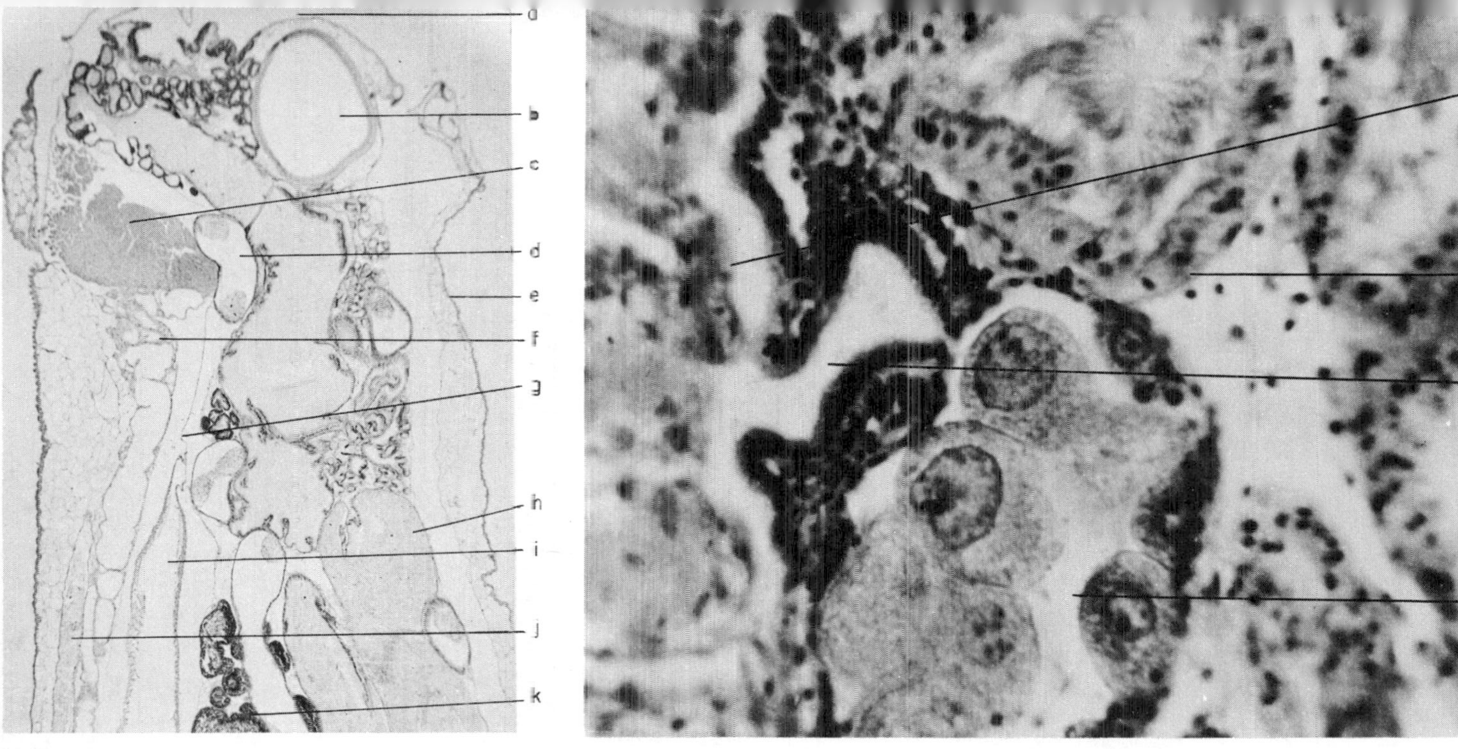

Fig. 9. Sagittal section through the anterior end of a male Teredo to show the heart and main efferent vessel. a, foot; b, style sac; c, adductor muscle; d, hindgut; e, mantle epithelium; f, kidney; g, pericardium; h, caecum; i, heart; j, anal duct; k, testis . x 31

Fig. 10. Cross section of Teredo to reveal the oviduct. a, gill; b, kidney tubule; c, oviduct; d, ovary . x 680

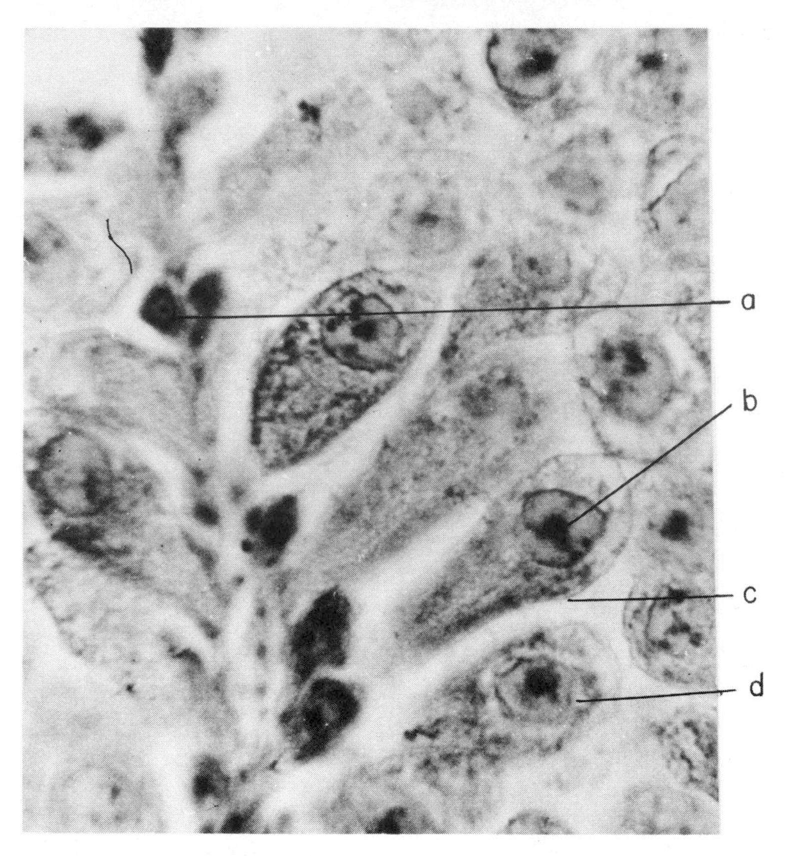

Fig. 11. Teredo ovary. a, germinal epithelium; b, nucleolus; c, oocyte; d, nucleus . x 715

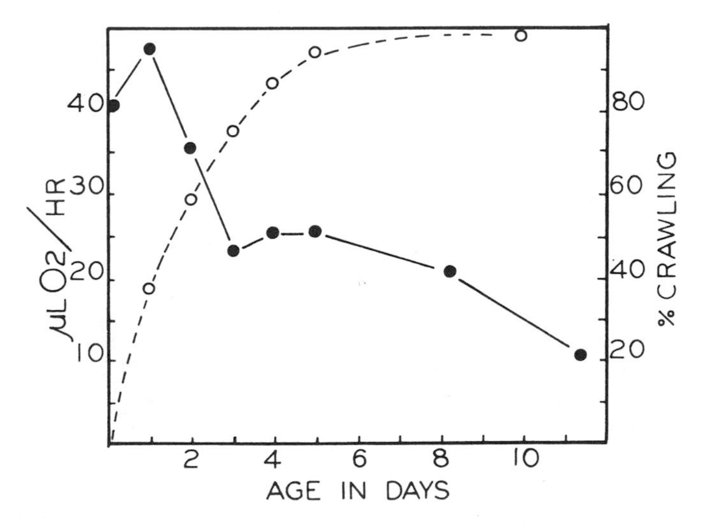

Fig. 12. Curve of normal respiration of unattached Teredo larvae (courtesy of Scientific Monthly)

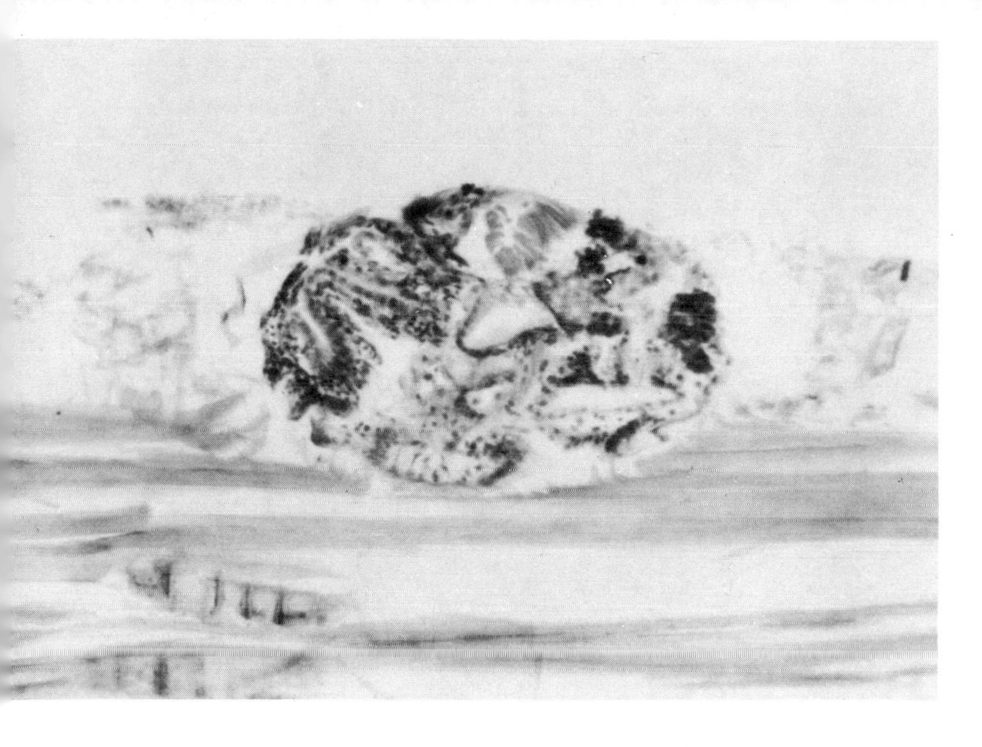

Fig. 13. Teredo larva *in situ* during initial penetration of wood. x 250

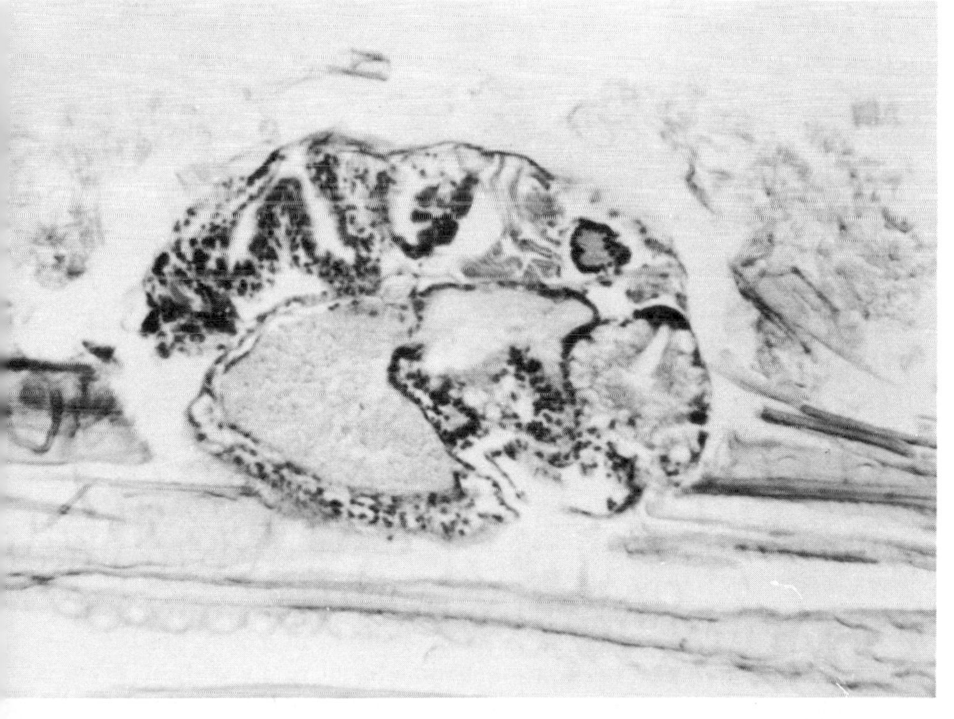

Fig. 14. Teredo larva at a later stage of penetration. x 290

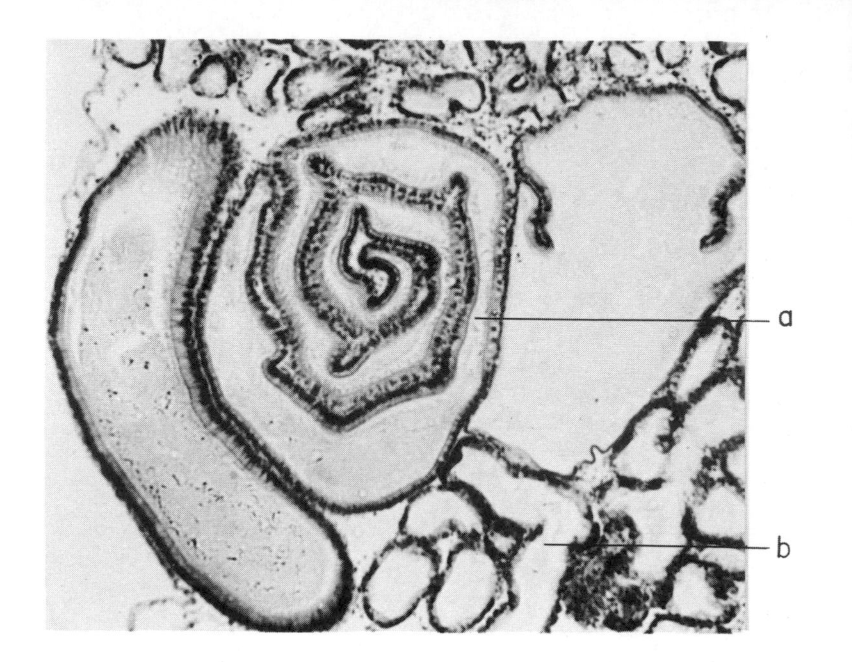

Fig. 15. Cross section of Teredo to illustrate the degree of development of the typhlosole in the foregut. a, typhlosole; b, foregut glands . x 140

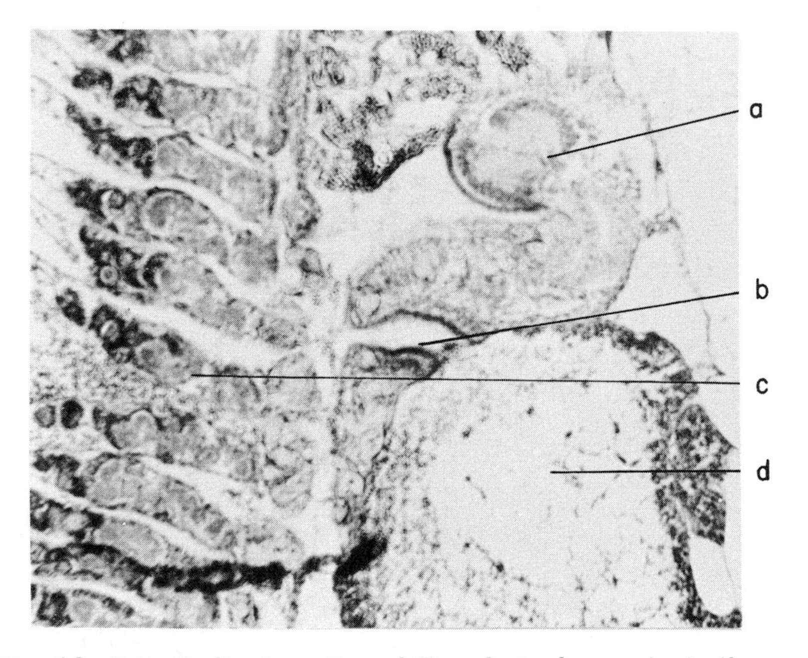

Fig. 16. Longitudinal section of Teredo to demonstrate the relationship between the male reproductive organs and the branchial chamber. a, visceral ganglion; b, testicular duct; c, gill; d, testis . x 133

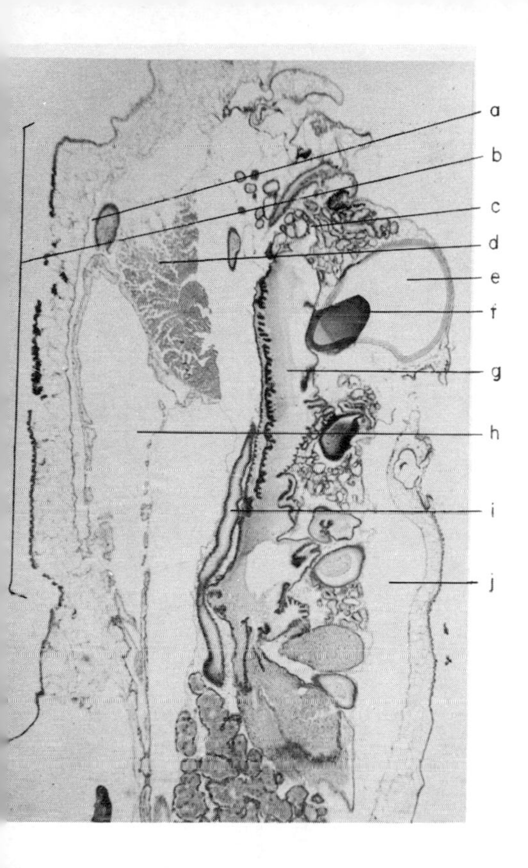

Fig. 17. Sagittal section through the anterior portion of a female Teredo. The considerable distension of the kidney system is probably a fixation artifact. a, anal duct; b, shell-secreting collar; c, foregut glands; d, adductor muscle; e, style sac; f, crystalline style; g, foregut; h, kidney; i, hindgut; j, mantle cavity . x 32

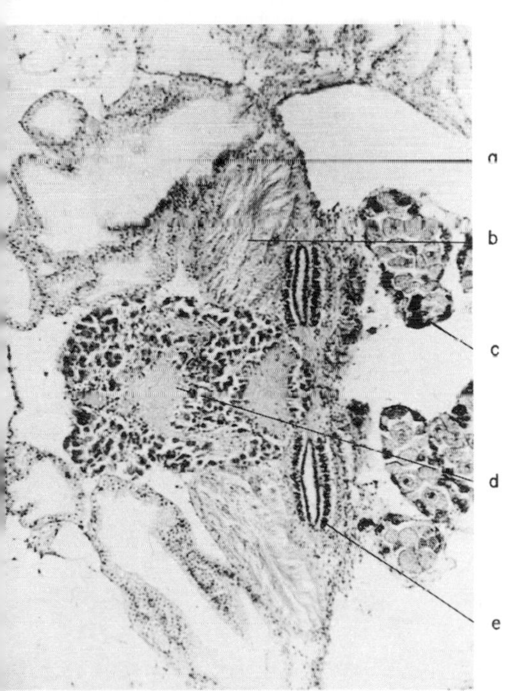

Fig. 18. Cross section of Teredo to show the relationship between the visceral ganglion and the orifices of both the kidney and the female reproductive systems. a, kidney tubule; b, kidney duct; c, ovary; d, visceral ganglion; e, oviduct . x 103

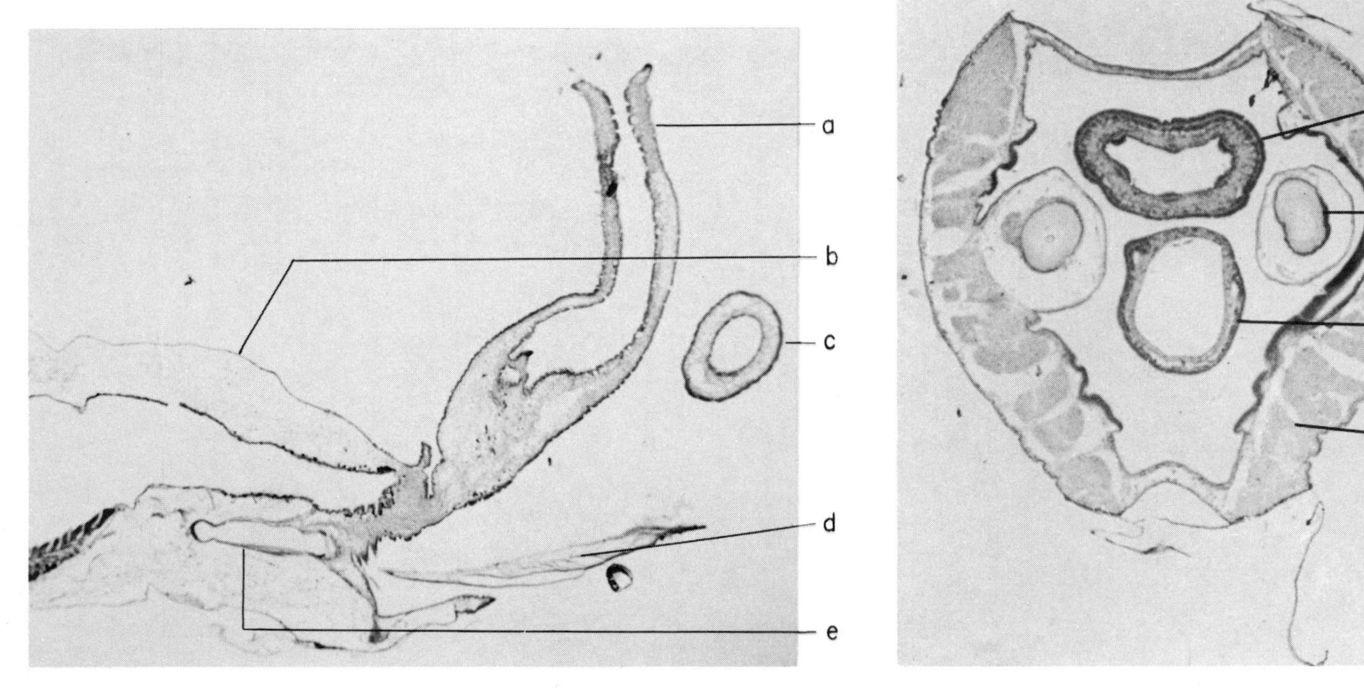

Fig. 19. Sagittal section through the posterior end of a small specimen that had been anesthetized prior to fixation. This section shows the relationship between the pallets and the siphons. a, incurrent siphon; b, mantle; c, excurrent siphon; d, pallet blade; e, pallet stalk . x 35

Fig. 20. Cross section through the posterior end of a small, narcotized Teredo. This section illustrates the structural relationships between various organs and parts in this portion of the animal. a, incurrent siphon; b, pallet stalk; c, excurrent siphon; d, pallet musculature. x 54

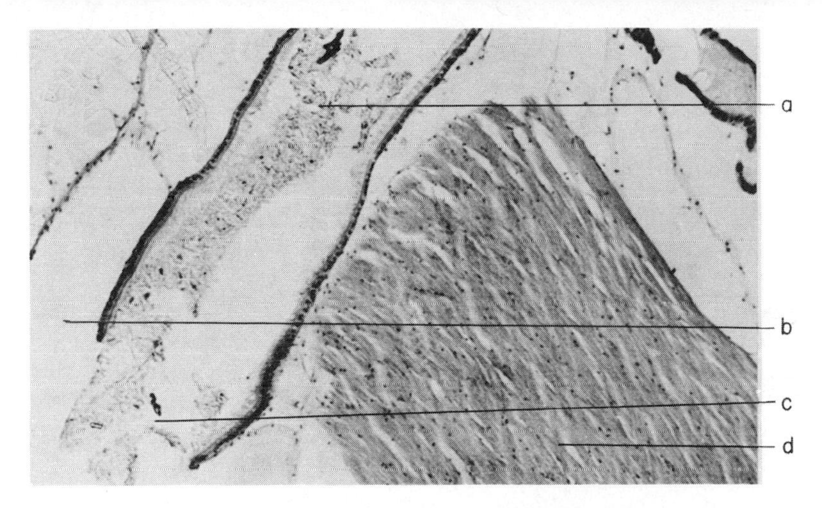

Fig. 21. Longitudinal section through adult Teredo to illustrate the relationship between the anus, the adductor muscle, and the mantle. a, preanal hindgut; b, anal duct; c, anus; d, adductor muscle . x 112

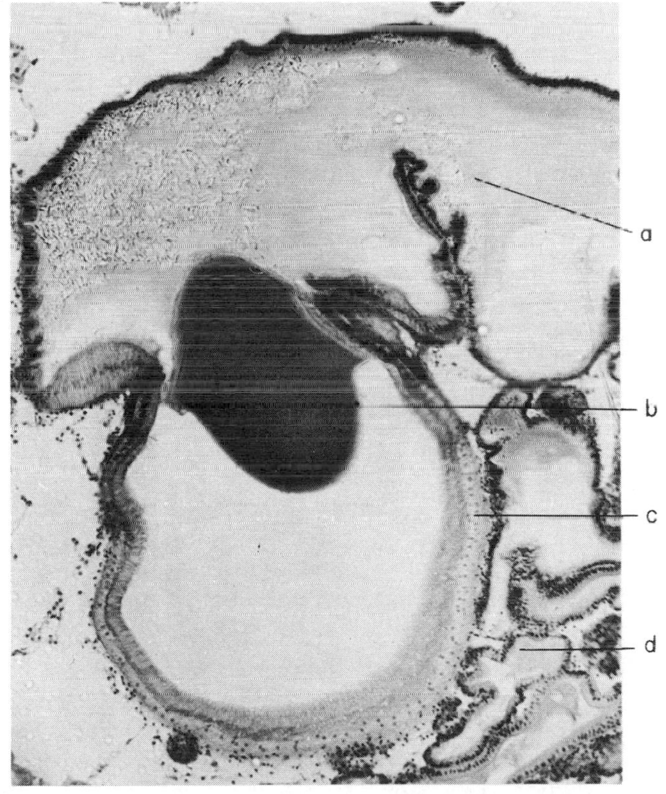

Fig. 22. Cross section to show structural relationships between the sac of the crystalline style and the foregut. The style does not appear too markedly distorted by fixation and later manipulation. a, foregut; b, crystalline style; c, style sac; d, foregut glands . x 100

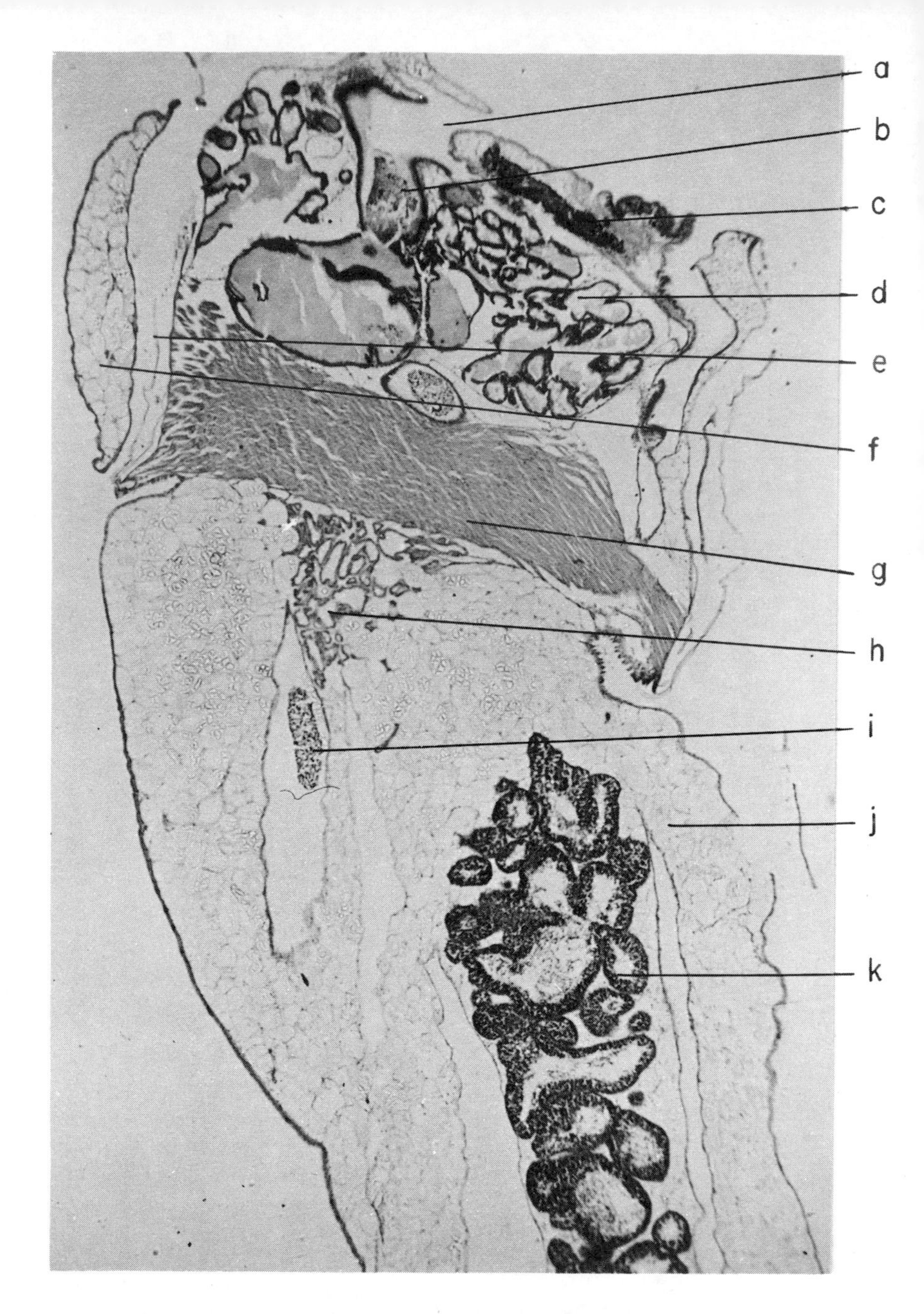

Fig. 23. Frontal section of male Teredo. This figure is included because it reveals the mouth and general foregut morphology better than a sagittal section can. Shown also is the relationship between the adductor muscle and the shells. a, mouth; b, esophagus; c, mucus glands of foot; d, glands of foregut; e, shell residue; f, cephalic hood; g, adductor muscle; h, kidney tubule; i, fecal pellet; j, mantle; k, testis . x 49

THE EARLY DEVELOPMENT OF *BANKIA SETACEA* TRYON

D. B. Quayle

Although the indigenous shipworm (*Bankia setacea* Tryon) of the north Pacific is of considerable economic importance because of its ability to destroy marine wooden structures and saw logs, no complete investigation of its life history has been made. Emphasis of the work done has been directed mainly to establishment of the breeding season by test block studies.

Fraser (1923, 1925, 1926) studied distribution and breeding with its possible relationship to temperature and salinity. The San Francisco Bay Marine Piling Committee (1927) devoted only a small portion of its efforts to *Bankia setacea.*

Boynton and Miller (1927) and Miller and Norris (1940) studied digestion and digestive enzymes in this species. White (1929, 1930) studied breeding, particularly at Departure Bay, British Columbia. Johnson and Miller (1935), during a general fouling survey, obtained information on breeding at Friday Harbor, Washington.

Coe (1941) studied sexual phases in *Bankia setacea.* Neave (1943), Black (1934), and Black and Elsey (1948) studied the breeding period in British Columbia by means of test blocks. Similar studies have been carried out by the W. F. Clapp Laboratories (Brown, 1955) in Washington waters. Miller and Dempster (1951) investigated the effects of low salinity and fresh water on *Bankia setacea,* and Miller (1951) gave an account of the distribution of Bankia and other shipworms in relation to ecological conditions. Trussell, Greer, and LeBrasseur (1956) studied the chemical control of *Bankia setacea.* The Larva has been described by Quayle (1953), who later (1956) gave a general account of the species.

The present observations are almost entirely the result of field studies, and were largely incidental to a program designed to investigate the larval life of the Pacific oyster (*Crassostrea gigas*) in British Columbia breeding areas.

Methods

Identification of the larva presented no problem, for it is the only species of Teredinidae that occurs in abundance in British Columbia waters. *Xylophaga washingtoni* is rather rare with limited distribution.

Quantitative plankton samples were taken by means of a 1 1/4-inch Jabsco pump powered by the main engine of the Shellfish Laboratory vessel *Saxidomus*. Samples to a depth of 30 meters were possible, but generally the maximum depth of sampling was limited by the vertical distribution of oyster larvae, or, in the case of the Ladysmith investigations, by the depth of the water.

The volume of the samples, normally 72 liters, was measured by a Leeds type water meter placed in the discharge line. The samples were concentrated in a plankton net of No. 20 or No. 25 bolting silk. At times a net of monel cloth (No. 20) was used. A plankton bucket permitted adequate washing of the net.

To count, the supernatant fluid in the samples was decanted and the remaining plankton was poured into a channeled plastic counting cell of 125 ml capacity, a size which permitted adequate rinsing of the sample jar. All of the larvae of concern (oyster and shipworm) in the sample were counted. Only immediately after a major oyster spawning was subsampling necessary, and this was never so with shipworm larvae, whose numbers did not approach those of the oyster larvae. These on occasion reached a density of 500 early stage larvae per liter.

Since during the summer in Pendrell Sound waters there is relatively little plankton other than lamellibranch veligers and protozoa (flagellates), counting presented no particular problem.

The Larva

Bankia setacea is an oviparous species and the larva is planktotrophic with a long pelagic life (Quayle, 1953). From the occurrence of size groups in successive plankton samples, the duration of the veliger period in Ladysmith Harbour appears to be about 3 weeks at a temperature of 12° to 15°C. Coe (1941) suggested a swimming period of 4 weeks in southern California waters.

The swimming and crawling activities of settling teredinid larvae have been described by Isham and Tierney (1953), and those of *Bankia setacea* are quite similar. The action of the foot and subsequent shell movements suggest the sequence of digging actions of mature sand-burrowing lamellibranchs (Quayle, 1949).

There is initial byssal attachment by *Bankia setacea,* although the byssus thread is very fine and weak compared to that of most lamellibranch species. A byssus was not observed in *Teredo pedicellata* by Isham and Tierney (1953) although Sigerfoos (1907) described one for *Xylotrya gouldi*.

The form of the valves immediately after settlement becomes asymmetrical; part of the posterior edge of the prodissoconch of only the right valve atrophies by what appears to be a process of erosion. The right valve at one period is little more than half the size of the left. When the newly settled animal is removed from its burrow and placed with the anterior gape down, the anteroposterior axis tilts markedly to the right, owing to the reduced length of the right valve. The cause of this atrophy is unknown, but it may have functional significance for it aids in forming the anterior gape, allowing development and operation of the foot. At this time (at the stage of a single row of teeth) there is no permanent posterior gape. Associated with settlement there must be quite drastic anatomical changes, particularly in musculature. Histological study of the process is required.

The first cutting teeth develop on the anterior edge of the shell, and on a specimen from a test block exposed for 48 hours a single row of 14 teeth had developed, the largest 14 μ in length. By the time a length of 20 mm is reached, 8 or 9 rows of teeth have been formed.

Observations indicate that *Bankia setacea* is able to bury itself completely about 24 hours after it has begun to burrow. The formation of the calcareous cone typical of the Teredinidea (Isham and Tierney, 1953) begins almost immediately. Within 4 days the 2 siphonal apertures separated by a calcareous bridge are well formed, and within 7 days the whole cone may be completed. Some cones may protrude slightly above the wood surface, but most are even with or below it. The diameters of the inhalant and exhalant apertures of the cone are approximately 80 μ and 50 μ, respectively. The bridge between the apertures disappears about one month after formation.

Pallets

Pallets are first formed when Bankia is about 0.5 to 0.6 mm in length. The initial section is spoon-shaped and not unlike the form of the pallet of *Teredo navalis*. The first pallet section measures about 90 μ wide and 200 μ in length.

Breeding

Most information on the breeding time in shipworms has been derived from the seasonal incidence of attack on test blocks. This method is of considerable practical importance and does provide a measure of breeding success, for there may be spawning without settlement. Breeding may also be studied by the occurrence of larvae in the plankton, and if sampling is adequate a measure of breeding intensity (but not necessarily success) may be obtained. Larval mortality may alter the relationship between breeding intensity and breeding success.

In 1951 in Ladysmith Harbour a general study was carried out on the seasonal abundance of planktonic organisms, among which were the larvae of *Bankia setacea.*

Samples of 283 liters each were taken at weekly intervals at a single station at depths of 0.91, 3.05 and 6.1 meters (bottom). The data on the mean monthly counts of shipworm larvae with mean monthly water temperatures at the 0.91 meter level are shown in figure 1.

In 1955, also in Ladysmith Harbour, plankton sampling specifically to determine the seasonal and vertical distribution of *Bankia setacea* larvae was carried out. Samples of 85 liters were taken at weekly intervals from a single station at depths of 0.91, 1.82, 3.64, 7.28, and 14.56 meters. These data are also graphed in figure 1.

A main spring spawning peak in the two series is demonstrated, although about a month apart, when the water temperature (3 feet below the surface) was 10°C or below. The salinity range is small in Ladysmith Harbour and varies between 24‰ and 28‰. The two succeeding minor spawning peaks are well separated in time, but the numbers are small for much significance to be attached to them. Fraser (1923, 1926) and White (1929, 1930) obtained data showing a similar main spring breeding. In contrast, Johnson and Miller (1935) concluded that *Bankia setacea* settles most intensively during October, November, and December with a cessation in January and February, beginning again in March and April and continuing off and on sporadically through the summer. They suggested that 7°C and 12°C are the limits for effective breeding in Puget Sound.

In San Francisco Bay, Miller (1926) and Kofoid and Miller (1927) considered the breeding season of *Bankia setacea* to begin in February when water temperatures are at a minimum for the year and to cease generally at the beginning of summer, with peak breeding in April or May. Coe (1941) stated that spawning occurs in this species in autumn and spring with resting periods in winter and summer in southern California.

Neave (1943), on the basis of test block studies, found maximum settling to occur during the months of September, October, and November in three different areas (including Ladysmith Harbour). He found no close relationship between the intensity of settlement and temperature or salinity. Black and Elsey (1948), also using test blocks in a number of areas along the British Columbia coast, found greatest infestation during the fall season, although at Shannon Bay, the area of heaviest attack, there was also a minor spring peak. Test blocks at Port Townsend, exposed during the period 1948-54, indicated breeding and settlement of *Bankia setacea* throughout the year according to Brown (1955). Trussel, Greer, and LeBrasseur (1956) found a fall and winter peak of settlement in Vancouver Harbour. The Pendrell Sound studies described in this account showed summer breeding.

These test block studies, which are widely spaced both in time and in place, show highly variable results and demonstrate the need for continuous long-term observations using both plankton and test blocks.

Vertical Distribution of Larvae

Ladysmith Harbour

In both the 1951 and 1955 investigations in Ladysmith Harbour, samples were taken at various depths to determine the vertical distribution of the larvae. These were day samples; consequently any diurnal or tidal effect would not be apparent.

In the 1951 series (193 samples), 4 per cent of the larvae were found at the 0.91 meter level, 32 per cent at the 3.05 meter level, and 64.4 per cent at the 6.1 meter depth. In 1955 (245 samples), 8 per cent of the larvae were found at the surface, 13 per cent at the 0.91 meter level, 16 per cent at the 1.82 meter level, 35 per cent at the 7.28 meter level, and 28 per cent at the 14.56 meter level, which is near the bottom.

Pendrell Sound

The Pendrell Sound oyster studies presented a further opportunity to observe the vertical distribution of *Bankia setacea* larvae. The intensity and time of sampling varied from year to year according to the abundance of oyster larvae, but the technique was uniform and all samples were taken at Station 2 (Quayle 1957).

The condensed data for the years 1952-56 are shown in table 1 together with typical temperature and salinity distributions taken on the days designated in the table.

In years when considerable numbers of Pacific oyster larvae were present, series of samples were taken at various depths at Station 2 (total depth 20 meters) to determine the possibility of a diurnal movement. Samples were usually taken every 3 hours, at times of slack water and halfway between, for periods of about 3 days. Both oyster and shipworm larvae were counted. The results of two such series are shown in tables 2 and 3, and those for a 3-day period in 1956 are shown in figure 2 together with the tidal fluctuations.

The 1956 series of samples (tables 1 and 3) indicated a tendency for *Bankia setacea* larvae to approach the surface during the hours of darkness and to be found at greater depths during the day. The fact that this tendency was not well marked in the 1952 series (table 2), apparently because of the selection of sampling depths, led to the statement that no diurnal movements had been observed (Quayle, 1953).

The data in table 2 show the larvae to be concentrated at the 7-meter depth and below in the day samples, while in the combined day and night samples, particularly those of 1956 (table 3), the upper level of larval concentration lies between 3 and 4 meters. In 1956 the night samples contained 4.5 times more larvae than did the day samples. No light intensity measurements were made, but Secchi disc readings during the summer in Pendrell Sound vary between 8 and 12 meters. A tendency for the larvae to be concentrated below the halocline is also indicated. Salinity stratification occurs in Pendrell Sound during late spring and summer as a result of river runoff from large rivers in Toba Inlet (Quayle, 1957). There is no evidence of a temperature barrier. That larvae are found at depths below those sampled in the described series is shown by their presence in samples taken

TABLE 1

BANKIA SETACEA: VERTICAL DISTRIBUTION OF LARVAE WITH TYPICAL TEMPERATURE AND SALINITY VALUES AT THE SAMPLING DEPTHS

Depth (meters)	‰ Larvae	Temp. °C	‰ Salinity
1952. Aug. 15			
0	...	22.5	17.3
.91	1	22.4	17.3
3.05	7	19.7	24.7
6.1	35	18.6	25.7
12.2	57	14.7	27.2
124 samples Aug. 12-Aug. 26 (day and night)			
1953. Aug. 9			
0	0	21.7	18.8
.91	0	21.6	18.8
3.66	3	18.1	25.3
7.32	50	13.7	27.2
10.98	47	12.2	27.8
60 samples Aug. 2-Sept. 8 (day)			
1954. Aug. 2			
0	0	21.6	17.1
.91	0	21.3	17.8
1.82	0	20.7	17.8
3.64	14	17.0	25.1
7.28	61	14.7	26.8
28.92	35	13.2	27.6
54 samples July 24-Aug. 30 (day)			

from 25 meters with a Clarke-Bumpus sampler, and in other pumped samples from the same depth.

As shown in figure 2, there appears to be some correlation between tidal level and the occurrence of *Bankia setaceà* larvae. In the July 26-28 series of samples, however, the tidal pattern was quite different from that during July 16-19, and there is no indication of a similar correlation in the later series.

TABLE 1 (cont.)

Depth (meters)	‰ Larvae	Temp. °C	‰ Salinity
1955. Aug. 4			
0	0	19.0	19.2
.91	0	19.0	20.0
1.82	0	19.0	19.3
3.64	0	18.8	20.6
7.28	10	17.8	24.4
10.92	7	.....	.....
14.56	36	15.5	28.2
18.20	47	.....	.....
53 samples July 27-Aug. 19 (day)			
1956. July 17			
0	0	24.2	20.3
.91	0	23.0	20.1
1.82	1	22.0	20.5
2.73	7	20.5	21.7
3.64	35	18.5	24.4
5.46	27	16.5	25.8
7.28	30	15.2	.....
137 samples July 16-19 (day and night)			
1956. July 26			
0	0	21.8	18.6
.91	0	21.6	18.6
1.82	0	21.6	18.6
2.73	1	20.4	23.8
3.64	8	17.1	25.6
4.55	28	14.9	26.5
5.46	63	14.2	26.7
161 samples July 25-28 (day and night)			

Also test blocks exposed at Station 2 in 1954 caught one *Bankia setacea* per 6.5 sq cm (1 square inch) at the 3.6 meter level, 10 per 6.5 sq cm at 7.3 meters, 25 per 6.5 sq cm at 12.7 meters, and 42 per 6.5 sq cm at 16.4 meters. On the other hand, in

TABLE 2
NUMBERS OF *BANKIA SETACEA* LARVAE IN SAMPLES OF 42 LITERS (3 CUBIC FEET), STATION 2, PENDRELL SOUND, AUGUST, 1952

Date	August 12		August 13			
Time	17:10	23:20	06:48	14:30	18:47	24:00
Depth in feet 3	2	3				
10	3	644				13
20	570	19	5	81	119	131
40	152	6	190	93	56	16
Total	727	672	195	174	175	160

Date	August 14				August 15		
Time	07:32	12:40	15:40	20:43	08:51	17:12	22:00
Depth in feet 3							
10		9	1			5	40
20	7	231	14	60	14	12	47
40	133	7	40	33	168	219	23
Total	140	247	55	93	182	236	110

Pendrell Sound oyster larvae are concentrated above a depth of 3 meters.

Discussion

Of interest is the apparent discrepancy between the planktonic larval and the test block studies in the breeding time of *Bankia setacea.* It is unfortunate that no test blocks were exposed during

TABLE 2 (cont.)

Date	August 16			August 23
Time	01:30	09:40	17:47	20:20
Depth in feet 3				0
10	32		3	5
20	86	19	26	15
40	13	64	115	16
Total	131	83	144	36

Date	August 24				August 25			
Time	02:30	08:30	14:20	20:40	03:00	09:30	15:00	21:00
Depth in feet 3	19	2		1	4			1
10	12	1	1	7	6	1	1	2
20	0	4	2	10	2	2	1	7
40	1	7	7	5	9	7	2	12
Total	32	14	10	23	21	10	4	22

the period of the larval studies in Ladysmith Harbour, for the relationship between numbers of larval shipworms and intensity of settling requires investigation similar to that which has been done with oysters. For instance, it has been found that in Pendrell Sound approximately one advanced stage Pacific oyster larva per 3 liters will provide a set of at least one spat per 6.5 sq cm (1 square inch) of collecting surface.

In connection with the cause of the difference in time of settling demonstrated by the two methods, the number of observations is insufficient to determine the degree of significance. With the test block method itself, which has been applied fairly

TABLE 2 (cont.)

Date	August 26			
Time	04:00	10:30	15:30	21:30
Depth in feet 3	1			4
10	2	4		4
20	6	1		3
40	3	0	6	5
Total	12	5	6	16

extensively, the annual variation both within and between areas is considerable. The only test block study in Ladysmith Harbour (Neave, 1943) showed fairly intensive settling in the period from June to October in 1941 with the greatest attack in June. In 1942 settling occurred mainly in the September-November period with the maximum in October. In addition, results from a single station may be misleading, and more stations might be used to advantage, together with longer periods of observation.

In general, however, the data indicate that *Bankia setacea* breeds most frequently during periods of low temperature, although the larvae may occur in strata with wide temperature differences. That there are exceptions to this is shown in the Pendrell Sound data.

Other species of Bankia whose breeding periods have been studied are generally high temperature spawners. Ralph and Hurley (1952) demonstrated that *Bankia australis* in New Zealand in 1951 had a breeding peak at a temperature of about 20°C. Sigerfoos (1907) concluded that *Bankia gouldi* spawned during periods of high water temperature in the Beaufort area. A similar conclusion was reached by Scheltema and Truitt (1954), who showed that the same species settles most intensively in Chesapeake Bay in summer when the water temperature is well above 20°C, although these authors believe that actual spawning may begin at lower temperatures.

The apparent diurnal movement demonstrated for *Bankia setacea* may be due to light. Isham, Smith, and Springer (1951)

TABLE 3

NUMBERS OF *BANKIA SETACEA* LARVAE IN SAMPLES OF 72 LITERS (3 CUBIC FEET), STATION 2, PENDRELL SOUND, JULY, 1956

Date	July 16	July 17							July 18								July 19					
Time	23:00	01:00	05:00	08:30	13:00	18:45	21:00	23:20	01:40	15:40	09:40	13:30	17:30	19:45	22:03	24:00	02:30	06:10	10:00	14:00	18:00	Total
Depth in feet 0	1	1																				2
3																						0
6	1																3					4
9								8	1								18	9				36
12	9	74	10			5	6	29	155	23						99	21	0				421
18	n. s.	n. s.	n. s.		3	4	12	21	46	81	1	4	15	93	75	27	108	30	0	3	0	523
24	29	40	41	61	n. s.	n. s.	n. s.	n. s.	n. s.	n. s.	n. s.	n. s.	n. s.	n. s.	n. s.	n. s.	n. s.	n. s.	n. s.	n. s.	n. s.	171
Total	40	115	51	61	3	9	18	58	211	104	2	4	15	93	75	126	150	39	0	3	0	

TABLE 3 (cont.)

Date	July 25		July 26							July 27							July 28						
Time	21:00	24:00	03:00	05:00	08:00	11:00	14:00	17:30	24:00	03:30	06:00	08:45	11:40	14 43	18:19	21:41	01:00	04:07	07:00	09:46	12:46	15:22	Total
Depth in feet 0																							0
3																							0
6																							0
9				2															3	1	1		7
12		31	1	0				1		2							3	6	0	2	0		46
15		42	9	7	3	1		10	15	1	12		1			20	26	38	0	0	4	2	151
18	4	54	59	22	0	0	1	27	13	24	21	2	9	0	6	36	20	62	12	6	21	14	417
Total	4	127	69	31	3	1	1	38	28	27	33	2	10	0	6	56	49	106	15	9	26	16	

have studied experimentally the effect of light on the movement of teredinid larvae (mainly *Teredo pedicellata*) and have concluded that under natural conditions the larvae will concentrate at the surface during darkness and during the day will be dispersed and tend to remain in the deeper, less illuminated water. Owen (1953) indicated that *Teredo norvegica* larvae tend to settle in increased numbers in dimly lit regions. Schwarz (1932) inferred that *Teredo navalis* exercises selection of the light-exposed north side of piling at Wangerooge. Vertical movements of other bivalve larvae have been indicated by Carriker (1951), Quayle (1952), and Yasuda (1952). Swimming rates of about 7.5 mm per second for *Teredo pedicellata* larvae found by Isham and Tierney (1953), if applied to *Bankia setacea*, would make it possible for this species to undergo the vertical movements indicated in the Pendrell Sound data.

LITERATURE CITED

Black, E. C. 1943. The Shipworm. Pac. Biol. Sta. and Pac. Exp. Sta. Prog. Rept., 21:7-9.

-------, and C. R. Elsey. 1948. Incidence of Wood-Borers in British Columbia Waters. Bull. Fish. Res. Bd. Can., LXXX.

Boynton, L. C., and R. C. Miller. 1927. The Occurrence of a Cellulase in the Shipworm. J. Biol. Chem., 75:613-18.

Brown, Dorothy J. 1955. Eighth Progress Report on Marine Borer Activity in Test Boards Operated during 1945. Compiled by Dorothy J. Brown, Wm. Clapp Laboratories, Inc., Duxbury, Massachusetts, Report No. 9440, Department of the Navy, 54 pp.

Coe, W. R. 1941. Sexual Phases in Wood-boring Mollusks. Biol. Bull., 81:168-76.

Carriker, M. R. 1951. Ecological Observations on the Distribution of Oyster Larvae in New Jersey Estuaries. Ecol. Monogr., 21:19-38.

Fraser, C. McLean. 1923. Marine Wood-Borers in British Columbia Waters. Trans. Roy. Soc. Can., 17:21-28.

-------. 1925. Marine Wood-Borers in British Columbia Waters. Trans. Roy. Soc. Can., 19:159-67.

-------. 1926. Marine Wood-Borers on the Pacific Coast of North America. Proc. Third Pan-Pacific Sci. Congr. (Tokyo), pp. 2270-75.

Isham, L. B., and J. Q. Tierney. 1953. Some Aspects of the Larval Development and Metamorphosis of *Teredo (Lydrodus) pedicellata* de Quatrefages. Bull. Mar. Sci. Gulf and Caribbean., 2:574-89.

Isham, L. B., F. G. Walton-Smith, and V. Springer. 1951. Marine Borer Attack in Relation to Conditions of Illumination. Bull. Mar. Sci. Gulf and Caribbean, 1:46-63.

Johnson, M. W., and R. C. Miller. 1935. Seasonal Settlement of Shipworms, Barnacles and other Wharf-Pile Organisms at Friday Harbor, Washington. Univ. Wash. Publ. Oceanogr., 2:1-18.

Kofoid, C. A., and R. C. Miller. 1927. Biological Section, pp. 188-343, in Marine Borers and Their Relation to Marine Construction on the Pacific Coast. Final Report, San Francisco Bay Marine Piling Committee.

Miller, R. C. 1926. Ecological Relations of Marine Wood-boring Organisms in San Francisco Bay. Ecology, 7:247-54.

-------. 1951. Distribution of Wood-boring Mollusks on the Pacific Coast of North America in Relation to Ecological Conditions. Rept. Mar. Borer Conf., U.S. Naval Civ. Eng. Res. and Eval. Lab., Port Hueneme, Calif., pp. G1-11.

-------, and R. P. Dempster. 1951. Effects of Low Salinity and of Fresh Water on the Northwest Shipworm, *Bankia setacea*. Rept. Mar. Borer Conf., U.S. Naval Civ. Eng. Res. and Eval. Lab., Port Hueneme, Calif., pp. F1-3.

-------, and E. R. Norris. 1940. Some Enzymes of the Northwest Shipworm, *Bankia setacea*. Proc. Pacific Sci. Congr. 6 (California, 1939), 3:615-16.

Neave, F. 1943. Seasonal Settlement of Shipworm Larvae. Fish. Res. Bd. Can. Prog. Rept. Pac., 54:12-14.

Owen, G. 1953. Vertical Distribution of *Teredo norvegica*. Nature, 171:484-85.

Quayle, D. B. 1949. Movements in *Venerupis (Paphia) pullastra* (Montagu). Proc. Malacol. Soc. Lond., 28:31-37.

-------. 1952. Structure and Biology of the Larva and Spat of *Venerupis pullastra* (Montagu)., Trans. Roy. Soc. Edin., 62: 255-97.

-------. 1953. The Larva of *Bankia setacea* Tryon. Rept. British Columbia, Dept. of Fisheries for 1951, pp. 88-91.

-------. 1956. The British Columbia Shipworm. Rep. British Columbia, Dept. of Fisheries for 1955, pp. 92-104.

-------. 1957. Summary of 1956 Oyster Breeding. Oyster Bull., British Columbia, Dept. of Fisheries, Shellfish Lab., Ladysmith, B.C., 8:1-32.

Ralph, P. M., and D. E. Hurley. 1952. The Settling and Growth of Wharf-Pile Fauna in Port Nicholson, Wellington, New Zealand. Zool. Publ. Victoria Univ. College, 19:1-22.

Scheltema, R. S., and R. V. Truitt. 1954. Ecological Factors Related to the Distribution of *Bankia gouldi* Bartsch in Chesapeake Bay. Publ. 100, Chesapeake Biological Laboratory, Solomons Islands, Md.

Schwarz, A. 1932. Der Lichteinfluss auf die Fortbewegung, die Einregelung und das Wachstum bei einigen niederen Tieren (Littorina, Cardium, Mytilus, Balanus, Teredo, Sabellaria). Senckenbergiana, 14:429-54.

Sigerfoos, C. D. 1907. Natural History, Organization and Late Development of the Teredinidae, or Shipworms. Bull. U.S. Bur. Fish., 28:191-231.

Trussell, P. C., B. A. Greer, and R. J. LeBrasseur. 1956. Protection of Saw Logs against Marine Borers. III, Storage Ground Study. Pulp and Paper Magazine of Canada, pp. 11-14.

White, F. D. 1929. Studies on Marine Borers. III, A Note on the Breeding Season of *Bankia setacea* in Departure Bay, British Columbia. Contrib. Can. Biol., N.S. IV, No. 3.

-------. 1930. The Pile Borer or "Teredo." Pac. Biol. Sta. and Pac. Exp. Sta. Prog. Rept., 5:3-6.

Yasuda, J. 1952. Tidal and Diurnal Changes of Occurrence of Molluscan Larvae at Maeshida, Aichi Prefecture. Bull. Jap. Soc. Sci. Fish., 17:6-8.

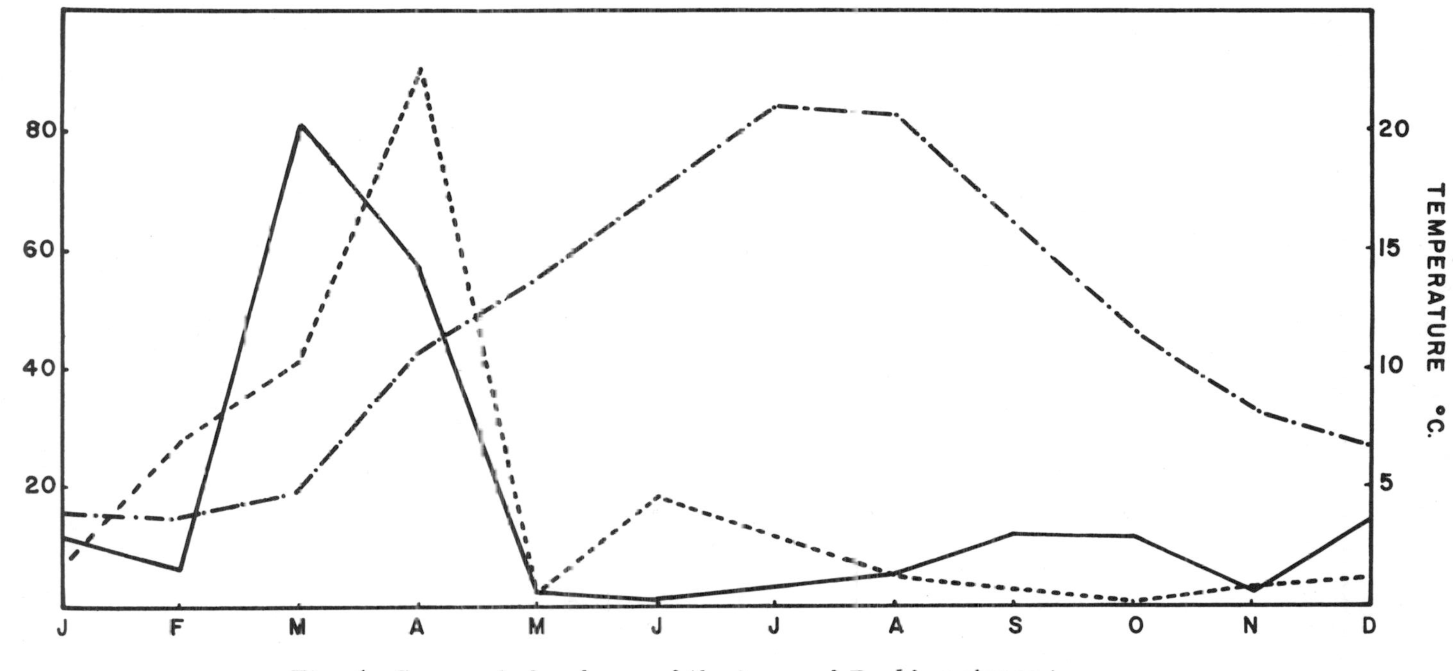

Fig. 1. Seasonal abundance of the larva of *Bankia setacea* in Ladysmith Harbour. (———) 1951; (-----) 1955; (–·–·–·–) thermograph temperature, 0.91 meters below surface

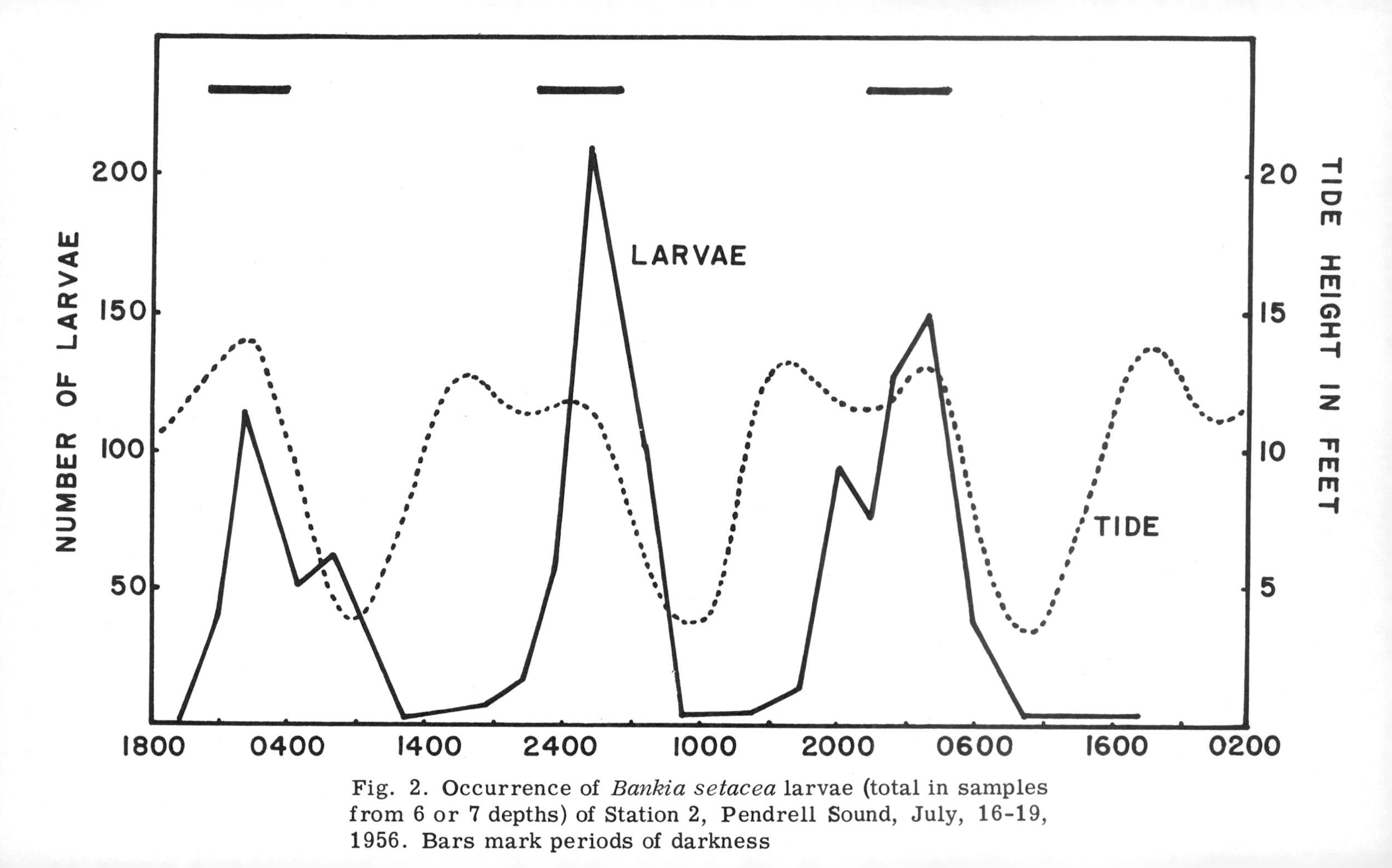

Fig. 2. Occurrence of *Bankia setacea* larvae (total in samples from 6 or 7 depths) of Station 2, Pendrell Sound, July, 16-19, 1956. Bars mark periods of darkness

THE GROWTH RATE OF *BANKIA SETACEA* TRYON

D. B. Quayle

Probably because of the difficulties involved in the determination of the rate of growth of marine wood borers, the literature on this subject is not extensive. Isham, Moore, and Smith (1951), who studied the growth of *Teredo pedicellata*, have adequately reviewed the literature to that date. Atwood and Johnson (1924) published a radiograph of shipworm-infested wood and thereby indicated the possibility of a technique for the study of the rate of growth of shipworms. Among those who have used this method are Ralph and Hurley (1952), who in New Zealand used X-rays in the determination of the rate of growth of *Bankia australis*. Crisp, Jones, and Watson (1953) demonstrated the use of X-ray stereoscopy for examining the infestation of shipworms *in vivo*, and Trussell, Greer, and LeBrasseur (1956) used radiographs to study short term (two months) growth of *Bankia setacea* in British Columbia. Taylor (1956) used the same technique to examine the effect of various preservatives in preventing shipworm and gribble attack on Douglas fir plywood.

Other studies on the growth of *Bankia setacea* were carried out by Kofoid and Miller (1927), who determined the rate in San Francisco Bay by measuring the burrows in test blocks exposed for known periods. The average rate ranged from 2.3 cm per month for specimens $1\frac{1}{2}$ months old to 6.3 cm per month for an individual 8 months old. Johnson and Miller (1935), in a fouling study at Friday Harbor, Washington, found an average rate of 1 cm increase in burrow length per month. To obtain more specific information on the growth rate of *Bankia setacea* in British Columbia waters, the present study was undertaken, and a preliminary report has already been given (Quayle, 1956).

Methods

The investigation was carried out at the Provincial Shellfish Laboratory, Ladysmith, British Columbia. Eight fir blocks 3/4 x 5 5/8 x 12 inches were covered with fiber glass except for the ends, so that attack would occur only in that area. Four more untreated blocks were interspersed among the fiberglass blocks, and the whole group was suspended in a frame at a depth of 3 feet below the surface from the laboratory float. No controls were used.

Following the first immersion in November, 1954, the blocks were removed at monthly intervals and X-rayed at a distance of 30 inches, with 43 K.V., 4 milliampere seconds, 14 x 17 Kodak B.B. film, Par speed screens, half wave rectification, on a General Electric D-38 25 M.A., 85 K.V. self-rectified tube.

Length measurements were made directly on the burrows as shown on the radiographs. The greatest diameter, taken usually just behind the valves, the length of the pallet stalk, the total length of the pallet, and the number of cones were recorded (table 1). The volume of the burrow, admirable though it is as a measure of growth, was not used.

Results

The first X ray was taken on December 15, 1954, and showed that twenty-seven shipworms had entered the ends of six of the twelve blocks. The mean length at this time was 10.5 mm, with a range of 5 to 7 mm, indicating a slight variation in the time of settling. Further attack did not occur until seven more shipworms were apparent on the April radiographs. The data in table 1 were calculated from the measurements of the initial group of twenty-seven specimens, which by July had been reduced to five by mortality. The shipworms in the block with the heaviest infestation (initially eleven, and later thirteen) died between the April and May measurements as shown by the disposition of the valves and pallets. By the June measurement all in this block were dead, indicating a life span of only 6 to 8 months under fairly crowded conditions.

Length

As shown in figure 1, with water temperatures below 10° C the growth rate is relatively low, and about 4 months were required to attain a length of 200 mm; at temperatures above that

TABLE 1

GROWTH RATE OF *BANKIA SETACEA* IN LADYSMITH HARBOUR 1954-55

Date	Mean Monthly Temperature °F	Mean Body Length in mm	Range	Mean Body Diameter in mm	Range	Total Pallet Length in mm	Range	Pallet Stalk Section Length in mm	Pallet Blade Section Length in mm	Stalk/Blade Ratio	Number of Cones	Range	Number of Animals
1954 December	45.0	10.5	5-17	2.2	1.5-2.5								27
1955 January	43.0	36.9	12-66	4.5	3.0-6.0								27
February	43.0	86.4	33-139	6.1	5.0-8.5	14.4	8-20	5.7	8.7	.660	5	4-8	27
March	44.0	139.6	61-207	7.6	5.5-9.0	19.0	12-25	7.6	11.5	.669	7	5-8	27
April	46.0	223.1	130-331	9.3	8.5-10.0	26.2	19-32	9.2	17.0	.543	9	7-11	27
May	54.0	340.3	255-465	10.1	9.0-12.0	34.4	32-40	12.8	23.0	.556	12	10-14	17
June	62.0	419.4	300-565	10.4	9.0-12.0	41.7	37-48	14.2	27.5	.517	15	14-17	13
July	66.0	505.0	430-570	12.0	12.0	50.0	48-59	16.0	34.0	.467	19	18-22	5

level a length of 300 mm was attained in 3 months. The greatest mean monthly increments (fig. 2) occurred in April, May, and June, with 120 mm in May. This may be owing to increased water temperature, to the relatively greater growth rate of larger animals, or to the increased availability of planktonic food. In connection with the latter possibility it should be noted that the spring phytoplankton bloom begins in Ladysmith Harbour normally in March, with maximum standing crop in April and May (Quayle, 1953).

The growth curves for three individual animals (1A, 7A, and 12B) that settled in the March-April period are shown in figure 3. Here, in spite of the smaller size, the initial growth rate is similar to that of the older groups, but it is greater thereafter. The second group showed a greater initial growth rate than the first, and this is no doubt due either to temperature or to availability of food. The mean length of 505 mm in July gave a mean monthly rate of 63 mm for this second group. The largest specimen (1A, fig. 3) was one of the second group and attained a length of 610 mm in about 5 months or an increment of 122 mm per month.

One of the blocks contained twelve shipworms, and, when the growth rate of those was compared to that of fifteen animals in five blocks, the former was found to have the highest rate. By April the single block group had a mean length of 50 mm more than the other group, although the initial size of the latter was slightly less than the group of twelve.

It should be mentioned that, since the burrowing for the most part was along the grain from the beginning, optimum conditions were favorable for maximum growth.

Diameter

In general, the rate of increase in the diameter of the burrow was similar to that of the length (fig. 1). When, however, the diameter is plotted against length rather than age, the relationship is somewhat different. There is at first a rapid increase in diameter up to a length of 50 mm, and from then until 250 mm the increase is only moderate. From a length of 250 mm to 400 mm the diameter increases more rapidly although the increase is not comparable to the initial rate. The greatest diameter attained was 12 mm.

Pallets

The data on pallet growth rate is not too satisfactory, for the degree of breakage at the blade end is unknown. The pallet stalk, blade, and number of cones, following the terminology of Clench and Turner (1946), were recorded. The pallet as a whole has a growth rate pattern similar to that of the length and diameter of the burrow. The length of the blade and the number of cones increase markedly with age while the stalk portion increases very slowly, although this is the growing area of the pallet where new cones are added. The largest number of cones found was twenty-four, but it is known that breakage occurred in handling during the measuring, so that the counts in older animals are low. As would be expected, the rate of increase in the number of cones follows closely that of the length of the blade.

Summary

The X-ray technique appears to be a simple and satisfactory method for the determination of growth rates (burrow dimensions) of shipworms. X-ray stereoscopy described by Crisp, Jones, and Watson (1953) may have assisted somewhat, but in this case little difficulty was experienced in identifying the separate burrows.

Either burrow length or size of pallets appears to be a good measure of growth, since these do not vary with the spawning condition of the animal, and the variability due to differential shrinkage is not involved.

The results indicate a rate of growth considerably in excess of previous growth measurements for this and other species of Bankia. A mean monthly increase of 122 mm was attained by one specimen, which reached a length of 610 mm (24 inches) in about 5 months.

Length, greatest diameter, and pallet length growth rates show the same general trend.

The degree of crowding experienced in this study did not appear to affect the growth rate, although it did seem to affect the life span.

A much more complex experiment than the one reported here is required in order to demonstrate the possible variations in growth of *Bankia setacea* with a range of age, size, temperature, feeding, and degrees of crowding.

LITERATURE CITED

Atwood, W. D., and A. A. Johnson. 1924. Marine Structures: Their Deterioration and Preservation. National Research Council. Washington, D. C.

Clench, J. W., and R. D. Turner. 1946. The genus *Bankia* in the Western Atlantic. Johnsonia, 2:1-28.

Crisp, D. J., L. W. G. Jones, and W. Watson. 1953. Use of Stereoscopy for Examinining Shipworm Infestation *in vivo*. Nature, 172:408.

Isham, L. B., H. B. Moore, and F. G. Walton-Smith. 1951. Growth Rate Measurement of Shipworms. Bull. Mar. Sci. Gulf and Caribbean, 1:136-47.

Johnson, M. W., and R. C. Miller. 1935. Seasonal Settlement of Shipworms, Barnacles, and Other Wharf-pile Organisms at Friday Harbor, Washington. Univ. Wash. Publ. Oceanogr., 2:1-18.

Kofoid, C. A., and R. C. Miller. 1927. Biological Section, pp. 188-343, in Marine Borers and Their Relation to Marine Construction on the Pacific Coast. Final Report, San Francisco Bay Marine Piling Committee.

Quayle, D. B. 1953. Condition in Oysters. Oyster Bull., British Columbia, Dept. of Fisheries, Shellfish Lab., Ladysmith, B. C., 4: No. 1.

-------. 1956. Growth of the British Columbia Shipworm. Pac. Biol. Sta. and Pac. Exp. Sta. Prog. Rept., 105:3-5.

Ralph, P. M., and D. E. Hurley. 1952. The Settling and Growth of Wharf-pile Fauna in Port Nicholson, Wellington, New Zealand. Zool. Publ. Victoria Univ. College, 19:1-22.

Taylor, F. H. C. 1956. The Effectiveness of Various Preservatives on Plywood in Preventing Attack by Shipworms and Gribbles. Pac. Biol. Sta. and Pac. Exp. Sta. Prog. Rept., 107: 13-18.

Trussell, P. C., B. A. Greer, and R. J. LeBrasseur. 1956. Protection of Saw Logs against Marine Borers. III, Storage Ground Study. Pulp and Paper Magazine of Canada, pp. 11-14.

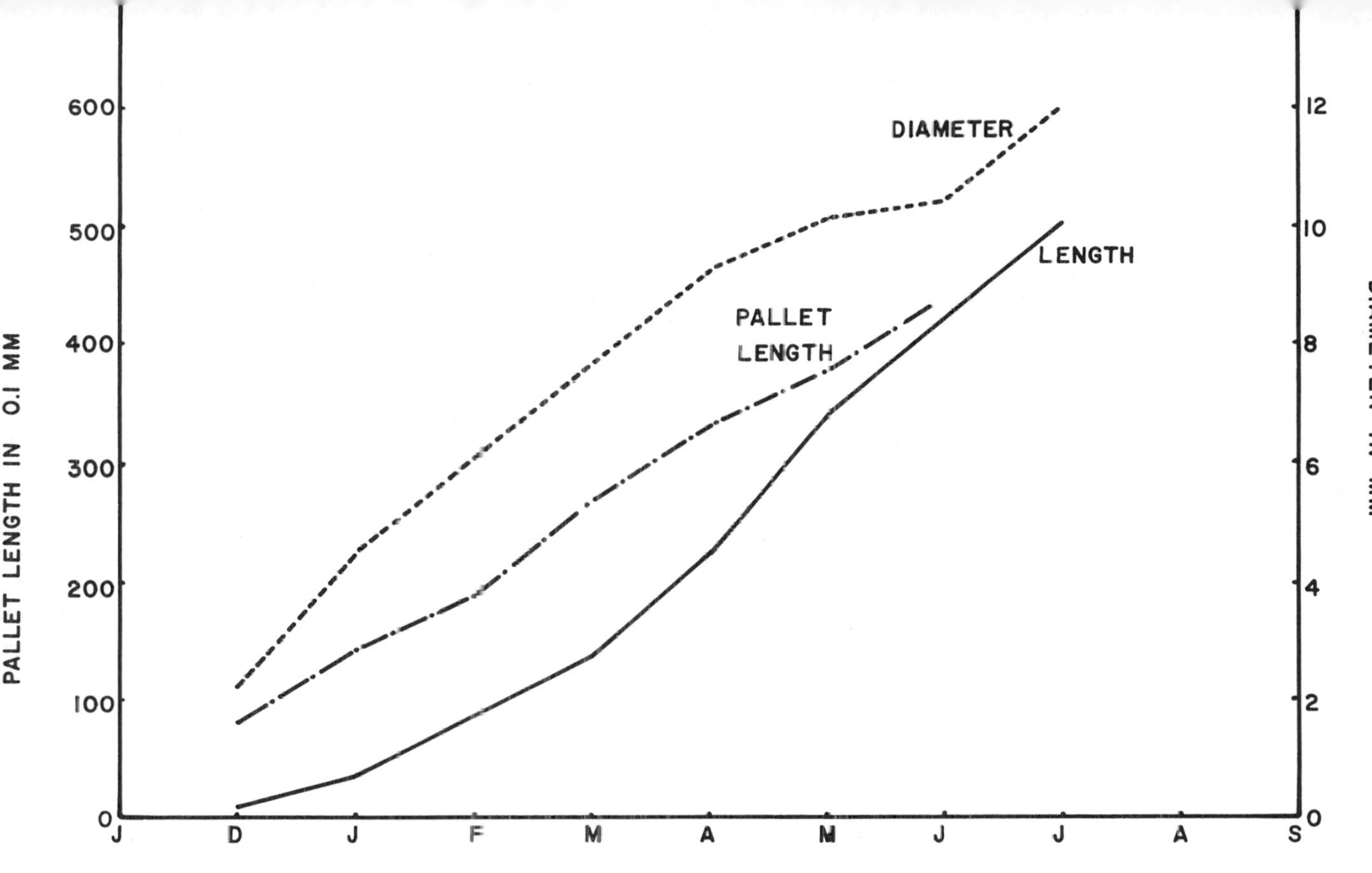

Fig. 1. Monthly growth in length, diameter, and pallet length of *Bankia setacea*, Ladysmith Harbour, 1954-55

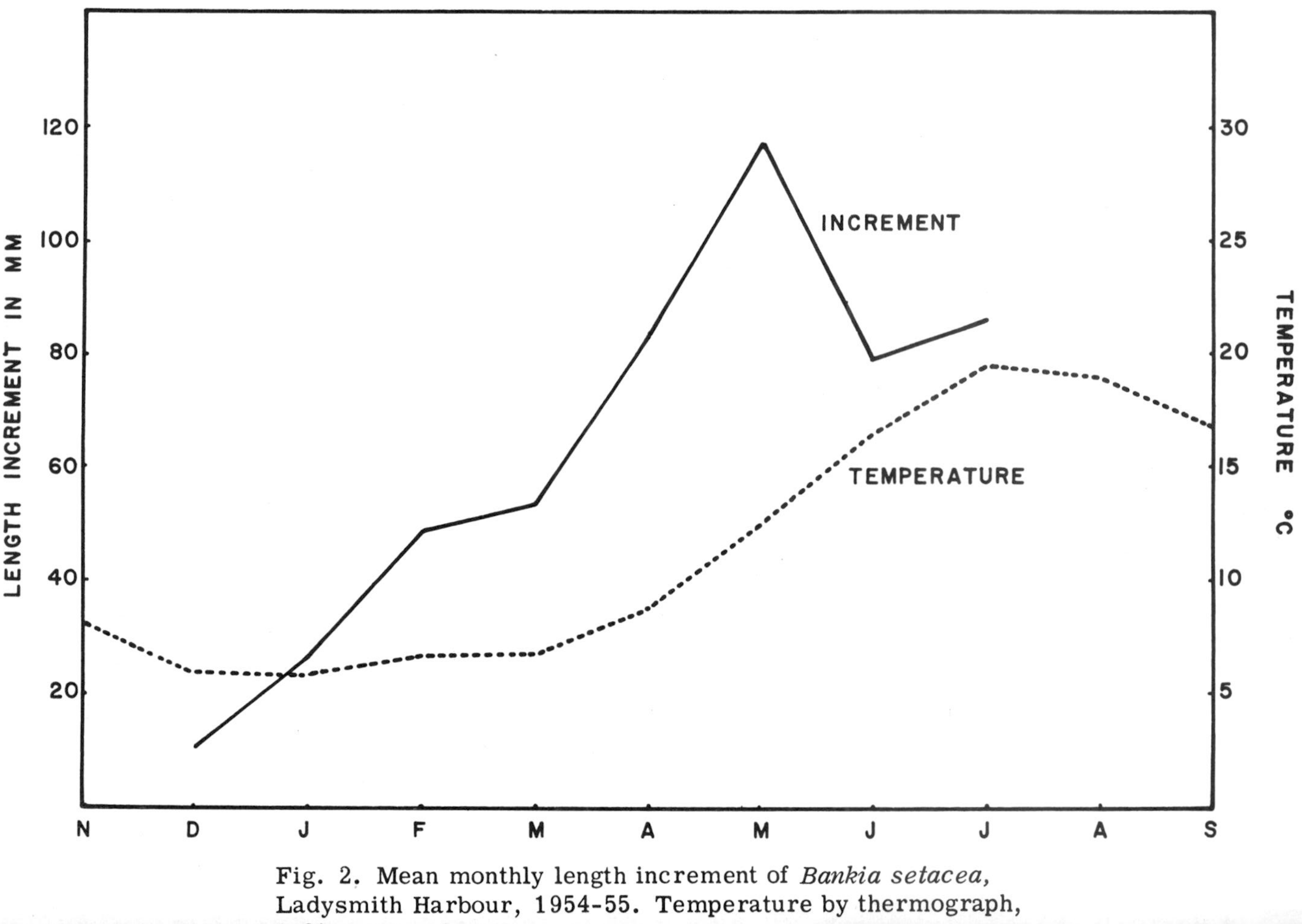

Fig. 2. Mean monthly length increment of *Bankia setacea*, Ladysmith Harbour, 1954-55. Temperature by thermograph,

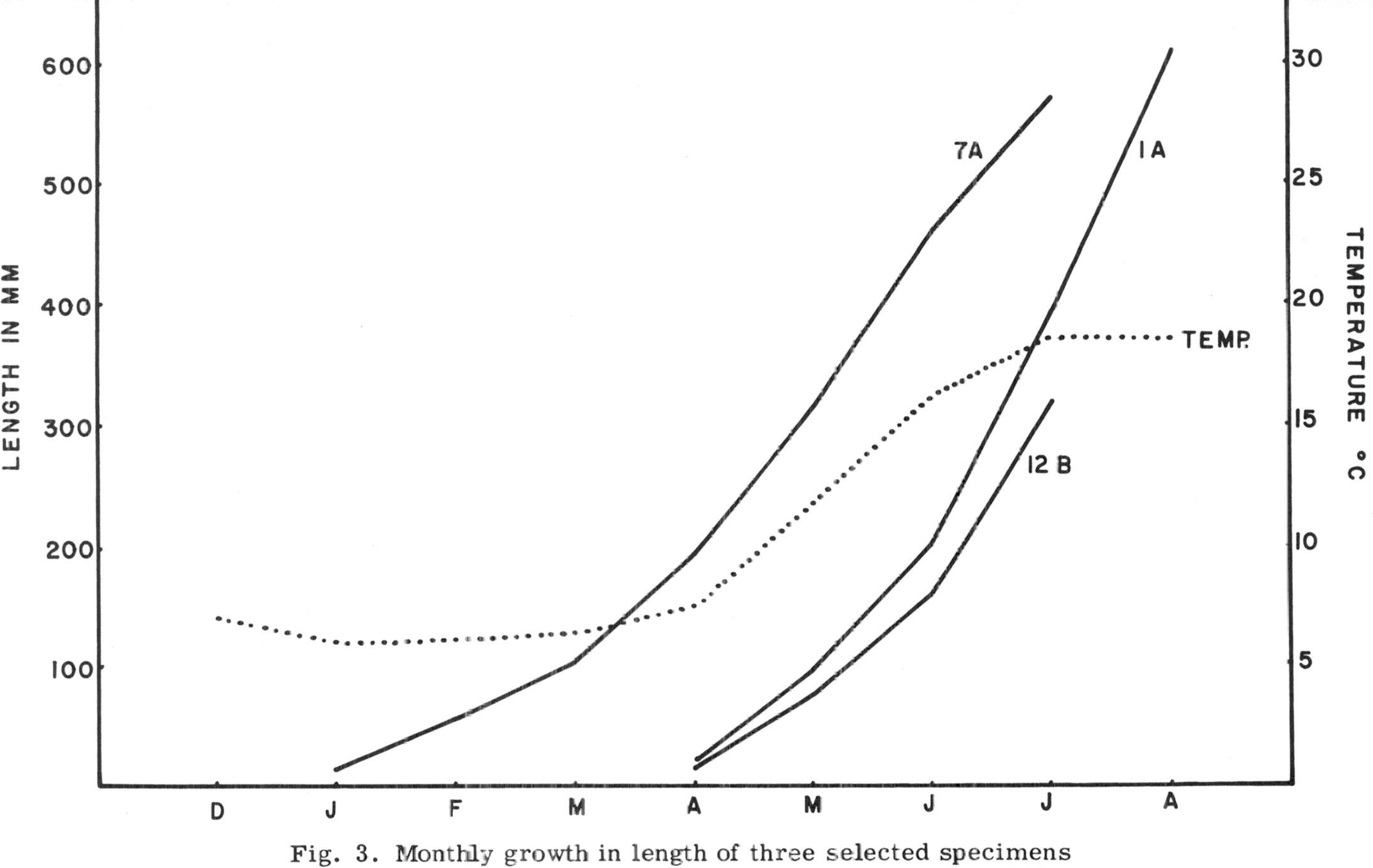

Fig. 3. Monthly growth in length of three selected specimens (1A, 7A, 12B) of *Bankia setacea*, Ladysmith Harbour, 1954-55. Temperature by thermograph, 0.91 meters deep

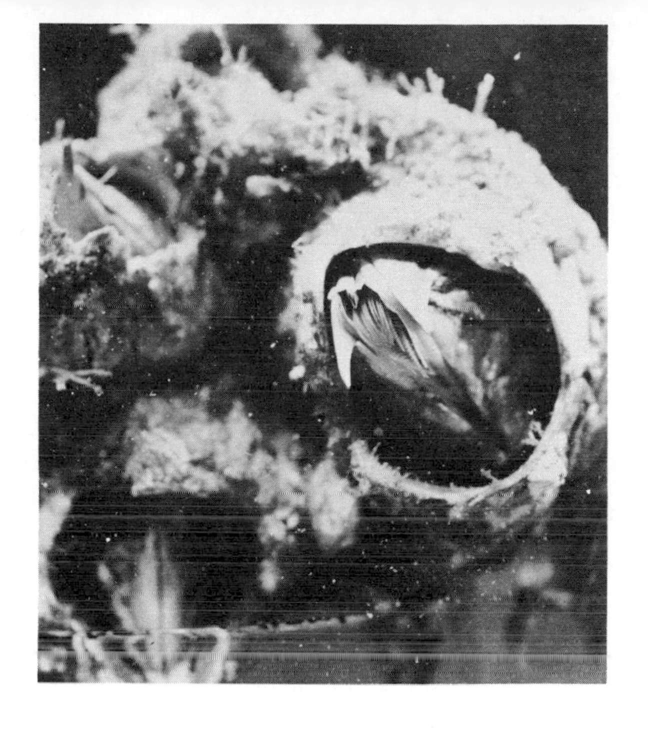

Barnacles

INTRODUCTION

Waldo Schmitt

Just being here is for me a very wonderful thing. I have passed San Juan Island on several occasions on the way to fisheries investigations in Alaskan waters, but this is the very first time that I have set foot ashore and have had the opportunity of visiting the university's fine facilities at Friday Harbor and meeting so many of the active staff and marine biologists sojourning here.

I am indebted to the Office of Naval Research, the University of Washington, and particularly to Dr. Arthur W. Martin and Dr. Dixy Lee Ray for inviting me and enabling my wife and me to attend.

The papers and discussions of the preceding sessions were highly informative and without question a great stimulus to the imagination. Each of us learned something he did not know before and, by the same token, envisioned new approaches to the solution of his own problems. May what you see and hear of barnacles today be equally useful to you.

There are more than 000 species of barnacles (not counting the parasitic Rhizocephala) distributed the world around from the high to the low latitudes, from spray-lashed shores to the ocean's great depths, from waters of high salt concentration to those of brackish estuaries. Over 100 of these species have thrust themselves upon man's attention because of their fouling proclivities--impeding the expedition of ocean-borne cargoes, and the proper functioning of navigational aids, buoys, and explosive mines. Some 2,000 other species of marine organisms have been recorded as associated with barnacles and in their turn may initiate, or compound, barnacle fouling problems.

Before we start today's session I should like to recognize and extend a special welcome to one of our visitors, Dr. Ira Cornwall. He is also a student of barnacles, who has come over from Nanaimo to be with us today. While he was serving with a British Columbia hospital unit in the first World War, the desire to name the barnacles he encountered along his coasts got him into the "game." I first became acquainted with him over thirty years ago as the result of some extremely interesting observations that he published on the respiratory adaptations of whale barnacles to the extremes of water temperature encountered in their journeyings from cold north Pacific waters to the breeding grounds of their hosts in equatorial, tropic seas. In the interim he has authored a number of other barnacle papers including a handbook, The Barnacles of British Columbia. Of late he has been investigating the curious but characteristic "figures" or patterns found by examining thin, ground sections of barnacle shells. These give hope that it may be possible to make specific identification from shell fragments of barnacles, both fossil and recent. It is a pleasure to have Dr. Cornwall with us today.

It is not inappropriate that this symposium is being held here in the Northwest. Our leading American authority on barnacles lives in Seattle and carries on her systematic researches in the Oceanography Department of the University of Washington, which has published a number of her papers on cirripedes. The United States National Museum has also been proud to publish the results of her study of collections of barnacles from all parts of the world sent to her for examination and report.

I have often wondered how Dora Henry ever got interested in barnacles. Since I now know, I'll tell you too: as a graduate student she undertook a study of gregarines found living in barnacles. Almost at once she realized that she would have to know the identities of the "host" animals that she was examining. At the time there was no easy way but "do-it-yourself," and she did. Now she is a barnacle specialist, and one in such demand that she never went back to those protozoan parasites of barnacles.

Dr. Henry's work illuminates what is involved in the identification of barnacles. Her admonition, that hasty identifications simply will not do if an investigator is not to learn too late that he has been working with two species instead of one, we should all take seriously to heart. Further, she justly emphasizes the handicaps under which the barnacle systematist operates because of the lack of up-to-date, comprehensive treatises, systematic monographs, and compilations of cirripede literature.

Those presently available were published a quarter of a century or more ago, long before today's pregnant knowledge of barnacle biology had been developed.

Foremost among the contributors of such knowledge are Dr. C. G. Bookhout and his associates at Duke University, Durham, North Carolina. His studies on the feeding, molting, and growth of barnacles have added much to our understanding of these animals. It is interesting to note that Dr. Bookhout bears witness to Dr. Henry's contention that a precise knowledge of the species involved is basic to any and all studies of barnacles, be they physiological, ecological, or of life histories. The originality displayed and the success achieved by Dr. Bookhout's "school" in rearing individual barnacles of definitely known species under controlled conditions marks one of the greatest advances made in recent years toward learning more about the biology of barnacles and no doubt also will contribute to the means of dealing with them as fouling organisms.

Further extension of our knowledge of barnacles comes from the studies of Dr. E. L. Bousfield of the National Museum of Canada, Ottawa. He has dealt primarily with conditions encountered in an estuarine environment that prove doubly interesting by way of comparison and contrast with the investigations of Dr. Joseph Connell carried out in a purely marine environment. Dr. Connell, now with the University of California at Santa Barbara, did not start using barnacles as subjects for biological research. He began with rabbits, but found that the things he wanted to do could better be done with animals that would stay put. Adult barnacles admirably filled the bill, and all students of barnacles have gained in knowledge of their biology by virtue of his choice.

Perhaps the outstanding investigator in the field of barnacle physiology is Dr. Harold Barnes, of the Marine Laboratory, Millport, Scotland. It was hoped that Dr. Barnes would be able to summarize and review all his researches on the many phases of barnacle physiology that he has studied, but these are so extensive that he has decided to limit his contribution to a consideration of the influence of one important environmental factor, temperature, on the life cycle of *Balanus balanoides* (L.).

This, then, is our "menu." It is now my privilege to declare the session on barnacles open, and to present today's speakers.

THE DISTRIBUTION OF THE AMPHITRITE SERIES OF BALANUS IN NORTH AMERICAN WATERS*

Dora Priaulx Henry

Balanus amphitrite Darwin is, with the possible exception of *Balanus tintinnabulum* Linnaeus, the most variable species in the genus and has a world-wide distribution in tropical and warm temperate waters. Many subspecies have been described, but in many cases the exact distribution is not known; in fact, the diagnoses of some of the subspecies are so inadequate that identification cannot be made with certainty. There are various reasons, in addition to its extreme variability, for our lack of knowledge of this species; some of the reasons are enumerated below.

Darwin, who described the species in 1854, named nine varieties but did not designate types or type localities, and some of the diagnoses are incomplete. Numerous subsequent workers reported the occurrence of some of Darwin's varieties, sometimes without any description or by using only color, a very unreliable criterion for identification in this species; or they reported the occurrence without naming the variety. Others described new varieties or subspecies using characteristics that were later found to be variable and, therefore, not of diagnostic value. Museum specimens, many collected from ships, have been used for most of the systematics of this species, and often locality data have been lacking or, for those from ships, the exact origin was not known. In additon, there is considerable evidence that this species has often been confused with other common fouling barnacles.

In North American waters, in addition to five subspecies of

---

*Contribution No. 214 from the Department of Oceanography, University of Washington.

B. amphitrite Darwin, the closely related species, *B. eburneus* Gould, *B. improvisus* Darwin, and *B. concavus pacificus* Pilsbry occur. As the readily available literature does not clearly differentiate these forms and as the known distribution has been extended, it is the purpose of this paper to give a key as an aid to identification and a table of the distribution. Revised diagnoses and remarks on variation are given for *B. eburneus* and *B. improvisus*, as well as some general remarks on the subspecies of *B. amphitrite*.

The *amphitrite* series is in the subgenus *Balanus* da Costa, balani in which the parietes but not the radii are porous; the basis is calcareous, either porous or solid; and the third cirrus is armed with teeth or spinules. The *amphitrite* series is differentiated from the other series with porous bases (*trigonus* and *perforatus*) by distinctive characters, except for *B. trigonus* Darwin. This species may be difficult to distinguish from colored subspecies of *B. amphitrite* if the opercular valves are lacking and the parietes are smooth. In the typical form, the walls of *B. trigonus* are ribbed and the scuta have one to six longitudinal rows of pits. Of the other members of the subgenus (those with solid or imperfectly porous bases), only *B. crenatus* may at times be confused with white subspecies of *B. amphitrite* or with *B. improvisus*. The scutum of *B. crenatus* lacks an adductor ridge.

Two other species of barnacles, although they do not belong to the subgenus *Balanus*, may superficially resemble some subspecies of *B. amphitrite*. *B. amaryllis* Darwin, which is a common fouling barnacle but does not occur in North American waters, may be identified by the poreless parietes. *B. tintinnabulum*, which is represented by several subspecies, has porous radii as well as porous parietes and bases.

Key to the Amphitrite *Series of* Balanus *in North American Waters*

1. Parietes white without colored longitudinal striae; epidermis usually persistent 2
1. Parietes white with colored or hyaline longitudinal striae or colored with white or darker colored longitudinal striae; epidermis usually not persistent 4
2. Radii narrow, arched with oblique, thin-edged summits; parietes sometimes with hyaline, longitudinal striae; epider-

mis usually more persistent on radii than on parietes
B. improvisus

2. Radii usually moderate, not arched, with thick-edged summits; epidermis usually more persistent on parietes than on radii 3
3. Tergum with spur over 1/2 own width from basiscutal angle; basal margin usually hollowed out on carinal side of spur; parietes sometimes with hyaline longitudinal striae in young
B. eburneus
3. Tergum with spur less than 1/2 own width from basiscutal angle; basal margin not hollowed out on carinal side of spur; parietes sometimes with traces of purple
B. a. pallidus
4. Tergum with spur 1/4 or less width of basal margin
B. a. saltonensis
4. Tergum with spur over 1/4 width of basal margin
5
5. Spur less than 1/2 own width from basiscutal angle
6
5. Spur over 1/2 own width from basiscutal angle
7
6. Scutum with pit below adductor ridge *B. a. inexpectatus*
6. Scutum without pit below adductor ridge; parietes with hyaline longitudinal striae or, less commonly, lavender striae
B. a. niveus
7. Scutum with prominent longitudinal striae; adductor ridge long *B. c. pacificus*
7. Scutum without longitudinal striae; adductor ridge short
B. a. denticulata

Balanus amphitrite Darwin*

Darwin (1854) recorded three of his nine varieties of *B. amphitrite* from North American waters; i.e., the typical form and *B. a. pallidus* from the West Indies, and *B. a. niveus* from the West Indies and Florida. Pilsbry (1916) redescribed *B. a. niveus* from specimens collected at various localities from Vineyard Sound to both the east and west coasts of Florida. He also described a new subspecies, *B. a. inexpectatus*, from the

---

*A paper is now in preparation on *B. amphitrite* with descriptions and photographs of the valid subspecies.

Gulf of California. Rogers (1949) described three new subspecies from California, *B. a. saltonensis* from the Salton Sea and two, which are here considered to be synonymous with *B. a. denticulata* Broch, from San Francisco.

The following subspecies have been found in North American waters: *B. a. niveus* from various localities on the east coast of the United States and the Gulf of Mexico; *B. a. pallidus* from two localities on the east coast and three on the Gulf of Mexico coast; *B. a. denticulata* Broch from Beaufort, North Carolina, three localities on the west coast of Florida, and four localities in California from La Jolla to Venice; *B. a. inexpectatus* from the Pacific coast of Baja California; and *B. a. saltonensis* from several localities in the Salton Sea and two localities in the Los Angeles Harbor Area.

Because of the great variation caused by age, corrosion, crowding, and difference in habitat, the key will be useful only for the typical forms. In many cases it is necessary to examine the mouth parts and the cirri, in addition to the shell and opercular valves, to identify specimens.

It should be pointed out, however, that the distribution of the teeth on the labrum is not of as great diagnostic value as previously thought. Pilsbry (1916) stated that *B. amphitrite* could always be differentiated from *B. eburneus* and *B. improvisus* by the fact that the labrum never has a series of graduated teeth extending down the sides of the notch. While this is true for most of the subspecies of *B. amphitrite*, including *B. a. niveus* and *B. a. inexpectatus*, there are several subspecies that cannot be distinguished by this character. Broch described *B. a. hawaiiensis* in 1922 and *B. a. denticulata* in 1927, both having numerous teeth extending into the notch of the labrum (see pl. 1, fig. 5), and Nilsson-Cantell (1938) named a new subspecies in which the labrum has similarly arranged teeth. Yet Bishop (1950) stated that *B. a. denticulata* can always be readily separated from other varieties of *B. amphitrite* by the denticulate labrum and suggests that it should perhaps be raised to specific status because of the character of the labrum. The labrums of *B. a. pallidus* (pl. 1, fig. 4) and *B. a. saltonensis* (pl. 1, fig. 6), which had not been previously examined, also have numerous teeth extending along the sides of the notch.

At the present time it is possible only to make a few general statements about the ecology of *B. amphitrite*. *B. a. niveus* and *B. a. inexpectatus* apparently are restricted to water with a relatively high salinity, while the other three subspecies are tolerant of a wide range in salinity. *B. a. denticulata,* which was first found in the Suez Canal by Broch (1927) in water with a salinity of about 50‰, is a common barnacle of the estuaries in South Africa but is not found on the marine coast (Millard, 1950). The salinities in the Knysna Estuary varied from below 21.7‰ at low tide to close to that of normal sea water. In North American waters this subspecies is present on the Atlantic coast, and on the open coast and in bays in the Gulf of Mexico, but was not found in the highly saline Laguna Madre and Baffin Bay where *B. eburneus* occurs. On the Pacific coast *B. a. denticulata* occurs in waters varying from brackish to nearly fresh, as well as on the open coast. *B. a. saltonensis* occurs in salinities varying from about 28‰ (Salton Sea) to about that of normal sea water (Los Angeles Harbor).

Whether the vertical distribution may be of use in differentiating the subspecies of *B. amphitrite* cannot be decided without further work. *B. a. niveus* and *B. a. inexpectatus* apparently occur, at least usually, below low tide level with the lowest limit about 40 fathoms for *B. a. niveus* (Pilsbry, 1953), and unknown for *B. a. inexpectatus. B. a. saltonensis, B. a. denticulata,* and probably *B. a. pallidus,* occur in the intertidal zone but their lower limits are not known.

Balanus eburneus Gould
Pl. 1, figs. 2-3; pl. 3, fig. 7; pl. 4, figs. 1-6

Balanus eburneus Gould, 1841, p. 15, pl. 1, fig. 6; Darwin, 1854, p. 248, pl. 5, figs. 4a-4d; Verrill, 1874, p. 579; Weltner, 1897, p. 266; Sumner, 1911. p. 129, 302, 645, Chart 84; Pilsbry, 1916, p. 80, figs. 14, 16b, pl. 24, figs. 1-1c, 2; Nilsson-Cantell, 1921, p. 309; Neu 1935, p. 93; 1939, p. 216, figs. 5a-5e, 6; Kolosváry, 1940, p. 36, figs. 3, 14, 18; 1947, p. 21, pl. 1, figs. 12-14, pl. 2, fig. 9, pl. 3, figs. i-i_2.

Conic or shortly cylindric with ovate orifice; white, smooth. Radii usually moderately wide with oblique irregular summits. Epidermis usually persistent, except on radii. Maximum diameter, about 40 mm. Scutum with prominent longitudinal striae; articular ridge very high, about 1/2 tergal margin; adductor ridge high but short, with a weak ridge below. Tergum broad, without a longitudinal furrow; spur long, width 1/3 or less that of basal margin, about 1/2 own width from basiscutal angle; basal margin hollowed out on carinal side of spur; carinal margin highly arched; crests for depressor muscles usually weak.

Labrum with numerous teeth continuing into notch. First maxilla with lower pair on a long prominence. First cirrus with very unequal rami; third cirrus with teeth on anterior margins; sixth cirrus with 7-9 pairs of spines.

Remarks. Young *B. eburneus*, before the development of the longitudinal striae on the scuta and the hollowing out of the basal margin of the terga, may be difficult to differentiate from young *B. improvisus* or *B. a. niveus*, especially if the longitudinal septa show through the wall. However, even in small specimens, the tergum (pl. 3, fig. 7) shows the characteristic arched carinal margin and the crests for the depressor muscles on a slight protuberance. Many of the young specimens have fewer than seven pairs of spines (five or six pairs) on the sixth cirrus so that the number of spines cannot be used to distinguished this species from *B. improvisus* or *B. a. niveus* as both Darwin (1854) and Pilsbry (1916) believed.

The greatest variation shown by the shell of *B. eburneus* is in the width of the radii. Specimens from Tampa Bay, Florida (pl. 4, figs. 4-6), are cylindrical with thinner walls than usual. Both the very narrow radii and the alae have extremely oblique summits so that the orifice is more toothed than in the typical form (pl. 4, figs. 1-3). The occurrence of narrow radii in these forms is surprising because in other species having both conic and cylindric forms the radii are wider in the cylindric than in the conic forms. The opercular valves of the Tampa Bay specimens are relatively narrower than in the typical form, but this is also true in cylindrical forms with wide radii. No differences in the mouth parts and cirri could be detected in the Tampa Bay specimens, and occasionally specimens with narrow radii were found with typical *B. eburneus* in other localities.

B. eburneus, unlike *B. improvisus* and *B. a. denticulata*, apparently never becomes corroded in brackish water. Specimens from the highly saline waters of the Laguna Madre and Baffin Bay, Texas attain a large size and do not differ from those in less saline waters.

Material examined. Maryland: Solomons Island, Chesapeake Biological Laboratory pier, with *B. improvisus*, October 1, 1944, U.S.N.M.

North Carolina: Beaufort, with *B. a. denticulata* and *C. fragilis*, John D. Costlow, Jr.; Wilmington, on test block, submerged November 21, 1940, removed July 8, 1941, with *B. improvisus*;

same, on test block, submerged January 27, 1941, removed August 16, 1941, William F. Clapp.

South Carolina: Sullivan's Island, on the jetty, with *B. a. niveus*, December 31, 1952, Charles S. Yentsch.

Florida: Jacksonville, on the jetty, with *B. improvisus, T. s. stalactifera* and *C. fragilis*; Tampa Bay, on a moored motor boat, with *B. a. denticulata* and *B. a. niveus,* December 29, 1952, Charles S. Yentsch; Boca Ciega Bay, with *B. a. denticulata* and *C. fragilis,* October 8, 1955; with *B. improvisus* and *B. a. niveus,* November 3, 1955; with *B. a. niveus,* November 4, 1955, U.S.N.M.; Saint Marks, with *B. a. denticulata*; Alligator Harbor, with *B. a. niveus*; Fort Walton, Charles S. Yentsch.

Gulf of Mexico: on a whistle buoy, with *B. a. niveus* and *B. improvisus,* June 26, 1952, Charles S. Yentsch.

Mississippi: Biloxi, on rock oysters, July 10, 1931, U.S.N.M.

Louisiana: on oyster shells, December 19, 1917, New Orleans, 1920, Shell Island, November 7, 1898, U.S.N.M.

Texas: Galveston, Offat's Bayou, March 24, 1940, U.S.N.M.; Mesquite Bay, with *B. improvisus*; Cedar Bayou, with *Ch. patula*; Rockport, with *B. improvisus*; Port Aransas, November 20, 1947; same, on old barge, with *B. improvisus*, December 1947; same, on driftwood, March 27, 1948; Riviera Beach, Baffin Bay, on piles, salinity 61‰, May 9, 1950; Laguna Madre, salinity + 50‰, March 11, 1950, Joel W. Hedgpeth.

Puerto Rico: San Juan, with *T. s. stalactifera* and *C. fragilis,* September, 1942, Naval Ordnance Laboratory, Washington, D.C.; Guayanilla, December 8, 1956, U.S.N.M.

Guadeloupe: Pte-à Pitre, on steel barge, August 1949, U.S.N.M.

Balanus improvisus Darwin
Pl. 3, figs. 1-8

Balanus improvisus Darwin, 1854, p. 250, pl. 6, figs. 1a-1c; Weltner, 1897, p. 266; Pilsbry, 1916, p. 84, figs. 16a, 17a-18f, pl. 24, figs. 3-3b, 5-5d; Nilsson-Cantell, 1921, p. 310; 1928, p. 33; Schaper, 1919-22, p. 227, figs. 22-30; Broch, 1924, p. 81, pl. 1, fig. 7; pl. 3, figs. 15, 16; Neu, 1932, p. 143, fig. 3; 1935, p. 94; 1939, p. 214, figs. 3, 4A-C; Ciurea, Monad, and Dinulesco, 1933, p. 2, figs. 1, 2a-2g, 3, 4a-4b; Henry, 1942, p. 110, fig. 1, pl. 2, figs. 12-13; Kolosváry, 1939, p. 92, fig. 2; 1943, p. 102, fig. 4.

Balanus improvisus var. *assimilis* Darwin, 1854, p. 250.

Balanus improvisus var. *gryphicus* Münter and Buchholz, 1869, p. 9.

Conic or cylindric, with diamond-shaped orifice; smooth, white, sometimes with hyaline longitudinal striae. Radii narrow with smooth arched, very oblique and thin-edged summits. Epidermis persistent on radii and sometimes on parietes. Maximum diameter 20 mm. Scutum with high articular ridge, a little over 1/2 tergal margin; adductor ridge high and long. Tergum with longitudinal furrow, varying in depth; spur short, width 1/4 basal margin, less than own width from basiscutal angle; basal margin straight on both sides of spur; crests for depressor muscles strong and numerous. Labrum with numerous teeth continuing into notch. First maxilla with lower pair of spines on a prominence; first cirrus with subequal rami; third cirrus with conic or spikelike teeth on anterior margins; sixth cirrus with 5 to 8 pairs of spines.

Remarks. None of the specimens (greatest diameter, 12 mm) are as large as the largest recorded in other parts of its range. Van Breeman (1933) gave 20.7 mm as the maximum diameter for specimens from Amsterdam, and Neu (1932, 1939) 17 mm for specimens from Helgoland and the Bosphorus. Perhaps this species does not attain as great a size in North American waters because of greater competition with other species of barnacles. In the estuaries of the Baltic and North Europe *B. improvisus* is usually the only barnacle present or, at most, competes with only one other species, i.e., *B. crenatus*.

The shells of young specimens may be difficult to distinguish from *B. eburneus*, *B. a. pallidus*, and *B. a. niveus*, but *B. improvisus* can be most easily differentiated from these species by the tergum (pl. 3, fig. 7), which even in young specimens is relatively wider and has a longer and narrower spur that is of a uniform width throughout its length.

Corroded specimens appear very different from the typical form (pl. 3, fig. 4, conic specimen) with narrow, thin-edged radii. In some corroded specimens (pl. 3, fig. 4) the radii cannot be differentiated, whereas in others (pl. 3, fig. 3) the radii are not arched, and therefore the edges appear thicker than usual. Such specimens may be confused with *B. a. pallidus* unless the opercular valves are examined, especially as the shells are often covered with debris which is difficult to remove without destroying some of the underlying shell.

The opercular valves of corroded specimens (pl. 3, figs. 5-6) also appear different from those of typical specimens (pl. 3, figs. 1-2). They are often irregular in shape, and there is a tendency for the ridges on the inner surfaces and the articulating margins to be worn. The longitudinal furrow of the tergum is usually much more prominent in corroded specimens.

When crowded, the shell may be pencil-shaped, and occasionally the basis is lengthened or the diameter at the apex may be greater than at the base. In some of the larger specimens there are 7 or 8 pairs of spines on the posterior cirri instead of 5 or 6 pairs as stated by both Darwin (1854) and Pilsbry (1916) (see discussion under *B. eburneus*). The larger specimens and some of the smaller have conic as well as spikelike teeth on the anterior margins of the third cirrus, and occasionally on the anterior ramus of the fourth cirrus.

B. improvisus has been used very frequently in various biological studies, both in connection with the control of fouling and in brackish water studies. Résumés of much of this work and an excellent bibliography may be found in Marine Fouling and Its Prevention (1952).

Material examined. Potomac River: Piney Point, with *B. a. pallidus,* May 31, 1937; Poseys Bluff, November 13, 1942; Kingcopsico Point, November 16, 1942; Ragged Point Bar, November 17, 1942; Blake Creek Bar, November 17, 1942; Heron Island Bar, November 19, 20, 1942; Coles Point Bar, November 26, 1942; Sheepshead Bar, December 1, 1942; same, December 7, 1942; Lower Cedar Point Bar, December 11, 1942; Swan Point Bar, December 15, 17, 1942; Flats Bar, January 11, 1943; Higgins Point Bar, January 26, 1943; Old Farms Bar, February 2, 1943; Piney Point Bar, June 23, 1943, U.S.N.M.

Maryland: mouth of Patuxent River, July 16, 1934; Solomons Island, Chesapeake Biological Laboratory pier, with *B. eburneus,* October 1, 1944, U.S.N.M.

North Carolina: Ft. Raleigh, Roanoke Island, with *B. a. pallidus,* October 1, 1943, U.S.N.M.; Wilmington, on test block, submerged November 21, 1940, removed July 8, 1941, with *B. eburneus,* William F. Clapp.

Florida: Jacksonville, on the jetty, with *B. eburneus, T. s. stalactifera,* and *C. fragilis,* Charles S. Yentsch; Daytona Beach, on test blocks, submerged January 5, 1942, removed March 5, 1942, William F. Clapp; Boca Ciega Bay, with *B. eburneus* and *B. a. niveus,* November 3, 1955, U.S.N.M.; Alligator Harbor, on eelgrass, Charles S. Yentsch.

Gulf of Mexico: on a whistle buoy, with *B. eburneus* and *B. a.*

TABLE 1

DISTRIBUTION OF THE AMPHITRITE SERIES OF BALANUS IN NORTH AMERICAN WATERS

Species	Atlantic Coast of North America	Pacific Coast of North America	Occurrence Elsewhere
B. a. niveus	Massachusetts to Gulf of Mexico		Caribbean Sea to southern Brazil; Mediterranean and Red seas to Madagascar and South Africa
B. a. pallidus	Chesapeake Bay to Gulf of Mexico		West Indies to Brazil; Red Sea to Madagascar; west coast of Africa
B. a. denticulata	North Carolina to west coast of Florida	San Francisco to La Jolla, California	Definitely known from Bermuda; southern England; Suez Canal; and South Africa
B. a. inexpectatus		Both sides of Baja California	Curaçao to Ecuador
B. a. saltonensis		Salton Sea, and Los Angeles Harbor, California	
B. eburneus	Massachusetts to the Gulf of Mexico		West Indies to Caribbean coast of South America; Mediterranean and Black seas; Hawaii
B. improvisus	Nova Scotia to the Gulf of Mexico		West Indies to Patagonia and Ecuador; Scotland and Norway to the ocean coast of France; Black and Red seas; West Africa (?)
			Panama to Peru
B. concavus pacificus		Northern California to Baja California	

niveus, June 26, 1952, Charles S. Yentsch; Heald Bank, with *B. a. niveus*, U.S.N.M.

Mississippi: Mississippi Sound, east end, September 1, 1947, U.S.N.M.

Louisiana: Lake Pontchartrain, 14 lots, 8 with *B. a. pallidus*, 1 lot on wood, rest on clam shells; 1 lot near South Point, with *B. a. pallidus*, May 9, 1923, U.S.N.M.

Texas: Mesquite Bay, on *B. eburneus*; same, on a diamond-back turtle, with *Ch. testudinaria*, July 15, 1946; Rockport, with *B. eburneus*; Harbor Island, on a crab and a rock oyster, March 18, 1948, Port Aransas, on laboratory intake pipe; same, on *B. eburneus* from old barge, December, 1947; same, on laboratory tank valve, March 26, 1948; same, on piles and on *B. galeatus*, April 13, 1948; same, on a buoy, April 14, 1948, Joel W. Hedgpeth.

Honduras: U.S.N.M.

Puerto Rico: Guayanilla, inside pipe, with *B. a. niveus*, December 8, 1956, U.S.N.M.

California: Bay View Park, San Francisco, February 25, 1939; San Pablo Bay, Frank L. Rogers.

Balanus concavus pacificus Pilsbry
Pl. 2, figs. 1-6

Balanus concavus pacificus Pilsbry, 1916, p. 104, fig. 25, pl. 23, figs. 1-2c; Giltnay, 1934, p. 1, figs. 1-4; Henry, 1942, p. 104, pl. 2, figs. 1-4; Cornwall, 1951, p. 328, pl. 4, figs. D and E.

Conic to cylindric with diamond-shaped orifice; smooth; white, yellow or pink with narrow to wide darker pink or reddish-purple longitudinal striae. Thin, transparent epidermis usually persistent. Maximum diameter, 34 mm. Radii wide with slightly oblique summits. Parietal tubes filled up above, no transverse septa. Opercular valves white with splotches of pink. Scutum with prominent, longitudinal striae. Articular ridge low, about 1/2 tergal margin; adductor ridge very prominent, extending nearly to the basal margin; a thin ridge bounds the pit for the lateral depressor muscle. Tergum broad with a narrow, infolded longitudinal furrow; spur long, separated by a little less than own width from the basiscutal angle; crests for depressor muscles moderate. Labrum with 0-3 teeth on each side of notch. First maxilla with lower, large pair of spines on a projection. First cirrus with slightly unequal rami; third cirrus and anterior ramus of fourth cirrus with conic or spikelike teeth on anterior margins in large specimens but only on anterior ramus of third cirrus in small; sixth cirrus with 3-4 pairs of spines.

Remarks. Specimens lacking opercular valves may be difficult to distinguish from *B. a. denticulata* or large forms of *B. a. saltonensis*; in fact, the latter was identified as this species when it was first reported from the Salton Sea. However, *B. c. pacificus* may always be differentiated from these two forms by the parietal tubes which lack transverse septa.

Material examined: California, locality unknown. Baja California, four localities (Henry, 1942, p. 104).

LITERATURE CITED

Bishop, M. W. H. 1950. Distribution of *Balanus amphitrite* Darwin var. *denticulata* Broch. Nature, 165:409-10.

Broch, H. 1922. Papers from Dr. Th. Mortensen's Pacific Expedition 1914-16. X, Studies on Pacific Cirripeds. Vid. Medd. naturh. Foren. Kjøbenhavn, 73:215-358.

-------. 1924. Cirripedia thoracica von Norwegen und dem norwegischen Nordmeer. Eine systematische und biologisch-tiergeographische Studie. Vid. Skript., I. Mat-naturw. Kl. 1924, 17:1-121.

-------. 1927. Zoological results of the Cambridge Expedition to the Suez Canal, 1924. VII, Report on the Crustacea Cirripedia. Trans. Zool. Soc. London, 22:133-37.

Ciurea, J., Th. Monad, and G. Dinulesco. 1933. Présence d'un cirripède operculé sur un poisson dulcaquicole européen. Bull. Inst. Océanogr. Monaco, 615:1-32.

Cornwall, I. E. 1951. The Barnacles of California (Cirripedia). Wasmann J. Biol., 9:311-46. -46.

Darwin, C. 1854. A Monograph on the Sub-class Cirripedia, with Figures of All the Species. The Balanidae (or Sessile Cirripedes); the Verrucidae, etc. Ray Society, London.

Giltnay, L. 1934. Note sur l'association de *Balanus concavus pacificus* Pilsbry (Cirripède) et *Dendraster excentricus* (Eschscholtz) (Echinoderme). Bull. Mus. Roy. Hist. Nat. Belg., 10:1-7.

Gould, A. A. 1841. Report on the Invertebrata of Massachusetts, Comprising the Mollusca, Crustacea, Annelida, and Radiata. Folsom, Wells and Thurston, Cambridge.

Henry, D. P. 1942. Studies on the Sessile Cirripedia of the Pacific Coast of North America. Univ. Wash. Publ. Oceanog., 4:95-134.

Kolosováry, G. von. 1939. Die Cirripedien (Subordo: Balanomorpha) des ungarischen Nationalmuseums. Ann. Mus. Nat. Hungar., 32:91-97.

-------. 1940. Les Balanides de l'Adriatique. Bull. Mens. Soc. Linn. Lyon, 9:35-38.

-------. 1943. Cirripedia Thoracica in der Sammlung des ungarischen Nationalmuseums. Ann. Mus. Nat. Hungar., 36:67-120.

-------. 1947. Die Balaniden der Adria. Ann. Mus. Nat. Hungar., 39:1-88.

Millard, N. 1950. On a Collection of Sessile Barnacles from Knysna Estuary, South Africa. Trans. Roy. Soc. So. Africa, 32:265-73.

Münter, J., and R. Buchholz. 1869. Ueber *Balanus improvisus* Darw. var. *gryphicus* Münter. Beitrag zur carcinologischen Fauna Deutschlands. Mitt. naturw. Ver. Griefswald, 1:1-40.

Neu, W. 1932. Das Vorkommen von *Balanus improvisus* Darwin auf Helgoländer Seetonnen. Zool. Anz., 99:143-49.

-------. 1935. *Balanus eburneus* Gould und *Balanus improvisus* Darwin als Bewuchs ausgehängter Platten im Goldenen Horn von Istanbul. Zool. Anz., 112:92-95.

-------. 1939. Bemerkungen über einige balanomorphe Cirripedien der Istanbuler Gewässer. Zool. Anz., 125:209-19.

Nilsson-Cantell, C. A. 1921. Cirripeden-Studien. Zur Kenntnis der Biologie, Anatomie und Systematik dieser Gruppe. Zool. Bidr. Upsala, 7:75-395.

-------. 1928. Studies on Cirripeds in the British Museum. Ann. Mag. Nat. Hist., Ser. 10, 2:1-39.

-------. Cirripeds from the Indian Ocean in the Collection of the Indian Museum, Calcutta. Mem. Ind. Mus., 13:1-81.

Pilsbry, H. A. 1916. The Sessile Barnacles (Cirripedia) Contained in the Collections of the U.S. National Museum: Including a Monograph of the American Species. U.S. Nat. Mus. Bull., 93.

-------. 1953. Notes on Floridan Barnacles (Cirripedia). Proc. Acad. Nat. Sci. Philad., 105:13-28.

Rogers, F. L. 1949. Three New Sub-species of *Balanus amphitrite* from California. Pomona Coll. J. Ent. and Zool., 41:3-12.

Schaper, P. 1919-22. Beiträge zur Kenntnis der Cirripedia

thoracica der Nord- und Ostsee. Wiss. Meeresunters., Abt. Kiel, N. F., 19:211-50.

Sumner, F. B., R. C. Osburn, and L. J. Cole. 1913. A Biological Survey of the Waters of Woods Hole and Vicinity. Pt. 2, Sect. 3: A Catalogue of the Marine Fauna. Bull. U.S. Fish. Comm., 1911, 31:549-794.

U.S. Naval Institute. 1952. Marine Fouling and Its Prevention. Prepared for Bureau of Ships, Navy Dept., by Woods Hole Oceanographic Institution. Annapolis, Md.

Van Breeman, L. 1934. Zur Biologie von *Balanus improvisus* (Darwin). Zool. Anz., 105:247-57.

Verrill, A. E., and S. I. Smith. 1874. Report upon the Invertebrate Animals of Vineyard Sound and Adjacent Waters, with an Account of the Physical Features of the Region. In Report of U.S. Comm. Fish and Fisheries, 1871-72, pp. 295-754.

Weltner, W. 1897. Verzeichnis der bisher beschriebenen recenter Cirripedienarten. Arch. Naturg., 1:227-80.

PLATE 1

Labrums, x 65

(Photographs by John Lincoln)

Fig. 1. *Balanus improvisus* Darwin. Lake Pontchartrain, Louisiana

Figs. 2, 3. *Balanus eburneus* Gould. Tampa Bay, Florida; specimen almost ready to moult, old membrane (fig. 2); new membrane (fig. 3)

Fig. 4. *Balanus amphitrite pallidus* Darwin. Potomac River

Fig. 5. *Balanus amphitrite denticulata* Broch. Huntington Beach, California

Fig. 6. *Balanus amphitrite saltonensis* Rogers. San Pedro, California

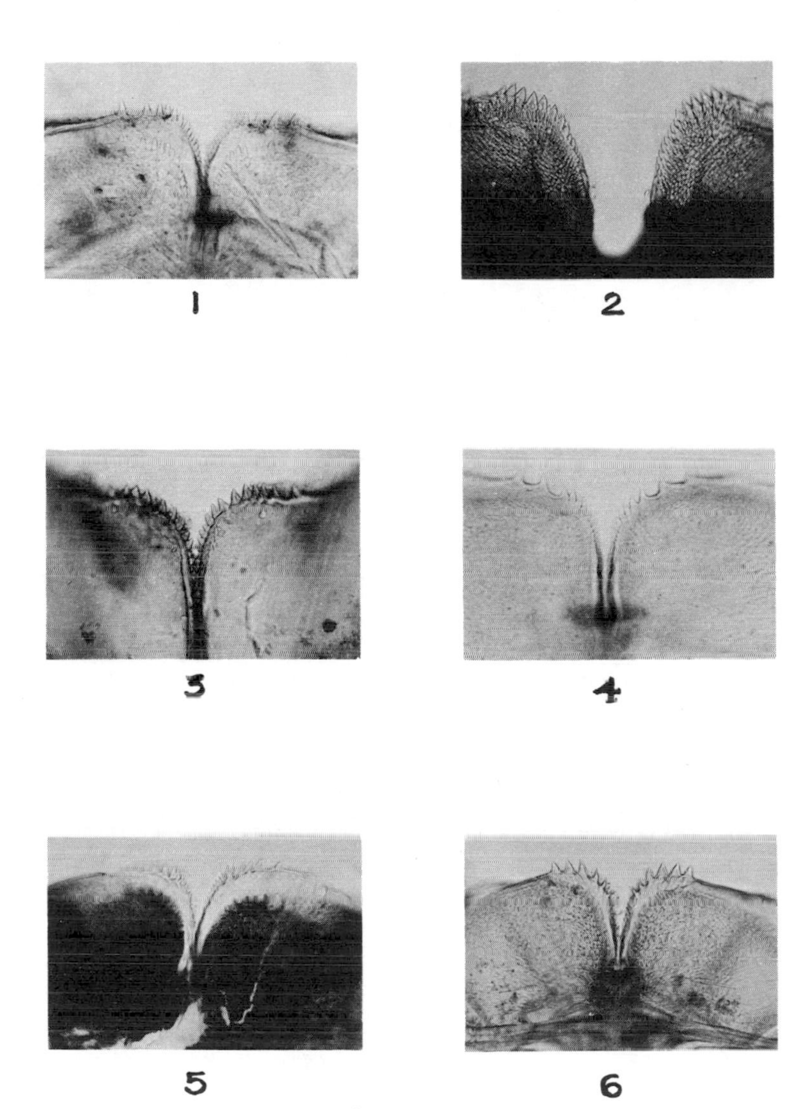

PLATE 2

Balanus concavus pacificus Pilsbry
Santa Maria Beach, Lower California
(Photographs by Elizabeth A. McGraw [figs. 1, 3-6]
and John Lincoln [fig. 2])

Fig. 1. Apical view, x 2

Fig. 2. Labrum, x 65

Figs. 3-6. Opercular valves, x 5

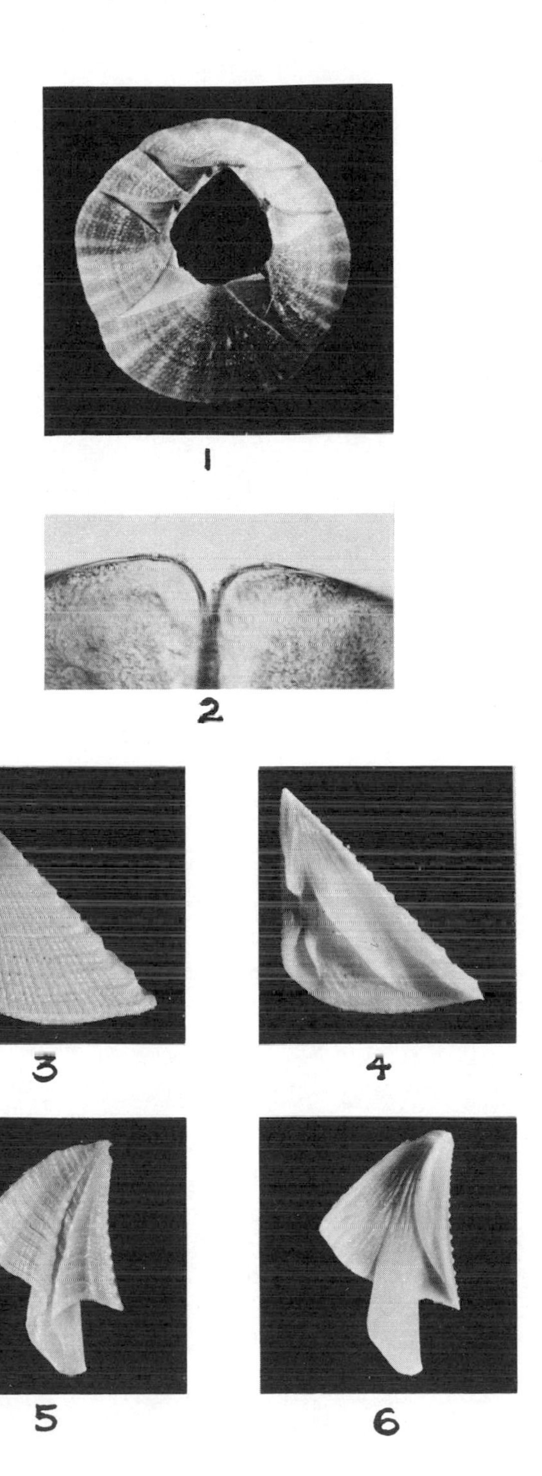

PLATE 3

Balanus improvisus Darwin

(Photographs by Elizabeth A. McGraw)

Figs. 1, 2. Opercular valves, x 5. Potomac River

Fig. 3. Lateral view of another specimen, x 2. Potomac River

Fig. 4. Corroded specimen with smaller, conical specimen on lateral compartment, x 2. Solomons Island, Maryland

Figs. 5, 6. Opercular valves of larger specimen shown in figure 4, x 5

Fig. 7. Terga of young specimens, x 5. Upper row: left, *Balanus amphitrite niveus* Darwin; right, *Balanus amphitrite denticulata* Broch; both from Tampa Bay, Florida. Lower row: left, *Balanus improvisus* Darwin; middle, *Balanus amphitrite pallidus* Darwin, shells shown in figure 8; right, *Balanus eburneus* Gould, Tampa Bay, Florida

Fig. 8. *Balanus amphitrite pallidus* Darwin on *Balanus improvisus* Darwin, x 2. Lake Pontchartrain, Louisiana.

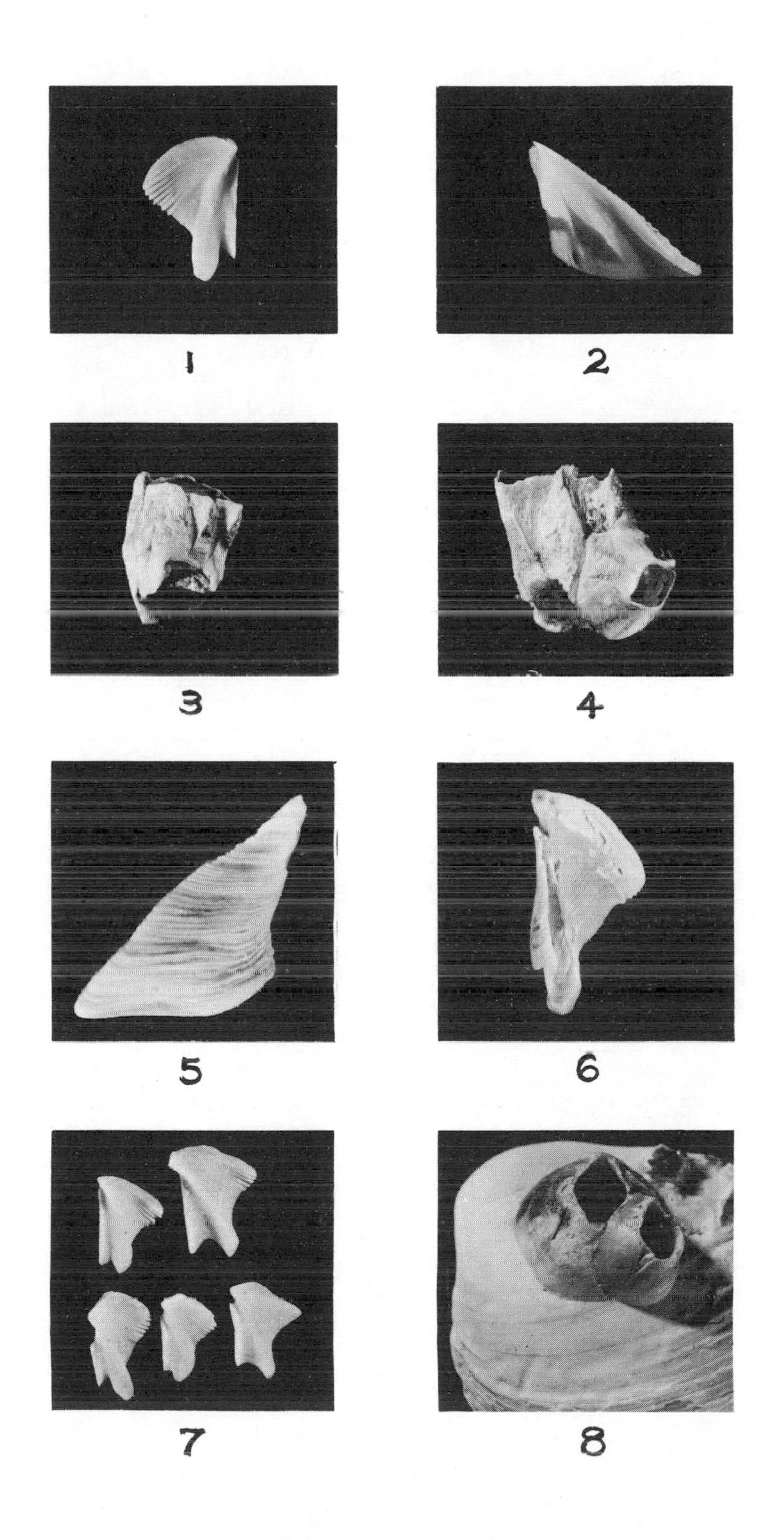
1
2
3
4
5
6
7
8

PLATE 4

Balanus eburneus Gould

(Photographs by John Lincoln)

Figs. 1, 2. Opercular valves of specimen shown in figure 3, x 5

Fig. 3. Lateral view of typical form, x 2. Saint Marks, Florida

Figs. 4, 5. Opercular valves of specimen shown in figure 6, x 5

Fig. 6. Lateral view of form with narrow radii, with young *Balanus amphitrite niveus* Darwin at apex, x 2. Tampa Bay, Florida

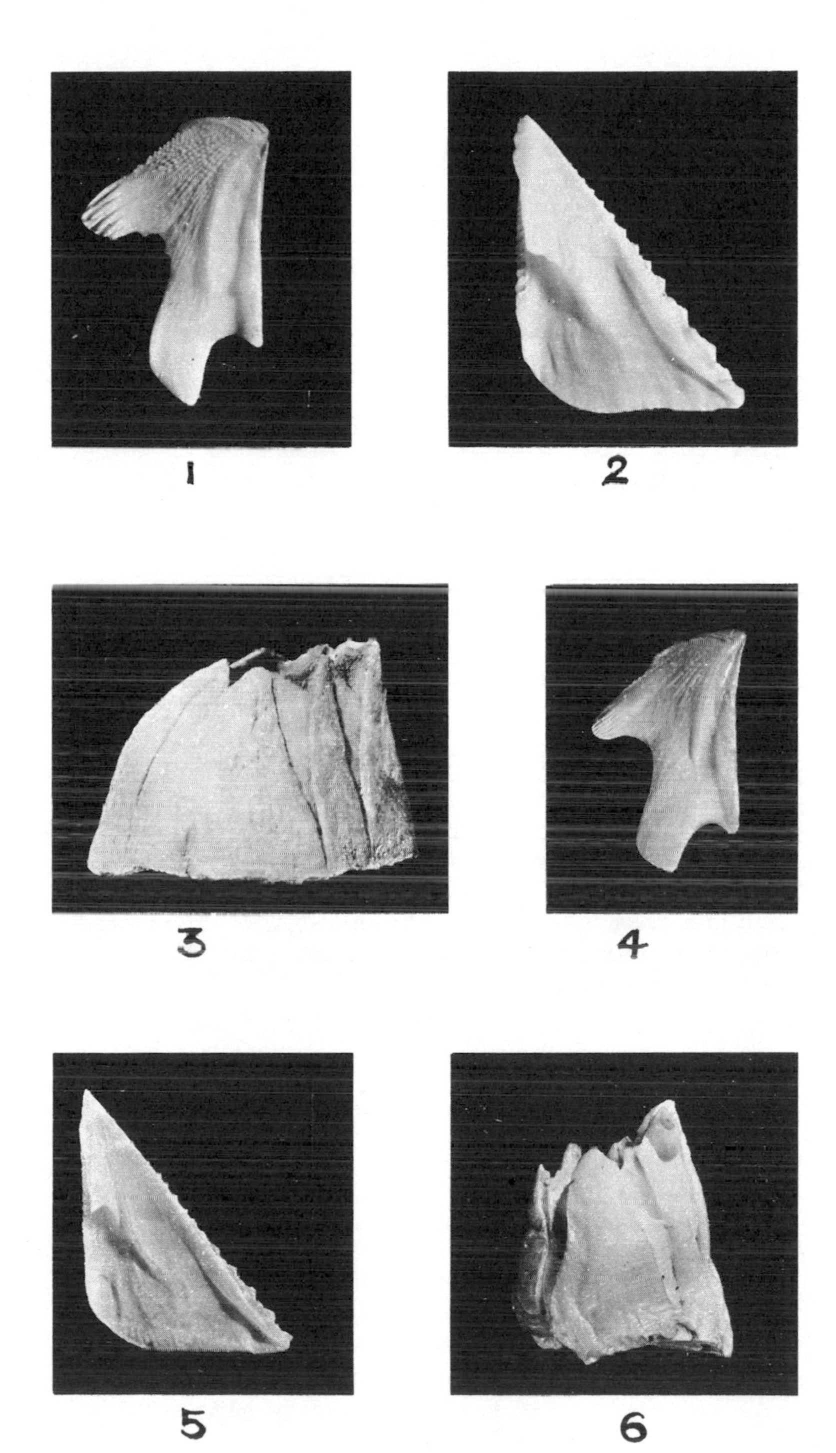

FEEDING, MOLTING, AND GROWTH IN BARNACLES*

C. G. Bookhout and John D. Costlow, Jr.

Most of the information we have today about growth and development of larval and adult barnacles has been based upon field studies. Although these have furnished and will continue to furnish us with a better understanding, it is only through successful rearing of barnacles from the egg through larval stages to the adult that reconstructed life cycles based upon planktonic studies can be verified. Nauplii of different species are similar, and cyprid larvae are almost impossible to separate into species; hence, it is not beyond the realm of possibility that some of the reconstructed life histories and ecological studies reported in the literature have been based upon larval stages of several species rather than one. If rearing studies could be made on all the species in a given area, the larval ecologist could proceed with some degree of assurance. Such studies are also prerequisite to physiological and genetical studies.

Larval Development in the Laboratory

Dr. Costlow and I have been interested in determining the food which would support complete development of acorn barnacles, the number and description of the larval stages, the frequency of molting, duration and mortality of each intermolt, and the length of time required for complete development in the laboratory.

---

*These studies were aided by a contract between the Office of Naval Research, Department of the Navy, and Duke University NR 163-194.

Until recently attempts to rear larvae from unhatched eggs and from naupliar stages have met with limited success. Groom (1894), for example, maintained *Balanus perforatus* nauplii only through the second stage, and Treat (1937) was unable to rear *Balanus balanoides* beyond the third stage. Sandison (1954) was also unsuccessful in maintaining 7 South African barnacles beyond the third naupliar stage (*Balanus algicola, Balanus amphitrite denticulata, Balanus maxillaris, Balanus trigonus, Chthamalus dentatus, Octomeris angulosa, and Tetraclita serrata).* Bassindale (1936) was able to rear *Verruca stroemia* to the cyprid stage by feeding the nauplii on *Nitzschia* sp., but the cyprids failed to settle. By using the same methods he could raise *Balanus balanoides* to the fifth naupliar stage only. Ishida and Yasugi (1937) followed the larval development of *Balanus amphitrite albicostatus*, but the cyprid stage did not settle and metamorphose.

There have been two published reports of barnacles successfully reared from egg to the settled form: that of Herz (1933) for *Balanus crenatus*, and that of Hudinaga and Kasahara (1941) for *Balanus amphitrite hawaiiensis*. All investigators to date have reared larvae in mass culture and have determined the time and stage of molting by daily sampling. While this may indicate the number of larval stages, it does not give accurately the molting frequency or the variations in intermolt periods within the population, nor does it take into account the per cent of mortality.

I should like to discuss briefly our study (Costlow and Bookhout, 1957a) of *Balanus eburneus*, which was selected because it had not been reared in the laboratory nor had its larval stages been described, even though it has a range from Massachusetts to South America (Pilsbry, 1916). Its larval history is known only through the report of Grave (1933), who stated merely that it passed through naupliar, metanaupliar, and cyprid stages in 7 to 10 days at Woods Hole, Massachusetts.

More recently, by the same methods, we have reared *Balanus amphitrite denticulata*, and we plan to rear the other barnacles in the Beaufort area.

Adult *Balanus eburneus* were removed from piles and cleansed of attached organisms. In the laboratory the basis was chipped away and the egg lamellae removed to finger bowls containing filtered sea water with a salinity of 28 ‰, *Chlamydomonas* sp., and 200,000 to 400,000 units of penicillin per liter. The developing ova obtained and successfully reared had attained the distinct median eyespot and gray color which correspond to the H stage of Groom (1894). The bowls were covered and maintained at 26° C in a constant temperature culture cabinet lighted with daylight fluorescent lamps. In order to obtain nauplii of known age, only those which were observed to hatch were used. At the time of removal, each nauplius was placed in a separate compartment of a rearing assembly of 100 wells with a capacity of 1.2 ml per well. The lucite assembly was then placed in a glass dish, covered, and maintained in a culture cabinet. The contents of the wells were checked two or three times per day with a binocular microscope. When an exuvia was found, it was removed and placed in 70 per cent alcohol, and the time, number of the molt, and mortality were recorded for each nauplius. After the second molt, freshly fertilized *Arbacia punctulata* eggs were introduced daily into the compartments in addition to *Chlamydomonas* sp. At this time the larval plutei, developed from the Arbacia eggs of the previous day, were removed.

The naupliar stages were determined from the number of molts under segregated conditions. These were drawn to scale on graph paper with the aid of a Whipple disc mounted in a compound microscope. The drawings were made from exuviae, from fixed specimens, and in a few cases from living nauplii. The setation formulae were obtained from the exuviae and dissected appendages from known stages. In addition to the 121 nauplii reared in individual compartments, hundreds of newly hatched larvae were maintained in finger bowls and plastic compartmented boxes. From these sources specimens were fixed daily in 70 per cent alcohol and Bouins fluid at 60° C. After the stages had been determined from the exuviae of segregated barnacles, the fixed specimens were staged and studied to determine the consistency in appendage setation.

The nauplii of *Balanus eburneus,* reared individually, pass through 6 naupliar stages and 1 cyprid stage, a conclusion which is consistent with Bassindale's (1936) rearing of *Verruca stroemia* and *Chthamalus stellatus* and the reconstructions of larval stages obtained from plankton by Groom (1894), Pyefinch (1948a), Knight-Jones and Waugh (1949), and Sandison (1954) for a variety of barnacles. Our results, Bassindale's (1936), and those based on reconstructions from the plankton do not agree with those obtained by culture methods reported by Herz (1933) and Hudinaga and Kasahara (1941). Herz found that *Balanus crenatus* passed through 8 naupliar stages and 1 cyprid stage; Hudinaga and Kasahara (1941) reported 7 naupliar stages and 1 cyprid stage for *Balanus amphitrite hawaiiensis*. The reason for the discrepancy between staged larvae of *B. crenatus* obtained from the plankton by Pyefinch (1949a) and those raised in mass culture by Herz (1933) is not known. The seventh stage of *B. amphitrite hawaiiensis* found by Hudinaga and Kasahara (1941) differed from the sixth only by the presence of completely developed paired eyes. The setation of the 3 paired appendages was the same for the sixth and seventh stages. Our study of *B. eburneus*, in addition to the reports of Pyefinch (1949a), Bassindale (1936), and Sandison (1954), shows that the change from rudimentary to completely developed paired eyes takes place in the sixth stage.

Setation and spine structure were considered in this study, but the number of naupliar stages was based solely on the exact number of molts through which an individual passed. Since Bassindale's (1936) introduction of the system of setation formulae, it has been applied to nauplii of many species of barnacles. Unfortunately, as asserted by Bassindale, setation formulae alone do not give a definite indication of stage or species. Knight-Jones and Waugh (1949) point out the extreme differences between types of setae and believe the formulae to be misleading. They note, however, a remarkable similarity between the setation of earlier stages of several species. Norris, Jones, Lovegrove, and Crisp (1951) found setation formulae to be of limited value in studies on *Balanus perforatus, B. improvisus, and B. amphitrite denticulata* and suggest that setation may be a developmental feature rather than a specific one for the separation of barnacle larvae. Pyefinch (1949a), however, examined thousands of staged nauplii of *Balanus crenatus* and found setation to be uniform within each stage. We (Costlow and Bookhout, 1957a) arrived at similar conclusions from the study of the development of *B. eburneus* under controlled conditions. We also found that the setation of the second to the sixth stages of *B. eburneus* is different from that described for other species. Since nauplii of different species of barnacles are known to have the same setation formulae and they may not be consistent in each stage of one species, we recommend that setation studies continue to be made but be used only as one of the characters for identification. By feeding varying amounts of the same diet to nauplii of known species and age, one could soon tell if development of the body and of setae in particular was speeded up or retarded. Now that barnacles can be reared in the laboratory, there is no reason why genetic experiments could not be used to determine which setae are constant for each stage and which are not.

Knight-Jones and Waugh (1949) used size as an initial means of establishing the six naupliar stages and corroborated it with setation analyses. Table II (Costlow and Bookhout, 1957a) gives the range in size of the 6 naupliar stages and 1 cyprid stage of *B. eburneus*. With setation and observable internal development, such as paired eyes and cirriform appendages, the carapace width and length would be more reliable than total length for laboratory-reared larvae and presumably also for planktonic material.

By rearing each nauplius in a separate compartment, we could determine the variation in duration of the intermolt periods for the six naupliar stages. The first naupliar stage had a duration of 15 minutes to 4 hours; the second from 1 to 2 days with an average of 1 day; the third from 1 to 4 days with an average of 1.5 days; the fourth from 1 to 4 days with an average of 2 days; the fifth from 1 to 5 days with an average of 2.6 days; and the sixth stage, 2 to 4 days with an average of 2.5 days. The cyprid stage ranges from 1 to 14 days, but successful attachment was observed only in those which settled within 1 to 3 days following the final naupliar molt. The over-all time of development in the laboratory ranges from 7 to 13 days.

One hundred per cent of the first-stage nauplii underwent ecdysis from 15 minutes to 4 hours after hatching whether food was available or not. The second-stage nauplii molt from 10 to 35 hours after hatching. During this stage the gut was green with *Chlamydomonas* sp. The third-stage nauplii may molt from the second to the fifth days. This is the stage when nauplii may die if proper conditions are not present. Therefore, it was at this time that we added fertilized Arbacia eggs to furnish food of animal origin. Even so, there was a mortality of 6 per cent. The fourth molt takes place from the third to the eighth days with the majority on the fourth day. It is accompanied by an increase in mortality. The molting of the fifth naupliar stage occurs from the fifth to the ninth days, with the majority on the seventh day. Mortality at this stage was about 33 per cent, greater than at any other. The sixth naupliar stage molts into a cyprid from the seventh to the twelfth days with a mortality of 22 per cent.

Under laboratory conditions we found that those cyprids which settled and underwent metamorphosis to the sessile stage did so within 1 to 3 days, whereas those that persisted as cyprids for longer periods, up to 14 days, failed to settle and died. Mortality in this stage was approximately 16 per cent.

The over-all time of development for *B. eburneus* is quite short when compared with the times given for other species. Unfortunately, comparison between species is not too reliable for the effect of temperature on individual species has not been determined and can only be inferred from studies on other species at different temperatures.

There is always the question of normality when organisms are reared in the laboratory, and of how survival compares with that in nature. In this study if a barnacle completed development, settled, and metamorphosed it was considered normal. In nature survival of barnacle larvae has been estimated by Pyefinch (1949b) to be between 1 and 9 per cent, and by Bousfield (1955) to be 10 per cent. Our observed survival of 16.3 per cent under laboratory conditions is higher than that estimated for larval barnacles in nature. It is undoubtedly true that the chief sources of mortality in nature are dispersal seaward and predation, whereas improper food and bacteria are the chief causes in the laboratory.

Growth and Molting of Attached Acorn Barnacles

The Cirripedia differ from other arthropods in that the body is enclosed by calcareous plates which are not shed with the chitinous layers of the animal body and the inner lining of the mantle. As early as 1854 Darwin reported that members of the family Balanidae and Chthamalidae underwent periods of exuviation. He reports that Mr. W. Thompson of Belfast kept 20 specimens of *Balanus balanoides* alive, presumably in the laboratory, and on the twelfth day all had molted once and one had molted twice. Thomas (1944) stated that *Balanus perforatus* molted every fifth day but did not give the age or the length of time the organisms were observed. Aside from these limited observations, little was known before 1953 concerning the relationship between growth of shell, ecdysis, and frequency of molting in any barnacle.

In order to answer these questions we had to find a method of rearing barnacles in the laboratory. Hence a series of experiments was run to determine if unicellular algae would support growth of barnacles from the time of settling over an extended period of time. It was found that *Balanus improvisus* could be reared on unialgal cultures of *Nitzschia closterium* or *Chlamydomonas* sp. or a mixture of these. *B. improvisus* exhibited a more rapid growth in a medium containing *Chlamydomonas* sp.

alone than in a mixed culture of *Nitzschia closterium* and *Chlamydomonas* sp. Although barnacles grown in unialgal cultures exhibited continuous growth, it was less than that found in the harbor. Barnacles maintained in the harbor and in cultures showed a slow initial increase in basis size and then a gradual increase which became more rapid, followed by a period of reduced growth. Even though the unialgal cultures did not maintain a growth rate equal to that under natural conditions where the food supply was more varied and presumably of greater magnitude, the resulting rate under artificial conditions was continuous, and the food source was considered adequate since mortality under these conditions amounted to only 0.5 per cent. The factors of social competition, crowding, and predation, which appear to account for high mortality observed in nature, were eliminated by complete segregation of individual barnacles in the laboratory. Recently we have found that animal food alone (i.e., Arbacia eggs) or in combination with *Chlamydomonas* sp. supports growth of older barnacles to a greater degree than unialgal cultures alone. Further nutritional studies should be made and refined to determine which sources of food will or will not support optimum growth. We do not know the essential foods necessary to maintain growth, or the role played by vitamins and trace elements.

The fact that *Balanus improvisus*, if it is not fed after it settles, will pass through one molt, but not two, implies that the animal possesses food which is carried over from the cyprid to the sessile barnacle. *Balanus amphitrite niveus,* on the other hand, must be fed or it will not pass through even the first molt.

Our experiments (Costlow and Bookhout, 1953) have shown that recently settled barnacles as well as large ones undergo frequent and regular ecdysis if they are fed. *Balanus improvisus* has been followed through 42 consecutive molts, extending over approximately 120 days, and the 2- to 3-day molting rate remained constant. Barnacles maintained on *Chlamydomonas* sp. alone showed more uniformity in molting than those fed on *Nitzschia closterium* alone or on a mixture of Nitzschia and Chlamydomonas. It is our belief that this is associated with the motility of *Chlamydomonas* sp., making it more readily available to the barnacles, whereas the diatom *Nitzschia closterium* tended to clump and settle to the bottom, becoming less readily available even though the culture medium was changed daily.

Numerous studies of the growth of various species of barnacles have been made in different parts of the world. However, the literature contains no reference to growth rates of barnacles raised in the laboratory. This phase will now be considered.

From daily measurements of rostrocarinal and lateral diameters, the daily increase in area of basis was computed. One might expect, as we did, that shell growth would be correlated with the molting cycle. However, we found no indication at the macrolevel of any correlation between shell growth and molting. Growth on the day of certain molts is less than on days of intermolt, but at other times in the same barnacle there is more growth at the day of molt than before or afterward. From this evidence it must be concluded that shell growth in *Balanus improvisus*, although continuous, is erratic. Since individual growth rates did not show a definite trend, it was thought that an average of a group of barnacles might. The results, shown by a histogram (Costlow and Bookhout, 1957b), reveal that there is no correlation between shell growth and the molting cycle. There is a general increase in the per cent of areal growth from the second molt through the ninth, and thereafter there is a gradual decline. In a histological analysis of shell-forming tissues in *Balanus improvisus*, Costlow (1956) looked for but found no correlation between the molting cycle and the cyclic activity of three cell types which form the various components of the shell.

Since *Balanus improvisus* molts more frequently than was suspected by Darwin (1854) and Thomas (1944), and is primarily a winter barnacle at Beaufort, North Carolina, it may or may not be typical of other species. Therefore, it was deemed advisable to make similar studies on what we thought was *Balanus amphitrite niveus* (now considered to be *Balanus amphitrite denticulata* by Dr. Dora Henry), a barnacle that settles during the summer and is commonly found in the intertidal region at Beaufort. The molting frequency, period of intermolt, and relationship of shell growth to the molting cycle were essentially the same as those found for *Balanus improvisus*. In spite of the uniform conditions provided by the culture cabinet, mortality in *Balanus amphitrite niveus* was high, even though the barnacles were maintained on the same diet that furnished the best growth in *B. improvisus* with practically no mortality (0.5 per cent). The question arose as to whether death could be due to intrinsic conditions, a particular time in the life cycle, or a specific period within the molting cycle. An examination of the graph in Costlow and Bookhout (1956) indicates that the greater percentage of mortality occurs during the first two molting periods, followed by a steady decrease in succeeding molts. Mortality of *Balanus amphitrite niveus* in the harbor is also very high, especially during the first week after settling. It may, therefore, be due to some genetic or environmental factors.

Costlow and Bookhout (1953) reared *Balanus improvisus* under continuous light without inquiring whether other light conditions might alter molting frequency. In 1956, however, they maintained several series of *Balanus amphitrite niveus* under three conditions of light: continuous light, continuous darkness, and 12 hours of light alternating with 12 hours of darkness. From an examination of the figure in Costlow and Bookhout (1956) it can be seen that each series exhibited essentially the same molting frequency. When accumulated daily areal increase of basis is plotted, it is noted that shell growth in continuous light was less than growth in continuous darkness. The growth under laboratory conditions, however, was never as rapid as that observed for barnacles grown in the harbor.

Body Growth versus Shell Growth in Balanus improvisus

The acorn barnacles are unique among the arthropods in that the body is permanently enclosed by, and separated from, the outer shell of calcareous plates. The rate of shell growth has been studied for several species of barnacles under both natural and experimental conditions. Barnes (1955) described the growth rate of *Chthamalus stellatus*, and Barnes and Powell (1953) studied the effect of varying conditions of submersion on shell growth in *Balanus balanoides* and *Balanus crenatus*. Crisp (1954) described morphological changes in the shell which are associated with differences in yearly growth rates of *Balanus porcatus*. The relationship between the daily increments of the continuously growing shell and the growth rate of the body, following a series of consecutive molts, is not known for any of the acorn barnacles. In order to obtain data over a series of consecutive molts of the same individual, it is theoretically possible to measure the exuvia, or some portion of it, following ecdysis. Since the shed exoskeleton tends to wrinkle, accurate measurements of the total length or width are not possible. The mandibles and maxillae, however, are sufficiently rigid to retain their shape and size and, as will be demonstrated, reflect the per cent increase of the entire body at molting.

The objectives of our study (Costlow and Bookhout, 1957b) were threefold: first, to determine whether or not body growth always accompanies an ecdysis; second, to follow the increase in body size, as represented by two mouth parts, through consecutive molting periods; and, third, to compare the relative increase in size of the body with that of the shell.

Balanus improvisus which had metamorphosed from the cyprid during the previous 24 hours were maintained individually in plastic compartmented boxes containing *Chlamydomonas* sp. at 25°C under daylight fluorescent lights. Each compartment was examined twice daily for the presence of molts. When exuviae were found they were mounted on glass slides, and the maxillae and mandibles were removed. The width of the mouth parts, at the base of the spines or teeth, was then measured with an ocular micrometer mounted in a compound microscope.

Daily measurements were also made of the barnacle shell, rostrocarinal length, and lateral width, using an ocular micrometer mounted in a dissecting microscope. The barnacles were observed and measured over a period of 60 days, an average of 20 molting periods.

A second series of twenty barnacles, of varying size and age, was obtained from the harbor. The entire body was removed from the shell and the width of the thorax, maxillae, and mandibles was measured. The thorax was measured to determine if, in living barnacles, a relationship exists between the size of the mouth parts and a major body dimension, or if growth of the mouth parts is differential.

The results obtained from measurements on whole bodies and mouth parts of 20 living barnacles taken at random from the harbor indicate that there is a definite correlation between the width of the thorax and the width of the maxillae. A ratio of 6.271 ± 0.104 was found in barnacles of various sizes and ages. A simple correlation coefficient of 0.922, significant at the 1 per cent level, was determined from the sample. This value exceeds that given by the standard tables at 18 degrees of freedom and indicates that 85 per cent of the variability in thorax size is associated with concomitant variability in maxillae size, leaving 15 per cent attributable to other factors including chance. Thus maxillary size, as well as per cent increase at molting, is not differential but reflects the increase in size of the body.

The mandibles, although they are more difficult to measure accurately, are approximately twice the width of the maxillae. This ratio is also maintained in barnacles of varying size and further supports the hypothesis that the size increase of the maxillae is not differential but reflects the size increase of the entire body.

A comparison of the per cent increase in size of the maxillae and mandibles of *Balanus improvisus* was made from the second to the twentieth molt. It was found that, if the mandible increases greatly following one molt, the maxillae increase approximately the same per cent, and, similarly, if one shows no increase at molting, no increase is found in the other. An increase in size of the maxillae does not necessarily accompany molting. There are several periods, ranging from 1 to 3 consecutive molts, when no measurable increase in size of maxillae occurred. Molting without growth was observed in each of the 20 barnacles studied, although not necessarily at the same time or molt. This was true even though the same quantity of food was supplied each day.

In contrast to the body, which may grow only at the time of molting, the calcareous plates of the shell grow continuously. As indicated previously, there were periods up to 3 molts when no increase in maxillae occurred, but during this same interval of time the shell continued to grow.

If the average increase in size of maxillae and shell for 20 barnacles is studied, it will be noted that the initial ratio between the maxillae and shell sizes is not maintained over any great length of time. As the body can grow only by molting, it might be expected to show an increase in size which would correspond with shell growth for that particular period. The per cent growth of maxillae, however, rarely equals the per cent accumulated shell growth during the intermolt period. If this trend were to continue, it would result in a large shell enclosing a relatively small body. The changing ratio is compensated for in part by changes in the opercular plate level at the time of molting. In a newly set barnacle the opercular plates occupy an extreme apical position. At each molt a new opercular hinge is secreted and the opercular plates gradually become more basal. The increase of relative shell size is accompanied by an enlarged mantle cavity which, at maturity, provides space for the egg lamellae. Although actual figures are not available, Bousfield (1955) has estimated that *B. improvisus* retains from 1,000 to 10,000 eggs during each breeding period. If the initial ratio between body and shell size were maintained, there would be insufficient space for eggs or larvae.

Some Problems for the Future

By use of the rearing techniques now available for larvae and adults, and by employment of others which will be perfected in time, physiological, genetic, and toxicity studies may be made. When large numbers of larvae and adults are available in the laboratory, the toxicity of various poisons can be investigated. It will also be of interest to subject breeding adults containing eggs and embryos to the same poisons to determine if larval development and settling will follow, and, if so, how they will be altered.

From field studies and from the geographical range given for various species of barnacles, we assume that larvae can exist over wide ranges of temperature and of salinity or that there are different physiological races. The breeding season in each area is, however, limited to a small portion of the year and may be governed in part by physical factors of the environment. Thorson (1950) states that the physiological limits of the larvae of marine forms are more closely confined than those of adults. To date all experimental work has been confined to one or two larval stages of the barnacle obtained by hatching eggs in the laboratory or from stages in the plankton. Therefore, we are particularly interested in determining the effect of controlled variations in temperature, salinity, light, and diet on the number of stages, the frequency of molting, duration of the intermolt periods, growth, and over-all time of development. The proposed program would lay a foundation for additional studies on the physiology and ecology of barnacles.

Knowles and Carlisle (1956), in their excellent review on endocrine control of crustacea, state that nothing is known about the endocrines of entomostraca to which Cirripedia belong, and little is known about the malacostraca except for the decapods. Whether entomostraca are affected by the hormones of the decapods is an open question which we hope to answer. We also plan to make a thorough histological analysis of larvae and adult barnacles to determine the location of secretory areas associated with the nervous system or glands which might be endocrine organs and affect the molting cycle. It is possible that barnacles have an endocrine gland comparable to the Y-organ of decapods which Gabe (1953) has found to secrete a molt-accelerating hormone. It is also reasonable to assume that there is another secretory area in barnacles which secretes a molt-inhibiting hormone, as has been described in decapods. If these areas can be found, and suitable extracts made, then the hormone can be added to the culture media of other barnacles to determine what effect it may have on the molting cycle and on other physiological functions.

LITERATURE CITED

Barnes, H. 1955. The Growth Rate of *Chthamalus stellatus* (Poli). J. Mar. Biol. Assoc, U.K., 35:355-61.

-------, and H. T. Powell. 1953. The Growth of *Balanus balanoides* (L.) and *Balanus crenatus* Brug. under Varying Conditions of Submersion. J. Mar. Biol. Assoc. U.K., 32:107-28.

Bassindale, R. 1936. The Developmental Stages of Three English Barnacles, *Balanus balanoides, Chthamalus stellatus,* and *Verruca stroemia*. Proc. Zool. Soc. Lond., 1:57-74.

Batham, E. J. 1945. *Pollicipes spinosus* Quoy and Gaimard. II, Embryonic and Larval Development. Trans. Roy. Soc. New Zealand, 75:405-18.

Bousfield, E. L. 1955. Ecological Control of the Occurrence of Barnacles in the Miramichi Estuary. Nat. Mus. Can., Bull. 137, Biol. Ser. 46.

Costlow, John D., Jr. 1956. Shell Development in *Balanus improvisus* Darwin. J. Morph., 99:359-416.

-------, and C. G. Bookhout. 1953. Moulting and Growth in *Balanus improvisus*. Biol. Bull., 105:420-33.

-------, and C. G. Bookhout. 1956. Moulting and Shell Growth in *Balanus amphitrite niveus*. Biol. Bull., 110:107-16.

-------, and C. G. Bookhout. 1957a. Larval Stages of *Balanus eburneus* in the Laboratory. Biol. Bull., 112:313-24.

-------, and C. G. Bookhout. 1957b. Body Growth versus Shell Growth in *Balanus improvisus*. Biol. Bull., in press.

Crisp, D. J. 1954. The Breeding of *Balanus porcatus* (Da Costa) in the Irish Sea. J. Mar. Biol. Assoc. U.K., 33:473-96.

Darwin, Charles. 1854. A Monograph on the Sub-class Cirripedia. Ray Society, London.

Gabe, M. 1953. Sur l'existence, chez quelques crustacés malacostracés, d'un organe comparable à la glande de la mue des insectes. C. R. Acad. Sci. Paris, 237:1111.

Groom, T. T. 1894. On the Early Development of the Cirripedia. Phil. Trans. Roy. Soc., 185:121-208.

Grave, B. H. 1933. Rate of Growth, Age at Sexual Maturity, and Duration of Life in Sessile Organisms at Woods Hole, Mass. Biol. Bull., 65:375-86.

Herz, L. E. 1933. The Morphology of the Later Stages of *Balanus crenatus* Bruguiere. Biol. Bull., 64:432-42.

Hudinaga, M., and H. Kasahara. 1941. Larval Development of *Balanus amphitrite hawaiiensis*. Zool. Mag. (Japan), 54:108-18.

Ishida, S., and R. Yasugi. 1937. Free Swimming Stages of *Balanus amphitrite albicostatus*. Botany and Zoology (Tokyo), 5: 1659-66.

Knight-Jones, C. W., and D. Waugh. 1949. On the Larval Development of *Elminius modestus* Darwin. J. Mar. Biol. Assoc. U.K., 28:413-28.

Knowles, F. G. W., and D. B. Carlisle. 1956. Endocrine Control in the Crustacea. Biol. Rev., 31:396-473.

Norris, E., L. W. G. Jones, T. Lovegrove, and D. J. Crisp. 1951. Variability in Larval Stages of Cirripedes. Nature, 167: 444-45.

Pilsbury, H. A. 1916. The Sessile Barnacles (Cirripedia) contained in the Collections of the U.S. National Museum: Including a Monograph of the American Species. U.S. Nat. Mus. Bull., 93.

Pyefinch, K. A. 1948a. Methods of Identification of the Larvae of *Balanus balanoides* (L), *Balanus crenatus* Brug., and *Verruca stroemia* O. F. Müller. J. Mar. Biol. Assoc. U.K., 27: 451-63.

-------. 1948b. Notes on the Biology of Cirripedes. J. Mar. Biol. Assoc. U.K., 27:464-503.

-------. 1949a. The Larval Stages of *Balanus crenatus*. Zool. Soc. London, 118:916-23.

-------. 1949b. Short-Period Fluctuations in the Numbers of Barnacle Larvae, with Notes on Comparisons between Pump and Net Plankton Hauls. J. Mar. Biol. Assoc., U.K., 28: 353-69.

Sandison, Eyvor E. 1954. The Identification of the Nauplii of Some South African Barnacles with Notes on Their Life Histories. Trans. Roy. Soc. S. Africa, 34:69-101.

Thomas, H. S. 1944. Tegumental Glands in Cirripedia Thoracica. Quart. J. Micro. Sci., 84:257-82.

Thorson, Gunnar. 1950. Reproductive and Larval Ecology of Marine Bottom Invertebrates. Biol. Rev., 25:1-45.

Treat, D. A. 1937. A Comparative Study of Barnacle Larvae. M.A. thesis (unpubl.), Dept. Biol., Western Reserve Univ.

STUDIES OF SOME FACTORS AFFECTING THE RECRUITMENT AND MORTALITY OF NATURAL POPULATIONS OF INTERTIDAL BARNACLES

J. H. Connell

The distribution and abundance of intertidal barnacles at any one place is determined by settlement and mortality, since immigration is impossible. I studied these aspects of the life history of a natural population of *Balanus balanoides* (L.) at Millport, Scotland, from 1952 to 1955. In the following general discussion, reference will be made to a detailed account of this study which is being prepared for publication.

Settlement

At Millport, small pieces of a basaltic stone, Bostonite, were attached to the intertidal rock by means of stainless steel screws. At low tide the stones were brought into the laboratory and the positions of newly attached cyprid larvae were mapped. At this time the metamorphosis or death of previously settled barnacles was also noted. The stones were always returned to the shore before the tide rose again, and when in the laboratory were kept outside on a window ledge except during the brief period when they were being examined. Thus the conditions were little different from those during a normal low tide. On one stone the cyprid mortality during the whole settlement season was only 5 per cent, indicating that the daily handling did not markedly increase mortality. Dr. C. G. Bookhout has informed me that attached barnacles can be shifted along smooth surfaces such as glass. Since my method for determining mortality depends on the immobility of barnacles, this shifting would cause an error. However, it is doubtful whether such movement occurred on the rough surface of the stones used.

In 3 successive spring settlement seasons, the settlement attained a density of 70 per sq cm on some of the stones. Small areas of each stone were kept clear by picking off the cyprids with a needle at each examination. On these cleared areas the settlement continued for several weeks after the maximum density had been reached on adjacent areas. Thus a great oversupply of cyprids had been produced in these years. Although variations between adjacent areas in the density of settlement at the end of the settlement season occurred, these variations could not be explained by a shortage of cyprids in the plankton. That this is not an isolated case is suggested by previous work. Hatton (1938), in his tables 12 to 15, gave the density of settlement in 2 or 3 successive years at St. Malo, France, for *Balanus balanoides* at several levels and locations. Annual variations at any one level and location were very slight. As at Millport, in no one year was the settlement consistently low on all the areas studied. Some variations in the planktonic phase occurred, since in one year the settlement started 3 weeks later than normal and lasted only 6 weeks, while in the next year the settlement season was 9 weeks long. The densities of settlement at the end of both seasons were similar, however.

Bousfield (1955) reported that the number of cyprids of *Balanus improvisus* Darwin surviving after the planktonic phase was 17 per sq inch; since only 5 per cent of the bottom was suitable for attachment, this meant that 340 cyprids were available per sq inch of suitable substratum. This was much greater than any densities actually reached on the substratum during or after settlement.

Barnes (1956), studying annual fluctuations in the planktonic development of *Balanus balanoides* at Millport, showed that, in 4 of the 9 years studied, the earliest group of larvae to be liberated failed to develop, so that few cyprids derived from this group were found in the plankton. Barnes stated that later liberations of larvae might develop normally in these years. One of the years when the early liberation failed was 1954, when the settlement occurred a month late on my experimental stones. The final density attained, however, was over 70 per sq cm, so that failure of part of the planktonic larvae to develop was not reflected in decreased settlement. Of the other three "failure" years, no quantitative data on settlement were given; in one of these years, 1946, the settlement was reported as "absent over a considerable area" by Pyefinch (1948). Thus, with the possible exception of the 1946 settlement at Millport, these studies indicate that the settlement density attained was not a function of the planktonic supply of larvae, probably because of a huge oversupply of cyprids. However, instances have been reported when the settlement may have been limited by the supply of planktonic cyprids. Hatton (1938) noted that the settlement in calm areas was much less than in wave-beaten places, possibly because the increased circulation brought more cyprids to the latter areas. Bousfield (1955) showed that the swimming activity of planktonic larvae determined the distribution of the adults in a large estuary. Finally, in certain years, as in 1946 at Millport, or in places where adults are few, as at the edge of a species' range, the planktonic phase might limit the density of settlement.

In such areas as Millport, where there was an oversupply of planktonic cyprids, variations in settlement density still occurred. These differences must have been due either to variations in the factors affecting initial settlement on the surface or to variations in mortality after settlement. With respect to the former, Crisp (1955) has shown that cyprids will attach more readily in moving than in still water; Barnes and Powell (1953) have suggested that settlement is stimulated by the draining of thin films of water. Many workers have noted that barnacles will first settle in grooves or hollows in the surface in preference to convexities. Knight-Jones (1953) has shown in a convincing series of experiments that the presence of previously settled individuals stimulates cyprids of the same species to settle.

With regard to mortality after settlement, Hatton (1938) observed that heat and/or desiccation cause great mortality; the settlement of *Balanus balanoides* at St. Malo was greater on north-facing slopes. He found that young barnacles could endure up to 6 days' emersion before they were all killed. By fixing a basin containing sea water above the level of the high-water spring tides and allowing the water to drip out through a tiny hole, he found that young barnacles, when kept moist, could exist above the level of the adults. In my work at Millport I found that, during an unusually warm period, cyprids suffered heavy mortality, but metamorphosed barnacles, only 3 to 5 days after attachment, withstood this weather. During gales, both groups suffered great mortality. By analyzing the mortality in hollows and on adjacent convexities, I found that in gales the barnacles on the convexities suffered damage, while those in hollows did not. Since the cyprids that arrive early settle in the hollows, the rate of survival of early settlers was much higher than that of later ones.

Hatton (1938) observed that the limpet *Patella vulgata* L. destroyed many young barnacles. This was also noted by Lewis

(1954) in Wales and by me at Millport. Other animals, such as fish, crabs, and birds, have been reported to feed on barnacles, but there is little evidence as to whether they prefer young barnacles. As will be shown later, the snail, *Thais lapillus* (L.), does not feed on young barnacles if older ones are present. The presence of algae affects settlement; Hatton (1938) found that the settlement on bare rock under large algae was less than on adjacent areas from which the algae had been removed. From all these studies it is clear that events occurring on the shore during settlement may be responsible for much of the variation in density observed at the end of the settlement season.

Mortality after Settlement

After the settlement becomes dense, growth causes the barnacles to touch each other. This results in various effects, collectively called crowding. Hatton (1938) noted that older barnacles displaced younger ones, and that, if two young settled adjacently, one was often displaced. He also observed that a young barnacle attached to an older one grew over the orifice and thus smothered the older one.

At Millport I observed all these occurrences. The most intense crowding was found between barnacles in the first months after settlement. In adjacent populations at the same level, the mortality was almost always higher at higher population densities. Where growth rates were high, as at low shore levels and in wave-beaten places, the barnacles grew into tall, trumpet-shaped forms when they were crowded. This was noted by Darwin (1854) and by many subsequent workers. Since the barnacles were attached only at the small base, they fell at the slightest shock, as noted by Hatton (1938). Barnes and Powell (1950) have described instances at Millport where entire areas have been stripped bare in this way; similar destruction was witnessed in each of the 3 years of my own study there. I have observed similar growth forms in *Chthamalus stellatus* Poli from southwest Ireland, in *Balanus balanoides* at Woods Hole, Massachusetts, and in *Balanus glandula* Darwin at Santa Barbara, California, and San Juan Island, Washington; it is, evidently, a common and widespread occurrence.

The mortality caused by this intraspecific crowding does not occur in every species. Hatton (1938) never observed displacement of one individual by another in *Chthamalus stellatus*. My observations of the same species at Millport support this except in a few instances in dense populations of young. Chthamalus grows more slowly than Balanus (cf. Hatton, 1938), and the settlement accumulates more gradually over a longer period.

Barnacles are often crowded and smothered by other species. Weiss (1948) observed tunicates and encrusting bryozoa growing over barnacles on panels suspended from a raft in Florida.

From the results of these studies of intraspecific crowding among barnacles, it is evident that a heavy settlement may subsequently suffer great mortality. In such cases fewer individuals, perhaps with a more unstable body form, may survive to maturity than from a lighter settlement.

Predation by *Thais lapillus* (L.)

Of the many reported predators of barnacles, few have been studied in detail. Probably the most common predators of intertidal barnacles in the northern hemisphere are gastropods of the genus Thais. In the North Atlantic, *Thais lapillus* is common on rocky shores, feeding mainly on mussels and barnacles. (This snail has been referred to as Nucella and Purpura by various authors.) At Millport and at Woods Hole I studied the effect of this whelk on populations of *Balanus balanoides*.

Some observations of *Thais lapillus* marked individually at Millport may be summarized as follows. During the summer they were actively feeding about half the time, while in winter they fed much less frequently. By placing whelks inside cages, I found that in summer they ate barnacles at the rate of about 1.0 to 1.3 per day, regardless of the size of the barnacle; in winter the feeding rate was about 0.4 per day. Great numbers occurred at mid-tide level and moderate numbers below, and, as observed by Moore (1938), they became much less abundant some distance below the top of the zone of *Balanus balanoides*. They often fed in groups and when not feeding were to be found densely packed in cracks or crevices; this was especially noticeable after a gale, when the whole population of whelks would be aggregated in sheltered places.

By attaching cages of stainless steel wire netting over small areas of barnacles, we could protect these populations from attack by Thais. Above mid-tide level during the first year after settlement, the mortality rates of protected and unprotected barnacles were about the same. In the second year, the mortality rate was much less for protected barnacles but continued high in areas open to predation. From this evidence it appeared that, when the whelks were feeding on populations having at least 2 year-groups of barnacles, they were ignoring the younger group.

At low shore levels, only a small proportion of the barnacles survived into the second year. At these levels, whelks began feeding on the younger year class about 6 weeks after settlement. By observing individual Thais feeding at low shore levels it was found that about 4/5 of them were feeding on barnacles 14 months of age, even though these older barnacles covered only about 1/5 of the area, and the rest was occupied by barnacles settled 2 months earlier. In addition, Thais feeding on the young year-group were selecting the larger of these as prey.

Thus direct observation supported the finding from experiments with cages that excluded predators: *Thais lapillus* was selecting larger barnacles as prey. If the result described previously is true, namely, that Thais can open only a certain number of barnacles per day regardless of the size of barnacle, then selection of larger barnacles would be advantageous.

In France, Hatton (1938) found that at the level of the high-water neaps the mortality rate after the age of 1 year was very low. My observations at Millport confirm this. Moore (1934) noted that the barnacle population at high shore levels was composed of a high proportion of large older barnacles, presumably of several year-classes. At low shore levels, these authors found that most of the barnacles were dead at the end of 2 years. At Millport, the age structure characteristic of high shore levels was reproduced at about mid-tide level on the areas protected from predation by cages. Since Thais does not occur at the upper levels of Balanus distribution, it is probably safe to say that the observed distribution of age structure of the *Balanus balanoides* population at Millport is determined in great part by predation, particularly by *Thais lapillus*.

One interesting side effect of the discovery of the selection of large-sized barnacles by Thais is seen in relation to some of the growth studies of *Balanus balanoides* made earlier. Moore (1935, 1936) noted at the Isle of Man and Plymouth, England, that in the first year of life the larger barnacles occurred at the lowest shore levels, but in the second year and later the larger ones were found sometimes at lower levels (in wave-beaten or muddy situations) and sometimes at high levels (in more protected or clearer waters). Hatton (1938) followed the growth of groups of barnacles for 3 years at various shore levels in various situations. The sizes during the first season were always greater at low levels; in the second season, however, the larger barnacles were found at the lower levels only in the most wave-beaten areas, whereas in protected areas the larger barnacles were found at the high levels. Barnes and Powell (1953) found that on panels attached to the pier at Millport the largest average size occurred at the lower levels at the end of the first and second growth seasons.

Moore (1935) observed that *Thais lapillus* was absent or uncommon where heavy sediment or great wave action occurred. In both his and Hatton's growth studies, the localities where these conditions obtain were also those where the larger barnacles were found at low levels in the second season. Thus where Thais was uncommon or, as in Barnes and Powell's study, absent, the barnacles at the lower levels were consistently larger. However, at places where Thais was more common, the larger average size occurred at the high-water neap level. Since Thais occurs only below high-water neaps, and selects the larger barnacles, the contradictory results obtained by Moore and Hatton might be explained by the action of Thais. In such studies of growth under natural conditions, it is necessary to follow individual animals in order to ascertain whether the inevitable mortality that occurs is random with respect to size.

Conclusions

From these studies it is evident that the distribution and abundance of intertidal barnacles is determined by many forces, interacting in often unexpected ways. Settlement may be affected by events both in the planktonic phase and on the shore; in my studies at Millport, the latter seem more important. Mortality in the first weeks of life results mainly from the action of the physical environment. Biological interactions, such as the crowding between barnacles which is most severe in the first year, and predation, which may fall heavier on the older groups, occurs mainly below the level of the high-water neaps. Once past the first year, barnacles at high levels may live for a yet undetermined number of years.

LITERATURE CITED

Barnes, H. 1956. *Balanus balanoides* (L.) in the Firth of Clyde: The Development and Annual Variation of the Larval Population, and the Causative Factors. J. Anim. Ecol., 25:72-84.

-------, and H. T. Powell. 1950. The Development, General Morphology and Subsequent Elimination of Barnacle Populations, *Balanus crenatus* and *B. balanoides*, After a Heavy Initial Settlement. J. Anim. Ecol., 19:175-79.

-------, and H. T. Powell. 1953. The Growth of *Balanus balanoides* (L.) and *Balanus crenatus* Brug. under Varying Conditions of Submersion. J. Mar. Biol. Assoc. U.K., 32: 107-28.

Bousfield, E. L. 1955. Ecological Control of the Occurrence of Barnacles in the Miramichi Estuary. Bull. Nat. Mus. Can., 137:1-69.

Connell, J. H. 1956. A Study of Some of the Factors Which Determine the Density and Survival of Natural Populations of the Intertidal Barnacle, *Balanus balanoides* (L.) Ph.D. thesis, Univ. of Glasgow (unpubl.).

Crisp, D. J. 1955. The Behaviour of Barnacle Cyprids in Relation to Water Movement Over a Surface. J. Exp. Biol., 32: 569-90.

Darwin, Charles. 1854. A Monograph of the Sub-class Cirripedia. The Balanidae, etc. Ray Society, London.

Hatton, Harry. 1938. Essais de bionomic explicative sur quel ques espèces intercotidales d'algues et d'animaux. Ann. Inst. Océanogr. Monaco, 17:241-348.

Knight-Jones, E. W. 1953. Laboratory Experiments on Gregariousness during Settling in *Balanus balanoides* and Other Barnacles. J. Exp. Biol., 30:584-98.

Lewis, J. R. 1954. Observations on a High-Level Population of Limpets. J. Anim. Ecol., 23:83-100.

Moore, H. B. 1934. The Biology of *Balanus balanoides*. I, Growth Rate and Its Relation to Size, Season and Tidal Level. J. Mar. Biol. Assoc. U.K., 19:851--68.

-------. 1935. The Biology of *Balanus balanoides*. IV, Relation to Environmental Factors. J. Mar. Biol. Assoc. U.K., 20: 263-308.

-------. 1936. The Biology of *Balanus balanoides*. V, Distribution in the Plymouth Area. J. Mar. Biol. Assoc. U.K., 20: 701-16.

-------. 1938. The Biology of *Purpura lapillus*. Part III, Life History and Relation to Environmental Factors. J. Mar. Biol. Assoc. U.K., 23:67-74

Pyefinch, K. A. 1948. Notes on the Biology of Cirripedes. J. Mar. Biol. Assoc. U.K., 27:464-503.

Weiss, C. M. 1948. The Seasonal Occurrence of Sedentary Marine Organisms in Biscayne Bay, Florida. Ecology, 29:153-72.

TEMPERATURE AND THE LIFE CYCLE OF *BALANUS BALANOIDES* (L.)*

H. Barnes

Introduction

There are three reasons for choosing this subject. First, there has been much misunderstanding and confusion with regard to the effects of temperature on the breeding of this common and widely distributed barnacle; it has not always been fully recognized that each phase of the reproductive cycle may respond in its own particular way to the temperature of the environment. As a result, established principles have been wrongly applied. In particular, concepts regarding the effect of temperature on the spawning of marine invertebrates have been applied to the so-called "spawning" of barnacles. However, whereas the term spawning is usually applied to the processes associated with the release of gametes, in the cirripedes the term is normally employed for the release of ripe embryos from the mantle cavity of the adults. There is little a priori reason to expect that these two phases of reproductive behavior, which are very different in character, will resemble one another in their responses to temperature changes of the environment.

Second, a discussion of this topic will provide an opportunity to consider how far the broad latitudinal pattern of distribution of the species may be interpreted in terms of the effect of tem-

---

*Contribution No. 937 from the Woods Hole Oceanographic Institution. The Woods Hole observations were made while the author was in receipt of a research fellowship from the institution. This fellowship, as well as leave of absence granted by the Council of the Scottish Marine Biological Association, is gratefully acknowledged.

perature on its reproductive physiology and breeding behavior. Third, the question of how far the results of such an inquiry are applicable to other common boreo-arctic species may be examined.

Distribution and the General Life Cycle

Balanus balanoides is a common barnacle widely distributed throughout the boreo-arctic (see Feyling-Hanssen, 1953a, for a review of the literature). It extends southward on the eastern side of the Atlantic from the coasts of the arctic to those of France and, after a break, to northern Spain (Bishop, Crisp, Fischer-Piette, and Prenant, 1957). Weltner (1900) has recorded the species from both Portugal and the Azores, but these records have been questioned by Southward and Crisp (1956). On the western shores of the Atlantic it reaches as far south as New Jersey (Darwin, 1854; Pilsbry, 1916; Richards, 1930). In the Pacific it is recorded from the Alaskan coast of the Bering Straits and as far south as Sitka; it is present in northern Japan (Hiro, 1935; Henry, 1942), and on the Siberian coast in Amur Bay and the Ochotsk Sea (Tarasov, 1937).

B. balanoides is an obligate cross-fertilizing hermaphrodite. The eggs, when ripe, are brought into the mantle cavity of the adult where, following upon copulation, fertilization takes place. The resultant fertilized egg masses, or ovigerous lamellae, are transferred to the base of the mantle cavity where they are retained throughout the period of embryonic development. The time of retention is termed the incubation period (Barnes, 1955; Barnes and Barnes, 1956). Gametogenesis, fertilization, incubation, and liberation of the fully developed nauplii from the mantle cavity are all well-defined phases of development, and any or all of them may be dependent upon temperature. Further, since the animal is littoral, the possible influence of both sea and air temperatures must be taken into account. It is also necessary to distinguish between temperature as it affects rate processes and temperature as a possible factor in any triggering mechanisms.

Influence of Temperature on the Life Cycle

It is generally recognized that the reproductive phases of the life cycle of many marine invertebrates are more sensitive to temperature than their nonreproductive phases, and consequently the effectiveness of temperature in determining the latitudinal distribution of a species has, on this account, been largely ascribed to its effect on breeding (Appellöff, 1912; Orton, 1920). The southern limit of a boreo-arctic species is, therefore, likely to be set by a maximum temperature in winter that does not exceed a certain critical value above which maturation of the gametes will not take place. Hutchins (1947) considered that the southern limit of *B. balanoides* was determined by a winter maximum sea surface temperature of 8°C--and indeed the limits are so approximated. He believed, however, that this low temperature was essential for an annual repopulation and associated it with the "spawning" period when nauplii are released; this "breeding" in the colder months was considered as an example of the temperature dependence stressed by Appellöff and Orton. Although there is now much evidence that certain phases of gametogenesis in invertebrates are markedly under temperature control (Loosanoff and Davis, 1950), there is little reason to suppose that the release of ripe embryos from the mantle cavity will be similarly dependent upon temperature. On the contrary, the evidence suggests rather that this phase of the life history is more closely synchronized with the spring diatom outburst (Barnes, 1957b).

The relation between the life cycle and temperature conditions may best be illustrated by a consideration of three widely separated localities where *B. balanoides* is common in the intertidal zone, namely, Spitzbergen (about 80°N. lat.), Millport, Scotland (55°45'N. lat.), and Woods Hole, Massachusetts (41°31'N. lat.). Data for the first of these localities have been given by Feyling-Hanssen (1953a), who has also provided some further details in correspondence. Data for the second two localities are largely unpublished personal observations. A summary of the life cycle and relevant temperature conditions will now be given.

At Spitzbergen, as elsewhere in the arctic, the animal's activities are crowded into the short summer season between the melting of the ice in May or June and its re-formation in November. In the far north the length of the open season is short so that, in spite of abundant food, the adult size at the end of the first season is small; the animals are not sexually mature until their second season. The reproductive products are developed and active growth takes place during the period of abundant food that follows the melting of the ice. Fertilization takes place in late summer or early autumn. Doubtless there is some embryonic development before the ice re-forms, but from November or December to the following spring (May or June) the animals are solidly frozen beneath the ice foot that covers the whole of the intertidal zone (Feyling-Hanssen, 1953b). Development during this period must be either arrested or extremely slow. With the disappearance of the ice foot and the onset of the diatom increase, the embryos are rapidly ripened, and nauplii are found in the plankton during June and July. The settlement of cyprids takes place during August. However, Feyling-Hanssen (1953b) has pointed out that in some years, when a late spring is followed by an early autumn, *B. balanoides* does not release its nauplii, so that subsequently a whole year-class will be missing.

The cycle is similar in other arctic regions (Bousfield, 1954, 1955; Barnes, 1957b). The animal is clearly well adapted to low temperatures. It has, therefore, been suggested (Barnes, 1957a) that the length of time of the ice-free period, rather than any deleterious effects of increasing low air or sea temperatures, is a most important factor in determining the northern limit of the species. This hypothesis is based upon the correlation that was shown to exist between the distribution pattern in northern seas and the extent and duration of ice formation. The correlation was believed to depend upon the fact that a minimum period is required for ripening of the embryos, development of the planktonic population to the cyprid stage, settlement of the cyprid, and growth of the spat to a point at which it can withstand the winter ice. If the ice-free period does not exceed this minimum, the species cannot exist.

In boreal regions where there is no winter ice formation, the various phases of the cycle extend over different periods. The development of reproductive products begins in late spring, almost as soon as the previous year's brood has been liberated, and continues during the summer at the time when the young animals are growing rapidly. At Millport, fertilization occurs early in November, and incubation extends throughout the winter months. The spring diatom increase normally takes place in March, and the nauplii are synchronously liberated at this time. Animals at all levels of the shore are sexually mature in their first year, during which growth is rapid. Indeed, under "optimal" growth conditions, namely, complete submersion and freedom from debris and smothering growths, maximum size is virtually attained five months after settlement, i.e., between April and September (Barnes and Powell, 1953). As in the arctic, there is a relatively long incubation period, but this cannot be ascribed to enclosure in ice. Rather, it has been suggested

(Barnes, 1957a) that it is largely due to the reduced activity of the animal during the winter months, which in most boreal regions are characterized by a paucity of plankton. Not only does this lack of plankton lead to reduced feeding activity and virtually no growth, but, since ventilation of the mantle cavity depends on the cirral activity associated with feeding, the supply of oxygen is inadequate for the maximum rate of respiration and development of the embryos that are contained within the mantle cavity. With only minor variations the life cycle is similar over the whole of the northern regions on both Atlantic and Pacific coasts (Bousfield, 1954; Runnström, 1925; Hatton and Fischer-Piette, 1932; Hatton, 1938.)

At Woods Hole and in other parts of the Buzzard's Bay region, where the species is approaching its southern limit, the planktonic regime is not typical of boreal regions, and as a result certain phases of the life history of *B. balanoides* become, so to speak, displaced. The reproductive products are formed during the summer, but fertilization takes place in early October; embryonic development is rapid, and nauplii are first released in midwinter. However, in some years the period of naupliar release may be more or less protracted with cyprids settling from February to early April. The rapid development of the embryos and their release without any delay may be ascribed to the presence throughout the winter of an abundant diatom population in this area (Fish, 1925). Feeding and growth are continued throughout the winter, and synchrony with a diatom outburst does not have to "await" a spring outburst.

It is very evident from the foregoing discussion that the various phases of the reproductive cycle take place under quite different temperature conditions. During much of the time when eggs and sperm are being developed, temperatures are relatively high--particularly near the southern limit of the species. There is further evidence (unpublished work) that temperatures up to 15°C have no deleterious effects on the general metabolic processes associated with gonad development, embryogenesis (incubation period), and adult growth. Further, as far as has been determined, the temperature coefficient (Q_{10}) of some of these processes approximates to that for many other biological systems, lying between 2 and 3. On the other hand, it must be pointed out that, unless there is extreme acclimation (which may be the case), the rapid development of the reproductive products in the arctic, where temperatures are low, suggests that other factors may be important. Of these the presence of abundant food is probably paramount.

Temperature Block to Maturation

The most striking feature of the reproductive cycle at Woods Hole is the delay in fertilization. This must be considered more fully. Examination of the shore population in this locality showed that by late summer the gonads were apparently ripe, i.e., the ovaries were prominent, thick in consistency, and creamy yellow in color. Nevertheless, fertilization was delayed until early October. This delay between the presence of apparently ripe sexual products and the onset of fertilization suggests that it is the final stages of gametogenesis that may be sensitive to temperature. Such a suggestion is in accord with the reproductive behavior of other invertebrates (see, for example, Loosanoff and Davis, 1950, on oysters). Examination of the temperature records showed that, although the sea temperature was relatively high (15°C at Woods Hole), the air temperature had dropped to 10° or below for about 10 to 15 days before fertilization. Preliminary laboratory experiments had previously shown that the fertilization of "ripe" eggs could be induced by maintaining animals at a temperature between 4° and 6°C (10°C has been more recently suggested by Crisp [1957] as the limit above which ripening of the gametes will not take place). In view of the fact that neurosecretory cells have now been shown to be present in cirripedes (Barnes and Gonor, 1958), attention may be drawn to the resemblance between the above results and those of Lubet (1956). He has shown that temperature shock, known to be effective in initiating liberation of gametes in certain mollusks, is mediated through neurosecretory activities. On the basis of the above evidence it is proposed that in *B. balanoides* there is a temperature block at a stage just preceding fertilization, i.e., in the final ripening of the gametes. Lowering of the temperature, either sea or air, can remove this block and initiate the processes leading to fertilization. (It should be pointed out that, although such a temperature block is perhaps the most important factor determining the southern limit of the species, other factors, in particular the effect of high summer temperatures on the adult, must also be taken into account.) If the above hypothesis is correct, then the minimum sea temperature suggested as a limiting factor by Hutchins (1947) and supposed by him to act on the "spawning" is not effective per se but is only empirically indicative that an appropriately low *air* temperature is reached earlier in the year.

Neither the mode of action nor the precise site at which this block is effective is known, but it seems to be more in the nature of a physiological trigger mechanism than an effect upon some rate process. It is not even known whether one or both gametes are affected. The final stages of maturation of the gametes are most probably involved, although copulation and fertilization may also be dependent upon temperature. However, copulation and fertilization both take place under water. At this time, although air temperatures have reached a value adequate to remove the block, *sea* temperatures are still high (15°C), and it would seem probable, therefore, that these processes (which must in any event be subsequent to the complete ripening of the gametes) are not involved in the temperature-dependent block. Even so, copulatory activity may be initiated as a result of factors arising from an earlier, temperature-dependent process.

The temperature requirement may be an upper critical limit, a critical departure from some mean level, or even oscillations of given magnitude and mean value. The behavior in the arctic, together with what is known of other invertebrates, tends to favor the first possibility.

Regulation of the Life Cycle

This temperature requirement for the final ripening of the gametes and the necessity for synchrony of the naupliar release with the spring diatom outburst are the two most important factors regulating the life cycle in relation to the environment. There are only limited times during the year when these requirements can be satisfied, and these will vary from place to place. Once maturation and naupliar release have been so determined, the remainder of the life cycle is readily adjusted to the very different environments.

Thus, in the arctic regions temperatures are probably rarely, if ever, high enough to prevent the ripening of the gametes; there is no block to this process which will, therefore, follow full development of the reproductive products. The life cycle in the absence of any temperature block is then determined by the synchrony of naupliar release and the vernal diatom outburst. As already noted, ripening of the reproductive products takes place directly after naupliar release during the short summer season, and fertilization takes place in early autumn. However, rapid development of the embryos is prevented by the onset of the winter ice in which the animals become frozen for several months.

Farther south, in typically boreal regions where ice does not form during the winter, the spring diatom increase occurs earlier in the year as does the release of nauplii. Reproductive products develop during the summer but here, in contrast to the arctic, must await an adequately low temperature before they can be finally ripened. Fertilization is, therefore, concomitant with the lowering of temperatures during the autumn. Even in boreal regions where the animals are not under ice, the incubation period is prolonged. Only when feeding is resumed at the time of the spring diatom increase are the embryos finally ripened and released from the mantle cavity.

At Woods Hole, because of the abundance of winter diatoms, feeding and ventilation of the mantle cavity continues after the egg masses have been formed. The incubation period is short, and, furthermore, naupliar release need not be delayed until spring.

It would appear that these two requirements, namely, synchrony of the naupliar release with an adequate food supply for the planktonic phases and a sufficiently low temperature for the final ripening of the gametes to take place, ensure that the species will produce only a single annual brood. In the arctic, where temperature conditions would not bar the production of a second brood, the season is too short for the redevelopment of reproductive products. In typically boreal regions the absence of a winter diatom population leads to a prolonged incubation period and reduces the time available for a second brood to be developed. In the more southerly boreal regions, further delay is caused by the fact that, although the reproductive products are well developed by late summer, final maturation must await the lower autumn air temperatures. At Woods Hole, although the incubation period is short and the subsequent development of new reproductive products rapid, the rise in temperature during the late spring and its maintenance at a high level during the summer does not allow the final stages of gametogenesis to take place until the autumn.

Discussion

No consideration has so far been given to the possibility that, quite apart from these reactions to environmental conditions, there may be some endogenous rhythmical control of breeding. The question may then be posed: suppose the required optimal conditions were established, either naturally or experimentally, would endogenous factors still limit the species to a single annual brood? If, as suggested below, endogenous factors are important, and if they are, as might be expected, under endocrine control, then it must be remembered that not only does a complex pattern frequently provide the stimulus for a specific response, but also the target organ itself may vary seasonally.

At least two other boreo-arctic balanids maintain a single annual brood wherever they are found; these are *B. balanus* and *B. hammeri*. These two species, both sublittoral, have a life history very similar to that already described for *B. balanoides*. Fertilization tends to be later, most probably because sea temperatures do not reach the appropriate low level until later in the year. The incubation period is shorter, and the release of nauplii during the spring may be somewhat earlier, than in the intertidal species (Pyefinch, 1948; Barnes and Barnes, 1954; Crisp, 1954). These two sublittoral species, particularly when in deeper water, are less subject to the environmental factors that have been invoked as largely determining the annual cycle of the intertidal species. This, in itself, adds point to the suggestion that endogenous control may, at least in part, regulate the production of only a single brood each year in typically boreo-arctic species.

In contrast, *B. crenatus* and *Verruca stroemia,* both mainly sublittoral species and abundant in northern waters, show a somewhat different breeding cycle from the species just discussed. The development of the major brood follows the boreo-arctic pattern: reproductive products develop during the summer, fertilization occurs during the colder months, and nauplii are released in the spring. However, small subsidiary broods appear to be produced during the summer, and in the case of *Balanus crenatus* there may be quite a moderate-sized autumn brood. The production of an autumn brood at Millport (Pyefinch, 1948; Barnes and Powell, 1953) is well documented in spite of statements to the contrary in Bishop, Crisp, Fischer-Piette, and Prenant (1957). These two species, therefore, have in part become independent of the restrictions that have been considered to characterize boreo-arctic animals.

It would appear that, further to elucidate the problems discussed above, something must be known of the endogenous factors involved. By analogy with other crustaceans such processes might be expected to be under hormonal control in which neurosecretory activities would be anticipated to play an important part. Since this paper was first presented it has been possible to demonstrate the presence of neurosecretory cells in the central nervous system of all the cirripede species so far examined. In the operculates the cells are by no means obvious, but this may in part be the result of the season when the animals were collected. In *Pollicipes (Mitella) polymerus* typical neurosecretory cells with heavily staining granules are present in all the major nerve ganglia (Barnes and Gonor, 1958).

Summary

1. A brief account is given of the life cycle of *Balanus balanoides* with particular reference to the phases of reproduction. The relation of the breeding cycle to temperature is discussed.
2. It is shown that many phases of the life cycle can take place at temperatures up to 15°C.
3. There is considerable evidence that the final stages of gametogenesis are inhibited at temperatures above 10°C; this temperature block to the final ripening of the gametes is one of the most important factors determining the southern limits of the species.
4. The possible modes of action of this temperature block are discussed.
5. Synchrony of naupliar release with the abundance of food and removal of any temperature block to gametogenesis are the prime factors determining the time sequence of the breeding cycle.
6. The question of the existence of endogenous rhythms is discussed and attention drawn to the breeding cycle of other northern species. The possible role of neurosecretion is considered.

LITERATURE CITED

Appellöff, A. 1912. Invertebrate Bottom Fauna of the Norwegian Sea and North Atlantic. In Murray and Hjört, The Depths of the Ocean, London.

Barnes, H. 1955. The Hatching Process in Some Barnacles. Oikos, 6:114-23.

-------. 1957a. The Northern Limits of *Balanus balanoides* (L.). Oikos, 8:1-14.

-------. 1957b. Processes of Restoration and Synchronization

in Marine Ecology: The Spring Diatom Increase and the "Spawning" of the Common Barnacle, *Balanus balanoides* (L.). Ann. Biol., 33:67-85
-------, and M. Barnes. 1954. The General Biology of *Balanus balanus* (L.) da Costa. Oikos, 5:63-76.
-------, and M. Barnes. 1956. The Formation of the Egg Mass in *Balanus balanoides* (L.). Arch. Soc. "Vanamo," 11:11-16.
-------, and M. Barnes. 1958. The Rate of Development of *Balanus balanoides* (L.) Larvae. Limnology and Oceanography, 3:29-32.
-------, and D. J. Crisp. 1956. Evidence of Self-fertilization in Certain Species of Barnacles. J. Mar. Biol. Assoc. U.K., 35:631-39.
-------, and J. J. Gonor. 1958. Neurosecretory Cells in Some Cirripedes. Nature, 181:194.
-------, and J. J. Gonor. 1959. Neurosecretory Cells in the Cirripede, *Pollicipes polymeris* J. B. Sowerby. Festschrift for Thomas G. Thompson, J. Mar. Res.
-------, and H. T. Powell. 1953. The Growth of *Balanus balanoides* (L.) and *B. crenatus* Brug. under Varying Conditions of Submersion. J. Mar. Biol. Assoc. U.K., 32:107-27.
Bishop, M. W. H., D. J. Crisp., E. Fischer-Piette, and M. Prenant. 1957. Sur l'écologie des Cirripèdes de la côte atlantique française. Bull. Inst. Océanogr. Monaco, 1099.
Bousfield, E. L. 1954. The Distribution and Spawning Seasons of Barnacles on the Atlantic Coast of Canada. Ann. Rept. Nat. Mus. Can. for fiscal year 1952-53, Bull., 132, 112-54.
-------. 1955. Cirripede Crustacea of the Hudson Strait Region, Canadian Eastern Arctic. "Calanus" Ser. 7., J. Fish. Res. Bd. Can., 12:762-67.
Crisp, D. J. 1954. The Breeding of *Balanus porcatus* (da Costa) in the Irish Sea. J. Mar. Biol. Assoc. U.K., 33:473-96.
-------. 1957. Effect of Low Temperature on the Breeding of Marine Animals. Nature, 179:1138-39.
Darwin, C. R. 1854. A Monograph on the Sub-class Cirripedia, Vol. II. Ray Society, London.
Feyling-Hanssen, R. W. 1953a. The Barnacle *Balanus balanoides* (Linné, 1766) in Spitzbergen. Norsk Polarinstitutt Skrifter, 98.
-------. 1953b. Brief Account of the Ice Foot. Norsk Geogr. Tidsskr., 14:45-52.
Fish, D. J. 1925. Seasonal Distribution of the Plankton of the Woods Hole Region. Bull. U.S. Bur. Fish., 41:91-179.

Hatton, H. 1938. Essais de bionomie explicative sur quelques espèces intercôtidales d'algues et d'animaux. Ann. Inst. Océanogr., N.S., 17:241-348.

-------, and E. Fischer-Piette. 1932. Observations et expériences sur le peuplement des côtes rocheuses par les Cirripèdes. Bull. Inst. Océanogr. Monaco, 592.

Henry, D. P. 1942. Studies on the Sessile Cirripedia of the Pacific Coast of North America. Univ. Wash. Publ. Oceanogr., 4:95-134.

Hiro, F. 1935. The Fauna of Akkeshi Bay. II, Cirrepedia. J. Fac. Sci. Hokkaido Univ. Zool., 4:213-29.

Hutchins, L. W., 1947. The Bases of Temperature Zonation in Geographical Distribution. Ecol. Monogr., 17:315-35.

Loosanoff, V. L., and H. C. Davis. 1950. Conditioning *V. mercenaria* for Spawning in Winter and Breeding its Larvae in the Laboratory. Biol. Bull., 98:60-65.

Lubet, P. 1956. Effets de l'ablation des centres nerveux sur l'émission des gamètes chez *Mytilus edulis* L. et *Chlamys varia* L. (Mollusques Lammellibranches). Ann. des Sc. Nat., Zool., 11e ser., pp. 175-83.

Orton, J. H. 1920. Sea-Temperature, Breeding and Distribution in Marine Animals. J. Mar. Biol. Assoc. U.K., 12:339-66.

Pilsbry, H., 1916. The Sessile Barnacles (Cirripedia) Contained in the Collections of the U.S. National Museum Including a Monograph of the American Species. U.S. Nat. Mus. Bull., 93.

Pyefinch, K. A. 1948. Notes on the Biology of Cirripedes. J. Mar. Biol. Assoc. U.K., 27:464-503.

Richards, H. G. 1930. Notes on Barnacles from Cape May County, New Jersey. Publ. Univ. Pennsyl. Contrib. Zool. Lab., 29.

Runnström, S. 1925. Zur Biologie und Entwicklung von *Balanus balanoides* (Linné). Bergens Mus. Aarbok. Naturv., 5.

Southward, A. J., and D. J. Crisp. 1956. Fluctuation in Distribution and Abundance of Intertidal Barnacles. J. Mar. Biol. Assoc. U.K., 35:211-29.

Tarasov, N. 1937. Contribution to the Fauna of Cirripedia Thoracica of the Arctic Ocean, III. Trans. Arctic Inst., Leningrad, 2:59-62.

Weltner, W. 1900. Die Cirripedien der Arktis. Fauna Arctica, 1:289-312.

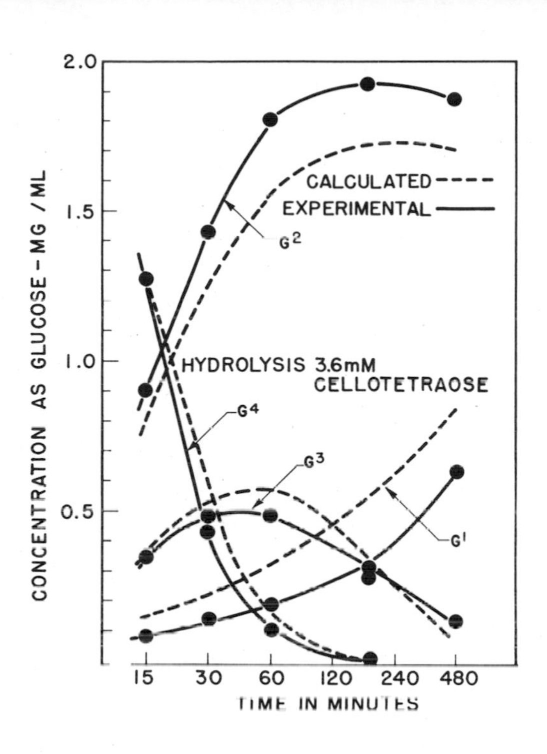
2.0
1.5
1.0
0.5
CONCENTRATION AS GLUCOSE - MG / ML
CALCULATED
EXPERIMENTAL
G2
HYDROLYSIS 3.6mM
CELLOTETRAOSE
G4
G3
G1
15
30
60
120
240
480
TIME IN MINUTES

Cellulases

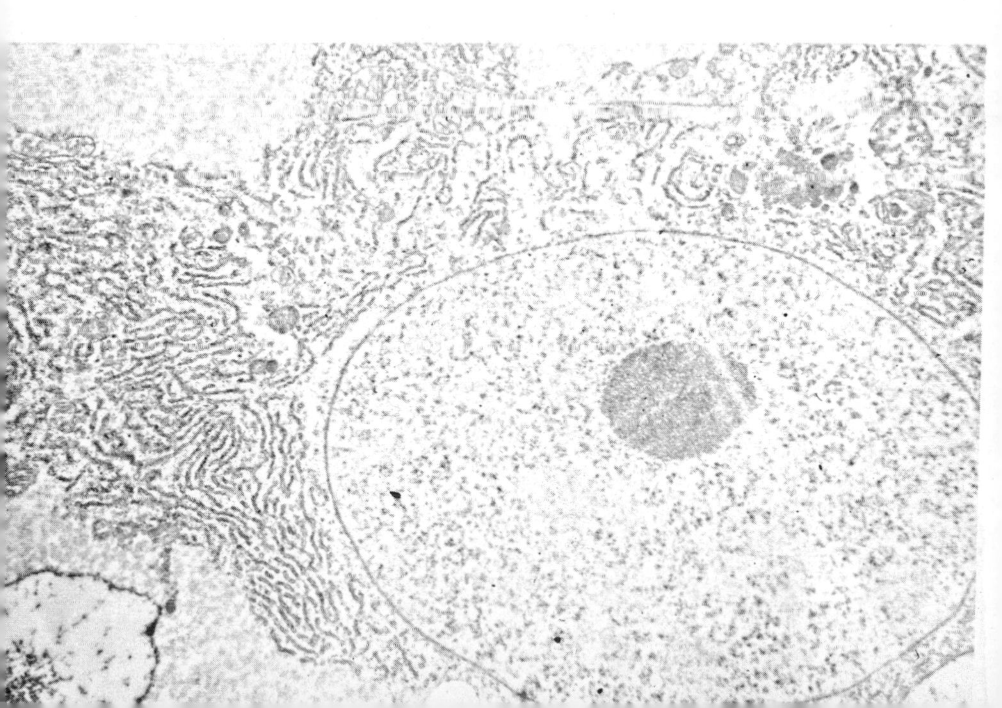

INTRODUCTION

C. B. van Niel

When knowledge in a particular field of biological investigation has reached the point where the organisms involved are fairly well known, we can begin to ask questions about details of their behavior. The previous sections of this symposium provide the basis for a general understanding of the major types of invertebrates that cause the deterioration of wooden structures in the sea. It is therefore logical that they be followed by a section in which is reviewed the currently available information on the chemical mechanisms by which these organisms act as cellulose-decomposing agents. More specifically, this part is intended to summarize our knowledge of the enzymes concerned in the primary degradation of cellulose, the cellulases.

It would have been possible to restrict this part of the program to a discussion of the cellulases of the Terediniens and Limnoria. That a more broadly conceived plan of organization was adopted can be attributed to the fact that investigations in this area have only barely begun to yield significant results. The reason for this is simple; it has been clearly expressed by Dr. Tracey in his comments on Dr. D. L. Ray's contribution on Cellulase from Limnoria, as follows:

"As one who has worked with cellulases, I should like to take this opportunity to say that not the least remarkable feature of Dr. Ray's results is that they have been achieved using what would seem to be such a very unsatisfactory source of enzyme. When it is realized that the great volume of work described has been done with an enzyme solution 1 ml of which requires the dissection of 600 animals, it will be obvious what courage and skillful planning must have been entailed. I know that the other workers in the field of cellulase research present here will join me in congratulating Dr. Ray on the splendid achievements outlined in her paper and envy her the determination and skill that made them possible."

Had the section been limited to this aspect, it would therefore have been very brief indeed. But the more elaborate program owes its inception to a consideration other than a mere desire to expand it to a respectable size. It has become increasingly clear that nearly every phase of scientific inquiry is greatly benefited by comparative studies. These often lead to the discovery of relationships between phenomena that are not always immediately apparent. And from such relationships new approaches to the problems frequently emerge. This aspect, too, may be illustrated by quoting Dr. Tracey. In the introductory chapter of his book, Principles of Biochemistry (Sir Isaac Pitman and Sons, London, 1954, p. 4), he stated:

"Our knowledge of biochemistry today has been severely conditioned by the domination of the subject by older disciplines. It has developed largely as a result of work done with a view to its application in other fields. These have been physiological medicine, nutrition, and the control of harmful organisms whether pests of man's stored products, his domestic animals and plants, or himself. Physiological medicine involves the study of one living thing, man, and the others emphasize the differences between organisms. It is the exploitation of differences in the biochemistry of organisms that is the basis of all attempts to kill one type of organism without killing another, whether in the control of a disease or in a search for an insecticide. This intensive investigation of the differences between organisms has tended to obscure the similarities that exist. Together with the fact that different techniques are often needed in dealing with widely different types of organisms it has led to the apparent fission of biochemistry into subjects such as bacterial biochemistry, agricultural biochemistry, and the biochemistry of man (this is usually simply 'Biochemistry'). Consequently biochemistry when compared with the older biological sciences such as botany and zoology is very unevenly developed. It is a fortunate fact that in the search for differences in the biochemistry of organisms there have been many failures, for these failures reveal similarities. In this book it is the similarities between organims that will be emphasized in the hope that if the subject is viewed as a whole some general principles of biochemistry will become apparent."

It will be evident why Dr. Tracey, who has himself contributed much to our knowledge of cellulases, was invited to prepare the general introduction to this part of the symposium.

In line with the above-quoted remarks it is also obvious that knowledge of the cellulolytic enzymes of organisms other than Limnoria and the Terediniens is important for a study of cellulose decomposition by the latter. During recent years a great deal of detailed information has been accumulated on the cellulases of various fungi, largely owing to the fact that these enzymes can be obtained with relative ease and in sufficient quantity from mold cultures. Consequently an attempt has been made to include reports of such studies on fungal cellulases by some of the most active workers in the field. A perusal of these papers shows that the research has led to the emergence of conflicting views, each one supported by certain kinds of experimental evidence. A situation such as this often arises in the course of scientific investigations; it usually indicates that rapid progress is being made in divergent directions but that there are still unbridged gaps preventing the reconciliation of apparently irreconcilable opinions. It may be confidently expected that further work will indicate the reasons for the present inconsistencies and thus focus attention on the general principles.

A truly comparative study of cellulases must, of course, also comprise those found in bacteria and protozoa. However, in this area our information is as yet very limited. Dr. Kadota has summarized the current knowledge of the marine cellulose-decomposing bacteria, thus providing the basis for future work on their enzymes. The review of termite cellulase shows that here also our comprehension does not yet extend to the enzyme level, and that the problem is complicated because different types of termites appear to digest cellulose by means of different agencies. Thus, while it is certain that in some termite species the primary attack on ingested cellulose proceeds under the influence of associated protozoa, in others the involvement of molds or bacteria has been invoked, and it is possible that yet other termites may produce their own cellulolytic enzymes. A parallel situation seems to exist in the case of the marine wood borers. Here, too, a close association with particular microorganisms is being advocated, while Dr. Ray's investigations have unequivocally demonstrated cellulase production by Limnoria itself.

Naturally, a discussion of the problem of cellulose decomposition in ruminants would have constituted a valuable contribution to this section of the symposium. Unfortunately, Dr. R. E.

Hungate, who had been asked to review this topic, could not resist the temptation to accept instead an invitation to go to Africa in order there to study cellulose digestion in some of the local herbivores.

The final paper deals with studies on the details of the mechanism of transglucosidations. At first sight its inclusion may appear somewhat surprising. But, if it is realized that the same enzymes are often involved in both the degradation and the synthesis of a particular substance, and that dextran, like cellulose, is a glucose polymer, the relevance of this contribution will be evident. Dextrans exhibit properties that render them especially favorable for the study of enzymatic polysaccharide synthesis, and eventually investigations of this sort are apt to contribute significantly also to our understanding of the mechanisms operative in the enzymatic decomposition and synthesis of cellulose.

THE ROLE OF CELLULASES IN NATURE

M. V. Tracey

Cellulose is a substance which is of prime importance to man as a raw material for the construction of his artifacts. Consequently it is natural that his first interest should be in cellulases produced by organisms that feed on things he has made of cellulose whether they be jetties, houses, newspapers, or viscose rayon garments. His second interest is in the nourishment of those domesticated animals for whom cellulose is a significant and probably essential component of the diet. Reliance on the results from these two fields of research does not and cannot lead to a balanced view of the place of cellulase in nature. Attention was not initially concentrated on these topics. In the late nineteenth and early twentieth century, work on cellulases was in effect a branch of the effort directed by German research workers toward a broadly based study of the physiology of digestion, and during this period a great deal of exploratory work on the occurrence of enzymes digesting cellulose in invertebrates was done. Most of it has not been repeated using modern methods and criteria and, until it is repeated, our knowledge of the distribution of cellulase in animals must remain sketchy and speculative. This body of knowledge could only have been produced by workers with complete freedom of choice of subject and can probably only be extended now by those in universities who enjoy a similar freedom. The enormous expansion of research, in large measure due to support by government and industry, has led in the cellulase field, as in others, to a perhaps premature focusing of effort onto restricted topics. The first of these was digestion in ruminants and in this progress has been slow owing to the complexity of the systems involved. Perhaps because modern technology was born in a temperate climate in which the tempo of decay is slow, the problem of fungal attack on cellulose (inherently simpler than the problem of ruminant digestion) had to wait until the recent war, much of it fought in hot and humid climates, made the problem acute. As a consequence, we know a good deal about how cellulose is broken down by fungi and are beginning to understand how it may be used as a source of energy by the ruminants. In nature, however, cows and sheep, molds and mildews are far outweighed by other organisms relying on cellulase for a variety of ends. We have enough information to be sure that much of the enormous weight of cellulose decomposed each year in the oceans and in or on the soil is the concern of organisms other than the fungi and bacteria and, even though these saprophytes may be quantitatively the most important, very little is known of their activities in these situations. We can, of course, extrapolate from the little detailed knowledge we have and flesh out the fragmentary skeleton of facts relating to the breakdown of cellulose in soils and the sea in a relatively convincing manner. But even so, one major field that must surely be of vital importance in any balanced view of the subject remains unexplored. All our knowledge of the breakdown of cellulose is concerned with organisms that use but do not make it.

It seems appropriate, therefore, to begin this survey by inquiring whether cellulases exist in green plants and whether they may have a general function or are limited to scattered specific roles of trivial importance compared to those they play in the chemotrophes. At first sight, it appears that their role is indeed trivial. Cellulase activity has been found in germinating seeds and also at what appear to be ridiculously low levels in many parts of the plant (Tracey, 1950). Though the levels are low, it would perhaps be unwise to dismiss them as insignificant for an extract of whole minced cow would also be expected to display little activity. The presence of cellulase in extracts of root, stem, and leaf argues a general function in the plant and the most likely such function would be in connection with the cell wall. The primary cell wall contains cellulose laid down in such a way that the molecular chains lie in a flat spiral and, after the bulk of cell growth is over, further layers of cellulose are added in the form of steeper intersecting spirals. There may be considerable angular dispersion among the microfibrils so that many primary cell walls present the appearance of oriented meshworks. Views on how this meshwork increases in area during growth are conflicting (for a recent summary, see Wardrop, 1955). Evidence that the mesh is loosened has been published in the form of electron micrographs (e.g., Frey-Wyssling, Mühlethaler, and Wyckoff, 1948) and this loosening may be thought of as a result of enzymatic degradation of some of the cellulose strands (Frey-Wyssling, 1949) or as a consequence of slipping between the components of the fabric as a result of internal pressure in the cell. It is difficult to reconcile the conflicting interpretations that have been advanced of observations on the submicroscopic structure of the cell wall during growth without assuming that the enzymatic dissolution of portions of the fabric occurs. In the very limited work done on the distribution of cellulase activity in the tobacco plant, it was observed that activities were higher in those regions of the plant in which growth was occurring more rapidly. If indeed cellulase does play a role in the remodeling of the primary cell wall during growth, then it might be expected to be present in cells in which growth was occurring and perhaps absent in cells no longer plastic. Thus, the very low levels of activity in extracts of plant organs may be an indication, not of the general presence of ineffective concentrations, but of the fact that most of the cells are not engaged in active remodeling of the cell wall and their contents dilute the extracts of those that are.

All living things contain the enzymatic seeds of their own decay, for complete or partial autolysis at death is a universal phenomenon. Many organisms nonetheless use dead tissues for protective, structural, or other purposes. Such tissues appear to be protected from attack by vagrant hydrolytic enzymes in a number of ways. Among these are the use of proteins rendered resistant to enzymatic breakdown by -S-S- bonding (keratins), by tanning with phenols (insect exoskeletons), or by induration with inorganic salts as in the bones of vertebrates and the shells of molluscs. In plants, the dead tissues include cork, in which the cell walls are protected by suberin, and the lignified schlerenchyma and xylem. In the formation of xylem from the cambium it is inevitable that cells in which elongation has ceased lie close to cells in which growth in length is actively proceeding. During the transformation of the fully grown cells into vessels, the walls are progressively lignified and the protoplasmic contents are lost. A similar growth process occurs during the formation of phloem from cambium, but in this tissue the protoplasmic contents of the cells are not lost and no lignification occurs. I think it reasonable to suppose that lignification of the xylem may be connected with the necessity of protecting the walls of empty dead cells from cellulase, and no doubt other enzymes, involved in the remodeling of nearby cell walls in the protoxylem. In suggesting that these methods of protecting dead tissues from hydrolytic attack by the organism's own enzymes had an evolutionary advantage, I would not wish to deny that subsequently this advantage was also operative in protection against the activities of other organisms and in conferring increased strength and other useful properties. The ability to protect the walls of empty cells from cellulase necessarily produced nearby may well, however, have been the prime cause of the establishment of lignin as a universally occurring constituent of vascular plants.

The cellulases of germinating seeds have been examined in more detail than those of the leaf, stem, and roots of the plant. They may, however, be regarded as a special case of the general need for cellulase in the plant's economy. The recent work of Sandegren and Enebo (1952) shows that barley cellulase is similar in most respects to cellulases from other sources. However, our general knowledge of the cellulases in the green plants is almost nil. We know that they are present, but of their properties and behavior we have only the information relating to those that may be of importance in brewing. Nothing is known of the presence of cellulase in other producers of cellulose. Among these are *Acetobacter* spp., some tunicates, and some fungi, members of the Oomycetes and Archimycetes, which have been shown to have a cellulose cell wall. It would be of great interest to examine these fungi for the presence of an intracellular cellulase during active growth.

The best known degraders of cellulose are also members of the plant kingdom--the saprophytic bacteria and fungi--and it is by taking advantage of the efforts of the former that many of the animal cellulose users operate. If the cellulases of bacteria and fungi are to be effective, they must hydrolyze cellulose in such a manner that the products of hydrolysis will pass in a metabolizable form through the cell wall of the organism concerned. There seems little doubt that some cellulases in the absence of β-glucosidase activity produce both cellobiose and glucose as sole soluble end products when acting on cellulose (Whitaker, 1953; Reese, 1956), and since *Trichoderma viride* (one of the most active cellulolytic fungi) produces no extracellular cellobiase activity (Reese and Levinson, 1952), it must be assumed that it is capable of absorbing either cellobiose or glucose. Whether those cellulolytic organisms which produce an extracellular cellobiase are incapable of transporting the disaccharide across the cell wall remains a matter for speculation. The cellulases of bacteria and fungi will be described by other contributors, but as background to the subject it should be remembered that cellulose, in the absence of man's activities, is not normally a substrate that is immediately available to saprophytes. In plant remains there are protective layers of cutin, suberin, or lignin associated with it and an abundance of more easily hydrolyzable materials. Moreover, cellulose itself is only slowly hydrolyzed by enzymes in the native state. Hence the users of cellulose tend to operate at the end of a chain of successive organisms, being satisfied with and owing their existence to their ability to subsist on the barer bones of the feast. Such successions in the fungi and the physiological adaptations they entail are discussed by Garrett (1955); similar sequences of attack are doubtless to be observed among the bacteria, and in these organisms mutual aid may also play a part. The occurrence of symbiotic phenomena in thermophilic cellulose fermentation was recognized in 1913 by Kroulík and a number of examples have been investigated in recent years involving the association of cellulase-producing bacteria with other bacteria or actinomycetes (Fahraeus, 1949; Enebo, 1951; Gyllenberg, 1952).

A very wide variety of animals can derive energy from cellulose that they have eaten. Our knowledge of the distribution of this ability in the animal kingdom is sketchy. In many instances, it rests solely on the presumption that the presence of cellulase in the gut implies that cellulose is hydrolyzed in significant amounts there and that the animal benefits therefrom. Most animals with a mouth or pharynx large enough to swallow bacteria will possess inhabited guts and cellulase producers can probably be isolated from all such communities. The situation is further confused because there is no reason to suppose that the gut of an animal making its own cellulase is free from microbial cellulase producers. Thus the demonstration that cellulase is present in an extract of a whole animal or in extracts of portions of its gut may be due to one or more of the following situations.

1. The animal produces cellulase and the products of its action represent a significant part of its food (some protozoa, insects, and shellfish).

2. The animal produces cellulase but uses it to gain access to other food and benefits mainly from the use of cellulase as a tool (some plant pathogenic eelworms).

3. The animal produces cellulase but the products of its action do not appear to form a significant part of its food (some protozoa, earthworms).

4. The animal does not produce cellulase and does not benefit significantly from breakdown of cellulose but its gut harbors a population of microorganisms, some of which produce cellulase. Any benefit from their activities is only significant indirectly, if at all, e.g., by virtue of vitamin synthesis (most vertebrates, herbivorous echinoderms?, many invertebrates?).

5. The animal does not produce cellulase but benefits from breakdown of cellulose by gut microorganisms. This situation often is indicated by major structural adaptations of the gut and sometimes by an altered metabolism of the host (low blood sugar, ability to use volatile fatty acids in bulk, etc.) (ruminants, macropod marsupials, many other vertebrate herbivores, some insects and ? some herbivorous mollusks).

The cellulase-producing protozoa living in the gut of animals are examples of group 1. Our knowledge of them is at present restricted to those in the gut of some ruminants and insects. More evidence seems to be required before other animals are assigned to group 1. The isopod Limnoria seems certainly to belong here for Ray and Julian (1952) showed not only that cellulase was present in the diverticula but that microorganisms could neither be seen in them nor cultured from them. Other animals with strong claims for inclusion are Hylotrupes larvae (Schlottke and Becker, 1942) and larvae of some other wood-eating insects (Mansour and Mansour-Bek, 1934), Teredo (Hashimoto and Onama, 1949; Greenfield and Lane, 1953), and Bankia (Nair, 1956). Inclusion of those mollusks in which cellulase has been demonstrated in style extracts should perhaps await clarification of the role played by the Spirochaetes that swarm in the styles (Lavine, 1946; Newell, 1953). The eelworms *Ditylenchus dipsaci* and *D. destructor* belong in group 2 (Tracey, 1952) as their pharynges are too narrow to admit bacteria and their guts appear free from them. Both are plant pathogens and appear to subsist on cell contents rather than cell walls. The free-living soil amoebae may belong in group 3. They produce cellulase and chitinase when feeding on pure cultures of bacteria that contain neither cellulase nor chitin but they may, of course, eat food containing either when in the soil (Tracey, 1955).

A distinction between groups 4 and 5 is difficult to draw in the absence of detailed knowledge of the nutritional physiology of the species concerned. Among the invertebrates, a tentative guess can sometimes be made on anatomical or physiological grounds. Thus, if digestion of cellulose is shown to occur in enteric diverticula whose lumina are too small to admit bacteria, then it is probable that cellulase is produced there by the animal. Similarly, if digestion of cellulose can be shown to occur intracellularly in the absence of intracellular symbionts (which are known to occur in some tissue of arthropods and mollusks), the same conclusion is justified. Earthworms provide a good example of the difficulty of interpretation that is often met with. Extracts of earthworms contain both cellulase and chitinase (Tracey, 1951). Their guts undoubtedly harbor soil organisms producing both these enzymes, and digestion is not intracellular. The gross structure of the gut is such as would suggest that there would not be adequate time during the passage of cellulose through it for significant digestion of cellulose to occur. These facts would suggest that they belong in group 4, but it was also found that extracts of washed gut wall of one species contained the enzymes and that concentrations were much higher in the first half of the intestine than the second while the portion from pharynx to gizzard contained little if any. These observations would suggest that earthworms do produce cellulase and should perhaps be placed in groups 1 or 3. The presence of an actively cellulolytic population in the gut, such as has been demonstrated by Florkin and Lozet (1949) in the snail *Helix pomatia*, is not enough to exclude the possibility that the host itself produces cellulase. Thus Nair (1956) has shown that in Bankia cellulase is to be found in style extracts and in extracts of the diverticula. It may be found that the style enzyme is of microbial origin while that of the diverticula, in which intracellular digestion occurs, is produced by Bankia.

Our knowledge of the anatomy and physiology of vertebrates is much more advanced than that of the invertebrates and it is easier to make guesses on these grounds. In the ox, in which efficient pregastric fermentation occurs, the surface area of the gut is three times that of the skin; in the horse, one of the most efficient postgastric fermenters, it has twice the area; while in carnivores, the ratio may be less than one. The increased gut area is usually due to increased sacculation in the walls of the stomach or colon and caecum. The extremes of specialization are seen in the sheep, ox, and horse, where the ratios of stomach contents to large intestine contents are respectively 1:0.12, 1:0.06, and 1:8.0, whereas in the pig and rabbit they are 1:1.3 (figures calculated from Elsden, Hitchcock, Marshall, and Phillipson, 1946). The insoluble nature of cellulose imposes a long period of fermentation if breakdown is to be substantially complete and the bulk of food normally takes two to three days to pass through a cow or sheep, but only one day in the horse or pig. These considerations are useful when applied to mammals but may not apply so obviously in other vertebrates. There appears little in the anatomy of the carp to indicate that it can use cellulose in its food as efficiently as a sheep (Bondi, Spandorf, and Calmi, 1957) while the *Tilapia esculenta* of Lake Victoria that feed exclusively on phytoplankton are unable to digest either green or blue-green algae (Fish, 1951).

That the efficient use of plant food by pregastric cellulose fermentation is of value in the exploitation of regions in which plant food may be scarce, is emphasized by the fact that it is a method that has been evolved twice in the mammalia. There seems no doubt that the digestive mechanism of the macropod marsupials, so similar to that of cows and sheep in all respects, owes its similarity to convergent evolution rather than derivation from any common ancestral form (Moir, Somers, and Waring, 1956). It seems odd, therefore, that cellulose digestion in vertebrates should be by means of enteric organisms and that no vertebrate is known to produce a cellulase.

One explanation of this lack would postulate the evolution of vertebrates from an echinoderm line that had irreversibly lost the ability to produce cellulase. It is unfortunate that there is so much obscurity at present surrounding the line of chordate evolution and that there is such a lack of information concerning digestion in the herbivorous echinoderms. It seems that *Strongylocentrotus purpuratus* is incapable of digesting the whole of the cell wall of *Iridophycus purpuratus* which may contain cellulose, though cultures isolated from the gut can do so. No proof that the sea urchin benefited from the activities of the enteric organisms was obtained (Lasker and Giese, 1954).

An alternative explanation of the absence of cellulase in vertebrates is that the production of cellulase by a vertebrate would not confer an evolutionary advantage. The most interesting fact about those cellulases that have so far been studied in a fairly pure state is the extreme slowness with which they act on cellulose. I have elsewhere (Tracey, 1953) argued that this is a direct consequence of the nature of the substrate which also imposes other peculiar features of their action. It is difficult to make a direct comparison to illustrate this slowness, but it will perhaps be enough to point out that amylase activity is normally assayed in terms of reducing sugar produced in 3 minutes at 20° while incubation periods when assaying cellulase with cellulose as substrate are normally 17-24 hours at 35°-40°. As a result of this slow digestion, efficient use of cellulose imposes a prolonged time of exposure to the enzyme. One consequence of this is the enormous spaces used in ruminant digestion coupled with a slow through-put. Another may be the elaborate mechanisms in herbivorous mollusks by which the particulate food is entrained in mucus and coiled round the style before proceeding to the intestine (Morton, 1952). In this way, orderly progression of the food is assured and all is evenly exposed to digestive enzymes for a similar time. Digestion in the seclusion of the diverticula after the more easily digested material has been brought into solution may also serve the same purpose of ensuring an adequate time of exposure to cellulase. The general pattern of digestion in vertebrates, insofar as the main sequence of enzymes to which food is exposed is concerned, seems as immutable as the possession of two limb girdles. If cellulase were to be produced by a vertebrate, we may guess therefore that it would accompany other polysaccharases either in the saliva or in the glands associated with the small intestine. If produced in the saliva, it could only be effective if the stomach were modified into a pregastric chamber of sufficient size to enable the long process of cellulose digestion to occur before the enzyme was inactivated by peptic secretions. If produced by the pancreas, then a voluminous colon or caecum would be required for the same purpose. All cellulases that we know of operate best at a pH at which microorganisms flourish, and it would be expected that a modified stomach would soon be a hotbed of fermentation, for no acid could of necessity be interposed to sterilize the food if salivary cellulase were used. Similar active fermentation would be expected in the colon and caecum if they were modified to allow prolonged action of pancreatic cellulase. The occurrence perforce of cellulose fermentation would impose modifications in the physiology of the host to deal with the fatty acids produced, and the animal would in this respect be in the position of a herbivore such as we know now that does not produce cellulase. It would, moreover, derive benefit from the fermentation by absorbing many of the synthetic products of the microorganisms. Thus production of its own cellulase might confer little, if any, advantage and the ability be lost.

Finally, it has been suggested (Pirie, 1953) that the production of cellulase might be a disadvantage owing to the presence in the animal's gut of a substrate other than cellulose but hydrolyzed by cellulase. If, for example, the integrity of the vertebrate gut depended in part on a substance containing a β-glucosidic link labile to cellulase then secretion of cellulase into the gut might be lethal. Cellulase produced by microorganisms could still be exploited subject to its never being free and active in solution. There is evidence (Halliwell, 1957) that little or no cellulase may appear in solution in the rumen liquor, and if such cellulase were inactivated or destroyed as soon as released by autolysis or digestion by the host, then the gut wall would be safe. A number of bacteria pathogenic to man produce a soluble chitinase (Clarke and Tracey, 1956) and probably a cellulase too. In view of Pirie's suggestion, it is interesting to speculate on whether they owe any of their pathogenicity to cellulase production. This last explanation of the absence of cellulases in vertebrates has the supreme advantage over the other two that I have advanced that it is open to experimental test. Though cellulase preparations (as what has not) were fed to man at one time for rather ill-defined medical ends, the experiments, though happily nonlethal, were hardly suitable as a basis for coming to a decision on Pirie's suggestion. It is to be hoped that the experiment will be done now that cellulase preparations are becoming a little more readily available.

This survey has been partial in both senses of the word; I have avoided those fields of which we shall shortly have expert accounts and I have given perhaps undue emphasis to those in which our knowledge is scanty or nonexistent. To formulate any satisfactory appreciation of the role of cellulase in nature much work remains to be done. I believe that some of the most serious gaps could best be filled by the work of laboratories such as Friday Harbor for we need knowledge of the digestion of marine herbivores more acutely perhaps than of the terrestrial.

LITERATURE CITED

Bondi, A., A. Spandorf, and R. Calmi. 1957. The Nutritive Value of Various Feeds for Carp. Bull. Fish Culture in Israel, 9:13.

Clarke, P. H., and M. V. Tracey. 1956. The Occurrence of Chitinase in Some Bacteria. J. Gen. Microbiol., 14:188.

Elsden, S. R., M. W. S. Hitchcock, R. A. Marshall, and A. T.

Phillipson. 1946. Volatile Acid in the Digestion of Ruminants and Other Animals. J. Exp. Biol., 22:191.

Enebo, L. 1951. On Three Bacteria Connected with Thermophilic Cellulose Fermentation. Physiol. Plant., 4:652.

Fahraeus, G. 1949. *Agrobacterium radiobacter* as a Symbiont in Cellulose Decomposition. Kgl. Lantbruchs. Hogskol. Ann., 16:159.

Fish, G. R. 1951. Digestion in *Tilapia esculenta*. Nature, 167: 900.

Florkin, M., and F. Lozet. 1949. Origine bacterienne de la cellulase du contenu intestinal de l'escargot. Arch. Internat. Physiol., 57:201.

Frey-Wyssling, A. 1949. Growth in Surface of Plant Cell Walls. lst. Int. Congr. Biochem. Abst., p. 505.

-------, K. Mühlethaler, and R. W. G. Wyckoff. 1948. Mikrofibrillenbau der pflanzlichen Zellwände. Experientia, 44:75.

Garrett, S. D. 1955. Microbial Ecology of the Soil. Trans. Brit. Mycol. Soc.,. 38:1.

Greenfield, L. J., and C. E. Lane. 1953. Cellulose Digestion in Teredo. J. Biol. Chem., 204:669.

Gyllenberg, H. H. G. 1952. Studies of Associative Populations in the Breakdown of Cellulose. Act. Agric. Scand., 2:183.

Halliwell, G. 1957. Cellulolytic Preparations from Microorganisms of the Rumen and from *Myrothecium verrucaria*. J. Gen. Microbiol., 17:166.

Hashimoto, Y., and K. Onoma. 1949. Digestion of Higher Carbohydrates by Molluscs. Bull. Jap. Soc. Sci. Fish., 15:253.

Kroulik, A. 1913. Ueber Thermophile Zellulosevergarer. Zentr. Bakt. Parasitenk. Abt. II, 36:339.

Lasker, R., and A. C. Giese. 1954. Nutrition of the Sea Urchin *Strongylocentrotus purpuratus*. Biol. Bull., 106:328.

Lavine, T. F. 1946. A Study of the Enzymatic and Other Properties of the Crystalline Style of Clams: Evidence for the Presence of a Cellulase. J. Cell. Comp. Physiol., 28:183.

Mansour, K., and J. J. Mansour-Bek. 1934. The Digestion of Wood by Insects and the Supposed Role of Micro-organisms. Biol. Rev., 9:363.

Moir, R. J., M. Somers, and H. Waring. 1956. Studies on Marsupial Nutrition. 1. Ruminant-like Digestion in a Herbivorous Marsupial (*Setonix brachyurus* Quoy and Gaimard). Austral. J. Biol. Sci., 9:293.

Morton, J. E. 1952. The Role of the Crystalline Style. Proc. Malacol. Soc. Lond.·, 29:85.
Nair, N. B. 1956. The Path of Enzymic Hydrolysis of Cellulose in the Woodboring Pelecypod *Bankia indica*. J. Sci. Indus. Res. (India), 15C:155.
Newell, B. S. 1953. Cellulolytic Activity in the Lamellibranch Crystalline Style. J Mar. Biol. Assoc. U.K., 32:491.
Pirie, N. W. 1953. Cellulose as a Subject for Speculation and Commercial Enterprise. Biochem. Soc. Symposia, 11:61.
Ray, D. L., and J. R. Julian. 1952. Occurrence of Cellulase in *Limnoria*. Nature, 169:32.
Reese, E. T. 1956. Enzymatic Hydrolysis of Cellulose. Appl. Microbiol., 4:39.
-------. and H. S. Levinson. 1952. A Comparative Study of the Breakdown of Cellulose by Micro-organisms. Physiol. Plant., 5:345.
Sandegren, E., and L. Enebo. 1952. Cell Wall Decomposing Enzymes of Barley and Malt. 1. Determination and Stability Investigations. J. Inst. Brewing, 68:198.
Schlottke, E., and G. Becker. 1942. Digestive Enzymes in the Intestine of *Hylotrupes* Larvae. Biol. Generalis, 16:1.
Tracey, M. V. 1950. Cellulase from Leaves and Roots of Tobacco. Biochem. J., 47:431.
-------. 1951. Cellulase and Chitinase of Earthworms. Nature, 167:776.
-------. 1952. Chitinase and Cellulase of Nematodes. 2nd Int. Cong. Biochem. Paris, Résumés des. Comm., p. 242.
-------. 1953. Cellulases. Biochem. Soc. Symposia, 11:49.
-------. 1955. Cellulase and Chitinase in Soil Amoebae. Nature, 175:815.
Wardrop, A. B. 1955. The Mechanism of Surface Growth in Parenchyma of *Avena* Coleoptiles. Austral. J. Bot., 3:137.
Whitaker, D. R. 1953. Purification of *Myrothecium verrucaria* Cellulase. Arch. Biochem. Biophys., 43:253.

CELLULOSE DECOMPOSITION: FUNGI

Elwyn T. Reese

How does an organism digest an insoluble substrate?

The Organism-Substrate Relationship

In order for a fungus to attack an insoluble substrate such as cellulose, it must produce and liberate the enzyme required for the hydrolysis of that substance. Such an action is quite remarkable since these enzymes are not produced in detectable amounts in the absence of the substrate. We are all familiar with the concept of enzymes produced in response to *soluble* inducers that enter the microorganism and react with the enzyme-forming system (adaptive enzymes). But how can an insoluble substrate, cellulose, induce the formation of an adaptive enzyme? There may be a reaction with cellulose at the surface of the organism, or traces of a soluble inducer may enter the cell. Our present data favor the idea of a soluble inducer. This requires the assumption that traces of adaptive enzymes are always being produced, even though the amounts are not detectable. These traces act on the insoluble substrate to produce soluble fragments which enter the cytoplasm and act as inducers. In one organism we have studied, *Trichoderma viride* (Mandels and Reese, 1957), cellulase is produced when the fungus is grown on the products of cellulose hydrolysis, e.g., glucose or cellobiose (or lactose which closely resembles cellobiose in structure).

Products of hydrolysis of other polysaccharides also induce the formation of the corresponding polysaccharase:

galacturonic acid induces polygalacturonase (Phaff, 1947)
xylose induces xylanase (Simpson, 1954)
N-acetylglucosamine induces chitinase (Mandels and Reese, 1957)

In our experiments with *T. viride,* conditions (addition of calcium and minor elements) are required for enzyme production that are not required for growth. To date, we have been unable to find other fungi that are induced to form cellulase in the presence of glucose or of cellobiose. Indeed, our evidence indicates that glucose is not a true inducer but is metabolized to an inducer, probably a β-glucoside. It appears that the β-1-4 glycosidic linkage must be present in soluble compounds that act as inducers of cellulase. We have no idea whether the monomeric inducer units of other polysaccharases undergo a similar change.

As the fungus penetrates a barrier, all components--not merely cellulose--are solubilized. Several enzymes, many of them adaptive, are involved. One of these is xylanase. A rather odd fact is that, while it is an adaptive enzyme, it is formed when the organism is grown on cellulose (in *T. viride* [Reese] and in *Stachybotrys atra* [Thomas, 1956]). An easy answer is that xylanase formation is induced by traces of xylan which are no doubt present in the wood cellulose (Solka Floc) that we commonly use. But xylanase is also formed on celluloses free of xylan, i.e., cotton, tunicate, and Acetobacter cellulose. Indeed, it is even produced under conditions where *cellulase* is induced in the absence of cellulose, i.e., when *Trichoderma viride* is grown on glucose under the proper conditions. Cellulase, however, is produced when the organism is grown on cellulose but not when it is grown on xylan.

Liberation of hydrolytic enzymes is an active function of living fungal cells (Mandels, 1956). During fungal growth on cotton or wood, enzymes are liberated, chiefly at the hyphal tip. They diffuse to the substrate and digest it. The hydrolysis products diffuse into the fungus cytoplasm. The hypha then grows into the digested region and maintains continual intimate contact with the substrate.

In attacking cotton, many fungi penetrate through the fiber wall into the lumen and do most of their digesting from within (fig. 2). Bacteria (and some fungi) appear to act from the outer surface, working slowly inward.

In the digestion of wood, one pattern of fungal attack is so unique and has received so much attention (Bailey and Vestal, 1937) that it has overshadowed other patterns. The growing fungus seems to follow the orientation of the cellulose fibrils (fig. 3c) and produces rather characteristic designs. Stanier (1942) found that some bacteria are oriented in a similar fashion on the cellulose of the cotton fiber. Perhaps such an action is due to spatial configurations in the substrate that permit more ready diffusion of enzyme, or perhaps the more readily hydrolyzable cellulosic fragments are to be found along the fibrils. The fungi producing this pattern were identified as Ascomycetes. Wood attacked by *Xylaria* spp. had such a pattern (Weston and Bailey, unpublished), as did cooling-tower wood attacked by *Chaetomium* spp. Savory, 1954). These are perithecial Ascomycetes.

However, Basidiomycetes, not Ascomycetes, are the important wood-destroying fungi, and these do not seem to follow the above pattern (Proctor, 1941). In general, the hyphae of the true wood rotters appears to grow in all directions, irrespective of the plane of fibril orientation. Indeed, their course is often perpendicular to the plane of orientation as the fungus grows from the lumen of one cell to that of the next. It may be that we are oversimplifying. Recently it has been shown that a single organism (*Halophiobolus rufus,* an Ascomycete) exhibits both patterns. The colored hyphae appear to grow in all directions, while the colorless hyphae followed "a steep spiral within the middle layer of the secondary wall" (Wilson, 1956). Further observations are required.

The amount of mycelium per wood cell required to give complete hydrolysis varies. Some published photos seem to suggest that one or two simple hyphae growing through a cell are sufficient to digest all of the cellulose therein. This would require movement of enzymes and hydrolysis products over appreciable distances. *Polyporus versicolor* is an example of such a fungus (Scheffer, 1936).

Filtrates containing both cellulase and xylanase act on holocellulose (spruce) to remove xylan much more rapidly than cellulose (unpublished data). The intact organism probably acts in a similar fashion, i.e., some components are digested before others. The possibility that xylan is digested very much in advance of cellulose should be kept in mind when an attempt is made to explain the presence of spiral cracks in the cell walls of decaying wood (fig. 4). The latter assume the angle of the fibrils (and so are reminiscent of the pattern of growth of the Ascomycetes described above). If this cracking is due to removal of some component, it would indicate a rather extensive diffusion of the digesting factor. If this is digestion, and if it is enzymatic, then it opens up the possibility that cellulase molecules may move along such channels and act at some distance from the fungus.

Solubilization as a Preliminary to Hydrolysis

I should like to bring together certain facts related to the early stages of decomposition of insoluble substrates by microorganisms. My interpretation of these facts must be regarded as incomplete, since it is based on insufficient data. This is an area, however, which I feel merits considerable attention.

Excellent work has been done in a related field, the early states of decomposition of structural proteins. An interesting summary has been presented by Linderstrøm-Lang (1952): "You are all familiar with the clothes moth, a nasty little insect whose larvae eat keratin of the hair in the clothes and thrive on it. Normally keratin is exceedingly stable and is not attacked by enzymes; but when these larvae eat the hair, it vanishes in their intestine and nice white crystals of uric acid appear at the other end. . . . We set out to investigate how the larvae do it. . . . The result of our investigation showed that the clothes moth larva has a very high alkalinity, pH 10, in its intestine, and secretes an enzyme, a proteinase, which acts at high pH, and further has the property of not being inhibited by SH groups. At the same time, a reducing agent seems to be secreted which reduces the hair so that the S-S- bonds are turned into -SH groups, whereby the protein is made soluble, and can be cleaved by the proteinase."

This account clearly distinguishes two steps, (1) solubilization and (2) hydrolysis. I suggest that all insoluble substrates must be solubilized prior to hydrolysis. In the above instance, solubilization is a nonenzymatic reaction brought about by a reducing agent which splits the bonds holding the chains closely together.

What are the properties of cellulose that may be involved in solubilization? Chemically, cellulose is for the most part a linear polymer of glucose and quite insoluble in water. Owing perhaps to greater attractive forces between chains, or to the intramolecular arrangements of the various groups, cellulose is much less soluble than other polymers of glucose (amylose, amylopectin, dextran) of equal chain length. Cross linkages have been reported in cellulose. Their frequency is usually placed at about 1 in every 500 anhydroglucose units, a value perhaps too low to make it likely that their rupture would have much effect on solubilization.

A few nonglucose units are present in the cellulose chain, or as polymers intimately associated with α-cellulose:

α-Cellulose	Xylose	Arabinose	Uronic acid anhydride
	%	%	%
Cotton	0.0	0.0	1.34
Softwood sulfite pulp	0.47	0.0	1.32
Hardwood sulfite pulp	0.33	0.08	1.0
Wheat straw	5.48	3.27	2.52

(Adams and Bishop, 1953)

The abundance of these, as well as close association with lignin and hemicelluloses, undoubtedly affects the ease and extent of solubilization of cellulose as it occurs in nature.

The physical nature of cellulose, as well as its chemical nature, may affect solubilization. One important factor is porosity. Cellulose is perforated by a continuous system of interconnected microcapillary spaces which may be filled with water or lignin or a wide variety of other organic compounds. The concentric layers seen in cross sections of cotton fibers and of secondary walls of wood cells are manifestations of variation in porosity. The pore size may be approximately 16-20 Å in cotton, and 40-60 Å in regenerated cellulose (Cooke, Dusenburg, Kienle, and Lineken, 1954).

A second important physical property of cellulose, from the standpoint of enzymatic hydrolysis, is hydratability, the ability to take up water. The greater the moisture uptake the greater the hydrolysis rate (Walseth, 1952). Any chain to be hydrolyzed must be in contact with water. The enzyme attacking the chain diffuses through water. Other properties of cellulose are important primarily in their effect on moisture uptake. "Crystalline" cellulose, resistant to enzyme action, has a lower moisture uptake than the more susceptible "amorphous" cellulose. Chemical modifications of cellulose fibers leading to decreased moisture uptake also lead to decreased susceptibility to enzymatic attack.

We suggest that the usual definition of solubility is too narrow. In polymers such as cellulose with tightly bound aggregates, it is possible that the loose chain ends can become completely hydrated, soluble for all purposes, except that they are unable to move freely (fig. 5). In the native, highly crystalline, celluloses, the "loose" chain ends would be extremely few. Solubilization, as here considered, would be the process of increasing the number of these.

Changes in Cellulose Accompanying Its Breakdown by Culture Filtrates

The properties of cellulose are modified both by the action of cellulolytic organisms and by the enzymes from these organisms. Many of these changes have been discussed before and will not be elaborated upon. Others are less well known and will receive more attention. In the following paragraphs, emphasis is on native cellulose, usually cotton. The changes are presented approximately in the order in which they occur.

Increase in alkali swelling

The cotton fiber is surrounded by a primary cell wall which has a restrictive effect on the swelling of the cellulose in 18 per cent alkali and on the adsorption of the dye, congo red. Damage to this wall by mechanical, chemical or biological action is detectable by an increase in alkali swelling and dye uptake (Marsh, Merola, and Simpson, 1953).

Alkali swelling provides our most rapid and most sensitive measure of the activity of filtrates of cellulolytic fungi on native cellulose. Changes in cotton treated with culture filtrates diluted 100x are detectable by this method in less than 30 minutes at 50° C (Reese and Gilligan, 1954). Under these conditions no changes are detectable by measurement of reducing sugars, or by any other method. The action is probably on the cellulose network of the primary wall.

Visible changes

Cotton fiber treated with cellulolytic fungus filtrates show visible changes of two types.

1. *Transverse cracking* occurs on short time (2 hour) exposure to the filtrates, provided the suspension is agitated (Marsh, 1957), but not in controls.

2. *Spiral fissures* appear on more prolonged incubation (5 days) in the culture filtrate. On swelling in cuprien, the hydrolyzed fiber "opened into a ribbon similar to an untwisted drinking straw" (Blum and Stahl, 1952). The time required to produce spiral fissures can be reduced to one hour by using undried fiber taken from the relatively mature but unopened boll (Marsh, 1957). It was suggested that the relatively larger spaces between cellulose aggregates in the undried fiber permitted greater diffusion of the enzyme molecules.

The spiral fissures in cotton are similar to those pictured by

Proctor (fig. 4) for similarly degraded woody cells and are supposedly related to the orientation of the underlying fibrils. In cotton, at least, the number of fissures per fiber is so low that any relationship to the fibril orientation does not appear highly significant.

Another interesting observation (Marsh, 1957) is that the undried cotton from the boll is attacked by *Aspergillus flavus* to produce the helical splitting. *A. flavus* does not degrade mature cotton. There is a possibility that the action responsible for the splitting is not cellulolytic, but that the organism is acting on materials other than cellulose occurring in the undried cotton.

The effects of transverse cracking and of spiral fissures on cellulose decomposition are not clear. It is uncertain to what extent "ribbon" formation, due to spiral fissures, occurs in the hydrolysis mixture and to what extent the opening up of the fiber is due to subsequent actions, such as the treatment with cuprien. It is quite well established that no detectable weight loss, change in DP, or change in crystallinity follows the cracking (Blum and Stahl, 1952). The primary wall seems to be an effective barrier to enzymatic degradation. From the fact that many fungi digest from within the fiber, we have assumed that the cellulose next to the lumen was more easily digested. However, if it can be shown that ribbon formation is the *usual* thing, our thinking will need revision for then the inner wall will be equally exposed and by inference must be equally resistant.

Loss in tensile strength

Perhaps the next detectable change in whole cotton fiber treated with enzyme is a loss in tensile strength (Blum and Stahl, 1952). This loss reaches a maximum (35 per cent) in 3 days with culture solution of full potency and does not increase thereafter. Transverse cracking and spiral fissures may account for the tensile strength losses.

Decrease in degree of polymerization (DP)

Enzymatic degradation of cellulose leads to a lowered DP. While this is difficult to demonstrate on whole fiber (see above), it is relatively easy to show with fibers put through the Wiley mill (40 mesh). In our tests (table 1) the changes in DP recorded were obtained in 2 days (or less) using a filtrate of *Trichoderma viride*. As native cellulose is chemically or mechanically "loosened up" the rate of hydrolysis is increased, as measured by change in DP or by production of reducing sugars. In degraded celluloses such as those used by Walseth (1952), the change in DP is quite rapid. On the other hand, in acid-prepared hydrocelluloses, DP changes accompanying enzyme (or acid) hydrolysis are either slight or absent (table 1).

TABLE 1
EFFECT OF ENZYME HYDROLYSIS ON ALKALI SOLUBILITY AND ON DP OF CELLULOSES AND HYDROCELLULOSES

Substrates	Solubility in Alkali* Before Enzyme Hydrolysis	Solubility in Alkali* After Enzyme Hydrolysis	Degree of Polymerization Before Enzyme Hydrolysis	Degree of Polymerization After Enzyme Hydrolysis	Loss in Weight during Enzyme Hydrolysis
Cellulóses (cotton)	%	%			%
A Kiered	4.9	7.6	4970	4200	20.
B Mercerized	6.8	12.3	5040	3040	31.
C Decrystallized	7.9	15.7	4670	3100	25.
D Amine decrystal.	11.2	35.8	3920	1630	69.
Hydrocelluloses					
AH	55.1	52.3	225	227	23.
BH	64.9	52.3	138	145	18.
CH	94.2	91.3	133	128	23.
DH	90.4	90.5	112	104	27.

*1 per cent of substrate in 10 per cent NaOH at 5°C for 30 minutes. Results are averages of two values. (Reese, Tripp, and Segal, 1957)

Increase in alkali solubility: the alkali soluble, water insoluble intermediate

When a fungus *(Chaetomium globosum)* attacks a cotton fiber (figs. 6, 1) it converts the adjacent cellulose to a form soluble in alkali (Abrams, 1950). No microscopic change is apparent during the early stages of digestion, but when the fiber is placed in alkali the affected portions are dissolved, leaving holes. As incubation with the fungus progresses, the number of holes increases (fig. 6). An important feature of this step is that the crystalline areas are being converted into a material which is susceptible to enzyme hydrolysis. This must be some sort of "loosening up" operation. Further hydrolysis to water soluble components is apparent by the later appearance of holes that are visible even without the addition of alkali (fig. 6). In wood, as in cotton, a marked increase in alkali solubility occurs during the early stages of attack (Von Pechmann and Schaile, 1950).

The hydrolysis of cellulose by culture filtrates proceeds in a similar manner, except that the enzyme acts on the whole surface rather than on restricted areas adjacent to the hyphal tips. There is a marked increase in alkali solubility (table 1), and this is accompanied by a decrease in degree of polymerization

(DP). Our results seem to indicate that, with enzyme hydrolysis, the action is on a small part of the total substance, i.e., on a thin surface layer.

The alkali soluble material is apparently made up of comparatively short chains since on their removal the residue has a much higher DP. This agrees with the generally accepted view

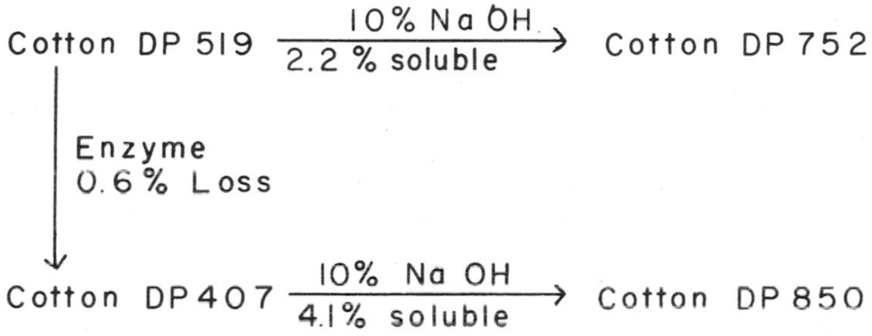

Cotton DP 519 —(10% NaOH; 2.2 % soluble)→ Cotton DP 752

↓ Enzyme 0.6 % Loss

Cotton DP 407 —(10% NaOH; 4.1% soluble)→ Cotton DP 850

that cellulose chains of low DP are alkali soluble. Alkali soluble cellulose, as might be expected of broken chains, does not contribute to tensile strength (Abrams, 1950). As a "loosened up," readily hydrated substance, the alkali soluble cellulose should be much more susceptible to further hydrolysis than the initial cellulose, but it actually appears to be more resistant. An interesting point that does not appear in the data (table 1) supports this. The alkali solubility has about doubled during the 2-day hydrolysis period. Yet, with this increase, the *rate* of hydrolysis has dropped to a very low level.

The alkali soluble fractions of naturally occurring celluloses and of chemically degraded celluloses may be quite different from each other and from the alkali soluble fraction resulting from microbial action. Celluloses A-D (table 1) have progressively higher alkali solubilities, and the hydrolysis rates are in the same order. On the other hand, the hydrocelluloses having very high alkali solubilities have comparatively low rates of hydrolysis by enzymes. Alkali solubility, per se, is therefore not a reliable index of susceptibility to enzyme hydrolysis.

Though there are no data on the thickness of the alkali soluble layer, the implication is that the factor responsible for its formation must have penetrated the native cellulose. Since the enzyme molecules seem unable to penetrate, is it possible that a nonenzymatic factor is responsible for the formation of the alkali soluble intermediate? Is this a part of the solubilization step?

Decrease in susceptibility to further enzyme action

The most rapid digestion of cellulose by culture filtrates occurs in the initial exposure. The decrease in rate of digestion and the eventual cessation of digestion are not due to enzyme inactivation. The rate falls off even though the enzyme solution is periodically renewed, and the proportion of active enzyme to substrate is kept constant. This is in marked contrast to the action of the fungus itself, which is able to digest the cellulose completely.

The rate of hydrolysis for filtrates of different organisms acting on cellulose (or wood or holocellulose, etc.) approaches zero at different per cent hydrolysis values (fig. 7). Filtrates of *T. viride* and *M. verrucaria* possessed nearly equal values when tested for activity against Walseth cellulose. Yet the filtrate of *T. viride* hydrolyzes ball-milled cotton and wood celluloses completely, while that of *M. verrucaria* is unable to digest more than a small per cent of each substrate, regardless of the time of incubation or the renewal of the enzyme solution.

A single filtrate acts differently against different cellulose preparations of the same particle size. The amount of residue remaining when the rate of hydrolysis by *T. viride* approaches zero is 75 per cent for Kiered cotton, 65 per cent for mercerized cotton, and 35 per cent for amine decrystallized cotton (materials shown in table 1; 40 mesh particles).

Finally, there are differences when a single filtrate acts against a single cellulose differing only in particle size. When *T. viride* filtrate acts on coarse cotton particles (fig. 8, E_4), the initial rate of hydrolysis is very low, and the rate approaches zero when 4 to 5 per cent of the cellulose has been hydrolyzed. With progressively smaller cellulose particles, curves E_3, E_2, and finally E_1 apply. For the small uniform particles obtained by ball-milling (E_1), the initially high hydrolysis rate remains nearly constant until the cellulose is completely digested. This is essentially the type of curve obtained by acid hydrolysis of cellulose. With acid hydrolysis, however, the initial particle size plays no important role for the acid is able to penetrate where the enzyme is not. Other cellulosic materials behave similarly to the cotton used in these experiments. Phosphoric acid prepared cellulose (Walseth, 1952) is rapidly hydrolyzed at first, but the rate falls off quickly. When the residue is retreated with phosphoric acid, it is reactivated to a high degree. The cellulose sol of Norkrans and Rånby (1956) similarly shows increasing resistance to hydrolysis as the more readily digested particles are removed.

The point we wish to make here is that filtrates of cellulolytic fungi cannot completely hydrolyze particulate cellulose, yet the intact fungi are able to do so. We do not know why. We have presented some observations that may or may not eventually lead to a solution. Our current belief is that readily hydrated chains are rapidly split off leaving bundles of chains (fig. 5) which resist further enzymatic hydrolysis. Inability of the large enzyme molecules to penetrate the bundles or to act on the outer chains of the bundle may be important factors. But, again, what does the fungus do that the filtrates are unable to do? Certainly it is in close contact with the cellulose and the enzyme activity locally may be quite high. It produces alkali soluble cellulose, but so, too, does the filtrate. It appears that the "necessary ingredient" is either closely associated with the surface of the fungus or is rapidly inactivated and so absent in filtrates being tested.

Decrease in moisture regain values

Enzyme hydrolyzed cellulose takes up less moisture than the original material (Walseth, 1952). This is generally considered evidence for the removal of the readily digestible cellulose, leaving behind less readily hydrated crystalline cellulose.

We have also observed lower "moisture regain" values on enzyme hydrolyzed samples. Oddly, however, the filtrate *(T. viride)* which gave greater loss in weight showed *less* of a change in "moisture regain" than did a filtrate *(M. verrucaria)* of lesser activity. The interpretation of our results based on alkali solubility appears to contradict those based on moisture regain. Our present data showing an increase in alkali solubility due to the action of culture filtrates (table 1) implies a "loosening" operation which should lead to *increased* moisture uptake.

Methods for Altering the Digestibility of Cellulose

Any change in cellulose that increases the moisture uptake should increase its digestibility. Conversely, any change that reduces moisture uptake should decrease digestibility.

Mechanical grinding of cellulose, increasing its surface by only 5 per cent, led to a 6-fold increase in its rate of hydrolysis by a filtrate of *Aspergillus luchuensis* (unpublished). It appears that this effect is related to the greater susceptibility of the cross-sectional areas to hydrolysis. Further reduction in particle size further increases digestibility (fig. 9, curve II).

Changes in the physical or chemical nature of cellulose lead to changes in its susceptibility to enzyme hydrolysis (fig. 9).

Swelling by strong acid (72 per cent sulfuric or 85 per cent phosphoric) or by strong alkali increases the susceptibility of the reprecipitated product to enzyme hydrolysis. Phosphoric acid treatment is preferable for most purposes, since it does not involve partial esterification of the cellulose as does the sulfuric acid treatment. Exposure of alkali cellulose to air leads to changes making the swollen cellulose more resistant to enzyme hydrolysis.

Treatment of native cellulose with dilute mineral acid, or with weak organic acids, reduces the susceptibility of the residual cellulose to enzyme, either by the removal of the more readily hydrolyzed (amorphous) areas or by the formation of substituted cellulose derivatives. While marked loss in tensile strength of cotton and a lower DP result from the treatment with dilute mineral acid, there is usually no increase in susceptibility to enzyme hydrolysis.

Irradiation with cathode rays (fig. 9) and with ultraviolet (Wagner, Webber, and Siu, 1947) makes cellulose more resistant to enzyme hydrolysis. Heavy dosages of cathode rays bring about a reversal in effect (at about 40 megareps), the residue *increasing* in susceptibility to enzyme action. At high dosage levels, the effect of depolymerization apparently outweighs the changes (probably oxidative) that make for resistance.

Chemical substitution of the cellulose molecule leads to an increase in solubility by keeping the cellulose chains apart, i.e., preventing aggregation, and thus increases digestibility. On the other hand, the addition of substituents to the anhydroglucose units increases resistance to enzyme action. For maximum reactivity a balance must be achieved between these two effects. The nature of the substituent and the nature of the reaction supplying the substituent contribute to the over-all effect. Since there are three possible sites for substitution on an anhydroglucose unit of cellulose, a degree of substitution (DS) of 3.0 is the maximum value attainable. The minimum DS conferring water solubility are (approximately) as follows:

Cellulose Derivative	Minimum DS for Water Solubility
Sulfoethyl cellulose	0.3
Cellulose sulfate	0.3-0.4
Carboxymethyl cellulose, Na, 50 T	0.5
Cellulose acetate	0.7-0.8

When every anhydroglucose unit has one or more substituents, the material is completely resistant. However, some products of DS 1.0 are susceptible to enzyme action since the DS is an average value, and some units escape substitution. Some workers appear to be using water-soluble cellulose derivatives without giving due consideration to the importance of these factors.

Chain length of cellulose derivatives is important in selection of an assay procedure. When enzyme activities are determined by sugar production, short chain lengths giving low viscosities are desirable. When activity is measured by decrease in viscosity, longer chains giving higher initial viscosities are preferable.

Conditions Affecting the Activity of Cellulase

We have discussed in earlier sections the importance of the source of the enzyme, i.e., the organism from which it was obtained, and the effect of cultural conditions on the production of cellulase. Conditions affecting cellulase activity are summarized briefly here.

pH:

The optimum pH for activity of fungus-derived cellulases is near pH 5.0; for bacterial cellulases, it is nearer to pH 7.0. Below pH 3.0 and above pH 8.5 activities are usually very low. Above pH 8.5 fungal cellulases appear to be relatively unstable and can be readily inactivated.

Temperature

There is great variation in optimum temperature for the enzymes of different organisms depending on stability. We use 50°C routinely for activity determinations (1- to 2-hour duration), but, where long incubation is involved and the heat stability unknown, 35-40°C is recommended.

Cofactors

No cofactors are known. Dialyzed solutions retain their activity. Resistance to inhibitors indicates that neither metals nor sulfhydryl groups are required for activity. Cellulase is active in phosphate, citrate, and all other buffers tested.

Within the limitations of the above we find fungal cellulases to be quite stable. Shaking for long periods may inactivate some cellulase. Addition of small amounts of protein appears to offer some protection against such inactivation (Basu and Whitaker, 1953). The enzymes can be precipitated with alcohol or acetone, adsorbed and eluted from solids, and lyophilized without appreciable loss in activity. They can be stored in a refrigerator for indefinite periods either as dried preparations or as solutions preserved with 0.01 per cent merthiolate.

Multiple Factors

At the outset I should like to say that I am reviewing some of our old data and interpreting them in the light of our past experiences. We have done little on this problem in the last three years. The concept of multiple factors, or of a multiple component cellulase, refers not to differences in cellulase from different organisms but rather to the presence of several enzymes in a single organism, all involved in the hydrolysis of cellulose. This concept has been developed within the past 10 years and is supported by the work of Jermyn (1952), King (1956), Miller (1956), and others (table 5). Exhaustive study is required before the ramifications are worked out. In addition to new techniques for enzyme separation, we sorely need well-defined substrate fractions representative of the various stages found during the enzymatic hydrolysis.

Other workers in the field have reached different conclusions. No attempt has been made here to evaluate the unienzymatic concept of cellulase. Its current champion, Dr. Whitaker, will undoubtedly so do.

When we began to study cellulose decomposition fourteen years ago, we accepted the then current view of cellulase as a single enzyme hydrolyzing cellulose to cellobiose. Of course, there were differences in cellulase depending upon the source; differences in pH and temperature relationships, for example. But, in general, the picture was one of simplicity. Then we found that some organisms which were unable to grow on native cellulose (cotton) did grow on some of the new water-soluble derivatives of cellulose (Siu and Reese, 1953). Earlier workers had also reported the growth of "noncellulolytic" organisms on variously degraded celluloses. This led us to propose a two-step process, the enzymes of which we designated C_1 and Cx. About the C_1 step, we had little to offer except that C_1 seemed to alter native cellulose in such a way that Cx could then complete the hydrolysis. Since "cellulase" covers both steps, the introduction of new terms was required. The subscripts were chosen to indicate a first step (C_1), followed by a possible series of steps (Cx).

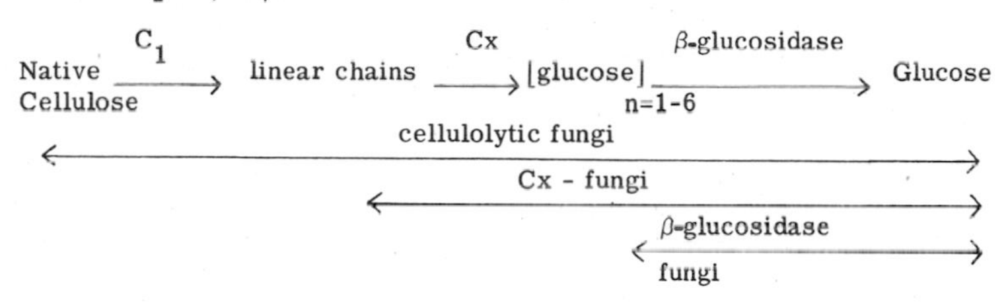

For purposes of simplification, we directed our attention to the extracellular enzymes. Culture filtrates (including those of the highly cellulolytic *Trichoderma viride* and the noncellulolytic *Aspergillus sydowi)* were compared for their relative activities on cotton cellulose and on carboxymethyl cellulose (CMC). When they were adjusted to the same activity on CMC and then tested for their activity on cotton, a 40-fold difference was found between them *(T. viride* being the more active). Thus it was shown that cellulolytic culture filtrates, like intact organisms, contain components for more than one step in cellulose breakdown.

Homogeneity of a substance is difficult to prove. Ninety-nine criteria may indicate the likelihood that a substance is homogeneous; the hundredth may prove otherwise. As no one criterion can establish homogeneity, so also might one question a single criterion of heterogeneity. The enzymes with which most of us have worked are far from being pure. The results of any procedure are open to more than one interpretation. When the application of x procedures indicates heterogeneity, and when no single interpretation can be applied to explain all the differences, then we feel that the case for "multiple factors" merits attention. Recognizing this we have set down some of the data that have accumulated.

Specificity.

The activities of different enzymes on a series of related substrates differ. Enzyme 1 may act on cellobiose faster than on salicin; enzyme 2 may do just the opposite. The identity of the linkage attacked puts both enzymes into the same class: β-glucosidase. The differences in relative activity single them out as different members of that class.

The types of cellulolytic activity which we have measured (table 2) have been selected both for the rapidity with which determinations can be made (maximum time, 2 hours) and for the variety of cellulose types represented by the substrates. All of the activities have been reduced to a unit basis by dilution to a selected value falling on the linear portion of the curve for activity vs. enzyme concentration. Changes in viscosity of CMC solutions (DS 0.7-0.8) have also been used as a measure of enzyme activity, but these results have not been expressed on a unit basis.

TABLE 2
TYPES OF CELLULOLYTIC ACTIVITY

Activity	Substrate	Enzyme + Substrate + Buffer		Time	Substrate Concentration	Determination*	Reference
Cx	CMC DS 0.52	1 ml	9 ml	1 hr.	0.5%	Reducing Sugar	Gilligan, Reese, 1954
W	Walseth Cellulose	1	1	2	0.5		Gilligan, Reese, 1954
SF†	Cotton fiber	← 15 ml →		1	1.6	Alkali uptake	Gilligan, Reese, 1954; Reese, Gilligan, 1954; Marsh, Merola, Simpson, 1953

*Assay conditions: pH 5.4; temperature 50°C.
†SF=Swelling factor.

Species differences. Filtrates from a wide range of cellulolytic organisms grown on cotton duck vary in their relative activities on the three substrates. The data recorded here (table 3) represent extremes. The differences are explainable on the assumption that several components are involved in cellulose hydrolysis, and that organisms differ in the amounts of each produced.

TABLE 3
VARIATION IN CELLULASES OF DIFFERENT ORGANISMS

Organism	Ratios	Activities in μ/ml		
		W	SF	Cx
Helminthosporium sp.	W/SF 9.4	75.0	8.0	7.5
Trichoderma viride	0.1	14.3	128.0	8.0
Fusarium roseum	W/Cx 10.4	83.0	120.0	8.0
Pestalotiopsis westerdijkii	0.9	29.0	44.0	30.0
Stachybotrys atra	SF/Cx 18.0	36.0	66.0	3.6
Streptomyces sp.	0.3	14.0	6.7	23.3
Myrothecium verrucaria		21.0	23.0	6.0

Digressing for a moment, I should like to comment briefly on the fungi studied most extensively in recent years.

Aspergillus oryzae (Jermyn, 1952) is not an *active* cellulolytic organism. It is unable to attack native cellulose (cotton). It does produce several components active against CMC and similarly degraded or modified celluloses. It is rich in β-glucosidases acting on simple glucosides. "Luizyme" (Luitpold-Werk, Munich, Germany), a commercial source of "cellulase," is probably derived from this species. *Stachybotrys atra* (Jermyn, 1953; Thomas, 1956) and *Myrothecium verrucaria* (widely used) degrade native cellulose (cotton) more rapidly than does *Trichoderma viride* (Reese, unpublished), but cell-free filtrates of the former are less active against *native* cellulose than is that of *T. viride*. *Chaetomium globosum* is an active cellulolytic organism used in deterioration tests of cotton fabrics. It is not used in enzyme studies because the enzymes do not accumulate in the culture fluids of shake flasks. It may be that they are inactivated rapidly or are adsorbed on the cellulose of the medium. The Basidiomycetes are the important wood-rotting fungi. The cellulases of these are being studied by Norkrans (1956, 1957) and by Jennison and his co-workers.

Differences during growth. During the early phases of growth of *M. verrucaria* on cellulose (fig. 10 B), the activity (W) measured against solid cellulose is high relative to the activity on soluble CMC (Cx). On longer incubation the activity against Walseth cellulose diminishes, while the Cx remains constant. Cultural conditions also affect the amount of each component present in the culture fluid. As the concentration of cellulose in the growth medium increases (fig. 10 A), the ratio W/Cx decreases. pH changes also affect the ratios. A reflection of the changes occurring during growth is seen in the variety of electrophoretic records obtained by Hash (1927) from different culture filtrates of *M. verrucaria.*

Differences resulting from enzyme purification. Cellulose adsorbs preferentially the factor which has high relative activity on the solid cellulose (table 4). The enzyme components active on the longest cellulose chains resist elution from cellulosic columns (Hash, 1957).

TABLE 4
CHANGES IN SPECIFICITY OF FILTRATES DUE TO ADSORPTION OF PART OF THE ACTIVITY ON CELLULOSE

Enzyme filtrate	RW/Cx		R SF/Cx	
	Before Adsorption	After Adsorption	Before Adsorption	After Adsorption
Stachybotrys atra	13.5	6.5	18.6	13.0
Fusarium roseum	9.5	6.5	14.8	7.2
Myrothecium verrucaria	3.9	2.9	6.5	2.5
Trichoderma viride	2.0	1.5	13.0	7.4
Pestalotiopsis westerdijkii	0.5	0.5	1.1	0.6

Method: To 20 ml filtrate pH 4.2, add 600 mg cellulose (Solka Floc). After 15 minutes at RT, centrifuge off cellulose. Assay original and supernatants by usual techniques. 47-54 per cent of the Cx of each solution was adsorbed under the test conditions.

Column separations have shown that various factors are adsorbed and eluted at different rates, resulting in the production of eluates differing markedly from each other (and from the original) in their specificities. We have eluted *T. viride* components from various adsorbents and have obtained eluates with the following extremes in the ratio W/Cx:

Adsorbent	RW/Cx Range
Cellulose	0.7- 3.0
Ca-PO_4 gel	0.4-14.0
Kaolin and celite	0.7- 6.6
Cellulose acetate phthalate	1.6- 3.2
$Fe(OH)_3$	0.5- 6.0
Cationic resin IR 50H	0.7- 3.0

Five electrophoretic fractions from *M. verrucaria* cellulase (supplied by Dr. Gail Miller) showed a progression in RW/Cx values from 1.0 to 2.6

Mobility

Proteins which move at different rates in column chromatography or under the influence of an electrical potential are considered to be different. In recent years, several highly purified proteins (including enzymes) have been separated into two or more independently moving fractions (lysozyme, ribonuclease, bovine plasma albumen, protamine-SO_4). Crystallinity no longer guarantees homogeneity.

In cellulase studies, several different techniques have shown the presence of components moving at different rates (table 5). The number of components was not always estimated by the authors. The values shown are our estimates based on the data seen. Typical data are shown in figures 11 and 12.

TABLE 5
DETECTION OF CELLULASES OF DIFFERENT MOBILITIES

Method	Components	Organism	Reference
Electrophoresis			
Paper	6	*Asp. oryzae*	Jermyn, 1952
Paper	7	*M. verrucaria*	Hash, 1957
Block (starch)	8	*M. verrucaria*	Miller and Blum, 1956
Block (starch)	5	Rumen of cow	King, 1956
Convection	6	*M. verrucaria*	Grimes, 1955
Moving boundary	3	*M. verrucaria*	Miller, unpublished
Chromatography			
Paper	3	*Asp. terreus*	Reese and Gilligan, 1953
Column			
Phosphate gel	5	*M. verrucaria*	Gilligan and Reese, 1954
Phosphate gel	3	*T. viride*	Gilligan and Reese, 1954

There is a surprisingly large number of components based on mobility. At present, Dr. Miller finds between 12 and 16 components in *M. verrucaria* cellulase subjected to block electrophoresis (unpublished). He is working also with fractions from calcium phosphate gel columns, and hopes to compare the fractions obtained by one technique with those from the other. When convinced that each component behaves as an entity, he will attempt to characterize them further. It is essential that other differences (e.g., substrate specificity) be correlated with differences in mobility.

Properties of enzyme fractions

In our fractionation of *T. viride* enzyme on calcium phosphate gel, we obtained 3 components (A, B, and C) which were eluted at different rates. When rechromatogramed on the same adsorbent they acted as separate entities.

Although many indications suggested that these fractions were not necessarily homogeneous, we determined to see what differences could be detected between fractions. In addition to mobility, there were marked differences in substrate specificities, and in the items listed below (table 6).

TABLE 6
COMPARISON OF VARIOUS *T. VIRIDE* FRACTIONS FROM CALCIUM PHOSPHATE GEL COLUMN

Fraction	R		Cellobiose Inhibition			Methocel Inhibition		Rel. Activity on Celluloses
			Viscosity			Viscosity		
	W/Cx	SF/Cx	0.5%	1.0%	2.0%	0.006%	.02%	$DP=\frac{10x}{x}$
			%	%	%	%	%	
A	0.48	13.0	+3.0	+5.0	2.0	(+4.)	0	1.1
B	0.34	1.1	25.0	36.0	42.0	35.0	64.0	1.6
CD	12.0	11.0	27.0	NT	51.0	33.0	63.0	0.66
Original	1.5	23.0	21.0	32.0	34.0	36.0	41.0	

*Values in ( ) are probably not significant.

Effect of inhibitors. Fraction A, acting on CMC, was either stimulated--or unaffected--by 1 per cent cellobiose; fractions B and C were inhibited by cellobiose. Similarly, methylcellulose (0.02 per cent) had no effect on fraction A, but strongly inhibited fractions B and C.

Mode of action. A rapid increase in the fluidity of CMC per unit increase in reducing value may indicate random splitting, while a more gradual increase per unit of reducing value may indicate endwise degradation. This parallels the analysis of α- and β-amylases by plotting starch-iodine color vs. reducing value (Hanes and Cattle, 1938). Fraction A, having the steeper slope, is considered to act more randomly than fractions B or C, which are "more endwise" in their actions (fig. 13). A Similar situation exists in *M. verrucaria*. Fractions of *M. verrucaria* cellulase were tested against CMC of different chain lengths. The first fractions eluted (from a cellulose column) were highly active on shorter chains. Later eluates were more active on longer chains (Hash, 1957).

The action of the three *T. viride* fractions on hydrolyzed celluloses differing in DP by a factor of 10 has been compared. Fraction B is more active on longer chains, fraction C against shorter, while fraction A showed about equal activity on long or short chains (table 6). Earlier workers (Siu, 1951, p. 274) found "no influence of the degree of polymerization of hydrocellulose on the rate of hydrolysis by enzymic preparations of fungi." It may be that such an effect can be found only when cellulase is separated into its individual components.

Synergistic effects. In following the recovery of activity from phosphate gel columns, we noted that, whereas nearly all the Cx could be accounted for, the recoveries of the W and SF activities were poor. Since protein recovery approached 100 per cent, it was unlikely that the W and SF activities were being left on the column. Either they were inactivated or a synergistic effect present in whole filtrate was lost on separation of the fractions.

In testing combinations of filtrates, we did indeed find stimulation (fig. 14). This was greatest between the first and last fractions from the column, but appreciable even between other pairs. The stimulatory effect was on W and SF activities, but not on Cx activity. The synergistic action was *not* due to nonenzymatic components, or to differences in amount of protein, or to the presence of noncellulolytic enzymes. The effect was not peculiar to fractions obtained from phosphate gel columns. A similar (but less pronounced) result was obtained with the first and last fractions from a kaolin column, in which the eluting agents were acetate and bicarbonate (rather than phosphate). Using the *T. viride* fractions successively rather than simultaneously we showed that the stimulation was due to action of fraction A in making the substrate more readily susceptible to fraction C.

Other indications of the multiple nature of cellulase

pH optima. Most pH-activity curves for cellulase have a single peak. The existence of more than one peak is an indication of more than one component. In our own work, we have found the two peaks only when measuring the SF activity. With Cx or W activity, only one peak was found.

Cellulase of	Acting on	Number of Peaks	Reference
Asp. oryzae	cuprophane	3	Freudenberg and Ploetz, 1939
Myrothecium verrucaria	cellulose sulfate	3	Grimes, 1955
Trichoderma viride, QM 6a	cotton (swelling factor)	2	Reese and Gilligan, 1954
Pen. pusillum, QM 137g	cotton (swelling factor)	2	Reese and Gilligan, 1954
Stromatium fulvum (larvae)	filter paper	2	Mansour and Mansour-Bek, 1937

Inactivation studies. From Dr. Norkrans' (1957) data on heat inactivation, we have derived the curves shown in figure 15. Leveling off of the inactivation at values lower than 100 per cent may indicate the presence of a heat-resistant component in the cellulase.

In our work on chemical inactivation of cellulase and of β-glucosidase (Reese and Mandels, 1957) we found occasional examples of inhibitors active against the cellulase of one or more organisms, but inactive against other cellulases. We also found several instances in which the curve for inactivation of cellulase of a single organism leveled off before reaching 100 per cent inhibition. As with heat inactivation these results suggest the presence of more than one component (fig. 16).

Activity curves. Curves for enzyme activity vs. concentration vary widely for enzymes from different sources and for different fractions from the same enzyme (fig. 17).

In concluding, we believe that a series of steps is involved in the breakdown of native cellulose by microorganisms. The first, or solubilizing, step may or may not be enzymatic. The later hydrolysis stages involve a series of enzymes. From a practical viewpoint, it is our inability to duplicate the solubilizing step that limits the production of sugars from cellulose by enzymatic means.

Mr. Robert Blum (Qm Lab) has offered an interpretation that agrees with the data presented. Enzymatic hydrolysis of the amorphous cellulose takes place. *This action results in cleavage of long chains where they pass through amorphous regions, leaving behind crystalline regions of shorter chain lengths.* These crystalline residues may be alkali soluble (note the high alkali solubility of the hydrocelluloses of DP 250, table 1). They would be expected to be more resistant to enzyme hydrolysis than the original material (as observed) and to have lower moisture regain values (which they do have). DP would also be lowered (as it is).

Where does this lead? The original cellulose particle presents a surface of mixed amorphous and crystalline areas. Enzymes would digest out the amorphous leaving exposed only the crystalline material. This action would be restricted to a surface layer because of the inability of enzyme molecules--because of their size--to penetrate into the particle. As a result, there remains a particle resistant to further enzyme hydrolysis. These particles do not spontaneously degenerate into susceptible ones. Yet the intact organism does digest them. The solubilization (C_1) step, then, remains as the conversion of tightly bound aggregates (crystalline) into loosely bound aggregates (amorphous) by an agent possessed by the organism but lacking in the filtrates. Blum believes this to be merely the result of high concentrations of cellulase acting locally in the case of the organism. However, if this were so, the much longer action of potent cellulolytic filtrates should show at least a little hydrolysis. And all organisms that can degrade water-soluble cellulose derivatives should be capable of attacking the native cellulose--but many can not.

The essential difference between Blum's interpretation and ours is on the nature of the alkali-soluble fraction, i.e., whether it is "crystalline" or "loosened up." Obviously the answer cannot be obtained by examination of the material dissolved in alkali. There is a possibility that enzymatic removal of amorphous cellulose might separate the crystalline particles from each other so that they can be mechanically removed from the parent substance (agitation). Or perhaps some physical method may be able to detect surface changes resulting from the enzyme action.

LITERATURE CITED

Abrams, E. 1950. Microbiological Deterioration of Cellulose during the First 72 Hours of Attack. Text. Res. J., 20:71-86.

Adams, G. A., and C. T. Bishop. 1953. Polysaccharides Associated with α-Cellulose. Nature, 172:28-29.

Bailey, I. W., and Mary R. Vestal. 1937. Significance of Certain Wood Destroying Fungi in the Study of the Enzymatic Hydrolysis of Cellulose. J. Arnold Arboretum, 18: 196-210.

Basu, S. N., and D. R. Whitaker. 1953. Inhibition and Stimulation of the Cellulase of *Myrothecium verrucaria*. Arch. Biochem. Biophys., 42:12-24.

Blum, R., and W. H. Stahl. 1952. Enzymic Degradation of Cellulose Fibers. Text. Res. J., 22:178-92.

Cooke, T. F., J. H. Busenbury, R. H. Kienle, and E. E. Lineken. 1954. Mechanism of Imparting Wrinkle Recovery to Cellulosic Fabrics. Text. Res. J., 24: 1015-36.

Freudenberg, K., and T. Ploetz. 1939. Über die Verschiedenheit von Cellulase und Lichenase. I, Mitteilung über den enzymatischen Abbau polymerer Kohlenhydrate. Zeitschr. für physiol. Chemie, 259:19-27.

Gilligan, W., and E. T. Reese. 1954. Evidence for Multiple Components in Microbial Cellulases. Can. J. Microbiol., 1: 90-107.

Grimes, R. M. 1955. Preparation and Study of Cellulases and Hemicellulases of *Myrothecium verrucaria*. Ph. D. Thesis, Mich. State Univ., East Lansing, Mich.

Hanes, C. S., and M. Cattle. 1938. Starch Iodine Coloration as an Index of Differential Degradation by the Amylases. Proc. Roy. Soc. London, Ser. B., 125:387-414.

Hash, J. H. 1957. Cellulolytic Enzyme Systems of *Myrothecium verrucaria*. Thesis, Va. Poly. Inst., Blacksburg, Va.

Jermyn, M. A. 1952. Fungal Cellulases. II, Complexity of Enzymes from *Asp. oryzae* that Split β-glucosidic Linkages, and Their Partial Separation. Austral. J. Sci. Res., Ser. B, 5:433-43.

-------. 1953. Fungal Cellulases. III, *Stachybotrys atra* Growth and Enzyme Production on Non-cellulosic Substrates. Austral. J. Biol. Sci., 6:48-49.

King, K. W. 1956. Basic Properties of the Dextrinizing Cellulases from the Rumen of Cattle. Va. Agr. Expt. Sta. Tech. Bull., 127.

Linderstrøm-Lang, K. 1952. Proteins and Enzymes. Stanford Univ. Press., Stanford, Calif.

Mandels, G. R. 1956. Synthesis and Secretion of Invertase in Relation to Growth of *Myrothecium verrucaria*. J. Bact., 71: 684-88.

Mandels, Mary, and E. T. Reese. 1957. Induction of Cellulase in *Trichoderma viride* as Influenced by Carbon Sources and Metals. J. Bact., 73:269-78.

Mansour, K., and J. J. Mansour-Bek. 1937. On the Cellulase and Other Enzymes of the Larvae of *Stromatium fulvum*. Enzymol. 4:1-6.

Marsh, P. B., G. V. Merola, and M. E. Simpson. 1953. An Alkali Swelling Test Applied to Cotton Fiber. Text. Res. J., 23:831-41.

-------. 1957. Microscopic Observations on Cotton Fibers Subjected to Enzymatic Degradation. Text. Res. J., 27:913-16.

Miller, G. L., and R. Blum. 1956. Resolution of Fungal Cellulase by Zone Electrophoresis. J. Biol. Chem., 218:131-37.

Norkrans, B. 1957. Studies of β-Glucoside--and Cellulose Splitting Enzymes from *Polyporus annosus* Fr. Physiol. Plant., 10:198-214.

-------, and B. G. Rånby. 1956. Studies of the Enzymatic Degradation of Cellulose. Physiol. Plant., 9:198-211.

Pechmann, H. von, and O. Schaile. 1950. Ueber die Aenderung der dynamischen Festigkeit und der chemischen Zusammensetzung des Holzes durch den Angriff holzzerstörender Pilze. Forstwiss. Centr., 69:441-65.

Phaff, H. J. 1947. Production of Extra-cellular Pectic Enzymes of *Pencillium chrysogemum*. I, On the Formation and Adaptive Nature of Polygalacturonase and Pectinesterase. Arch. Biochem., 13:67-81.

Proctor, P. 1941. Penetration of the Walls of Wood Cells by the Hyphae of Wood Destroying Fungi. Yale Univ. Sch. For. Bull., 47.

Reese, E. T. 1955. Enzymatic Hydrolysis of Cellulose. Appl. Microbiol., 4:39-45.

-------, and W. Gilligan. 1953. Separation of Components of Cellulolytic Systems by Paper Chromatography. Arch. Biochem. Biophys., 45:74-82.

-------, and W. Gilligan. 1954. Swelling Factor in Cellulose Hydrolysis. Text. Res. J., 24:663-69.

-------, and Mary Mandels. 1957. Chemical Inhibition of Cellulases and β-Glucosidases. Res. Rept., Pion. Res. Div. Quartermaster Res. and Dev. Center, Natick, Mass. Microbiol. Ser., 17:1-42.

-------, V. Tripp, and L. Segal. 1957. Effect of Cellulase on Celluloses and Hydrocelluloses. Text. Res. J., 27:626-32.

Savory, J. G. 1954. Breakdown of Timber by Ascomycetes and Fungi Imperfecti. Ann. Appl. Biology, 41:336-47.

Scheffer, T. C. 1936. Progressive Effects of *Polyporus versicolor* on the Physical and Chemical Properties of Redgum Sapwood. USDA Tech. Bull., 527.

Simpson, F. J. 1954. Microbial Pentosanases. I, A Survey of Microorganisms for the Production of Enzymes that Attack the Pentosans of Wheat Flour. Can. J. Microbiol., 1:131-39.

Siu, R. G. H. 1951. Microbial Decomposition of Cellulose. Reinhold Publ. Co., N.Y.

-------, and E. T. Reese. 1953. Decomposition of Cellulose by Microorganisms. Bot. Rev., 19:377-416.

Stanier, R. Y. 1942. The Cytophaga Group. A Contribution to the Biology of Myxobacteria. Bact. Rev., 6:143-96.

Thomas, R. 1956. Fungal Cellulases. VII, *Stachybotrys atra*. Production and Properties of the Cellulolytic Enzyme. Austral. J. Biol. Sci., 9:159-83.

Wagner, R. P., H. H. Webber, and R. G. H. Siu. 1947. Effect of Ultraviolet Light on Cotton Cellulose, and Its Influence on Subsequent Degradation by Microorganisms. Arch. Biochem., 12:35-50.

Walseth, C. S. 1952. Influence of Fine Structure of Cellulose on the Action of Cellulases. Tappi, 35:233-38.

Weston, W. H. 1957. Personal communication.

Whitaker, D. R., J. R. Calvin, and W. H. Cook. 1954. Molecular Weight and Shape of *Myrothecium verrucaria* Cellulase. Arch. Biochem. Biophys., 49:257-62.

Wilson, Irene M. 1956. Some New Marine Pyrenomycetes on Wood or Rope: *Halophiobolus* and *Lindra*. Trans. Brit. Mycol. Soc., 39:401-15.

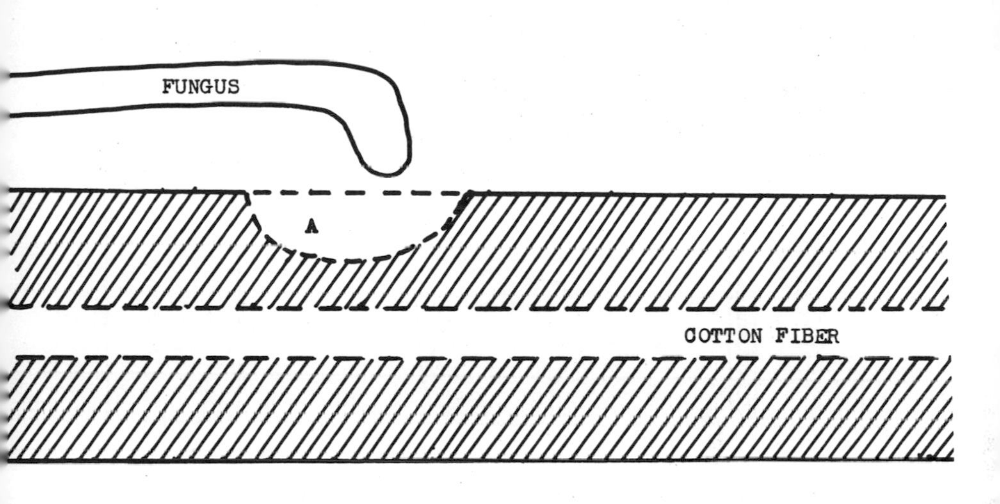

Fig. 1. Fungus attacking a cotton fiber. A, First change is conversion to an alkali soluble intermediate

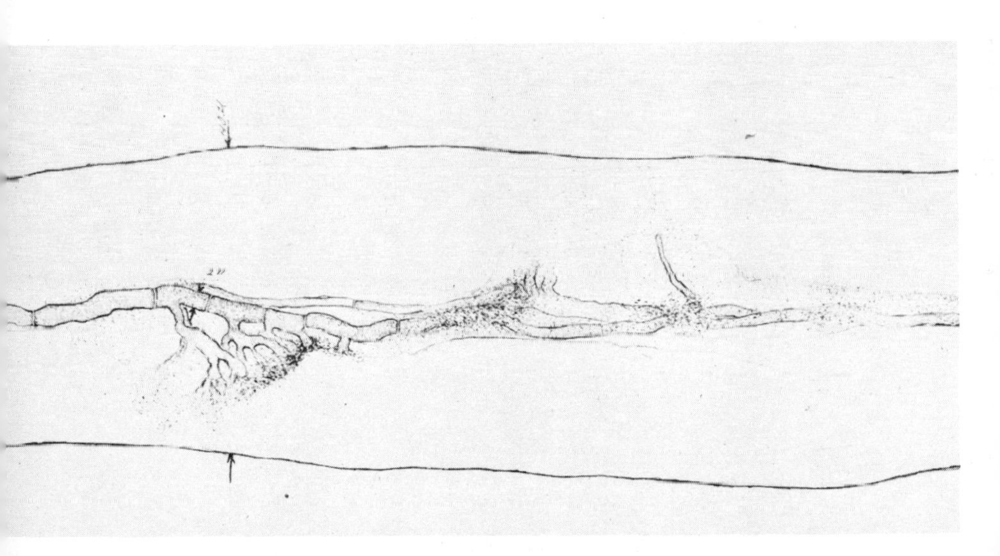

Fig. 2. Destruction of cotton fiber from within by *Memnoniella echinata*. (Drawing by W. L. White)

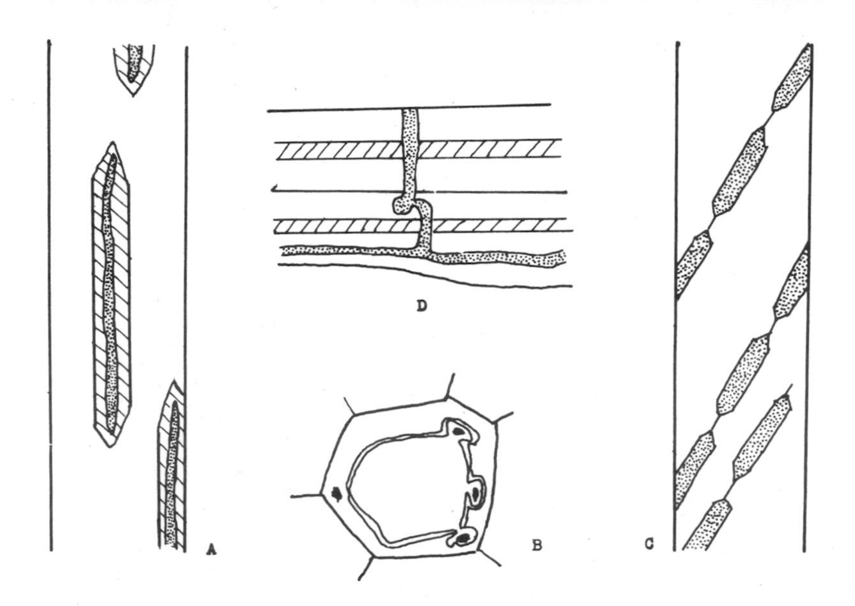

Fig. 3. Growth of fungi in wood. A, Fungus inside of cavity which it produced; B, Cavities produced by fungus, cross section; C, Cavities produced by fungus apparently following orientation of cellulose fibrils; D, Fungus penetrating through walls of wood cells. (A, B, C, after Bailey and Vestal, 1937)

Fig. 4. Spiral cracking in Douglas fir caused by *Fomes pini*, believed to assume the angle of the fibrils of the cellulose of the secondary wall. (After Proctor, 1941)

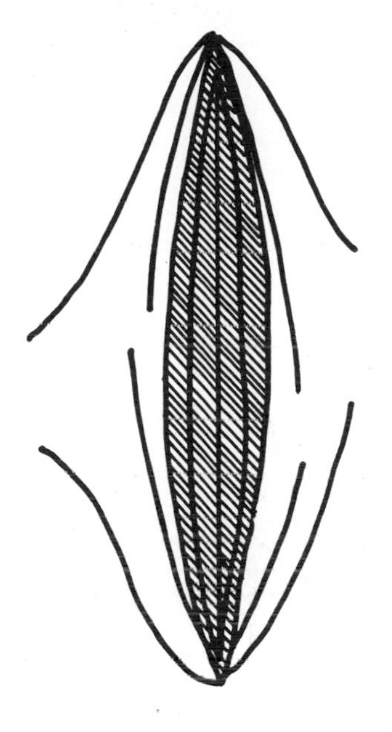

Fig. 5. Hydrocellulose particle (schematic) showing free, hydrated chains attached to an enzymatically resistant core

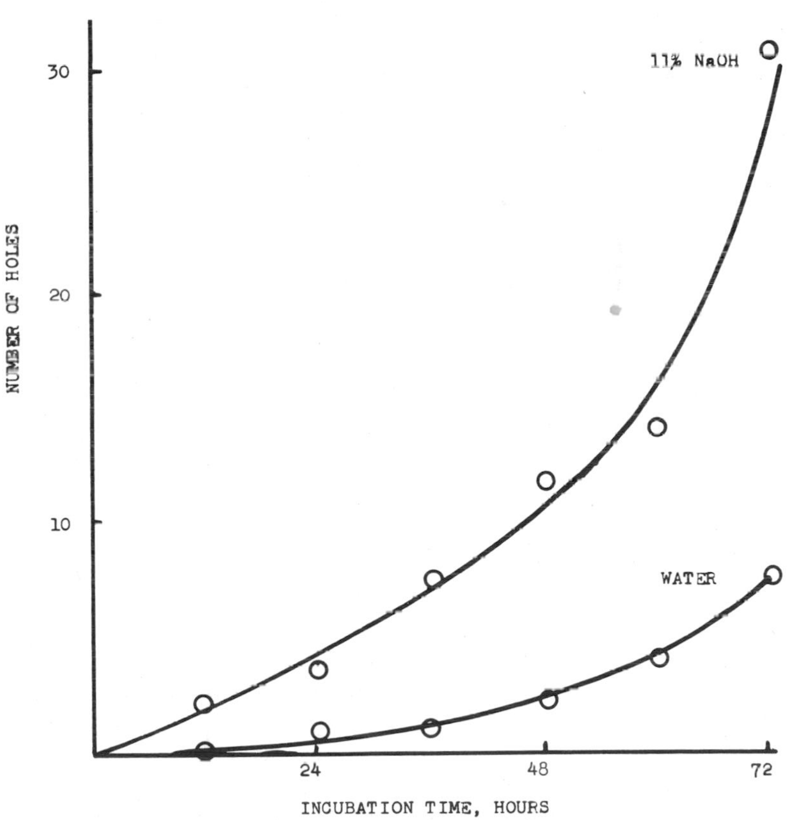

Fig. 6. Attack of cotton fiber by *Chaetomium globosum*. (Abrams, 1950)

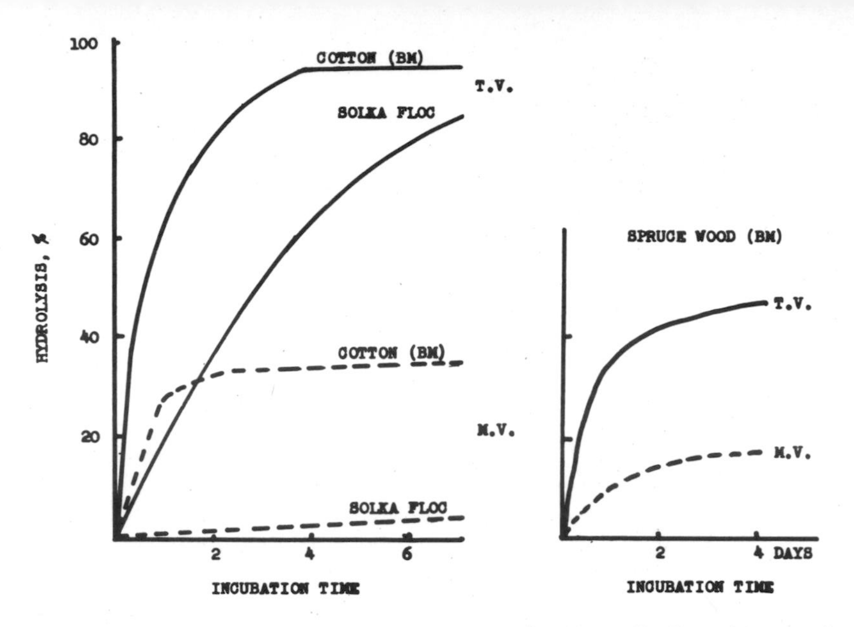

Fig. 7. Enzymatic hydrolysis of cotton (ball-milled), of a wood cellulose (Solka Floc), and of spruce wood (ball-milled). T. V., filtrate of *Trichoderma viride* QM 6a; M.V., *Myrothecium verrucaria* QM 460

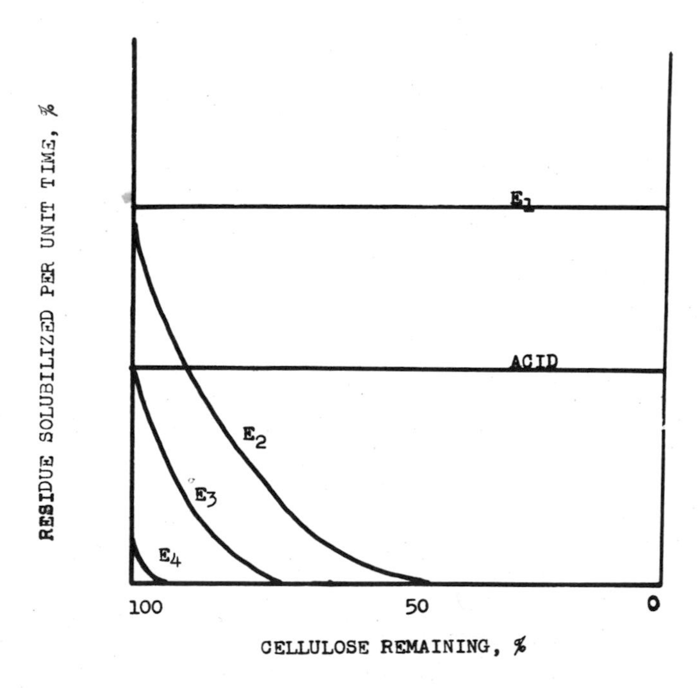

Fig. 8. Rate of hydrolysis of cellulose as affected by particle size (diagrammatic); E_4, cotton fiber; E_3, coarse; E_2, medium; E_1 ball-milled. Acid: relatively independent of particle size

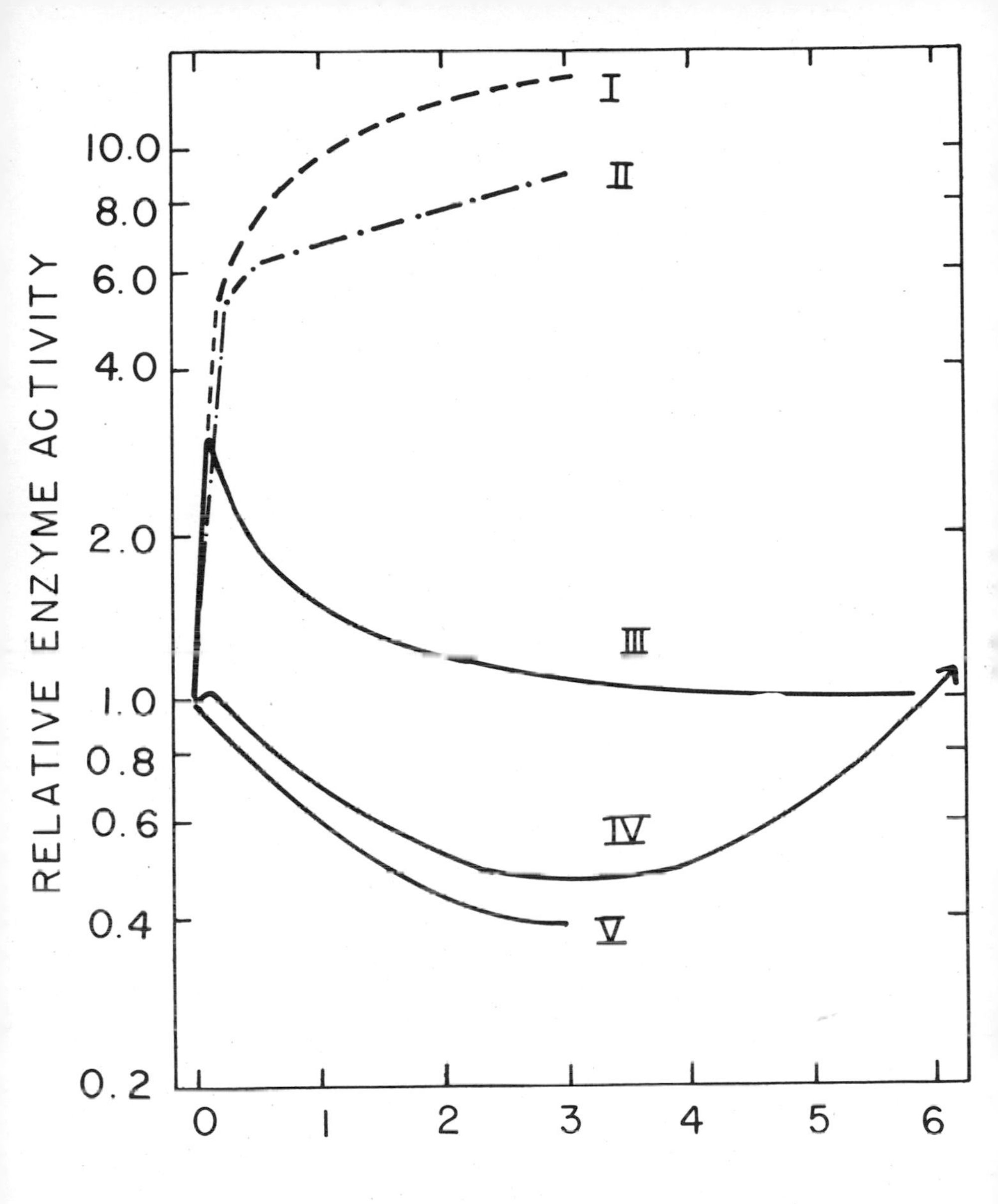

Fig. 9. Effect of various treatments on the susceptibility of cotton cellulose to enzymatic hydrolysis. I, Effect of 85 per cent H_3PO_4, or of 72 per cent H_2SO_4, abscissa in hours; II, Effect of ball milling (vibratory), abscissa in hours; III, Effect of 30 per cent NaOH, abscissa in days exposure to alkali swollen material to air; IV, Effect of irradiation with cathode rays (van de Graaf), abscissa 0-60 megareps; V, Effect of weak organic acids, abscissa in days of refluxing (Loss in weight = 5 to 6 per cent in 3 days). (Reese, 1955)

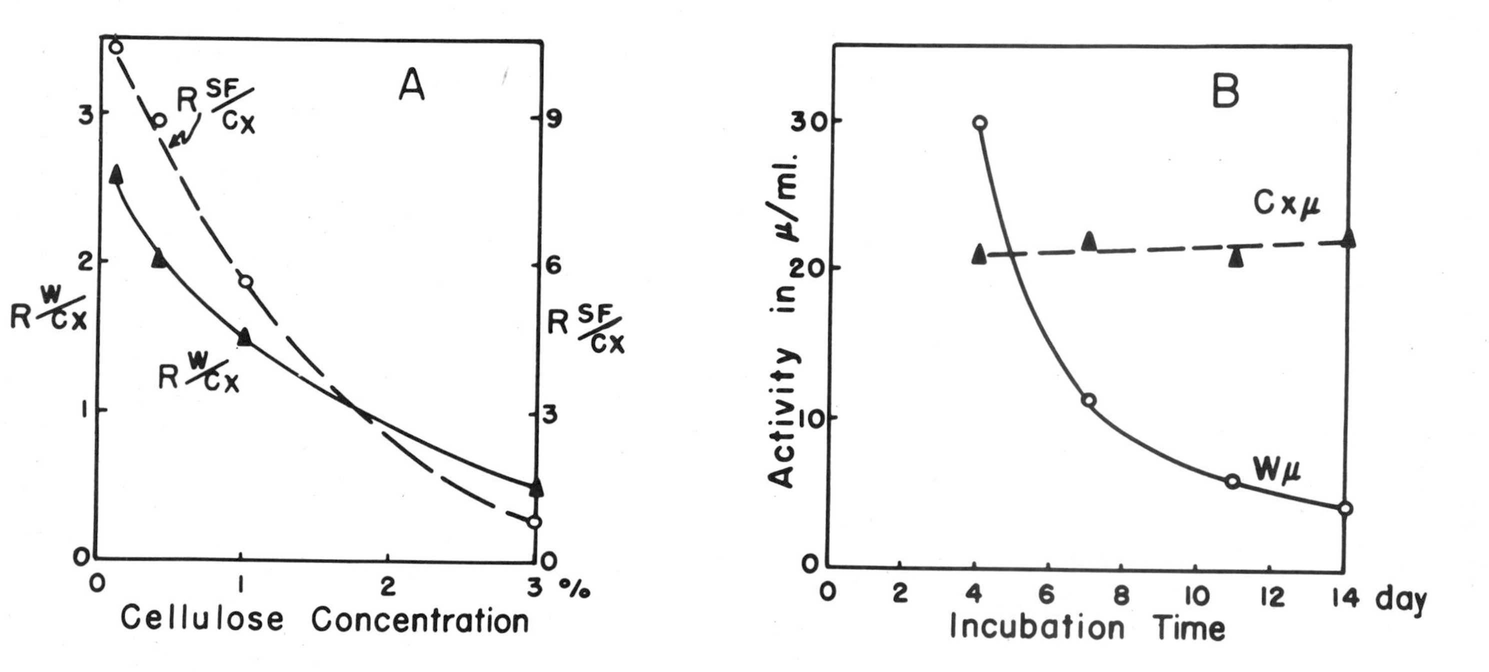

Fig. 10. Effect of growth conditions on nature of the cellulolytic components. Organism is *M. verrucaria*. A, Effect of cellulose concentration in the medium; B, Effect of age of culture (grown on 3% cellulose). (Gilligan and Reese, 1954)

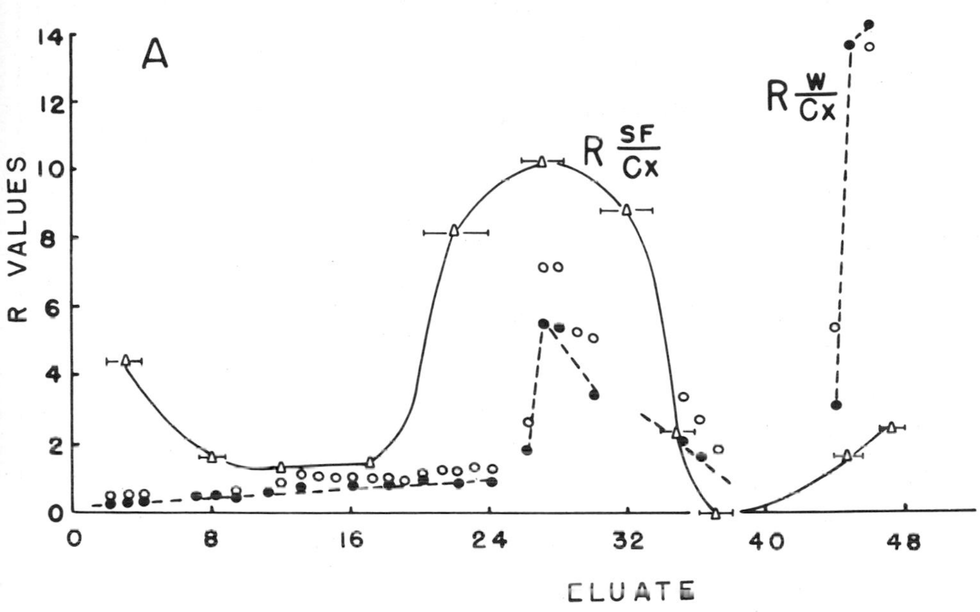

Fig. 11A. Separation of components of *M. verrucaria* cellulase on calcium phosphate gel column. Nature of fractions, based on their relative activities. (Gilligan and Reese, 1954)

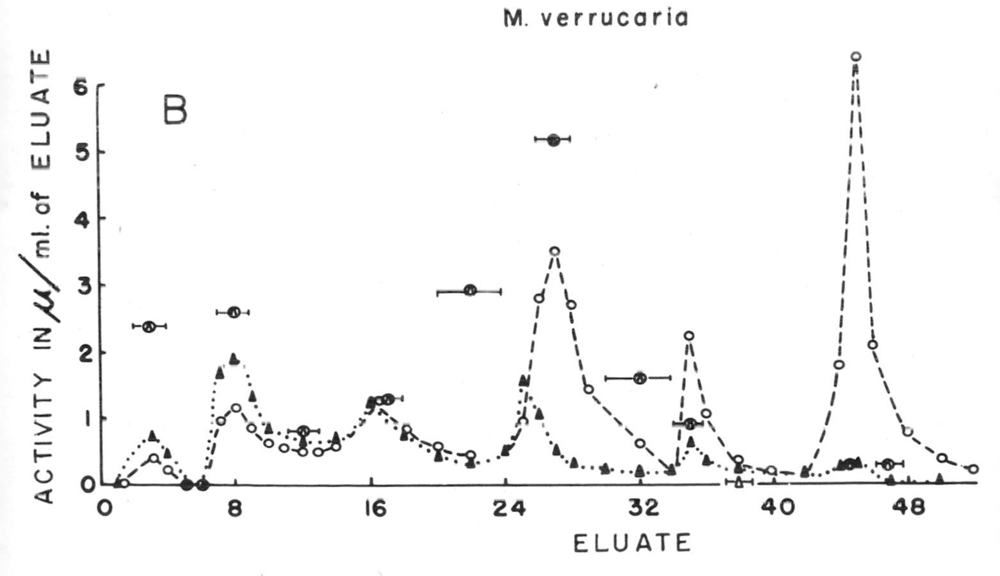

Fig. 11B. Separation of components of *M. verrucaria* cellulase on calcium phosphate gel column. Distribution of activities o - - - o W on Walseth cellulose; ▲ . . . ▲ Cx, on CMC; |⊘| SF, on cotton fiber. (Gilligan and Reese, 1954)

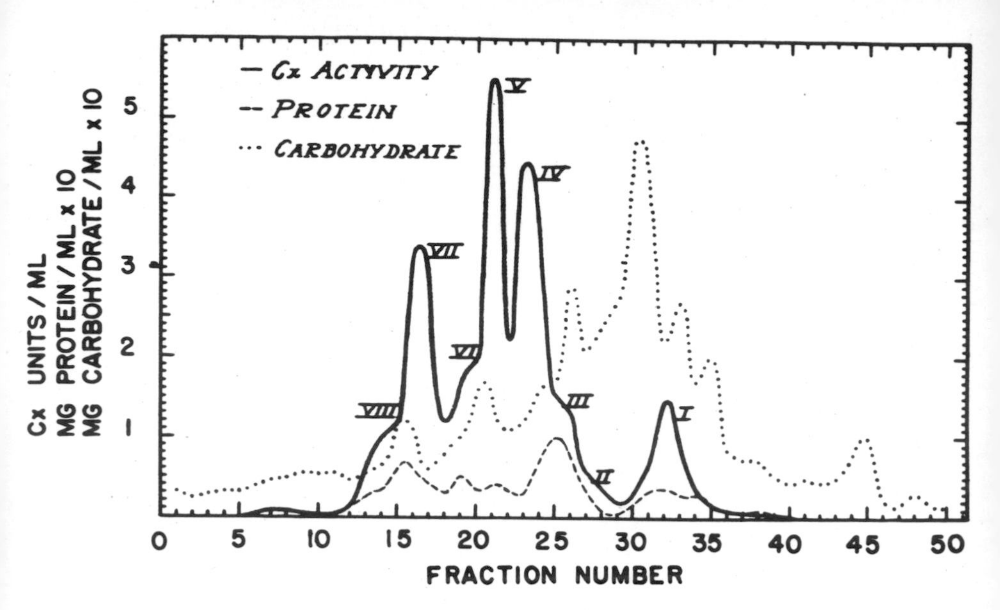

Fig. 12. Separation of components of *M. verrucaria* cellulase in zone electrophoresis. Distribution of protein, carbohydrate and Cx activity. pH 7.2, 42 hours, 400 volts. (Miller and Blum, 1956)

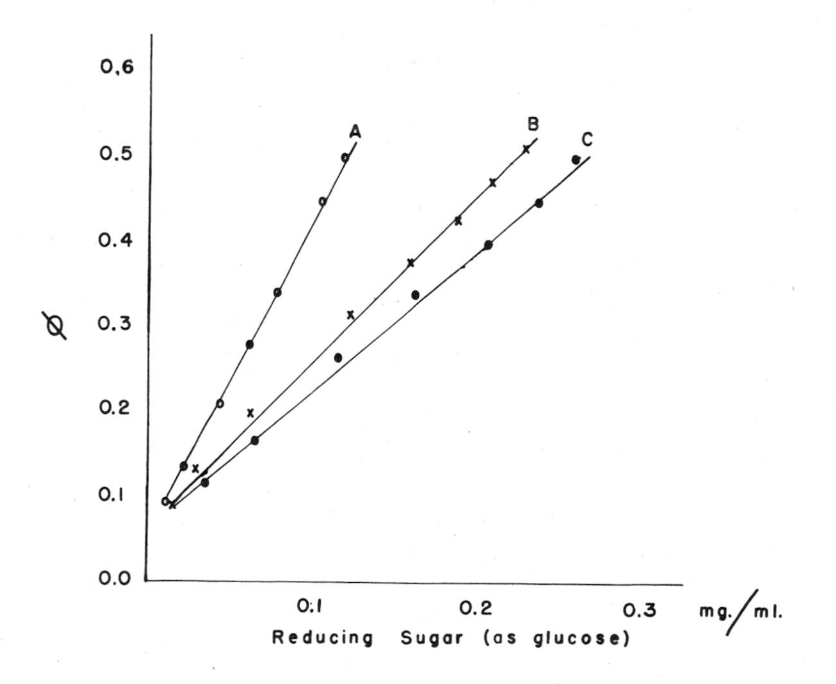

Fig. 13. Hydrolysis of CMC by three components of *T. viride* cellulase. Fluidity vs. reducing groups. Fractions are from Ca-PO_4 gel column. (Gilligan and Reese, 1954)

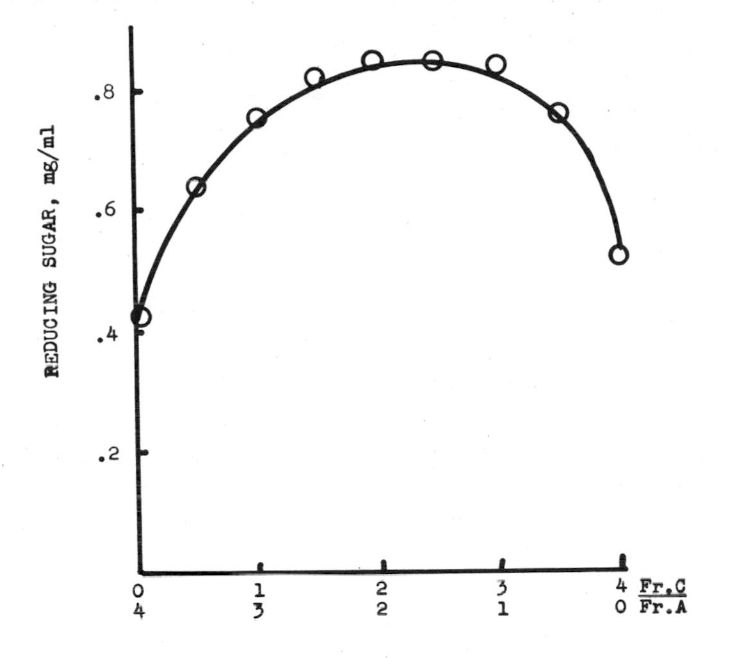

Fig. 14. Synergistic effect of one *T. viride* component on another. Substrate = Walseth cellulose; enzyme fractions are first and last eluted from Ca-PO_4 gel column

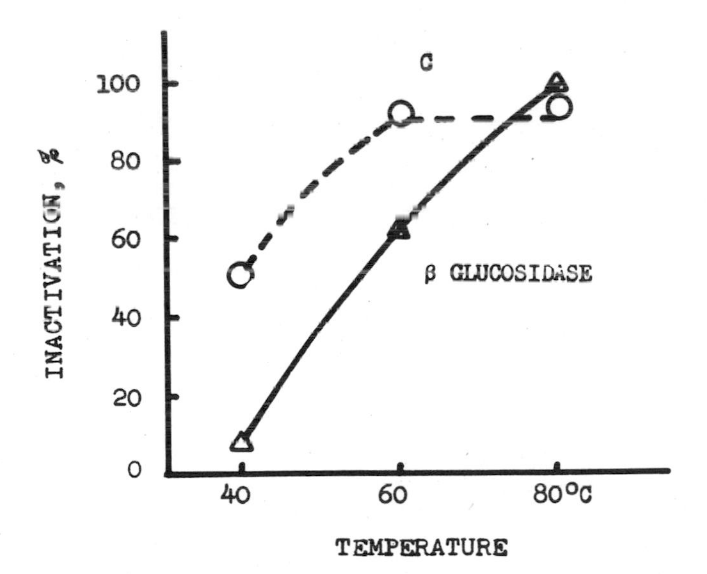

Fig. 15: Heat inactivation of cellulase and of β-glucosidase of *Polyporus annosus* Fr. Inactivation time 20 minutes; cellulase (C) determined on cellulose sol; β-glucosidase by hydrolysis of p-nitro phenyl β-glucoside. (Derived from data of Norkrans, 1957)

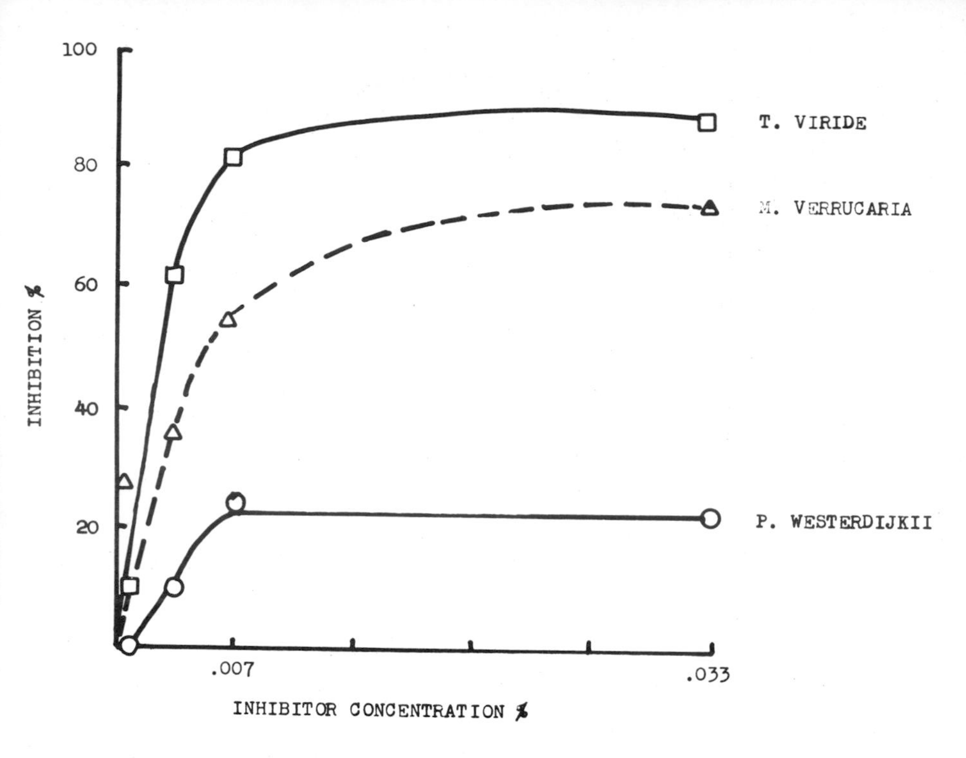

Fig. 16. Inhibition of cellulase (Cx) by disodium ethylene bis-dithiocarbamate. Filtrates of three cellulolytic fungi. (Reese and Mandels, 1957)

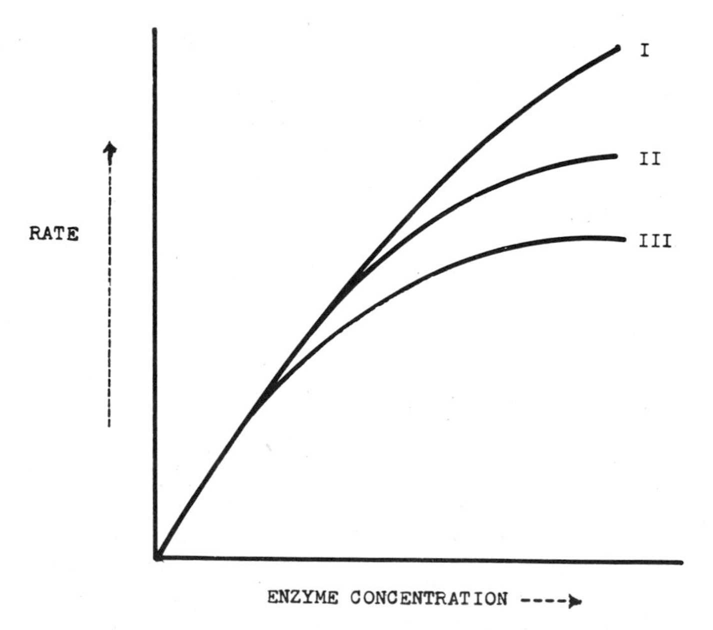

Fig. 17. Enzyme activity curves (diagrammatic)

SOME PROPERTIES OF THE CELLULASE OF *MYROTHECIUM VERRUCARIA**

D. R. Whitaker

Introduction

Cellulose has probably been studied more intensively than any other natural product, and the economic importance of its biological degradation is well recognized. However, until quite recently, cellulases have been a rather neglected group of enzymes. Thus before 1945 only two cellulases had been studied in any detail: the cellulase of the snail, *Helix pomatia,* and the cellulase of Luizym, a commercial product containing enzymes of the mold, *Aspergillus oryzae*. This period of neglect appears to be over; laboratories in at least eight countries are now engaged in research on cellulases, and, if the present interest continues, the next few years should see a very rapid advance in knowledge of these enzymes.

The subject of our own investigations at the National Research Laboratories in Ottawa has been a cellulase obtained from culture filtrates of the mold, *Myrothecium verrucaria* (Alb. and Schw.) Ditm. ex Fr., United States Department of Agriculture strain no. 1334.2. These investigations were carried out in collaboration with Drs. S. N. Basu, J. R. Colvin, W. H. Cook, E. Merler, and R. Thomas. The following discussion will review some of our findings.

Preparation and Purification of the Enzyme
(Whitaker, 1953; and Thomas and Whitaker, 1958)

Our original method of growing the mold for cellulase pro-

---

*Contribution from the Division of Applied Biology, National Research Laboratories, Ottawa. Issued as N.R.C. No. 5374.

duction has been modified to simplify the procedure for larger scale preparations. The main features of our present method are illustrated in figure 1. The mold is grown at 30°C in 10-gallon flasks containing 5 gallons of a dispersion of ground, extracted cotton linters in a solution of mineral salts. The flasks are mounted on a rotary shaker operating at a speed just above that required to keep the cellulose in suspension. A current of sterile, scrubbed air passes over the surface of (not through) the medium. The cellulase activity of the culture medium reaches a maximum within 7 to 10 days.

To isolate the enzyme from the culture medium, the residual cellulose and mold mycelium are filtered off and the filtrate, after clarification in a Sharples supercentrifuge, is concentrated in a Mojonnier low temperature evaporator and then saturated with ammonium sulfate at 1°C by a dialysis procedure which further concentrates the filtrate while it is undergoing saturation. The crude enzyme precipitated from solution contains virtually all the protein originally present in the culture medium. Its electrophoretic pattern is shown in figure 2. Cellulase activity is associated with the main central peak. The components responsible for the other peaks can be removed by a sequence of fractionations with ethanol at a low ionic strength and fractionation with polymethacrylic acid. The purified enzyme then shows only one peak on electrophoresis and one peak in the ultracentrifuge (fig. 3). Balance sheets of "protein accounted for" and "enzymatic activity accounted for" showed high recoveries at each stage of the purification procedure, and the final product showed similar enrichments in enzymatic activity toward a series of substrates, ranging in degree of polymerization from two (cellobiose) to several thousand (cotton linters). These results suggest that the cellulase activity of the filtrate was due to one enzyme.

Figures 4 and 5 illustrate some results of zone electrophoresis in Whatman 3MM and in glass fiber filter paper which confirm these findings in a simple and direct way. The dotted arrows in the figures mark the initial position of the enzyme preparations; the solid arrows mark the position to which an electrically neutral component would be transported by electro-osmotic flow during the period of electrophoresis. The only indicator requiring comment is p-anisidine. This indicator gives a typical aldohexose color reaction after electrophoresis of the enzyme on cellulose paper at pH 7 and pH 5. As the enzyme preparation was free of carbohydrate (no color reaction with anthrone) and gave no color reaction with p-anisidine on glass fiber paper (fig. 5B, strip 3) or after electrophoresis in Whatman 3MM paper at pH 3 (a pH at which cellulase activity is suppressed), the reducing sugar indicated by p-ansidine evidently arises from the action of cellulase on the paper during electrophoresis.

Electrophoresis of the crude enzyme in Whatman 3MM paper (fig. 4) resolved 7 components reacting with ninhydrin; the major components were nos. 4 and 5, and the latter trailed as though absorbed by the paper. The p-anisidine color reaction was displaced slightly in the direction of electro-osmotic flow from component 5 but conformed to its shape. Other tests for cellulase activity showed similar correlations between cellulase activity and component 5. The separations on the glass fiber paper (fig. 5) exclude the possibility that the apparent electrophoretic homogeneity of the purified enzyme in Whatman 3MM paper is an effect of adsorption on the supporting medium. They also demonstrate electrophoretic homogeneity both with respect to the wide range of chemical components for which alkaline permanganate is an indicator and with respect to enzymatic activity toward substrates covering a wide range of chain lengths.

Physical Properties of Purified Enzyme
(Whitaker, Colvin, and Cook, 1954)

Certain physical properties of the purified enzyme are listed in table 1. According to our calculations, a cigar-shaped protein molecule with a length of roughly 200 Å and a maximum breadth of about 30 Å would have equivalent hydrodynamic properties. This length, which for purposes of comparing relative orders of magnitude can be taken as an approximation to the real length of the enzyme molecule, is equal to that of a cellulose chain with a degree of polymerization (DP) of about 40.

The cellulose series thus comprises chains with lengths much less than, of the same order as, and much greater than the length of the enzyme molecule. These categories of relative chain length are illustrated in figure 6 by representative chains of the soluble oligoglucosides with DP's between 2 and 6, a cellodextrin with a number-average DP of 24, and a swollen cellulose with a weight-average DP of 1,000, these being the substrates to be considered in the following survey of the

TABLE 1
PHYSICAL PROPERTIES OF PURIFIED MYROTHECIUM CELLULASE

Diffusion coefficient ($D_{w, 20}$)	$5.6 \pm 0.1 \times 10^{-7}$
Sedimentation constant ($S_{w, 20}$)	$3.7 \pm 0.05 \times 10^{-3}$
Intrinsic viscosity ($[\eta]$)	$8.7 \pm 0.2 \times 10^{-2}$*
Frictional coefficient (f/f_o)	1.44 ± 0.02
Molecular weight	$63,000 \pm 1500$†

*Assuming a nitrogen content of 16 per cent.
†Assuming a partial specific volume of 0.74 cc/g.

influence of chain length on the interaction between substrate and enzyme.

Hydrolysis of Soluble Oligoglucosides

The solubility of the first members of the cellulose series varies inversely with their degree of polymerization, and a DP of about 10 is often considered to mark the transition between chains that give true solutions in water and chains that give colloidal dispersions. The substrates to be considered in this section are the first five members of the series and certain derivatives of them. Cellobiose is readily available commercially; the others--cellotriose, cellotetraose, cellopentaose, and cellohexaose--were obtained by acetolyzing cellulose, deacetylating the products, and fractionating them on a column of Darco G-60 charcoal (Whitaker, 1954b).

Rates of hydrolysis

(Whitaker, 1956a, and unpublished data)

At a sufficiently low enzyme concentration, [E], and substrate concentration, [S], the rate of hydrolysis, dS/dt, of the substrate is directly proportional to both enzyme and substrate concentration and can be expressed by the usual equation for a second order reaction:

$$-dS/dt = K\,[E]\,[S] \qquad \ldots\ldots.1$$

The second order velocity constant, K, is numerically equal to the rate of the reaction when substrate and enzyme are at the unit concentration of 1 mole/liter. If rates are to be compared under second order conditions at one enzyme concentration, $[E]_1$, then, since enzyme is not consumed during the reaction, the product $K[E_1]$ can be replaced by one constant, K', to give an equation as for a first order reaction:

$$-dS/dt = K' [S] \qquad \ldots\ldots 2$$

This pseudofirst order velocity constant is numerically equal to the rate of the reaction when the substrate is at unit concentration and the enzyme is at concentration $[E]_1$ and can be evaluated in the usual way from the equation:

$K' = \frac{1}{t} \ln [S]_0/[S]$ where $[S]_0$ is the initial substrate concentration and $[S]$ is the substrate concentration after hydrolysis for the time, t.

A comparison of such pseudofirst order velocity constants for an enzyme concentration of 20 μg/ml is given in table 2.

TABLE 2
ENZYMATIC HYDROLYSIS OF β-1:4-OLIGOGLUCOSIDES: PSEUDOFIRST ORDER VELOCITY CONSTANTS*

Substrate	K' $sec.^{-1} \times 10^5$
Cellobiose	1.2
Cellotriose	16
Cellotetraose	83
Cellopentaose	500

*Solvent: acetate buffer of pH 5.0; temperature: 29.4°C; enzyme concentration: 20 μg protein/ml.

Cellobiose, the first member of the series, is hydrolyzed very slowly; cellopentaose, the last member, is hydrolyzed very rapidly, and within the series each increase in DP results in a sharp increase in the velocity constant. The increase is greatest at the start: raising the DP from 2 to 3 increases the rate constant by about 1,500 per cent; raising it from 3 to 4 increases the rate constant by about 500 per cent. A similar but much more marked trend is shown when certain other substituents are added to the chains. Thus, if a β-linked sorbitol group is added to the reducing end of cellobiose, the increase in rate constant is about 900 per cent; if it is added to cellotriose, the increase is about 50 per cent; if it is added to cellotetraose, the increase is only 10 per cent. Similarly, although methyl β-glucoside is not hydrolyzed by the enzyme, methyl

β-cellobioside is hydrolyzed about seven times as fast as cellobiose, i.e., at a rate approaching that for cellotriose.

Some indications of the cause of this extreme dependence of rate of hydrolysis on DP can be obtained by measuring the initial rate of hydrolysis as a function of initial substrate concentration and analyzing the data in terms of the simple reaction scheme of Michaelis and Menten. This reaction scheme:

$$E + S \underset{k_{-1}}{\overset{k_1}{\rightleftharpoons}} ES \overset{k_2}{\rightarrow} E + \text{products}$$

postulates two steps for the reaction; in the first, enzyme and substrate combine reversibly to form an enzyme-substrate complex, in the second, the complex breaks down to give the products of the reaction catalyzed by the enzyme; k_1, k_{-1} and k_2 are the rate constants for the individual steps. The overall rate of the reaction in this scheme is given by the equation:

$$-dS/dt = \frac{k_2 \bar{K} [S] [E]}{1 + \bar{K} [S]} \quad ....3$$

$$\text{where } \bar{K} = \frac{k_1}{k_{-1} + k_2} \quad ....4$$

At sufficiently low substrate concentrations, $1 + \bar{K} [S] \approx 1$ and equation 3 assumes the form of equation 1 with K replaced by $k_2 \bar{K}$,

$$\text{hence } K \approx k_2 \bar{K} \quad ....5$$

At sufficiently high substrate concentrations, $1 + \bar{K} [S] \approx \bar{K} [S]$, and at a given enzyme concentration, E_1, the rate of the reaction should level off at a maximum value of $k_2 [E_1]$ and become independent of substrate concentration thereafter.

As shown in figure 7, this latter expectation is not realized: the rate passes through a maximum and then decreases as substrate concentration is increased. Similar relations between substrate concentration and the velocity of the reactions catalyzed by certain other hydrolytic enzymes have been attributed to the formation of a second enzyme-substrate complex containing two molecules of substrate, the second molecule entering at an essential "water site" on the enzyme surface which must be reoccupied by water molecules before hydrolysis of the first molecule of substrate can occur (Morton, 1956). To account for inhibition at higher substrate concentrations, the Michaelis-Menten reaction scheme must be modified, the simplest modification being:

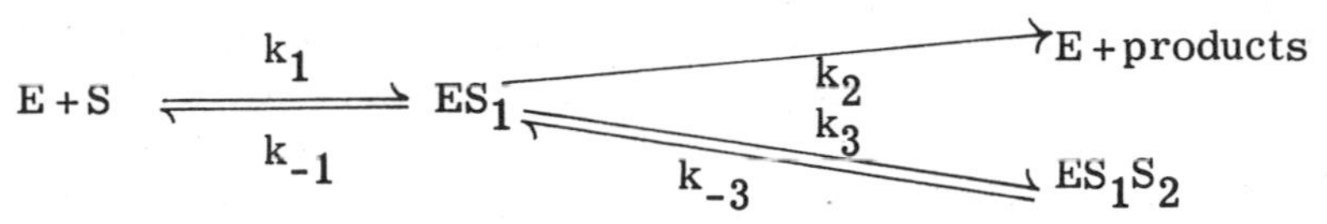

With respect to the problem under consideration, the cause of the trend shown in table 1, this complication is a side issue and the main question raised by it is whether the magnitude of k_3 relative to k_2 is such as to preclude an approximate evaluation of k_2 and $\overline{K}$ by methods, like that of Eadie (1952), which assume Michaelis-Menten kinetics. As an indication that it is not, (1) Eadie plots of the data up to near the maxima in figure 7 show little deviation from a straight line; and (2) as shown in table 3, the values of k_2 and $\overline{K}$ given by the slope and intercepts of the lines satisfy the condition expressed by equation 5. The following preliminary analysis is more concerned with trends in the orders of magnitude of k_2 and $\overline{K}$ than with their exact numerical values; the values of k_2 and K in table 3 will therefore be accepted as good enough for the purpose in view.

TABLE 3
ENZYMATIC HYDROLYSIS OF β-1:4-OLIGOGLUCOSIDES: k_2 and $\overline{K}$*

Substrate	k_2 sec^{-1}	$\overline{K}$ liters/mole	$k_2\overline{K}$ liters/mole/sec.	K† liters/mole/sec.
Cellotriose	18	230	4.2×10^3	5.2×10^3
Cellotetraose	48	870	4.2×10^4	2.6×10^4
Cellopentaose	122	2680	3.3×10^5	1.6×10^5

*Temperature: 24.9°C; pH: 5.0.
†Calculated from K' in table 2.

These values indicate that both k_2 and $\overline{K}$ increase with the DP of the substrate, and hence, as indicated by equation 5, both can contribute to the increase in K, the second order velocity constant. However, the extent of their contributions depends on the relative magnitudes of the two terms in the denominator of $\overline{K}$. If (1) $k_{-1} >> k_2$, then $\overline{K} \approx k_1 / k_{-1} = k_{eq}$, the equilibrium constant of the enzyme-substrate complex, and hence $K \approx k_2 \overline{K} \approx k_2 k_{eq}$; if (2) $k_{-1} << k_2$, then $\overline{K} \approx k_1 / k_2$ and $K \approx k_2 \overline{K} \approx k_1$. If condition (1) holds throughout the series, both factors contribute; if condition (2) holds, i.e., if the reaction is essentially unidirectional, only one factor contributes directly to the value of K. These are, of course, extreme conditions, and neither may hold. The uncertainty can be resolved in various ways: e.g., if the initial course of hydrolysis can be followed over sufficiently brief time intervals, k_1 can be estimated directly. As yet, such measurements have not been carried out.

Measurements of K over a range of temperatures give an Arrhenius activation energy of about 12,000 calories for all these hydrolyses. This result is of interest in two respects. It indicates, first, that increases in the entropy of activation rather than decreases in the energy of activation are primarily responsible for the increase in the rate constants with DP. Second, it shows the extent to which energy barriers are reduced by the enzyme as a catalyst: the corresponding activation energies when the hydrogen ion is the catalyst are about 29,000 calories (Freudenberg and Blomqvist, 1935).

The increase in $\overline{K}$ with DP may be accounted for without difficulty: irrespective of the limiting conditions discussed above, it follows from the definition of $\overline{K}$ (equation 4) that, if both $\overline{K}$ and k_2 increase with DP, the equilibrium constant, k_1/k_{-1}, must increase with DP, i.e., the tendency to form an enzyme-substrate complex increases with the chain length of the oligoglucoside. Explanations are less obvious for the increase in k_2 and for the rapid falling off with increasing DP of the increase in rate constant from addition of a β-sorbityl or β-methyl group. Some evidence of the source of these effects is discussed in the following section.

Method of degradation

(Whitaker, 1954a and b; Whitaker and Merler, 1956.)

Data on the method of cleavage were obtained by carrying out hydrolyses and at various time intervals adding samples of hydrolyzate to boiling ethanol to stop the reaction; the constituent sugars in the sample were then separated by paper chromatography as shown in figure 8, eluted from the chromatogram and analyzed. The method of analysis employed was to measure the reducing value before and after acid hydrolysis; the ratio of the two values gives the DP of the oligoglucoside and hence provides a check on the identity of the sugar.

An example of data obtained in this way is shown in figure 9. The substrate, cellotetraose, may be represented as follows:

$$G^{1}_{NR} \underset{a}{-- 0 --} {}^{4}G^{1} \underset{b}{-- 0 --} {}^{4}G^{1} \underset{c}{-- 0 --} {}^{4}G_{R}$$

where - 0 - represents a glucosidic linkage, G_R the anhydroglucose unit with a free reducing group at carbon atom 1, G_{NR} the anhydroglucose unit with a free (nonreducing) hydroxyl group at carbon atom 4, and G the anhydroglucose units which are substituted at both positions. Hydrolytic cleavage at either of the two terminal linkages a and c will give a molecule of glucose and a molecule of cellotriose which can undergo further hydrolysis; cleavage at the central linkage b will give two molecules of cellobiose. All three sugars, cellotriose, glucose, and cellobiose, appear as initial hydrolysis products in figure 9. The first two indicate that cleavage of terminal linkages does occur; furthermore it occurs at a moderately fast rate. However, it is apparent from the amount of cellobiose formed simultaneously that the preferred point of cleavage is at the central linkage, the rate of hydrolysis at this linkage being nearly twice the combined rates of cleavage at the two terminal linkages.

Similarly with the next sugar, cellopentaose, cleavage at the terminal linkages takes place fairly rapidly, but the predominance of cellobiose and cellotriose in the initial stages of hydrolysis indicates a substantial preference for cleavage at the two nonterminal linkages. With cellohexaose, the main initial hydrolysis products are cellotetraose, cellotriose, and cellobiose, the three sugars given by cleavage of the three nonterminal linkages.

The remaining substrates in the series are cellobiose and cellotriose. Since the first has only one glucosidic linkage, the question of discrimination between linkages does not arise. Cellotriose has two terminal linkages, one adjacent to the reducing end of the chain, the other adjacent to the nonreducing end, but cleavage at either linkage gives the same hydrolysis products, glucose and cellobiose. However, a suitably labeled cellotriose was obtained by adding the enzyme to a solution of cellopentaose containing a high concentration (approximately 10 per cent w/v) of C^{14}-labeled glucose. The cellotriose isolated from the reaction mixture had incorporated about 0.3 per cent of the C^{14} supplied and, as it proved to be labeled virtually exclusively in the anhydroglucose unit at the reducing end of the chain, can be represented as follows:

$$G_{NR} \underset{a}{--\,0\,--} G \underset{b}{--\,0\,--} G^{*}_{R}$$

When this labeled cellotriose was partially hydrolyzed by the enzyme, about 84 per cent of the C^{14} split off appeared as cellobiose, the remainder as glucose. It follows that linkage a, the linkage adjacent to the nonreducing end of the chain, is hydrolyzed about five times as fast as the other linkage.

The trend shown by these results may be illustrated as follows, with the preferred points of cleavage indicated by arrows.

(The bar joining certain arrows is used as a symbol for "and/or.")

Cellobiose	$G_{NR} - 0 - G_R$
Cellotriose	$G_{NR} - \underset{\uparrow}{0} - G - 0 - G_R$
Cellotetraose	$G_{NR} - 0 - G - \underset{\uparrow}{0} - G - 0 - G_R$
Cellopentaose	$G_{NR} - 0 - G - \underset{\uparrow}{0} - G - \underset{\uparrow}{0} - G - 0 - G_R$
Cellohexaose	$G_{NR} - 0 - G - \underset{\uparrow}{0} - G - \underset{\uparrow}{0} - G - \underset{\uparrow}{0} - G - 0 - G_R$

Effects, such as the above, which are associated with the proximity of a linkage to the ends of a chain are commonly referred to in polymer chemistry as "end effects." The preferences shown above indicate that the linkages adjacent to either end, particularly linkages adjacent to the reducing end, are less rapidly hydrolyzed by the enzyme. However, their rates of hydrolysis are not determined by their position in the chain alone. Thus, at least one of the terminal linkages of cellotetraose is hydrolyzed more rapidly than either linkage of cellotriose, and each terminal linkage of cellotriose is hydrolyzed more rapidly than the single linkage of cellobiose. The increase in rate constant from converting the chain to a β-methyl or β-sorbityl glycoside--a conversion which could be indicated above by adding -0-X to the reducing end of each chain--shows a similar dependence on the length of the chain undergoing substitution. This substitution should free hydrolysis of the linkage adjacent to the reducing end from any effects due to the presence of a free reducing group or the absence of a glucosidic linkage at its end of the chain. When this linkage is the only linkage of the chain undergoing substitution, the resulting increase in rate constant is relatively large, but each increase in the DP of the chain adds a linkage farther removed from the influence of the reducing end, and the increase in the over-all rate constant from glycoside formation falls off rapidly.

This evidence that the rate of hydrolysis of a linkage depends on both its position in a chain and the length of the chain suggests that the complex formed between enzyme and substrate is not confined to the immediate vicinity of the glucosidic linkage undergoing cleavage but extends over several anhydroglucose units of the chain. The increase with DP of k_{eq}, the equilibrium constant for the enzyme-substrate complex, is also in line with an increase in DP increasing the number of functional groups contributed by the substrate to the enzyme-substrate complex. It would follow that the sites responsible for this attraction are distributed over a relatively large surface on the enzyme.

This suggestion is of interest relative to some consequences of modifying the properties of the enzyme by partial denaturation (table 4, Whitaker, 1956a). None of the treatments listed caused any precipitation of the enzyme. The results show, first,

TABLE 4
EFFECTS OF VARIOUS TREATMENTS ON ENZYME ACTIVITY

Treatment A: 5 ml samples of enzyme solution (240 μgm protein/ml) added to 25 ml volumetric flasks in thermostats at temperatures specified. Flasks withdrawn after 5 minutes and cooled in ice bath.

Temp. °C.	Exposure Time	Residual Activity (%) Toward			
		Cellobiose	Cellotriose	Cellotetraose	Cellopentaose
46.2	5 minutes	100	100	100	100
53.0	5 minutes	87	87	85	102
58.0	5 minutes	46	46	45	98
65.1	5 minutes	4	3	21	90
69.6	5 minutes	1	0	15	75
75.6	5 minutes	0	0	10	19
80.7	5 minutes	0	1	5	13

Treatment B: 4 ml samples of enzyme solution (200 μgm protein/ml) added to 10 ml volumetric flasks in thermostat. Flasks removed at times specified. Cold buffer blown in to promote foaming.

Temp. °C.	Exposure Time	Residual Activity (%) Toward			
		Cellobiose	Cellotriose	Cellotetraose	Cellopentaose
60.1	30 seconds	97	96	94	96
60.1	60 seconds	67	65	69	75
70.1	30 seconds	10	27	50	82
70.4	60 seconds	1	5	49	81
79.3	30 seconds	3	37	45	64
79.7	60 seconds	2	18	30	32

that a pretreatment which results in a certain loss of activity toward one substrate can result in a significantly different loss toward another; second, that, wherever such differences result, the greater loss is shown toward the substrate of lower DP, i.e., toward the substrate which is less readily cleaved by the enzyme. The nature of the changes in structure responsible for these effects cannot be profitably discussed in any detail at present: heating and foaming can alter the properties of proteins in too many different ways, and the data necessary to limit the field of speculation have not yet been obtained. For example, much depends on whether the changes in structure were irreversible, as a long chain capable of binding the enzyme at several different sites might be expected to be more effective in restoring the configuration of the enzyme than a short chain. In any case, a lower rate of cleavage appears to be correlated with a more critical dependence on the integrity of the enzyme molecule.

The data in table 5 have a bearing on another aspect of the mechanism of hydrolysis. Hydrolysis of a glucosidic linkage

TABLE 5
OPTICAL ROTATIONS DURING HYDROLYSIS OF CELLOPENTAOSE*

Reactants	Time (minutes)	Optical Rotation		
		Before Addition of Ammonia	After Addition of Ammonia	Increase
Enzyme		-0.003 ± .006	0.000 ± .001	0.003 ± .006
Enzyme + Substrate	0	+0.276 ± .002	+0.276 ± .002	0.000 ± .004
	30	+0.324 ± .003	+0.391 ± .005	0.067 ± .006
	100	+0.431 ± .006	+0.492 ± .004	0.061 ± .007
	220	+0.548 ± .006	+0.597 ± .004	0.049 ± .007

*Conditions: enzyme concentration: 200 μgm/ml; substrate concentration: 22 mgm/ml; pH: 5.0; temperature; 24.3°C; solvent: 0.05 M sodium chloride.

gives a sugar with a free reducing group which, depending on the mechanism of hydrolysis, can be in either α or β configuration. Exchange reactions with water subsequently convert either stereoisomer to an equilibrium mixture of the two forms; if the newly formed reducing groups were in α configuration, this conversion reduces optical rotation; if they were in β configuration, it increases the rotation. The nature of the change in rotation due to this mutarotation can be distinguished from other changes that may be occurring simultaneously by measuring the change in rotation from addition of alkali, since under alkaline conditions the conversion to an equilibrium mixture is almost instantaneous. The increases shown in table 5 are small--the optical rotations of the sugars involved are such that only small changes can be expected--but statistically significant. Hence the enzyme cleaves a β-glucosidic linkage to give a reducing group in β configuration, i.e., the β configuration is retained.

As discussed by Koshland (1953), since a single displacement reaction at an asymmetric carbon atom generally inverts the configuration (e.g., $\beta \rightarrow \alpha$), retention of configuration in the type of reaction under consideration suggests either (1) a double displacement reaction at carbon atom 1 with two inversions of configuration ($\beta \rightarrow \alpha \rightarrow \beta$), or (2) a double displacement reaction at carbon atom 4 of the sugar on the other side of the glucosidic linkage. Assuming (1), the cleavage of a glucosidic linkage by the enzyme can be represented by some such double displacement reaction as the one shown in figure 10. In this particular reaction an electron donor, D, on the enzyme displaces oxygen from the linkage and breaks the chain (displacement 1; the electron donor is now in α configuration with respect to carbon atom 1 or ring a); another electron donor, ROH, then displaces D to give a free sugar with a reducing group in β configuration. Ordinarily this electron donor will be water (R = H), but it has been generalized to illustrate the following point. If glucose, when present at sufficiently high concentration, can compete appreciably with water as an electron donor, then, provided the displacement reactions take place at carbon atom 1 as assumed in figure 10, the glucose group transferred should appear at the reducing end of the sugar formed. If the displacement reactions take place at carbon atom 4 on the other side of the glucosidic linkage, the glucose should appear at the nonreducing end. It will be recalled that the cellotriose prepared from cellopentaose and C^{14}-labeled glucose was labeled with C^{14} in the anhydroglucose unit at the reducing end. The labeling of this sugar is thus consistent with the type of mechanism assumed in figure 10.

In terms of this aspect of its mechanism, the enzyme thus resembles the α-amylases, which hydrolyze 1:4-α linkages with retention of configuration, rather than the β-amylases, which hydrolyze such linkages with inversion of configuration.

Time course of hydrolysis
(Whitaker, 1954b)

The course of hydrolysis of cellotetraose and cellopentaose when the initial substrate concentration is moderately low is illustrated in figures 9 and 11. The oligoglucosides formed as initial hydrolysis products break down in turn to give a mixture of glucose and cellobiose, and as the latter is hydrolyzed very slowly it accumulates throughout the periods of hydrolysis shown. The preference of the enzyme for cleaving nonterminal linkages has the effect that in these periods of hydrolysis greater amounts of cellobiose and smaller amounts of glucose are formed than would be expected from strictly random cleavage.

If the substrate concentration is low enough and the enzyme

concentration is high enough to give first order kinetics throughout the hydrolysis, the course of hydrolysis can be described by a set of equations such as the following for a hydrolysis of cellotetraose:

$$dG^4/dt = -k4\ G4 = -(k^4_{2:2} + k^4_{3:1})\ G^4 \qquad ....6a$$

$$dG^3/dt = \frac{3}{4}\ k^4_{3:1}\ G^4 - k^3\ G^3 \qquad ....6b$$

$$dG^2/dt = k^4_{2:2}\ G^4 + \frac{2}{3} k^3\ G^3 - k^2 G^2 \qquad ....6c$$

where G^n is the concentration, expressed in terms of glucose, of the oligoglucoside of the nth degree of polymerization, and k^n is the corresponding velocity constant; $k^4_{2:2}$ denotes the velocity constant for cleavage of cellotetraose at its central linkage, and $k^4_{3:1}$ denotes the sum of the velocity constants for cleavage at its two terminal linkages. These equations simply state in terms of first order kinetics that the rate of change in concentration of an oligoglucoside is equal to the difference between the rate at which it is formed by cleavage of higher oligoglucosides and the rate at which it is itself hydrolyzed. They can be integrated in turn to give a corresponding set of equations:

$$G^4 = G^4_0\ e^{-k^4 t} \qquad ....7a$$

$$G^3 = \frac{3}{4}\ \left(\frac{k^4_{3:1}}{k^4 - k^3}\right)\ G^4_0\ \left(e^{-k^3 t} - e^{-k^4 t}\right) \qquad ....7b$$

etc.

where G^4_0 is the initial concentration of cellotetraose and t is the time of hydrolysis. A time course calculated in this way is shown in figure 12.

As the initial substrate concentration increases, deviations from the conditions assumed in these equations should become increasingly serious. A general treatment, valid at all substrate and enzyme concentrations, would have to allow for the effects of transfer reactions, for competition between oligoglucosides for the enzyme when the latter is not present in excess, and for the rates of hydrolysis having the type of dependence on substrate concentration shown in figure 7. Needless to say it would be hopelessly complicated.

The time course of these reactions, while adding nothing new on the mechanism of hydrolysis of the oligoglucosides, has an important bearing on evidence relating to the method of degradation of longer chains. If a chain is so long that virtually all its linkages are well removed from the ends, the influence of end effects on its degradation should be either all-important or unimportant: all-important if--as has been postulated for some cellulases--longer chains are degraded by cleavage of cellobiose units from the ends of the chains; unimportant if cleavage can take place at any interior linkage, in which case the degradation will approximate to a process of random cleavage. These two mechanisms would appear to be readily distinguished by the nature of the soluble hydrolysis products formed during a hydrolysis, e.g., an apparent absence of higher oligoglucosides and a preponderance of cellobiose would appear to indicate endwise cleavage. However, for reasons discussed later, the enzymatic hydrolysis of cellulose is often a very slow process, and higher oligoglucosides may be undetectable simply because their rates of formation are too slow to compensate for their rapid rates of hydrolysis. Evidence of a preponderance of cellobiose is equally inconclusive: preferences in cleavage similar to those already discussed as well as endwise cleavage can be responsible for it. In general, other evidence is needed to rule out the alternative mechanism. The following section provides an illustration.

Hydrolysis of a Cellodextrin
(Whitaker, 1956b)

This cellodextrin has been referred to previously as an example of a substrate with chains of the same order of length as the enzyme molecule. It was prepared by acetolyzing absorbent cotton and, after deacetylation, freeing the product of soluble sugars and coarse aggregates by repeated washing and sedimentation, first in a Sharples supercentrifuge, then in a preparative ultracentrifuge. The resulting product was a translucent gel which dispersed readily in water to give a colloidal suspension.

The data in table 6 are from a hydrolysis carried out in the presence of a small amount of C^{14}-labeled cellobiose; this sugar was prepared from labeled cellulose obtained by growing *Acetobacter xylinum* in a medium containing C^{14}-labeled glucose. The cellodextrin itself was degraded quite rapidly: within 60 minutes the dry weight of residual cellodextrin fell from an initial value of 18 mgm/ml to 8 mgm/ml. The small decrease in number-average DP during the hydrolysis is, in itself, consistent with either degradation by random cleavage or degradation by endwise cleavage of cellobiose units since, when the DP is as low as 24, small changes in DP can be accounted for by

TABLE 6
HYDROLYSIS OF A CELLODEXTRIN*

Time of Hydrolysis (minutes)	DP of Residual Cellodextrin	Concentration of Soluble Hydrolysis Products: mgm/10 ml			
		Cellobiose	Glucose	Cellotriose	Cellotetraose
0	24	1.5	0	0	0
10	18	6.5	0.7	4.9	3.7
20	21	11.3	0.9	8.6	6.0
30	19	16.1	1.5	10.5	6.3
60	21	23.7	2.5	15.2	6.7
		Specific Activity of Soluble Hydrolysis Products counts/min./mgm			
0		624	....	....	....
10		128	27	8	13
20		74	28	3	12
30		58	17	4	9
60		32	13	7	3

*Conditions of hydrolysis: enzyme concentration: 40 μgm/ml; initial substrate concentration: 18.2 mgm/ml; pH 5.0; temperature: 39°C. Weights of all oligoglucosides taken as the weight as glucose after complete hydrolysis.

either mechanism. However, the amounts of cellotriose and cellotetraose formed are quite substantial in comparison with the amounts of cellobiose formed. Thus, during the first 10 minutes, at least seven molecules of cellotriose and four of cellotetraose are formed for every ten molecules of cellobiose formed. This result not only is consistent with random cleavage but in conjunction with the other data in the table rules out endwise cleavage as the mechanism of degradation. The argument is as follows.

If endwise cleavage is assumed, the formation of cellotriose and cellotetraose can be accounted for in one of two ways. First, these oligoglucosides could represent the ends of partially degraded chains. In this case, in view of the 10:7:4 ratio mentioned above, virtually all the chains that underwent hydrolysis must have been short chains with DP's less than about 10. The selective removal of such short chains, assuming they were present, should lead to an increase in the average DP of the chains that remain. Hence the decrease in the average DP rules out this possibility. A second possibility is that the higher oligoglucosides represent products formed from cellobiose by transfer reactions. If so, their specific activity in C^{14} should be not less than that of cellobiose. The specific activities of the higher sugars are so low that they approach the experimental error of the measurement. Thus this explanation is ruled out also.

Hydrolysis of a Swollen Cellulose (Whitaker, 1957)

Before considering experimental data on the hydrolysis of this substrate, it is desirable to consider one respect in which the representation in figure 6 is incomplete.

The enzyme appears in this figure as a relatively minute object beside a very long chain. However, in degrading cellulose an enzyme is confronted not by well-separated, individual chains, but by a closely packed assemblage of chains; and relative to certain dimensions of the assemblage the enzyme is still a large molecule. It has been shown by electron microscopy that most of the cellulose in plant fibers occurs in flattened threads called microfibrils. These microfibrils can be extremely long but are only about 100 Å wide and less than that in thickness (Balashov and Preston, 1955), and they are closely packed together. The chains making up the microfibril are held together by a combination of hydrogen bonds and van der Waals forces, and throughout much of their lengths their arrangement with respect to one another approaches that of a crystal lattice.

This supermolecular structure has an important bearing on the enzymatic degradation of native cellulose. Within the so-called "crystalline" regions, adjacent chains are separated by distances of only a few Å. Hence even a fairly small molecule cannot move through the lattice; the only chains accessible to it are those on the outside of the crystallite. A large molecule is not only subject to this restriction but will also have difficulty in moving through the narrow spaces between the microfibrils. Consequently, the enzymatic hydrolysis of native cellulose tends to be a slow, topochemical reaction in which, as shown for example by Walseth's (1952) data, the course of hydrolysis is completely dominated by accessibility factors. A substrate presenting at each stage in a hydrolysis a relatively small number of accessible chains, which can be subject to repeated cleavage while the bulk of the remaining chains are virtually unattacked, is obviously unsuitable for a study of the mechanism of degradation at the molecular level. More accessible celluloses can be obtained in various ways: swelling agents reduce crystallinity; regeneration eliminates microfibrillar structure; and grinding opens up the fibrous structure.

None of these treatments give a substrate with all its chains equally accessible to enzymes, but they reduce the complication sufficiently to make possible an analysis of the mechanism of degradation during, at least, the initial stages of a hydrolysis.

Figure 13 shows the loss in weight and the change in $\overline{DP_N}$, the number-average DP, and the $\overline{DP_W}$, the weight-average DP, during the hydrolysis of a ground, swollen cotton cellulose. $\overline{DP_N}$ is the average DP given by osmotic pressure or reducing end group measurements; $\overline{DP_W}$ is the average given by suitably calibrated viscosity measurements. The initial stages of the hydrolysis show a marked drop in $\overline{DP}$ before the loss in weight becomes appreciable. This combination of effects excludes endwise cleavage as the mechanism of degradation. As hydrolysis continues, both the rate of fall in DP and, as shown by other data not included in the figure, the net rate of hydrolysis decreases. By the final stages, any tendency of further cleavage to lower the DP is nearly balanced by the passage of short chains into solution.

Table 7 gives the results of a test for random cleavage based

TABLE 7
FIRST AND SECOND MOMENTS DURING HYDROLYSIS OF SWOLLEN CELLULOSE

Time (minutes)	First Moment $\times 10^{-2}$	Second Moment Exptl. $\times 10^{-4}$	Second Moment Calc. $\times 10^{-4}$
0	5.08	50.8	....
5	3.63	25.2	27.1
15	3.12	16.7	19.6
30	2.46	10.5	11.3
60	2.20	7.6	7.8
120	1.81	5.4	5.6

on Beall and Jorgensen's (1951) method. The statistical moments referred to in the table are related to the average DP's as follows:

$$u_1 = \overline{DP_N}$$

$$u_2 = u_1 \cdot \overline{DP_W}$$

where u_1 and u_2 are the first and second moments, respectively. The agreement between the experimental value of the second moment and the value calculated on the assumption of random cleavage is about as good as can be expected.

Trends in the Method and Rate of Hydrolysis

The two main trends shown in the hydrolysis of the soluble oligoglucosides are for the degradation to become increasingly random and increasingly rapid as chain length increases. The preceding evidence on the method of degradation in the next two ranges of chain length shows that the first trend is maintained. It also extends the resemblance to the α-amylases suggested by the optical configurations of the hydrolysis products of cellopentaose. The second trend is not maintained. The rate probably reaches a maximum within the series of soluble oligoglucosides itself, but, in any case, once the chains become insoluble their rates of hydrolysis are determined by their accessibility rather than by their length. The colloidal cellodextrins probably represent the closest approach to an insoluble substrate with completely accessible chains, and their rates of hydrolysis are relatively high. The more slowly degraded swollen or regenerated celluloses have low but not negligible crystallinities and fine capillary spaces, and, once the more accessible chains are degraded, their rates of hydrolysis tend to fall off rapidly. Native celluloses, with their high crystallinity and their microfibrillar structure, have the lowest accessibilities to enzymes and are degraded very slowly.

Some Other Factors

The properties and reactions considered in the preceding sections were those of a purified enzyme and chemically defined substrates. Much of the literature on Myrothecium cellulase deals with the properties of culture filtrates and crude enzyme preparations and in many respects is extremely confused, conflicting reports on certain properties tending to be the rule rather than the exception. Some factors contributing to this condition are discussed below.

Cellulases are extracellular enzymes, and, depending on the conditions of culture, their properties can be modified in several ways during the period in which they are accumulating in the culture medium. Some cultural conditions, e.g., aeration of the medium in such a way as to cause foaming, can favor complete or partial denaturation of the enzyme. Other conditions can favor the production of various metabolic products which can form complexes with enzymes and other proteins. The complexing agents that we have encountered are partly responsible for the yellowish-orange color of the culture filtrate. They tend to precipitate with the enzyme at each step in the purification procedure and show little tendency to dissociate from the enzyme during dialysis. They appear to have acidic properties and can be removed by ion exchange if precautions are taken to avoid partial denaturation of the enzyme on the exchange column. The data of Miller and Blum (1956) suggest that under other cultural conditions other complexing agents are produced. Their crude enzyme preparations were resolvable by zone electrophoresis at pH 7 into at least eight components with enzymatic activity toward carboxymethylcellulose. They regard this as proof of the formation of eight different enzymes. However, as their enzyme preparation contained not only protein but an approximately equal amount of an electrophoretically heterogeneous polysaccharide, their evidence may simply indicate that Myrothecium cellulase, like *Stachybotrys atra* cellulase (Thomas, 1956), can form complexes with certain fungal polysaccharides. The effect of such complexing agents on cellulase activity is unknown, but one property of cellulases which can be expected to be influenced by complex formation is thermostability. Many complexes between enzymes and natural products are much more stable than the free enzyme itself. The varying reports on the thermostability of Myrothecium cellulase, ranging from evidence of moderate thermolability (Saunders, Siu, and Genest, 1948) to evidence of extreme thermostability (Kooiman, Roelofsen, and Swerris, 1953), are therefore not necessarily incompatible.

Another variable factor in culture filtrates of Myrothecium is the presence of certain other enzymes. Some culture filtrates show very little hydrolytic activity toward cellobiose, and, in the case of our filtrates at least, this activity is inhibited by moderately high concentrations of cellobiose (Whitaker, 1953). However Kooiman, Roelofsen, and Sweeris (1953), using another strain of the mold, obtained filtrates with a high cellobiase activity which was not inhibited by high substrate concentrations. Furthermore, in contrast with the extreme thermostability associated with their filtrate's enzymatic activity toward substrates of DP greater than 2, this cellobiase was relatively thermolabile. They also obtained evidence of a thermolabile glucotransferase which gave some conversion of cellobiose to cellotriose and glucose. They suggested that these two enzymes might be identical, but Aitken, Eddy, Ingram, and Weurman (1956) concluded that they were not, on the grounds that the same strain of the mold sometimes gave filtrates with low cellobiase activity but appreciable transferase activity. Unfortunately most of these data are qualitative only. Crook and Stone (1957) have extended these observations on transferase activity with evidence that sugars other than those of the cellulose series, e.g., laminaribiose, are also formed at high cellobiose concentrations. These enzymes introduce obvious complications--e.g., if a separate cellobiase is present, assays of cellulase activity based on the amount of reducing sugar formed from cellulose under standard conditions can be in serious error. Kooiman *et al.* took advantage of the extreme thermostability of the cellulase in their culture filtrates to eliminate this complication. However, their technique of exposing filtrates to high temperatures is inapplicable to filtrates in which the cellulase itself is thermolabile and, in any case, is a rather drastic treatment which, as indicated by the data in table 4, may give rise to other problems.

The substrates employed in the study of cellulases represent another variable factor. Apart from properties previously discussed, cellulose and all its insoluble hydrolysis products have adsorbent properties which vary with the accessibility of the cellulose and the treatment to which it has been subjected; associated with these adsorbent properties are the properties of cellulose as a weak ion exchanger. The adsorption of certain compounds on the substrate can have marked effects on the activity of Myrothecium cellulase. Among those we have observed are stimulation of cellulase activity by adsorbed proteins, polypeptides, and basic dyes, and inhibition by adsorbed acid dyes (Whitaker, 1952; Basu and Whitaker, 1953). These stimulations and inhibitions were all strongly pH dependent and varied considerably with the nature of the substrate. They indicate that the relation between pH and enzymatic activity toward an insoluble substrate is not determined by the properties of the enzyme alone but depends also on a variety of other factors: the nature of the substrate, adsorbable impurities in the enzyme preparation, etc. Such effects probably account for the diversity of pH activity curves which have been reported for this enzyme.

To avoid these complications of insoluble substrates, various soluble derivatives of cellulose have been employed as substrates for cellulases. This practice was introduced by Ziese

(1931), and the type of derivative he employed--hydroxyethylcellulose, i.e., a neutral derivative--is useful for many purposes. However, Jermyn's (1952) evidence suggests that certain β-glucosidases as well as cellulases can degrade them. Derivatives with ionizable groups such as carboxymethylcellulose figure more prominently in the literature on Myrothecium cellulase, but such derivatives introduce a very serious complication, particularly when pH is a variable. Less highly substituted carboxymethylcelluloses are, in fact, used extensively as ion exchangers for protein chromatography.

Naturally occurring fibers represent the ultimate complication since, although some fibers, e.g., cotton fibers, may contain large amounts of cellulose, their physical structure and chemical composition continue to reflect the fact that they were once living cells. One recent use of them as substrates for cellulases is based on the observations of Marsh, Bollenbacher, Butler, and Guthrie (1953) that the extent to which cotton fibers were swollen by an 18 per cent sodium hydroxide solution was markedly increased if the fibers were first immersed in culture filtrates of various species of fungi. These species included Myrothecium and other cellulolytic fungi as well as some species usually regarded as noncellulolytic. They designated the factor or factors responsible for this effect as "S-factor." Reese and Gilligan (1954) use the term "swelling factor" and have employed this test to characterize cellulases. The latter term is not too apposite since it suggests the action of a swelling agent, whereas the action involved is damage to the outer primary wall of the fiber, a structure that tends to restrain the swelling of cotton fibers. The primary wall is the most chemically heterogeneous structure in the fiber: cellulose accounts for about half its weight; the rest is a mixture of waxes, pectins, proteins, etc. (Tripp, Moore, and Rollins, 1954). The present evidence suggests that cellulases contribute to S-factor activity, but, in view of the nature and complexity of the effect, its significance for purposes of characterizing enzymes remains obscure.*

---

*The writer wishes to express his thanks to the publishers of Archives of Biochemistry for permission to reproduce figure 3; to the publishers of Nature for permission to reproduce figures 4 and 5; and to the Canadian Journal of Biochemistry and Physiology for permission to reproduce Figures 8 and 13.

LITERATURE CITED

Aitken, R. A., B. P. Eddy, M. Ingram, and C. Weurman. 1956. Action of Culture Filtrates of the Fungus *Myrothecium verrucaria* on β-glucosans. Biochem. J., 64:63-70.

Balashov, V., and R. D. Preston. 1955. The Fine Structure of Cellulose and Other Microfibrillar Substances. Nature, 176: 64-65.

Basu, S. N., and D. R. Whitaker. 1953. Inhibition and Stimulation of the Cellulase of *Myrothecium verrucaria*. Arch. Biochem. Biophys., 42:12-24.

Beall, G., and L. Jorgensen. 1951. The Mechanism of the Hydrolytic Cleavage of the Inter-unit Bonds in the Cellulose Molecule. Text. Res. J., 21:203-14.

Crook, E. M., and B. A. Stone. 1957. Enzymic Hydrolysis of β-glucosides. Biochem. J., 65:1-12.

Eadie, G. S. 1952. Evaluation of the Constants Vm and Km in Enzyme Reactions. Science, 116:688.

Freudenberg, K., and G. Blomqvist. 1935. Hydrolysis of Cellulose and Its Oligosaccharides. Berlin, 68B:2070-82.

Jermyn, M. A. 1952. Fungal Cellulases. II, The Complexity of Enzymes from *Aspergillus oryzae* That Split β-glucosidic Linkages, and Their Partial Separation. Austral. J. Sci. Res., 5:433-43.

Kooiman, P., P. A. Roelofsen, and S. Sweeris. 1953. Properties of Cellulase from *Myrothecium verrucaria*. Enzymologia, 16:237-46.

Koshland, D. E. 1953. Stereochemistry and the Mechanism of Enzymic Reactions. Biol. Rev. Camb. Phil. Soc., 28: 416-36.

Marsh, P., K. Dollenbacher, M. L. Butler, and L. R. Guthrie. 1953. S-factor--a Microbial Enzyme Which Increases the Swelling of Cotton in Alkali. Text. Res. J., 23:878-88.

Miller, G. L., and R. Blum. 1956. Resolution of Fungal Cellulase by Zone Electrophoresis. J. Biol. Chem., 218:131-37.

Morton, R. K. 1956. The Group-transfer Activity of Certain Hydrolytic Enzymes. Disc. Faraday Soc., 20:149-56.

Reese, E. T., and W. Gilligan. 1954. The Swelling Factor in Cellulose Hydrolysis. Text. Res. J., 24:663-69.

Saunders, P. R., R. G. H. Siu, and R. N. Genest. 1948. A Cellulolytic Enzyme Preparation from *Myrothecium verrucaria*. J. Biol. Chem., 174:699-703.

Thomas, R. 1956. Fungal Cellulases. VII, *Stachybotrys atra:*

Production and Properties of the Cellulolytic Enzyme. Austral. J. Biol. Sci., 9:159-83.
-------, and D. R. Whitaker. Unpublished procedure.
-------, and D. R. Whitaker. 1958. Zone Electrophoresis of *Myrothecium* Cellulase. Nature, 181:715-16.
Tripp, V. W., A. T. Moore, and M. L. Rollins. 1954. A Microscopical Study of the Effects of Some Typical Chemical Environments on the Primary Wall of the Cotton Fiber. Text. Res. J., 24:956-70.
Walseth, C. D. 1952. The Influence of the Fine Structure of Cellulose on the Action of Cellulases. Tappi, 35:233-38.
Whitaker, D. R. 1952. An Effect of Proteins and Proteoses on the Cellulase of *Myrothecium verrucaria*. Science, 116:90-92.
-------. 1953. Purification of *Myrothecium verrucaria* Cellulase. Arch. Biochem. Biophys., 43:253-68.
-------. 1954a. Mutarotation after Hydrolysis of Cellopentaose by *Myrothecium verrucaria* Cellulase. Arch. Biochem. Biophys., 53:436-38.
-------. 1954b. Hydrolysis of a Series of β-1, 4'-Oligoglucosides by *Myrothecium verrucaria* Cellulase. Arch. Biochem. Biophys., 53:439-49.
-------. 1956a. The Steric Factor in the Hydrolysis of β-1, 4'-Oligoglucosides by *Myrothecium* Cellulase. Can. J. Biochem. and Physiol., 34:102-15.
-------. 1956b. The Mechanism of Degradation of a Cellodextrin by *Myrothecium* Cellulase. Can. J. Biochem. and Physiol., 34:489-94.
-------. 1957. Mechanism of Degradation of Cellulose by *Myrothecium* Cellulase. Can. J. Biochem. and Physiol., 35: 733-42.
-------. Unpublished data.
-------, J. R. Colvin, and W. H. Cook. 1954. The Molecular Weight and Shape of *Myrothecium verrucaria* Cellulase. Arch. Biochem. Biophys., 49:257-62.
-------, and E. Merler. 1956. Cleavage of Cellotriose by *Myrothecium* Cellulase. Can. J. Biochem. and Physiol., 34: 83-89.
Ziese, W. 1931. The Action of Enzymes of the Gastric Juice of *Helix pomatia* and Those of Barley Malt on Cellulose Glycol Ether. Z. Physiol. Chem., 203:87-116.

Fig. 1. Method of growing mold for cellulase production

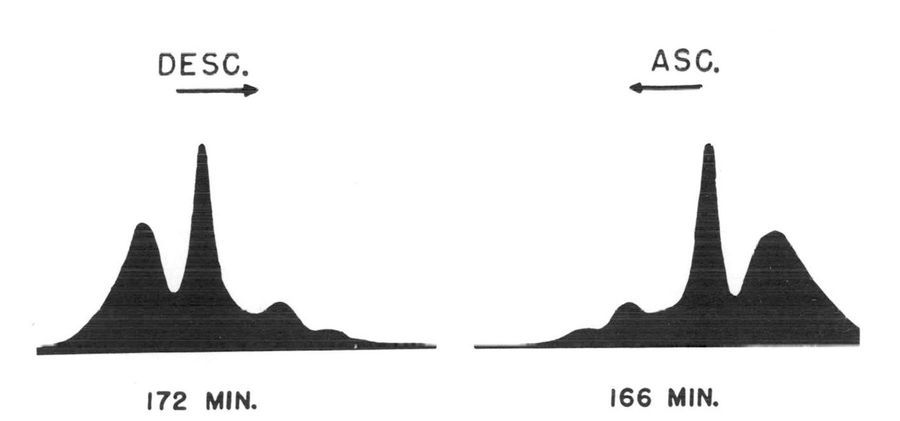

Fig. 2. Electrophoretic pattern of crude enzyme at pH 7

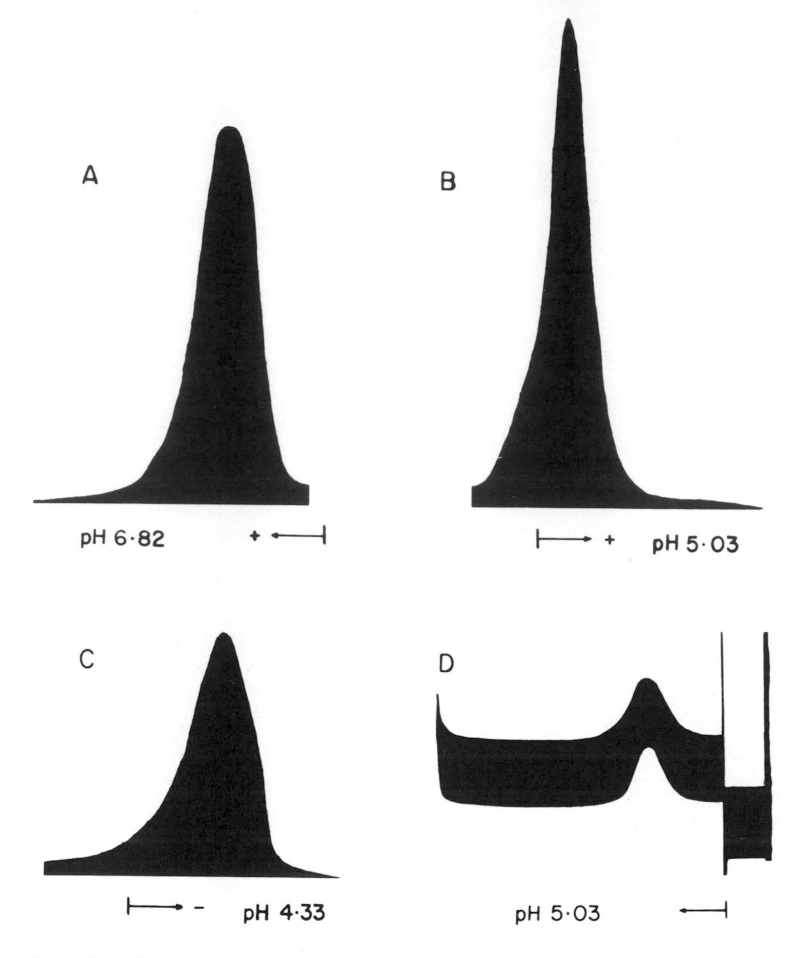

Fig. 3. Electrophoretic (A, B, C) and sedimentation pattern (D) of purified enzyme

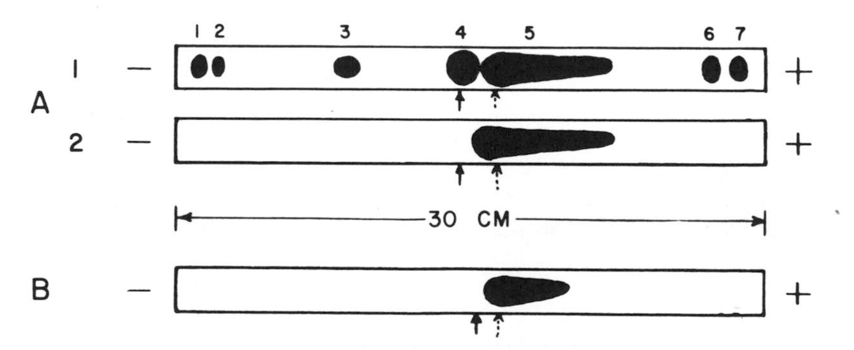

Fig. 4. Zone electrophoresis at pH 7 in Whatman 3 MM paper. A, crude enzyme. Indicators: 1, ninhydrin, 2, p-anisidine hydrochloride; B, purified enzyme. Indicator: ninhydrin

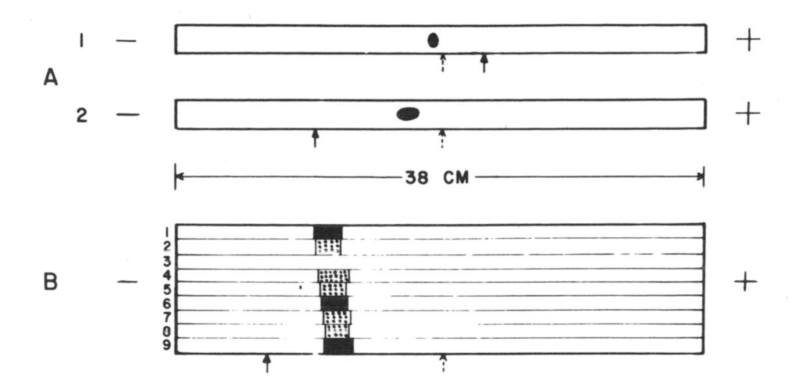

Fig. 5. Zone electrophoresis of purified enzyme on glass fiber paper. A, 1, at pH 3.0; 2, at pH 6.7. Indicator: alkaline permanganate; B, at pH 6.7. Indicators: (a) alkaline permanganate (strips 1, 6, and 9; (b) p-anisidine: (i) after spraying strip with acetate buffer of pH 5.0 (strip 3); (ii) after spraying strip with buffered solutions of methyl β-cellobioside (strip 8), methyl β-cellotetraoside (strip 5), cellopentaose (strip 7), and carboxymethyl cellulose (strip 2); (iii) on Whatman No. 1 paper after strip 4 had been sprayed with buffer and then left in contact with the Whatman paper

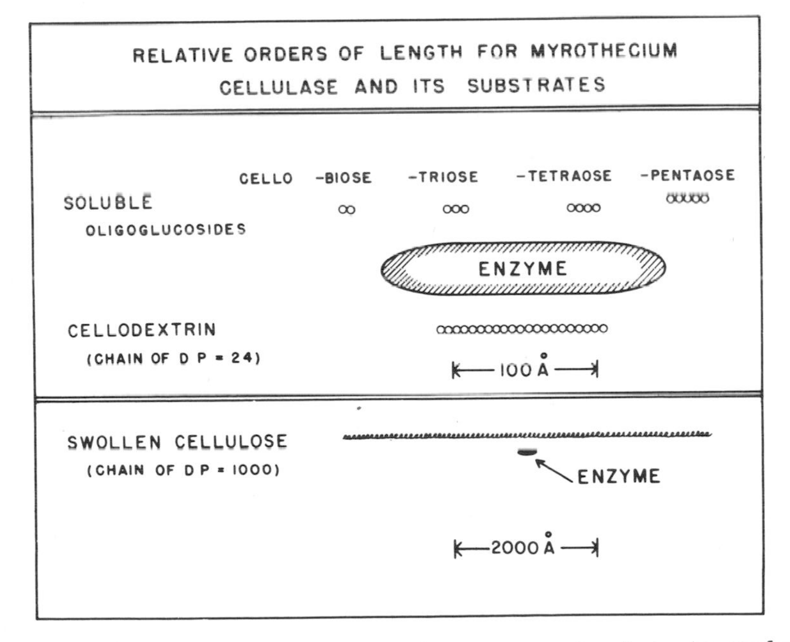

Fig. 6. Comparison of lengths of enzyme and substrate molecules

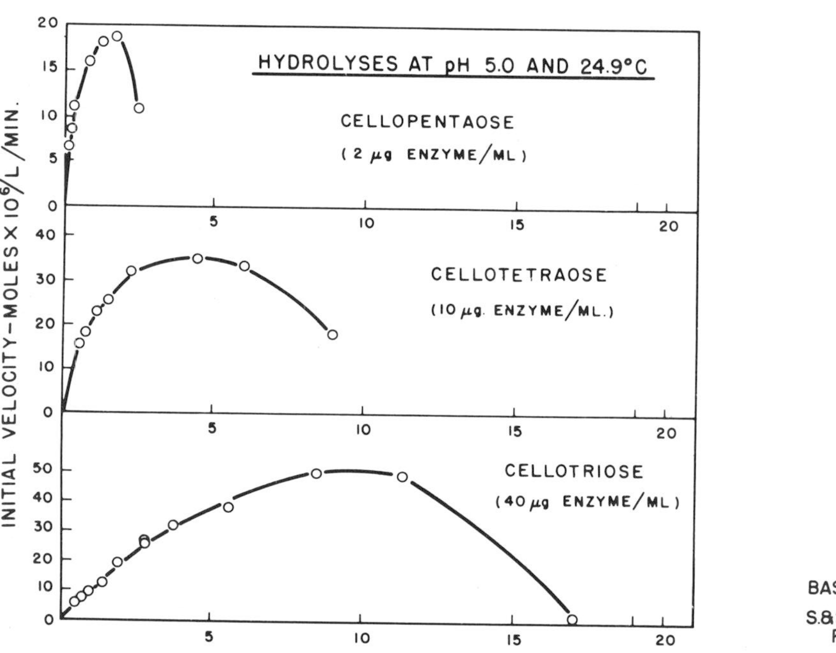

Fig. 7. Variation of initial rates of hydrolysis of cellotriose, cellotetraose, and cellopentaose with initial substrate concentration

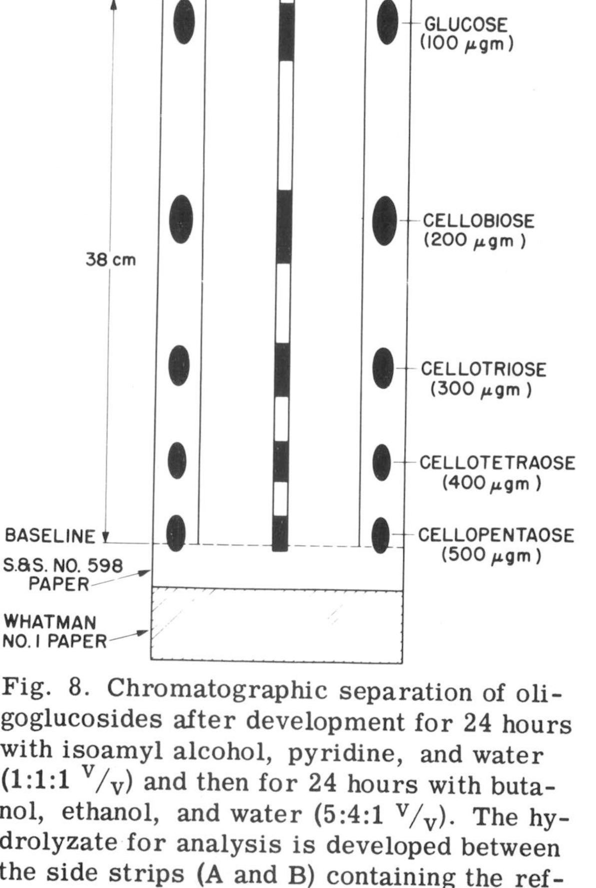

Fig. 8. Chromatographic separation of oligoglucosides after development for 24 hours with isoamyl alcohol, pyridine, and water (1:1:1 $^{V}/_{V}$) and then for 24 hours with butanol, ethanol, and water (5:4:1 $^{V}/_{V}$). The hydrolyzate for analysis is developed between the side strips (A and B) containing the reference sugars; a section of it is shown at C

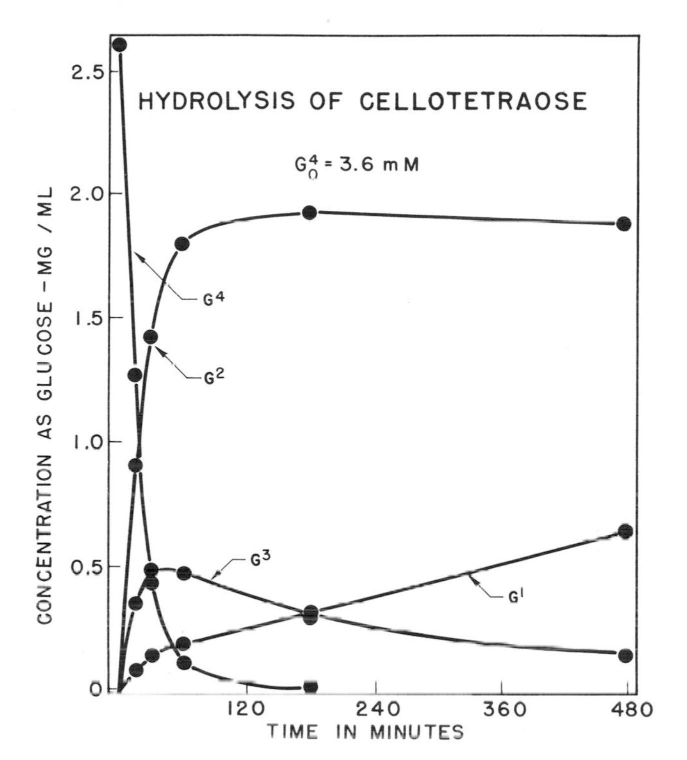

Fig. 9. Time course of hydrolysis of cellotetraose at pH 5. Temperature: 35°C, enzyme concentration: 20 μgm/ml. G^4 denotes cellotetraose; G^3, cellotriose, etc.

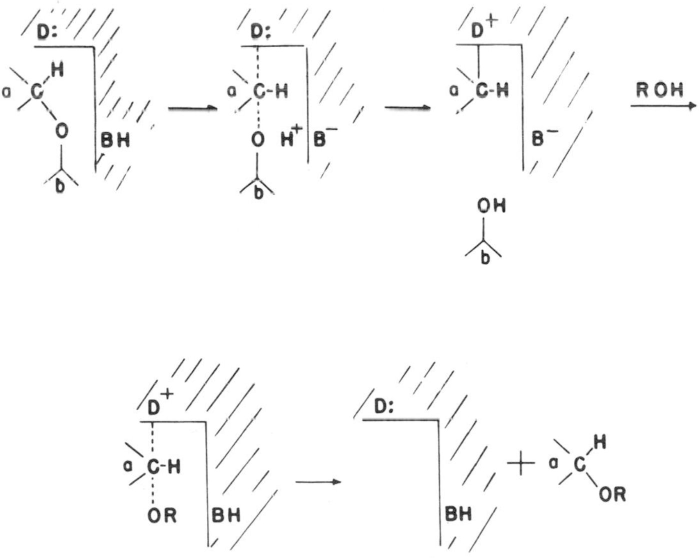

Fig. 10. Cleavage of a glucosidic linkage by a double displacement reaction at the active site of an enzyme

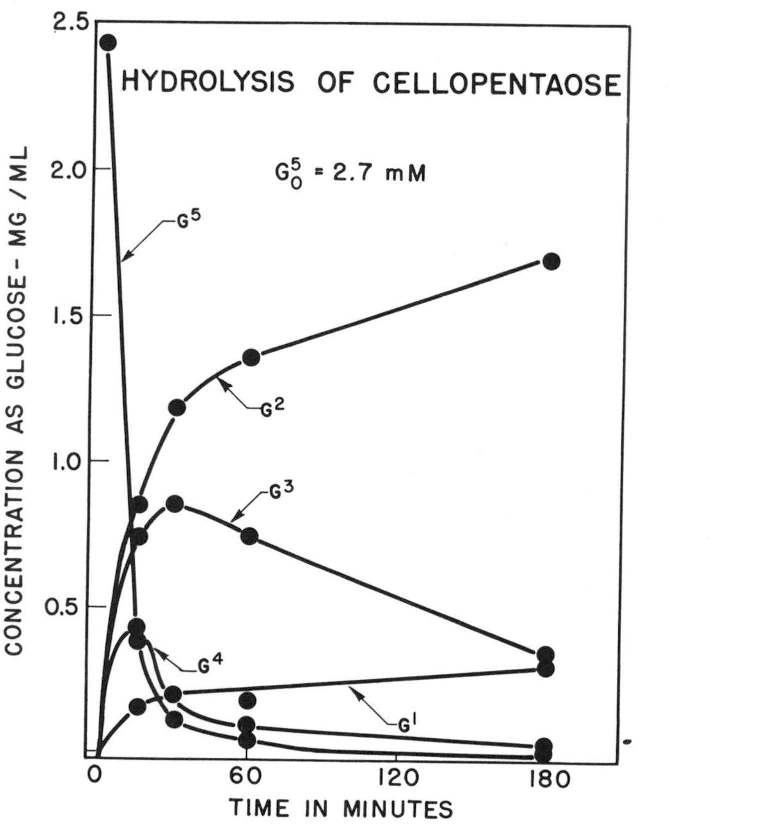

Fig. 11. Time course of hydrolysis of cellopentaose. Conditions and notation as for figure 9

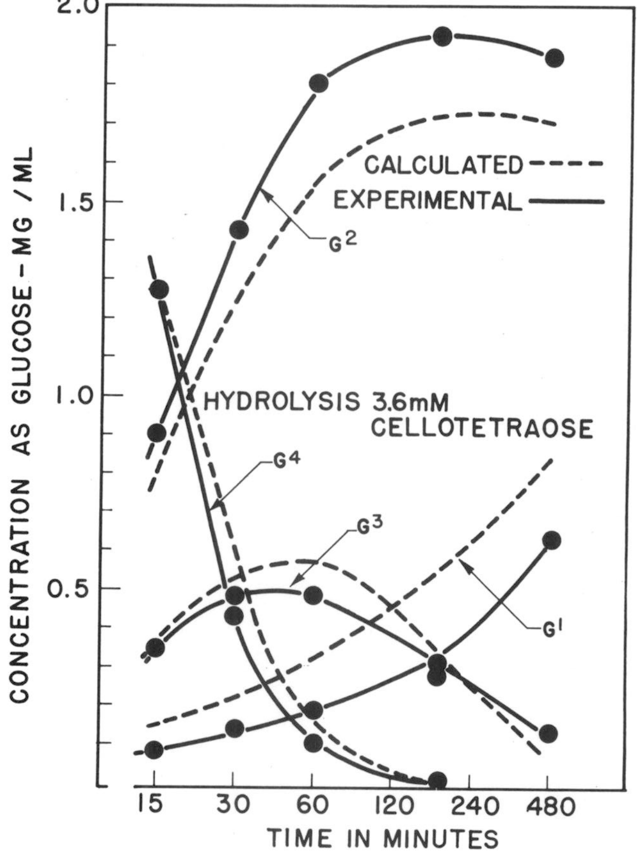

Fig. 12. Comparison of calculated and observed time course for cellotetraose

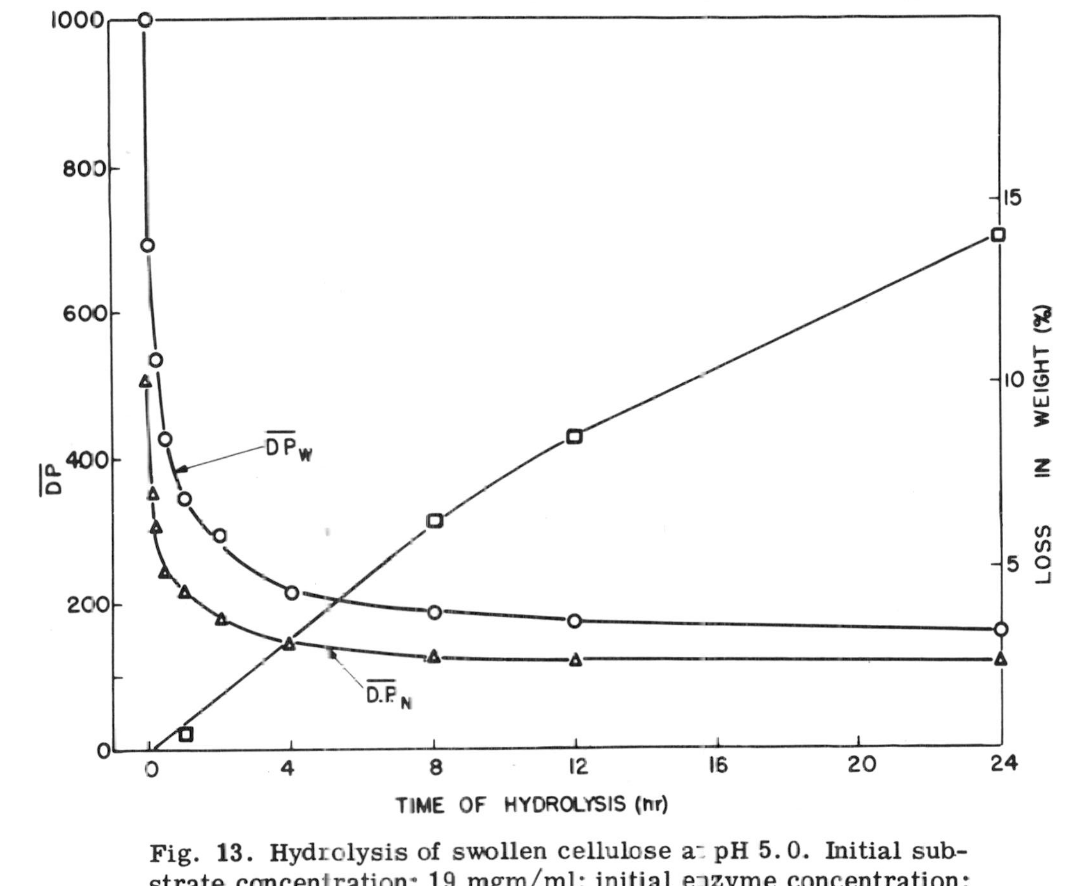

Fig. 13. Hydrolysis of swollen cellulose at pH 5.0. Initial substrate concentration: 19 mgm/ml; initial enzyme concentration: 20 μgm/ml

CELLULOSE-DECOMPOSING BACTERIA IN THE SEA

Hajime Kadota

Cellulose-decomposing bacteria are widely distributed in the sea, where they play an important role as mineralizers of organic matter and influence the productivity of the sea. Besides influencing the carbon cycle in the sea, these organisms are of economic importance in that they cause the deterioration of fishing nets and ropes and play a part in the fouling of ships' bottoms and other wooden structures exposed to the sea. In the latter case these organisms may work symbiotically with shipworms.

In the present paper an attempt has been made to summarize and correlate the literature on cellulose-decomposing bacteria in the sea, with particular reference to the distribution of these organisms and to the symbiosis of these bacteria with shipworms.

Distribution of Cellulose-decomposing Bacteria in the Sea

At first cellulose decomposition in the sea was studied primarily from the economic standpoint. Most of the early works were mainly concerned with the deterioration of fishing nets in the sea.

Nishimura and Yamazoe (1914) and Kawai (1914) observed that fishing nets made of cotton and hemp deteriorated rapidly in raw sea water but never deteriorated in sea water that had previously been sterilized. On the basis of these data, Nishimura, Yamazoe, and Kawai believed that the deterioration of cellulosic fishing nets immersed in the sea was caused by marine microorganisms and not by abiogenic factors. From the cotton cords that had rotted from prolonged immersion in sea water, Nishimura and Yamazoe (1914) isolated several crude cultures of microorganisms that might be active in deterioration. Dorée (1920) also found that the deterioration of cotton textiles in the sea was caused by microorganisms. Kotani (1937) reported that, in sea water supplemented with inorganic nutrient salts, cotton textiles rotted rapidly when they were inoculated with marine bacteria but did not rot under sterile conditions. Atkins and Warren (1941), Barghoorn (1942), Barghoorn and Linder (1944), and several other workers also reported that microorganisms were responsible for the deterioration of cotton, hemp, jute, and sisal cordage as well as pilings and other wooden structures in the sea.

As mentioned above, it was early recognized that cellulose-decomposing bacteria were mainly responsible for the deterioration of fishing gear, timbers, wooden pilings, and other cellulose-containing structures exposed to the sea. However, the distribution of cellulose-decomposing bacteria had never been studied in detail over a long period and from a purely scientific standpoint.

Issatchenko (1921) first isolated from the mud of the salt lake Ssaky in Crimea two aerobic bacteria, one of which decomposed cellulose without forming a pigment (*Bact. cellulosae album*), whereas the other colored the paper yellow (*Bact. cellulosae flavum*). From the deep water of Lake Tambukan (Kaukasus), Waledinsky and Ssweschnikoff (1926) obtained crude cultures of aerobic cellulose-decomposing bacteria. Rubentschik (1928) isolated from salt estuaries in southern Russia an actinomycete (*Actinomyces melanogenes*) capable of slow decomposition of paper. Later (1933) he found that anaerobic cellulose digesters were also distributed in such aquatic environments. The decomposition of cellulose by these organisms was unimpaired by a salt concentration as high as 15 per cent NaCl. Waksman, Carey, and Reuszer (1933) found in the Gulf of Maine and George's Bank that cellulose-decomposing bacteria were generally present in sea water, and particularly abundant in bottom deposits and diatom tows. They observed that species of Cytophaga, Cellvibrio, and Cellfalcicula were present in the sea. All these bacteria were aerobic, and some of them were capable of attacking both cellulose and agar, while others decomposed cellulose only. Some of these bacteria were also able to decompose a number of other polysaccharides, as well as various mono- and disaccharides. In the natural environment these bacteria were invariably accompanied by numerous protozoans, including flagellates, ciliates, and amoebae. Bavendam (1932) found aerobic cellulose digesters in all marine mud samples, and anaerobic cellulose-decomposing bacteria in a few of the samples he examined. By use of the minimum dilution method, ZoBell (1938) demonstrated 1,000 cellulose digesters per gram of some marine mud samples. In Marine Microbiology ZoBell (1946) stated that marine microorganisms endowed with the ability to utilize cellulose "may be demonstrated in most 10 to 100 ml samples of sea water and in nearly all one gram samples of bottom deposits." From marine materials, Stanier (1941) isolated and described *Vibrio fuscus, Pseudomonas iridescens, Cytophaga krzemieniewskae,* and *Cytophaga diffluens*; all are new species that digest cellulose and agar. According to Stanier, *Vibrio granii* and *Pseudomonas droebachense,* which had been named previously by Lundestad (1928), digest both agar and cellulose. Using fresh water media, Kotani (1937) obtained from sea water several cultures of aerobic cellulose-decomposing bacteria which he described but failed to name. In an investigation of marine agar-digesting bacteria of the south Atlantic Coast of North America, Humm (1946) isolated and described *Vibrio agar-liquefaciens, Pseudonomas droebachense, Pseudonomas iridescens, Vibrio notus, Vibrio stanierii,* and *Pseudomonas elongata*; again, all of these digest both cellulose and agar.

About 20 species of marine bacteria were reported by the afore-mentioned workers as cellulose decomposers. However, the greater number of these organisms have not been precisely studied from the ecological standpoint.

Recently Kadota (1956) made a study of marine cellulose-decomposing bacteria in Maizuru Bay and Hiroshima Bay to determine the identity of the species involved and their relative abundance under various ecological conditions. Some of the environmental factors influencing the activity of marine cellulose-decomposing bacteria, the important physiological activities of these bacteria, their behavior toward cellulose and cellulose derivatives, and the disinfecting effects of several chemical agents upon these bacteria were also investigated in this study. Using various kinds of media suitable for the growth of both aerobic and anaerobic marine cellulose-decomposing bacteria, Kadota observed that aerobic cellulose-decomposing bacteria were distributed widely in sea water and in bottom deposits, and that anaerobic cellulose digesters existed hardly at all in sea water and only sparsely in bottom deposits. These data suggested that the deterioration in sea water of fishing nets made of cellulose fibers was caused primarily by the activities of aerobic cellulose-decomposing bacteria. From various marine materials Kadota isolated 252 cultures of marine aerobic cellulose-decomposing bacteria, and classified and identified these organisms into 3 genera, 18 species, and 2 varieties, as follows: *Pseudomonas marinocellulosae, Vibrio aquamarinus, Vibrio purpureus, Vibrio purpureus* var. *albus, Vibrio gilvus, Vibrio albogilvus, Vibrio macerans, Vibrio marinoliquefaciens, Vibrio simplex, Vibrio ferrugineus, Vibrio brevis, Vibrio fulvus, Vibrio aurantiacus, Vibrio marinotypicus, Vibrio euryhalis, Vibrio marinoflavescens, Vibrio xanthus, Cytophaga haloflava, Cytophaga haloflava* var. *nonreductans*, and *Cytophaga rosea.*

It was found from the population counts continuously carried out in Maizuru Bay that there were no definite seasonal cycles in the total abundance of aerobic cellulose-decomposing bacteria in sea water or in marine muds. However, the composition of the flora of cellulose-decomposing bacteria did vary somewhat seasonally; i.e., the kinds of cellulose-decomposing bacteria predominating in sea water varied seasonally. For example, *Vibrio purpureus* and *Vibrio aquamarinus* predominated in summer, and *Vibrio marinoflavescens* and *Vibrio fulvus* in winter. The former group could grow at temperatures ranging from 20° to 35°C and developed most rapidly at 30°C. The latter group, on the other hand, could not grow at temperatures higher than 30°C and grew most rapidly at 20° to 25°C. Therefore, the seasonal variations in the kinds of cellulose-decomposing bacteria found in sea water were thought to be attributable to the optimum growth temperatures of the organisms.

It seems that during the summer months the deterioration of cellulosic fishing gear immersed in sea water is caused primarily by the *Vibrio purpureus* group, and during the winter months by the *Vibrio marinoflavescens* group. Since cellulose-decomposing activity of the *Vibrio purpureus* group at 25° to 30° C was generally much higher than that of the *Vibrio marinoflavescens* group at 15° to 25°C, seasonal fluctuation of the deterioration rate of cotton fishing nets exposed to the sea may be attributable partly to the seasonal variation in the kinds of cellulose-decomposing organisms in sea water.

From the results of the extensive observations on vertical and geographical distribution of aerobic cellulose-decomposing bacteria in Hiroshima Bay and Maizuru Bay, it was found that the population of these bacteria was always larger in bottom deposits (10 to 100,000 cells per gm mud, in general) than in overlying water (1 to 100 cells per ml water, in general), in which marked stratification of these organisms did not exist.

In some cases, it was observed that the abundance of aerobic cellulose-decomposing bacteria in bottom muds decreased with the distance from the shore. It seems well to consider, however, that this may be caused by the quantity of the available organic matter present in the mud rather than by the distance from the shore or the depth of the overlying water.

According to Kadota (1956), the distribution in the sea of cellulose-decomposing bacteria belonging to genus Cytophaga did not always correspond to that of the total aerobic cellulose-decomposing bacteria. In the marine materials examined, cytophagas constituted from 0 to 33.3 per cent of the total number of aerobic cellulose-decomposing bacteria. The abundance of cellulose-decomposing cytophagas in sea water seemed to be closely related to that of suspended matter. Since suspended matter provides solid surfaces for the attachment of epiphytic bacteria, it is not surprising that the abundance of cytophagas is closely related to the abundance of suspended solids. Kadota also found that, on the cotton fishing nets which had been immersed for 10 days in sea water, cytophagas constituted from 22.2 to 40.5 per cent of the total number of aerobic cellulose-decomposing bacteria. This percentage was more than twice as large as that found in the sea water where the nets had been immersed. It is especially interesting that a large number of cytophagas were found on the nets immersed in the sea water in which cytophagas were never found at all. This result clearly indicates that cytophagas developed better on submerged surfaces such as fishing nets than when freely floating in the sea.

Changes in flora of cellulose-decomposing bacteria occurring on cotton fishing nets during prolonged immersion in sea water were also observed by Kadota. From the results of those observations it is suggested that after about 10 days' immersion the proportion between the number of cytophagas and that of true bacteria remained practically constant throughout the prolonged period of immersion, although the absolute number of both groups might increase in the course of the immersion period.

Although cellulose-decomposing cytophagas are generally less abundant in sea water than cellulose-decomposing true bacteria, the former may also play an important role in the deterioration of cellulosic fishing gear submerged in the sea because of their tenacious attaching property.

Besides deteriorating during immersion in the sea, cellulosic fishing gear is known to rot during storage after fishing. Kadota found that a large number of cells of marine aerobic cellulose-decomposing bacteria that had grown on the cotton fishing nets immersed in the sea survived or developed for a considerable period in storage under dry as well as wet conditions, and that under such conditions the survival period of the cytophagas was much longer than that of the true bacteria.

It seems, therefore, that, in the deterioration of cellulosic fishing gear stored after fishing or immersed again in sea water, the cytophagas play a more important role than the true bacteria, although, in the deterioration of gear newly immersed in sea water, the true bacteria play a more important role than the cytophagas.

Cellulose-decomposing Bacteria Associated with Shipworms

It is generally thought that cellulose-decomposing microorganisms may often work symbiotically with shipworms or other wood borers and may be instrumental in the destruction of timbers, wooden pilings, and other cellulose-containing structures exposed to the sea. ZoBell (1946) stated in Marine Microbiology that the bacteria may convert the cellulose and lignin into products more readily assimilated by the wood borers. He said that the gut of engorged *Teredo* as well as its burrows contained large numbers of cellulose- and lignin-digesting bacteria. According to Barghoorn and Linder (1944), cellulose-decomposing fungi in the sea may make the wood more susceptible to attack by wood borers.

From the intestinal organs of *Teredo navalis*, Hidaka (1954) isolated cellulose-decomposing bacteria that rapidly decomposed cellulose. These bacteria closely resembled *Vibrio aurantiacus* Kadota and developed rapidly in sea water media containing phosphates, nitrates, and sulfates at 27°C. Hidaka and Saito (1956) studied the cellulose-digesting activities of the above-mentioned bacteria and the biochemical behavior under various conditions of a crude preparation of cellulase obtained from the culture of these bacteria by precipitation with acetone. The results of their experiments suggested that the optimum temperature for the cellulase from these bacteria was 37°C, and the optimum pH for it lay within the range of 6.2-6.4. This enzyme was activated by Mn^{++} to a certain extent, and inhibited by Cu^{++} in very low concentrations.

Saito and Hidaka (1954) compared the biochemical behavior of crude cellulase prepared from the cellulose-decomposing bacteria that were isolated from the intestinal canals of *Teredo navalis* with that of cellulase obtained from the liver of *Teredo navalis*. Both these cellulases decomposed regenerated cellulose more easily than native cellulose and attacked the wood preparations in proportion to the degree of delignification. Crude cellulase preparations obtained by Saito and Hidaka (1954) from both the bacteria and the organs contained xylanase, amylase, and mannase besides cellulase. Influence of pH value upon the activities of crude cellulase preparations obtained from the bacteria, the liver of *Teredo navalis*, and the whole body of this animal was as shown in figure 1.

According to these data, the pH optimum of cellulase from the bacteria was about 6.4, and that of cellulase from the liver of *Teredo navalis*, 5.8. It was also found from these data that the mixture of cellulase from the bacteria and from the liver of *Teredo navalis* showed both pH optima (at 5.8 and 6.6). The pH optima of the mixture of cellulases from the bacteria and the liver accorded approximately with those of cellulase preparation obtained from the whole body of *Teredo navalis*: viz., the pH optima of cellulase from the whole body of this animal were 5.6 and 6.6.

From the afore-mentioned experiments, Saito and Hidaka (1954) concluded that the cellulose-decomposing bacteria in the intestinal tract of *Teredo navalis* play an important role in the digestive processes of the host animals. It is thought that these bacteria may help the digestion of shipworms by partially digesting cellulose as is the case with the cellulose-digesting bacteria in the intestine of herbivores.

Kadota (1951 and 1953) also found that *Vibrio purpureus*, which easily digested agar and cellulose and produced reducing sugars, was commonly distributed in the intestinal tract of sea slugs.

LITERATURE CITED

Atkins, W. R. G., and F. J. Warren. 1941. The Preservation of Fishing Nets, Trawl Twines and Fibre Ropes for Use in Sea Water. J. Mar. Biol. Assoc., 25:97-107.

Barghoorn, E. S. 1942. The Occurrence and Significance of Marine Cellulose-destroying Fungi. Science, 96:358-59.

-------, and D. H. Linder. 1944. Marine Fungi: Their Taxonomy and Biology. Farlowia, 1:395-467.

Bavendamm, W. 1932. Die mikrobiologische Kalkfällung in der tropischen See. Arch. f. Mikrobiol., 3:205-76.

Dorée, C. 1920. The Action of Sea Water on Cotton and Other Textile Fibers. Biochem. J., 14:709-14.

Hidaka, T. 1953. On the Cellulose Decomposing Bacteria Found in the Digestive Organs of *Teredo* (in Japanese). Memoirs of the Faculty of Fisheries, Kagoshima Univ., 3:149-57.

-------, and K. Saito. 1956. Studies on the Cellulose Decomposing Bacteria in the Digestive Organs of Ship-worm (*Teredo navalis*). II, On the Bacterial Cellulase (in Japanese). Memoirs of the Faculty of Fisheries, Kagoshima Univ., 5:172-77.

Humm, H. J. 1946. Marine Agar-digesting Bacteria of the South Atlantic Coast. Bull. Duke Univ. Mar. Lab., 3:45-75.

Issatchenko, B. L. 1921. Ber. d. Russisch. Hydrolog.

Kadota, H. 1951. Studies on the Biochemical Activities of Marine Bacteria. I, On the Agar-decomposing Bacteria in the Sea. Memoirs of the College of Agriculture, Kyoto Univ., 59: 54-67.

-------. 1953. Studies on the Biochemical Activities of Marine Bacteria. II, On the Properties of Agar-digesting Enzyme of *Vibrio purpureus*. Memoirs of the College of Agriculture, Kyoto Univ., 66:31-38.

-------. 1956. A Study on the Marine Aerobic Cellulose-decomposing Bacteria. Memoirs of the College of Agriculture, Kyoto Univ., 74:1-128.

Kawai, K. 1914. An Experiment on the Deterioration of Textiles (in Japanese). Suisan Koshujo Shiken Hokoku (J. Imp. Fisheries Inst.), 9:235-50.

Kotani, K. 1937. Preliminary Notes on the Cellulose-decomposing Bacteria in the Water (in Japanese). Suisan Kenkiu-shi (J. Fisheries), 32:435-38.

Lundestad, J. 1928. Ueber einige an der norwegischen Küste isolierte Agar-spaltende Arten von Meerbakterien. Centralbl. f. Bakt., II Abt., 75:321-44.

Nishimura, T., and H. Yamazoe. 1914. A Study on the Microorganisms Causing the Deterioration of Fishing Nets (in Japanese). Suisan. Koshujo Shiken Hokoku (J. Imp. Fisheries Inst.), 9:209-34.

Rubentschik, L. 1928. Zur Frage der aeroben Zellulosezersetzung bei hohen Salzkonzentration. Centralbl. f. Bakt., II Abt., 76:305-14.

-------. 1933. Zur anaeroben Zellulosezersetzung in Salzseen. Centralbl. f. Bakt., II Abt., 88:182-86.

Saito, K., and T. Hidaka. 1954. Studies on the Cellulose Decomposing Bacteria in the Digestive Organs of Ship-worm (*Teredo navalis*). III, Relation between the Enzyme of Digestive Organs and Bacterial Enzyme (in Japanese). Memoirs of the Faculty of Fisheries, Kagoshima Univ., 3:50-55.

Stanier, R. Y. 1941. Studies on Marine Agar-digesting Bacteria. J. Bact., 42:527-59.

Waksman, S. A., C. L. Carey, and H. W. Reuszer. 1933. Marine Bacteria and Their Role in the Cycle of Life in the Sea. I, Decomposition of Marine Plant and Animal Residues by Bacteria. Biol. Bull., 65:57-79.

Waledinsky, J., and I. Ssweschnikoff. 1926. Arb. a. d. Balneolog. Instit. d. Kaukasischen Mineralwässer, 3 (quoted in Rubentschik, 1928).

ZoBell, C. E. 1938. Studies on the Bacterial Flora of Marine Bottom Sediments. J. Sediment. Petrology, 8:10-18.

-------. 1946. Marine Microbiology. Chronica Botanica Co., Waltham, Mass.

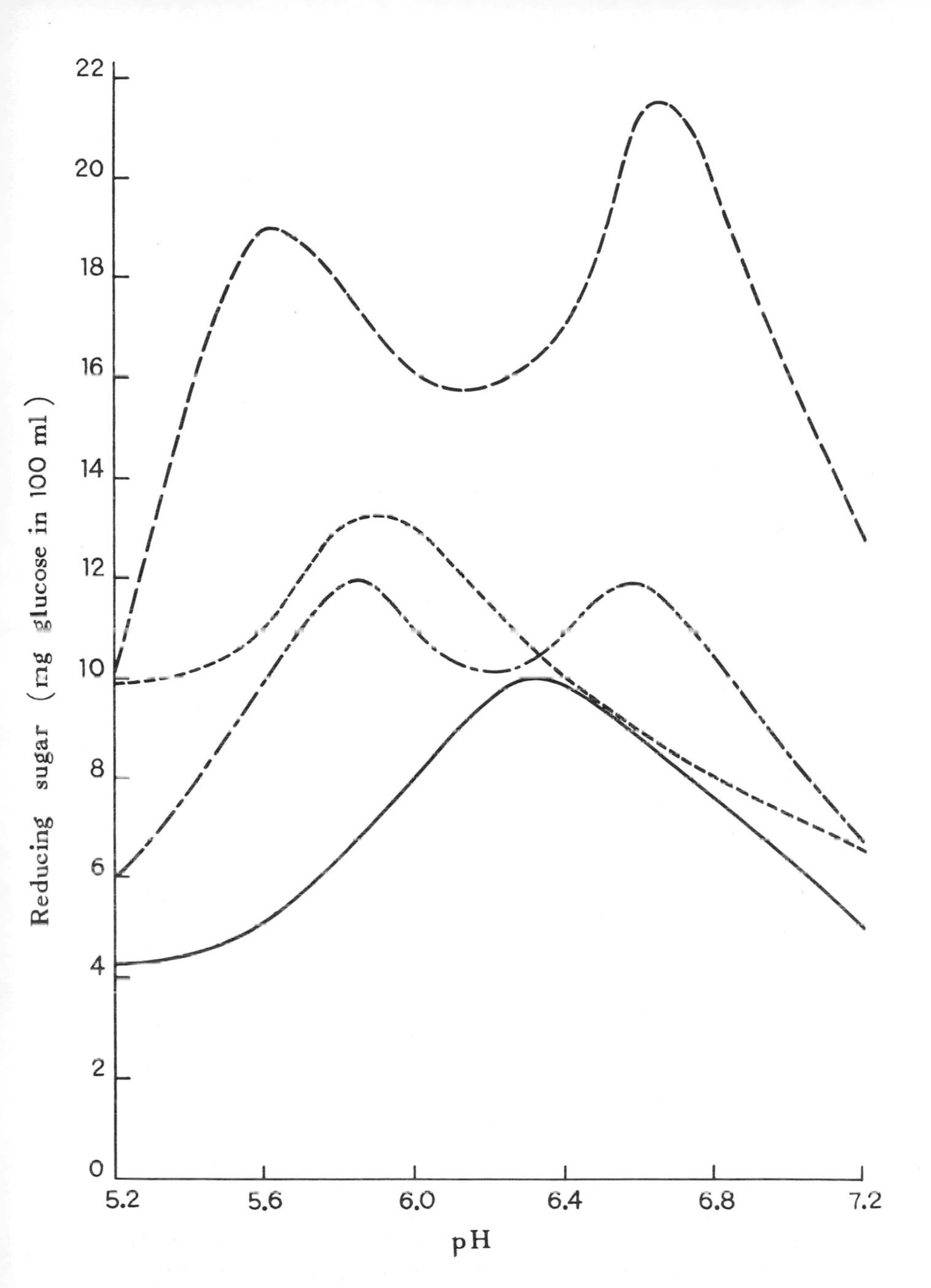

Fig. 1. Influence of pH value upon the activity of crude cellulase preparations obtained from the bacteria, the liver of *Teredo navalis*, and the whole body of *Teredo navalis*. Cellulase from the bacteria (———); cellulase from the liver of *Teredo navalis* (-----); mixed cellulase preparations from the bacteria and the liver of *Teredo navalis* (–·–·–·–); cellulase from the whole body of *Teredo navalis* (– – – –) (from Saito and Hidaka, 1954)

TERMITE CELLULASE

R. H. McBee

The relatively rapid rate at which the wood-eating termites can riddle a log or a substantially constructed wooden building has attracted the attention of biologists for generations. One of the most remarkable things noticed by early investigators was the large numbers of protozoans in the hindgut of the wood-eating termites. This led to speculation regarding a possible symbiotic association, but it was not until the 1920's that the situation was truly analyzed.

Cleveland (1924) noted that in the following families of the wood-eating termites: Kalotermitidae, Rhinotermitidae, and Mastotermitidae, intestinal protozoa were always present in the specimens he examined, whereas in the higher or fungus-raising termites of the family Termitidae there were no protozoa except in 3 of the 75 species examined, and these 3 fed on wood.

Cleveland (1925c, 1925d) observed that specimens of the termite commonly found in the eastern United States, *Reticulitermes flavipes,* could be defaunated by exposure to a temperature of 36°C for 24 hours. Such termites starved to death on a diet of wood, whereas refaunated forms survived normally. Similar experiments were performed in which termites were defaunated with an atmosphere of 0_2.

Cleveland (1925b) convinced himself that most of the flagellates in the termite gut are symbionts digesting wood for the termites, although he did not postulate a mechanism by which the termite profited.

Not all of the protozoans were beneficial, however. Those of the genera Trichonympha, Trichomonas, and Leidyopsis digested cellulose; Streblomastix did not. Of the wood-digesting protozoa, Trichonympha appeared to be the most effective.

Trichonympha campanula could be observed with 1/3 to 1/2 of its body filled with solid wood particles which had been ingested (Cleveland, 1925a). Although some of Cleveland's conclusions were based on what other investigators believed to be inadequate experimental evidence, they have substantially withstood reinvestigation.

The actual cellulolytic ability of the protozoans was demonstrated by Trager (1932, 1934). He was able to cultivate *Trichomonas termopsidis* from *Zootermopsis angusticollis* and to grow it in a cellulose medium for over three years, free from all contaminants except a small bacillus that was not cellulolytic. It was necessary to have the cellulose finely divided. Trager used cellulose precipitated from 50 per cent LiCl solution. The cellulase could be purified from homogenates of the protozoa by adsorption on alumina from which it could be partially eluted with 3 per cent KH_2PO_4. Trager also observed the effects of a cellobiase in the protozoan contents.

The presence of large numbers of bacilli and spirochaetes in the termite gut was noted by Dickman in 1931. He postulated that abundant cellulose-digesting bacteria were taken in with the food by the termites, but he was unable to recover them by culturing the intestinal contents. It is doubtful if his culture methods were adequate, and his original premise would not hold for termites consuming sound wood.

Beckwith and Rose (1929) found a few cellulose-decomposing bacteria in the gut contents of termites but concluded that they were probably contaminants.

The possibility of cellulose digestion by bacteria in the gut of the termites has not been an easily downed theory, and possibly it is well that this has been the case. Baldacci and Verona (1939) found large numbers of Cytophaga in the gut of the termite and considered them significant, but Ghidini (1940) concluded that they could not be important because the RQ (0.82) of the defaunated termite with its Cytophaga intact was the same whether or not it fed on cellulose.

Hungate (1936, 1938) reinvestigated the roles of various microorganisms in the nutrition of Zootermopsis and concluded that cellulose digestion was principally the role of the protozoans and that bacteria and fungi played no significant part in this process. The defaunated termite could digest some of the more readily soluble portions of the wood and could possibly by its own digestion satisfy up to 1/3 of its energy requirements for maintenance. Therefore, the protozoans must be necessary to supplement the digestive ability of the termite living on wood.

A possible role of bacteria living inside the protozoans and producing cellulase there was suggested by Pierantoni (1936), but there has been no good evidence presented to support this theory.

The work of Hungate (1939) and others has established without much doubt that in some of the termites the protozoans are the principal agents of cellulose digestion. This raised the question of the mechanism by which the termite benefited from this digestion. Early workers looked for an increase in reducing sugars in the gut of the termite due to the cellulose digestion by the protozoans. This idea that glucose would be the end product of such a digestion has been difficult to dispel, even though on a physiological basis it is not very logical. We find that Pochon still believed this theory worthy of mention in 1949 when considering the role of cellulolytic microorganisms in the rumen of cattle.

It would probably be more nearly correct to speak of cellulose fermenting protozoa rather than to call them cellulose digesters. It has been shown repeatedly that the termite protozoa are anaerobic, being killed by the presence of atmospheric oxygen. They ferment the sugars formed by cellulose hydrolysis, producing a variety of products, the most important of which appear to be CO_2, H_2, and acetic acid in variable concentrations. Hungate (1943) was able to show that the hindgut of the termite Zootermopsis is permeable to acetate. In the 1930's and the early 1940's, this idea of an organism living primarily upon acetate which it might absorb from its digestive tract was not as easy to believe as it is now with our more recent knowledge of the role of acetate in both energy yielding and synthetic reactions in all types of living creatures. The finding of a similar situation in the ruminant animals, where it is easier to measure absorption of the acids through the gut wall than it is in the termite, has made this hypothesis more credible.

A question which is so frequently asked that I believe it must be answered here, although I assume that all of you know the answer, is: How can both organisms live off the same food? The answer is relatively simple when we consider the metabolisms of the two creatures. The protozoan is anaerobic, living without benefit of oxygen. Therefore, there is available to it only about 1/20 of the energy in the food it ferments. The remainder of the energy available from the complete oxidation of the food to CO_2 and H_2O is partially available to the termite, which is an aerobic organism. The hydrogen produced by the protozoan probably is a waste product not available to any creature except the hydrogen-oxidizing bacteria.

We thus obtain a picture of the wood-eating termite that looks something like this: the termite collects the wood, comminutes it, moistens it with its body fluids, and then passes this food on to its captive protozoa in the hindgut with very little predigestion. The protozoan mass may be from 16 to 36 per cent (Katzin and Kirby, 1939) of the total weight of the termite. The protozoans ingest and ferment the wood particles, giving off CO_2, H_2, and a variety of organic compounds, chiefly acetate. The acetate is absorbed through the termite gut wall where it is metabolized to yield the energy for growth of the termite and the gathering of more food for the protozoans.

The role of fungi in termite nutrition has also been investigated especially with respect to the nitrogen economy of the termite. Hungate (1944) was not able to show any significant contribution to the nitrogen economy of wood-eating termites by the wood rot fungi.

This picture shows that the termite cellulase is of protozoal origin and that fermentation is involved instead of simply hydrolysis of the cellulose. If this were the whole picture it would be simple to discuss the cellulase of termites. But it appears that the digestion of cellulose in termites may also proceed by two other mechanisms.

Holdaway (1933) observed that *Eutermes exitiosus* has no intestinal protozoa, yet digests wood. This may be due either to bacteria or to a termite cellulase.

Misra and Ranganathan (1954) observed that the workers of *Termes (Cyclotermes) obesus* (Dr. Becker believes this organism is actually in the genus Odontotermes), a mound-building termite which maintains fungus gardens, will eat wood but have no flagellates in the gut. Cellulose-decomposing bacteria were isolated from the gut contents, and it was concluded that these bacteria, through the production of an extracellular cellulase, were responsible for the cellulose decomposition by the termite. The published data are very incomplete, and I believe that one could use the same information to show that the bacteria are not responsible for the cellulose digestion.

As a final touch, Dr. Tracey and Dr. Jermyn have told me at this symposium that there is in Australia another queer creature of the animal world, a termite with its own cellulase.

These data have not appeared in print, and therefore an evaluation will have to wait until a later date.

At present the existence of a true termite cellulase has not been well established, nor has the bacterial decomposition of cellulose for the termite been adequately investigated. These two possibilities cannot be ignored, however, and further work may show either one or both situations to exist. The cellulase of the termite protozoans has been studied sufficiently well so that there is little doubt as to its source or the mechanism by which the termite benefits by the presence of its symbionts. Therefore any discussion of termite cellulase ends up by being primarily concerned with the protozoan cellulase and with the mutualism existing between the termite and its intestinal protozoa.

LITERATURE CITED

Baldacci, E., and O. Verona. 1939. Isolamento di schizomiceti del G. *Cytophaga,* dall' intestino delle termiti. Soc. Ital. Biol. Sper. Bol., 14:156.

Beckwith, T. D., and E. J. Rose. 1929. Cellulose Digestion by Organisms from the Termite Gut. Proc. Soc. Exp. Biol. and Med., 27:4-6.

Cleveland, L. R. 1924. The Physiological and Symbiotic Relationships between the Intestinal Protozoa of Termites and Their Host, with Special Reference to *Reticulitermes flavipes* Kollar. Biol. Bull., 46:178-227.

-------. 1925a. The Method by Which *Trichonympha campanula,* a Protozoöan in the Intestine of Termites, Ingests Solid Particles of Wood for Food. Biol. Bull., 48:282-88.

-------. 1925b. The Feeding Habit of Termite Castes and Its Relation to Their Intestinal Flagellates. Biol. Bull., 48:295-308.

-------. 1925c. The Effects of Oxygenation and Starvation on the Symbiosis between the Termite, *Termopsis,* and Its Intestinal Flagellates. Biol. Bull., 48:309-26.

-------. 1925d. Toxicity of Oxygen for Protozoa *in vivo* and *in vitro*: Animals Defaunated without Injury. Biol. Bull., 48:455-68.

Dickman, A. 1931. Studies on the Intestinal Flora of Termites with Reference to Their Ability to Digest Cellulose. Biol. Bull., 61:85-92.

Ghidini, G. M. 1940. Ricerche sulla attività cellulosolitica della flora e fauna intestinale di *Reticulotermes lucifugus,* Rossi. Soc. Ital. Biol. Sper. Bol., 15:220-21.

Hungate, R. E. 1936. Studies on the Nutrition of *Zootermopsis.* I. The Role of Bacteria and Molds in Cellulose Decomposition. Zentralbl. Bakter., Ser. 2, 94:240-49.

-------. 1938. Studies on the Nutrition of *Zootermopsis.* II. The Relative Importance of the Termite and the Protozoa in Wood Digestion. Ecology, 19:1-25.

-------. 1939. Experiments on the Nutrition of *Zootermopsis.* III. The Anaerobic Carbohydrate Dissimilation by the Intestinal Protozoa. Ecology, 20:230-45.

-------. 1943. Quantitative Analyses on the Cellulose Fermentation by Termite Protozoa. Ann. Entomol. Soc. Am., 36: 730-39.

-------. 1944. Termite Growth and Nitrogen Utilization in Laboratory Cultures. Proc. and Trans. Texas Acad. Sci., 1943, 27:01 08.

Holdaway, F. G. 1933. The Composition of Different Regions of Mounds of *Eutermes exitiosus* Hill. Austral. J. Council Sci. Indus. Res., 6:160-65.

Katzin, L.I., and H. Kirby, Jr. 1939. The Relative Weights of Termites and their Protozoa. J. Parasitol., 25:444-45.

Misra, J.N. and V. Ranganathan. 1954. Digestion of Cellulose by the Mound-building Termite *Termes (Cyclotermes) obesus* (Rambur). Indian Acad. Sci. Proc. sec. B, 39:100-13.

Pierantoni, U. 1936. La simbiosi fisiologica nei termitidi xilofagi e nei loro flagellati intestinali. Arch. Zool. Ital., 22:135-73.

Pochon, J. 1949. Anaerobies cellulolytiques. Ann. Inst. Pasteur, 77:419-33.

Trager, W. 1932. A Cellulase from the Symbiotic Intestinal Flagellates of Termites and of the Roach, *Cryptocercus punctulatus.* Biochem. J., 26:1763-71.

-------. 1934. The Cultivation of a Cellulose-digesting Flagellate, *Trichomonas termopsidis,* and of Certain Other Termite Protozoa. Biol. Bull., 66: 182-90.

CELLULOSE DIGESTION IN INSECTS

Reuben Lasker

A search through the literature for references to cellulose digestion by insects is disappointing. Considering the advances made in the study of mold and bacterial cellulases, it seems unfortunate that insect cellulases have been neglected. However, a further analysis of the situation reveals that the probable reason for this lack of progress has been the failure to demonstrate that any insect can truly digest cellulose by its own enzymes. Nevertheless, some published work of a preliminary nature has been done which shows that insects can digest cellulose, leaving the question of how they do it open to speculation.

The earliest of these studies showed that an element of the plant cell wall, presumably cellulose, disappeared after ingestion by the insect under investigation. For example, Haberlandt (1918) compared the food and feces of two caterpillars, *Cemiostoma laburnella* and *Zeuzera pyrina,* and reported that the cell walls of the ingested plant material were softened after passage through the intestine of these animals. Similarly, Biedermann (1919) observed the weakening of grass cell walls in the gut of two grasshoppers and concluded from this evidence that a "cytase" must exist to destroy the cellulose wall. Belehrádek (1922) used a somewhat more experimental approach and studied the effect of saliva of the stick insect *Dixippus morosus* on thin slices of almond endosperm. He observed microscopically that the cell walls dissolved under the action of the saliva.

A few early workers were interested in the commercially important termites, wood-boring beetles, and their larvae. The line of experimental approach was a rough quantitative "before and after" analysis. The wood into which these insects bored was analyzed for its cellulose content, and the results were compared to a similar analysis of the ingested wood after it had passed through the animal. Cellulose digestion by the insect was inferred if there was a significant reduction in the cellulose content of the wood as it appeared in the excreta. Probably the most notable study of this nature was that by Oshima (1919), who showed for the first time that termites were capable of digesting wood. It was some years later before any cognizance was was taken of the possible role of symbiotic microorganisms. Falck (1930) also did this kind of study on the wood-boring house beetle, *Hylotrupes bajulus*, which attacks pine wood. Schlottke and Becker (1942) found cellulase in this animal. Ripper (1930) investigated several wood-boring insect larvae and found that two of them, *Cerambyx cerdo* and *Xestobium rufovillosum*, reduce the cellulose content of wood after ingestion and also give positive results for a cellulase in the gut contents. However, he did not consider bacteria as the source of the cellulase since their occurrence was sporadic and the size of the microbial population varied widely. Mansour and Mansour-Bek (1934) found a cellulase in the gastric juice of *Macrotoma palmata* larvae. These authors stated than an extensive search was made for cellulose-decomposing microorganisms but with negative results.

The concept of symbiosis between microorganisms and the cellulose-eating insects has been well established and documented by the work on termites by Cleveland (1924, 1925, 1928) and on the termite protozoans by Trager (1932) and by Hungate (1946). Their findings should caution us against the interpretation of cellulose digestion in terms of an animal cellulase alone. At about the same time Cleveland was studying termites, further enforcement of the symbiosis concept was given by Werner (1926) and Wiedemann (1930), who found that the larvae of lamellicorn beetles had a much dilated hindgut, containing an abundant population of bacteria and flagellates. These larvae feed chiefly on rotting wood. According to Wiedmann, the bacteria digest the wood, the flagellates eat the bacteria, and the animal benefits by digesting the flagellates and bacteria. This example is reminiscent of the analogous situation in the cow, which uses its intestinal bacterial flora as a source of dietary protein. Both Werner and Wiedemann isolated cellulose-decomposing bacteria from the excreta and intestine of these animals.

This brief review of the literature illustrates that cellulose does disappear when it passes through the intestine of some insects. However, except for the case of termites, where the protozoa are involved in cellulose digestion, we are not sure whether any insect has a cellulase or not.

In each case where an attempt has been made to define the role of microorganisms in cellulose digestion, the difficulties in obtaining positive evidence become apparent. That is to say, when there is only negative evidence that microorganisms may be responsible for the digestion of cellulose, it can be argued that they are still important in digestion but that the investigator is unable to demonstrate their existence. Thus, if cellulolytic activity is found in an extract of tissues or in the secretion of a gland, the question remains whether or not this activity is attributable to microorganisms; conversely, finding a cellulose-digesting microbe in the intestine of an animal does not establish conclusively its benefit to the host.

In 1954 an investigation was begun at Stanford University to study the paper-eating insects commonly known as silverfish. Three features of this animal's biology favored us in a study of its cellulose digestion: (1) the diet of the silverfish contains large quantities of cellulose; (2) these animals are easy to rear in the laboratory since they are already domesticated; and (3) the eggs are heavily cutinized and relatively impermeable to water and many dissolved substances. Much of the work I am going to discuss has appeared in detail elsewhere (Lasker and Giese, 1956; Lasker, 1957) but it is summarized briefly here for its relevance to the general and comparative aspects of cellulose digestion.

Stanford University, like other institutions in a temperate climate, has been, and is still, plagued by silverfish. A typical example of the damage done by these animals is shown in figure 1.

Cellulose digestion in the silverfish, *Ctenolepisma lineata,* was shown in several ways. Digestibility coefficients were determined for a number of animals fed on starch-free filter paper cellulose for a month or longer (table 1). The data indicate that 70 to 85 per cent of the cellulose ingested is actually digested. Not only did the silverfish digest a considerable amount of cellulose eaten, but in many cases they gained weight on a diet of cellulose alone (fig. 2).

Respiratory quotients (CO_2/O_2) determined on silverfish fed cellulose alone for a month or longer, were, in three experiments, 0.91, 1.09, and 0.92. These may be compared with the results from an experiment done with gelatin-fed animals where an RQ of 0.75 was obtained. The results suggest that silverfish are metabolizing almost pure carbohydrate when fed on cellulose alone.

TABLE 1
DIGESTIBILITY COEFFICIENTS FOR SILVERFISH FED CELLULOSE FOR SEVERAL MONTHS

No.	Initial Wet Wt. Silverfish (mg)	Dry Wt. Cellulose Consumed (mg)	Dry Wt. of Feces (mg)	Digestibility Coefficient (%)
1	22·21	9·39	1·37	85·4
2	25·21	7·46	1·94	74·2
3	25·38	3·35	0·99	73·2
4	20·01	1·58	0·38	75·9
5	23·51	4·49	0·58	87·0
6	27·25	2·33	0·63	71·7
7	15·67	5·99	0·84	85·8

The most decisive evidence of cellulose digestion was obtained by feeding silverfish with uniformly labeled C^{14}-cellulose. The high radioactivity of the carbon respired as carbon dioxide indicates extensive digestion and metabolism of radioactive cellulose (table 2).

Incorporation of the cellulose-derived carbon was shown by the high radioactivity of the animal's tissues after the gut and its contents had been removed.

Tests were made for bacteria and other alimentary microorganisms. The results of these tests may be summarized as follows. Enrichment cultures containing filter paper and Hungate's mineral medium (1950) were unsuccessful in isolating cellulose-decomposing microbes from silverfish intestine. In 7 out of 20 cases, molds were obtained that did decompose the filter paper, but no molds were ever found to be present in the intestine of hundreds of organisms examined. Additional tests were made in which cellulose was supplemented with organic nutrients in various combinations and concentrations as suggested by the demands of various microorganisms, particularly those digesting cellulose. Although many bacteria developed in such media, none was cellulolytic. To exclude the possibility that anaerobic bacteria might decompose the cellulose for the silverfish, additional tests were made using the critical methods devised by Hungate (1950) for the detection of anaerobic cellulolytic microorganisms. These tests were also negative, and it soon became evident that anaerobic conditions were unfavorable for the silverfish microflora, since only small populations developed under

TABLE 2
RADIOACTIVITY OF $Ba^{14}CO_3$ DERIVED FROM THE RESPIRATORY CO_2 OF SILVERFISH FED ^{14}C CELLULOSE

Planchet No.	mg $BaCO_3$	Counts* per min.	*d*/min./mg $BaCO_3$	*d*/min./0·1 mM $BaCO_3$
		Experiment 1		
1	4·0	3710	3840	$7{\cdot}58 \times 10^4$
2	6·5	5150	3380	$6{\cdot}69 \times 10^4$
3	1·3	1100	3230	$6{\cdot}37 \times 10^4$
4	2·0	1730	3420	$6{\cdot}75 \times 10^4$
5	2·1	1750	3300	$6{\cdot}51 \times 10^4$
6 background	...	59	....	....
		Experiment 2		
1	2·9	1510	2560	$5{\cdot}05 \times 10^4$
2	2·0	1200	2830	$5{\cdot}58 \times 10^4$
3	1·2	855	3220	$6{\cdot}36 \times 10^4$
4 background	...	45	....	....

*Counts per minute are presented after correction for self-absorption, coincidence, and the efficiency of the counter which was determined in each run by counting a radioactive standard sample; *d* = disintegrations.

anaerobic conditions; yet, when aerobic plates were made from the same samples, many additional colonies appeared.

Protozoa (parasitic sporozoans) occurred in the silverfish intestine only sporadically and were few in number. Considering the rare occurrence of these protozoans and their parasitic mode of life, they cannot be implicated in cellulose digestion by the silverfish. Intracellular symbiotic microorganisms were sought by various diagnostic staining procedures, but none was found.

If the silverfish does not depend upon microorganisms to digest its cellulose, digestion should be possible in the complete absence of microorganisms. By washing silverfish eggs (fig. 3) in a mixture of mercuric chloride and ethanol, it was possible to obtain nymphs which were bacteria-free. When these nymphs were fed radioactive cellulose under aseptic conditions, the carbon dioxide respired was radioactive . These data offer convinc-

TABLE 3
RADIOACTIVITY OF $Ba^{14}CO_3$ DERIVED FROM THE RESPIRATORY CARBON DIOXIDE OF BACTERIA-FREE SILVERFISH*

Animal no. 1, fed ^{14}C cellulose with no additional food; nos. 2 and 3, fed ^{14}C cellulose and oats soaked in yeast extract and liver extract.

Planchet No.	mg $BaCO_3$	Counts per min.	*d*/min./mg $BaCO_3$	*d*/min. 0·1 mM $BaCO_3$
		Animal no. 1		
1	1·7	458	1480	$2{\cdot}92 \times 10^4$
2	1·2	600	2710	$5{\cdot}34 \times 10^4$
3	0·6	277	2360	$4{\cdot}65 \times 10^4$
4	0·5	379	3600	$7{\cdot}10 \times 10^4$
5 background	...	33	....	....
			Average	$5{\cdot}12 \times 10^4$
		Animal no. 2		
1	1·9	255	604	$1{\cdot}19 \times 10^4$
2	2·8	384	627	$1{\cdot}23 \times 10^4$
3	0·3	74	740	$1{\cdot}46 \times 10^4$
4 background	...	38	...	....
			Average	$1{\cdot}29 \times 10^4$
		Animal no. 3		
1	2·5	164	327	$6{\cdot}44 \times 10^3$
2	1·8	134	342	$6{\cdot}73 \times 10^3$
3	0·5	68	446	$6{\cdot}78 \times 10^3$
4	0·3	52	456	$6{\cdot}97 \times 10^3$
5 background	...	29	...	....
			Average	$7{\cdot}73 \times 10^3$

*The animals weighed less than 1 mg each.

ing proof that the silverfish can digest cellulose by virtue of its own enzymes.

It was also possible to demonstrate by reducing sugar analyses that a cellulase was present in extracts of the midgut tissue of nonsterile silverfish. Figures 4 and 5 are pH activity curves for silverfish midgut cellulase and cellobiase. The activity of the midgut mash on the two substrates was not different enough to determine whether or not a different enzyme was involved in the splitting of the cellobiose. The bimodal character of these curves indicates that a number of enzymes may possibly be involved. The peak pH activities shifted 1/2 of a pH unit from experiment to experiment. Possibly in a mash there is some interaction between different proteins. Some preliminary attempts were made to fractionate silverfish cellulase extract into a more pure form. It was possible to isolate an active cellulase preparation in the 60 per cent and 70 per cent saturated ammonium sulfate fractions of the soluble proteins from the midgut. This preparation was three to four times as active as the original cell-free extracts.

Evidence has been presented here which shows that the silverfish *Ctenolepisma lineata* can digest cellulose without the aid of alimentary microorganisms. This information was obtained by a fruitful combination of several methods which the author believes may prove equally enlightening if applied to other animals. Radioactive cellulose is a useful tool which might be used further to elucidate some of the puzzling mechanisms of cellulose degradation in many animals, and particularly in insects.

LITERATURE CITED

Belehrádek, J. 1922. Expériences sur la cellulase et l'amylase de la salive chez *Dixippus morosus*. Arch. Internat. Physiol., 17:260-65.

Biedermann, W. 1919. Beitrage zur vergleichenden Physiologie der Verdauung. VIII, Die Verdauung pflanzlichen Zellinhalts im Darm einiger Insekten. Arch. ges. Physiol., 174:392-425.

Cleveland, L. R. 1924. The Physiological and Symbiotic Relationships between the Intestinal Protozoa of Termites and Their Host, with Special Reference to *Reticulotermes flavipes* Kollar. Biol. Bull., 46:178-227.

-------. 1925. The Effects of Oxygenation and Starvation on the Symbiosis between the Termite, *Termopsis*, and Its Intestinal Flagellates. Biol. Bull., 48:309-26.

-------. 1928. Further Experiments on the Symbiosis between Termites and Their Intestinal protozoa. Biol. Bull., 54:231-37.

Falck, R. 1930. Die Scheindestruktion des Koniferenholzes durch die Larven des Hausbockes (*Hylotrupes bajulus* L.). Cellulosechemie, 11:89-91.

Haberlandt, G. 1918. Mikroskopische Untersuchungen über Zellwandverdauung. Beitr. allg. Bot., 1:501-35.

Hungate, R. E. 1946. The Symbiotic Utilization of Cellulose. J. Elisha Mitchell Sci. Soc., 62:9-24.

-------. 1950. The Anaerobic Mesophilic Cellulolytic Bacteria. Bacteriol. Rev., 14:1-49.

Lasker, R. 1957. Silverfish, a Paper-eating Insect. Sci. Monthly, 84:123-27.

-------, and A. C. Giese. 1956. Cellulose Digestion by the Silverfish *Ctenolepisma lineata*. J. Exp. Biol., 33:542-53.

Mansour, K., and J. J. Mansour-Bek. 1934. On the Digestion of Wood by Insects. J. Exp. Biol., 11:243-56.

Oshima, M. 1919. Formosan Termites and Methods of Preventing Their Damage. Philippine J. Sci., 15:319-83.

Ripper, W. 1930. Zur Frage des Celluloseabbaus bei der Holzverdauung xylophager Insektenlarven. Z. vergleich. Physiol., 13:314-33.

Schlottke, E., and G. Becker. 1942. Verdauungsfermente im Darm der Hausbockkäfer-Larven. Biol. Gen. (Vienna), 16:1-11.

Trager, W. 1932. A Cellulase from the Symbiotic Intestinal Flagellates of Termites and of the Roach, *Cryptocercus punctulatus*. Biochem. J., 26:1762-71.

Werner, E. 1926. Der Erreger der Zelluloseverdauung bei der Rosenkäferlarve (*Potosia cuprea* Fbr.), *Bacillus cellulosam fermentans* n. sp. Zentr. Bakteriol. Parasitenk. Abt. 11, 67:297-330.

Wiedemann, J. F. 1930. Die Zelluloseverdauung bei Lamellicornierlarven. Z. Morphol. Okol. Tiere, 19:228-58.

Fig. 1. Typical damage done to paper by feeding silverfish

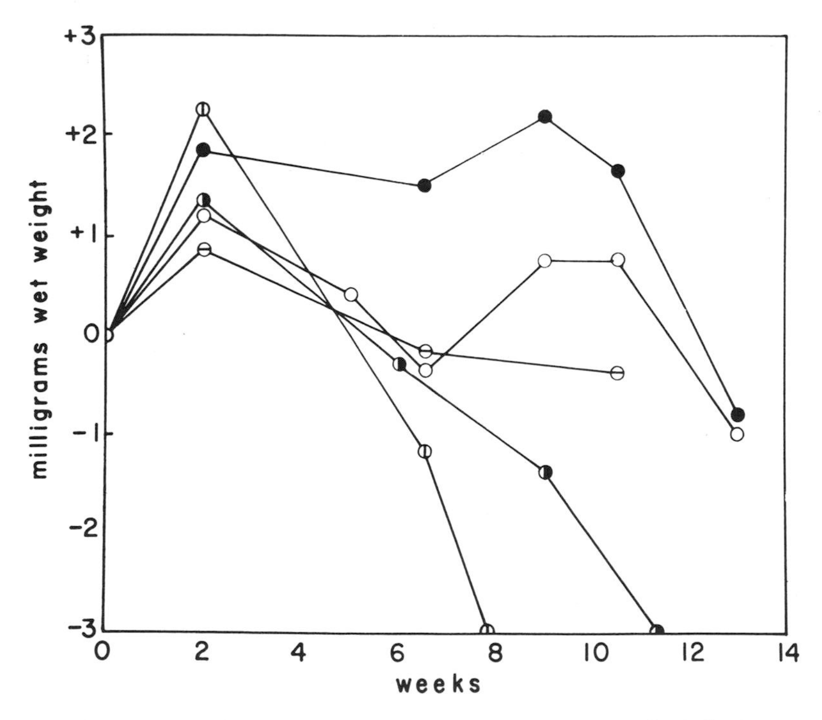

Fig. 2. Gain and loss of weight by several silverfish fed only cellulose for several weeks

Fig. 3. An egg, a newly hatched nymph, and an emerging nymph of the silverfish, *C. lineata*. The egg is 1 mm long

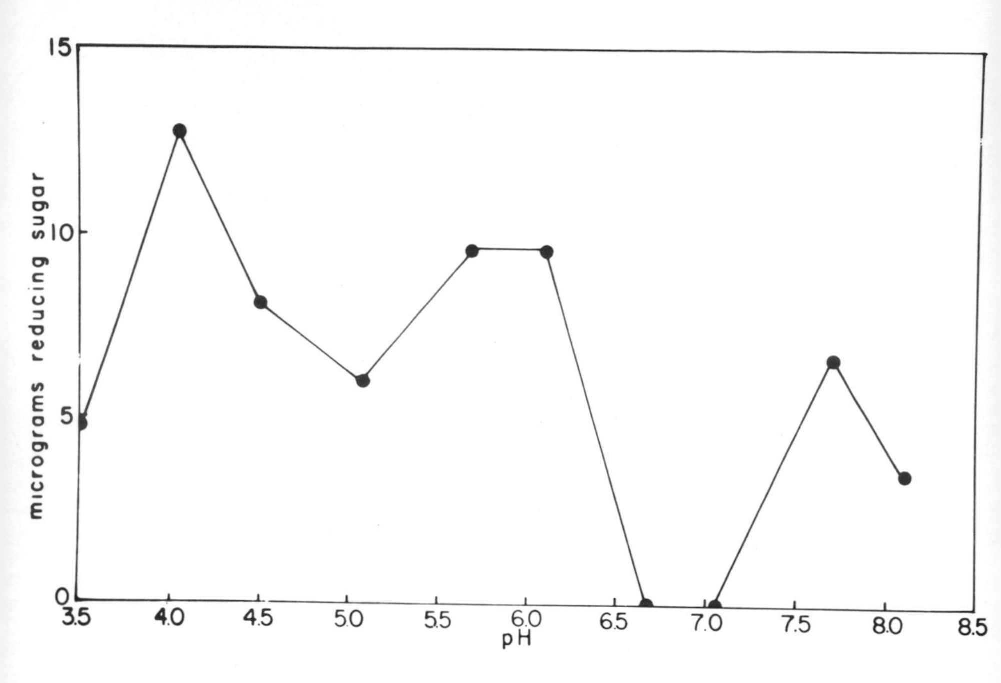

Fig. 4. The pH activity curve of silverfish cellulase

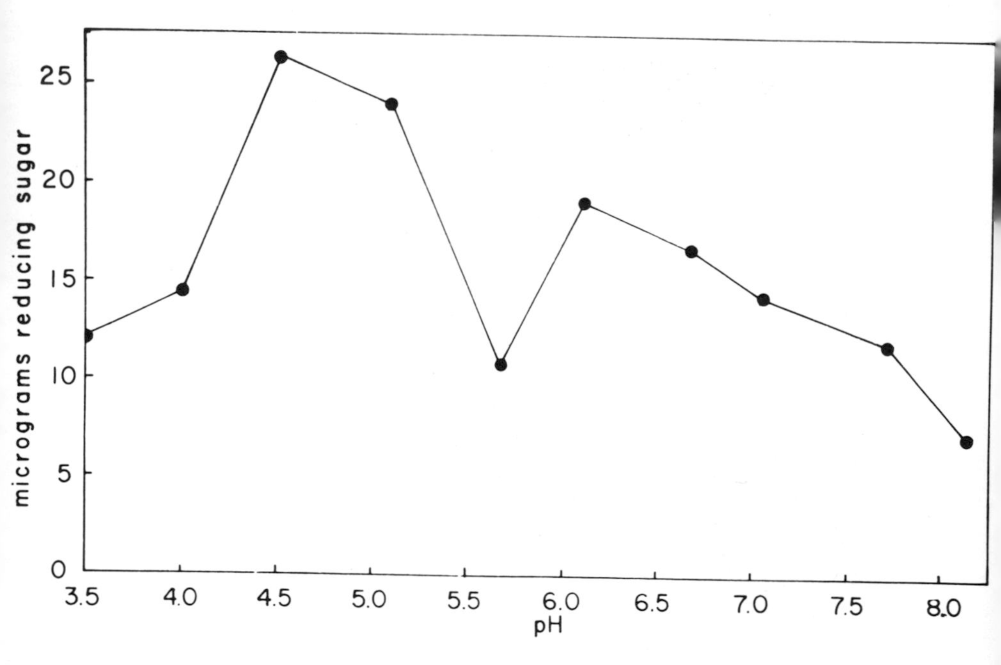

Fig. 5. The pH activity curve of silverfish cellobiase

ON CELLULOSE AND β-GLUCOSIDE-SPLITTING ACTIVITY IN ENZYME PREPARATIONS FROM SOME HYMENOMYCETES

Birgitta Norkrans

Within the systematic group of Hymenomycetes, we will find many genera and species empirically and experimentally known as cellulose-decomposers, decomposers of litter as well as of wood. In comparison with Ascomycetes and Fungi Imperfecti, however, they have been used only in relatively few cases as enzyme producers for cellulase studies. Of course they offer some disadvantages. They do not form asexual spores, which could be easily handled as inoculation material, and they are, generally, slower growing than the Ascomycetes.

The two Hymenomycetes which I have used in recent studies are the root-rotting fungus on spruce, *Polyporus annosus*, and different strains of *Collybia velutipes*, a stump-rotting fungus. The crude enzyme preparations have been obtained by treating culture filtrates with three volumes of cold acetone, and the vacuum-dried precipitate has been dissolved in water amounting to a twentieth of the filtrate volume. On culture media with glucose as a sole carbon source, no cellulolytic activity has been proved in that manner; on media with cellobiose, however, a weak activity has been induced, and, finally, on cellulose the highest activity has been obtained. In general, I have tested the cellulolytic activity turbidimetrically with a precipitated cellulose as enzyme substrate. The difference between the initial extinction value and that measured during the experiment has been taken as a measure of cellulolytic activity. The precipitated cellulose is very far from a native cellulose, but it is an unsubstituted cellulose. It has an average DP of about 350. The initial enzymatic depolymerization proceeds rapidly until a DP limit of about 50. Electron micrographs of such cellulose residues reveal a homogeneous mass of distinct particles about 300 Å long and about half as wide. These are in clear contrast to those of the untreated cellulose sols containing larger aggregates of longer, disordered fibers that form an irregular network of unresolved lumps. This part of the investigation was made in collaboration with Dr. Bengt Rånby, formerly at the Institute of Physical Chemistry, Uppsala (Norkrans and Rånby, 1950). The enzymatic degradation apparently occurs in easily accessible regions of the cellulose aggregates leaving more resistant particles as a residue. The higher resistance of the particles surely depends upon a higher crystallinity since an X-ray diagram of the precipitated cellulose, after an enzyme treatment (in this early investigation by Norkrans, 1950a and b, a Tricholoma preparation was used), seems to show a relative increase in crystallinity. It shows not only more distinct lines but additional lines which were invisible in the untreated cellulose sol.

DP values for cellulose sol treated with different enzyme preparations for various times illustrate the above-mentioned DP limit of 50. This limit was not exceeded in spite of an addition of fresh enzyme after 168 hours (tested with *Collybia velutipes,* strain L 1). Furthermore, the breakdown seems to give, maximally, 6.5 to 6.7 mg reducing substances calculated as glucose, amounting to a breakdown of 65 to 67 per cent of the initial cellulosic material. This limit is probably related rather to the morphological structure of the cellulose sol particles than to enzyme inactivation.

I should like to draw attention also to another fact, the effect of heat-treated enzyme preparations. They have been heated at 120°C by autoclaving to 2 atm, and still the inactivation was not complete. A certain activity was demonstrated in all test methods used: a certain increase in mg reducing substances, an increase in mg glucose (determined manometrically with the aid of glucose dehydrogenase), and a marked drop in DP. By the way, a comparison between the DP values and the corresponding values for reducing substances lends support to the theory of random splitting along the cellulose chain. An enzyme attack proceeding from the endgroup, resulting in such a drop in DP, would have given much higher values for reducing substances.

The above-mentioned facts constitute the main data about the cellulose residue. The breakdown products obtained and demonstrated by paper chromatography are mainly glucose, cellobiose, and some slower moving substances, one of which may be cellotriose; whether these latter substances are primary products or not may be left undecided.

And then what about the activities of the enzyme preparations against β-glucosides? Three β-glucosides have been tested as enzyme substrates, p-nitro-phenyl-β-glucoside, called "niphegluc," and salicin, as well as cellobiose. The extracellular activity produced by mycelia of varying age has been followed. In all tests the hydrolysis rate for β-niphegluc seemed to be somewhat higher than that for the two natural β-glucosides, salicin and cellobiose. The niphegluc test gives the amount of aglucon split off, the two other tests, however, the glucon part. Since the β-glucosidases have been shown to act in transferase reactions giving a number of glucose oligosaccharides, the lower value in the salicin and cellobiose tests might depend upon that fact.

And then I have tried to find out whether one, two, or more extracellular enzymes are responsible for the activities against cellulose and the β-glucosides. The extremely high heat stability of the cellulolytic activity suggested heat treatments as a possible way to fractionate the enzyme preparations. Thus, enzyme preparations have been treated for 20 minutes at different temperatures, at 40°, 60°, 80°, and 100°C. At first the activities have been tested against niphegluc. The main result seems to be that the niphegluc-splitting activity was completely, or nearly completely, eliminated by a treatment at 60°C, whereas a certain cellulase activity could be detected even after a treatment at 120°C. This means we should have a separate niphegluc-splitting β-glucosidase *besides* the cellulase.

Treatment at 80°C or 100°C caused hardly any further reduction of the cellulase activity than a treatment at 60°C. All strains of Collybia behave in approximately the same way. This fact, that the treatment at 60°, 80°, or 100°C seems to reduce the activity against the cellulose sol to about the same level, might suggest the idea of two separate cellulase components in the Collybia preparations, one heat stable one, stable at least in this crude enzyme preparation (Norkrans, 1957a and b).

Some enzyme preparations have been tested also for their salicin- and cellobiose-splitting activity. A 60°C treatment eliminated the salicinase activity from the preparation in the same way as the niphegluc-splitting activity, whereas a cellobiase activity was obtained even after a treatment at 100°C (in the case of *Collybia velutipes*, strain Mi 460). This shows that the cellobiase activity resisted the heat treatment in about the same way as the cellulase activity did. This may suggest that the heat stable cellulase component(s) also have a cellobioase activity.

Furthermore, zone-electrophoresis experiments have been

performed in collaboration with Dr. Jerker Porath at the Institute of Biochemistry, Uppsala. The electrophoresis tests were run (in acetate buffer of pH 4.7) in a vertical column. The supporting medium has been a semisynthetic, insoluble, hydrophilic substance which cannot be attacked by the enzyme preparation. The fractions show adsorption in ultraviolet light.

The adsorption values, measured at a wave length of 270 mμ, reached a maximum in the same fraction as the cellulase activity, whereas the maximum niphegluc-splitting activity was found in another adjacent fraction. The fact that the activity maxima appear in different fractions should prove the presence of two different extracellular enzymes or enzyme complexes, one active against β-niphegluc, one active against cellulose. The cellulase curve is wider, a fact pointing to an enzyme complex. Preliminary experiments (*Collybia velutipes* Mi 460), however, show that the maximum cellobiase activity was obtained in this niphegluc-splitting fraction, too, not in the cellulase fraction. This means that, even if the above-mentioned heat stable component should have cellobiase activity, a separate β-glucosidase with cellobiase activity in addition to the cellulase is present in the enzyme preparation. The maximum salicin-splitting activity has also been found in this niphegluc-splitting fraction. Whether or not the niphegluc-splitting enzyme, the salicin-splitting enzyme, and the extracellular cellobiase are identical cannot be decided by these experiments. I think, however, it has been shown that two extracellular enzymes or enzyme complexes are involved here in the total breakdown of cellulose: an extracellular β-glucosidase in addition to a cellulase.

LITERATURE CITED

Norkrans, B. 1950a. Studies in Growth and Cellulolytic Enzymes of Tricholoma. Symb. Bot. Ups., XI:1.

-------. 1950b. Influence of Cellulolytic Enzymes from Hymenomycetes on Cellulose Preparations of Different Crystallinity. Physiol. Plant., 3:75.

-------. 1957a. Studies of β-glucoside and Cellulose-splitting Enzymes from *Polyporus annosus*. Fr. Physiol. Plant., 10: 198-214.

-------. 1957b. Studies of β-glucoside and Cellulose-splitting Enzymes from *Collybia velutipes*. Physiol. Plant., 10:454.

-------, and B.G. Rånby. 1956. Studies of the Enzymatic Degradation of Cellulose. Physiol. Plant., 9:198-211.

CELLULOLYTIC ACTIVITY IN TEREDO

Leonard J. Greenfield

Utilization of wood for its potential nutritive value usually is associated with cellulolytic activity on the part of the feeding organism. Among the Terediniens, this type of activity has been studied from two major viewpoints. The first involves indirect or presumptive evidence based on analyses of wood particles during or after their passage through the Teredo's digestive tract. The second is directly concerned with production of reducing substances by means of cellulose digesting enzymes.

In one of the earlier studies of this problem, Harrington (1921) extracted a material from the digestive diverticula or "liver" of *Teredo norvegica* which partially hydrolyzed sawdust, but negative results were obtained when filter paper was used as a substrate. This author suggested that the results might be inconclusive because shortage of materials forced him to use small concentrations of enzyme and prevented him from running sufficient controls.

A subsequent series of experiments by Dore and Miller (1923), resulting in presumptive evidence of wood digestion in *Teredo navalis*, involved analyses of woody matter in the feces ejected from the excurrent siphon of the organism. These were compared with similar determinations on uningested wood fragments (Douglas fir). By eliminating the animal substance from the fecal material, results were calculated on a basis of wood constituents. It was then found that about 80 per cent of the cellulose was lost during the passage of wood through the digestive tract. The suggestion was made that the disappearance of this constituent was due to its hydrolysis to simple carbohydrates which the animal utilized as food. This conclusion was further substantiated as a result of experiments with the Northwest shipworm, *Bankia setacea* by Miller and Boynton (1926). The caeca from a number of specimens of this species were removed and their contents, consisting of wood borings, were analyzed for free reducing sugar content. Comparative determinations were performed on pulverized samples of the original wood. The amount of reducing sugars from the caecum was about four times as great as that of the pulverized controls. Under the conditions of the experiment, this quantitative figure did not indicate the true amount of sugar formed since in all probability wood was still being digested at the time when the specimens were sacrificed, in addition to which unknown amounts of sugar may have been absorbed by the caecal typhlosole. On the other hand, the increased quantity of reducing sugar in the digestive tract was believed to be associated with cellulose loss by the wood, which was characteristic of the previously analyzed ejected particles.

Following the presumptive evidence of cellulose digestion and adhering to the line of investigation as set by Harrington, Boynton and Miller (1927) tested the enzymatic action of various extracts from *Bankia setacea* on sawdust and on filter paper. Three types of extracts were prepared with the objective of at least partially delimiting the site of enzyme action. The first consisted of esophagus, stomach, crystalline style, digestive diverticula, anterior end of the caecum, and some portions of foot, mantle, kidney, etc.; the second was composed of the crystalline styles; and the third contained the material in the first extract minus crystalline styles. In all cases material was ground and kept in 30 per cent alcohol (to prevent bacterial action) at pH 6 for three days. The supernatants obtained by centrifugation were utilized as enzyme sources and were allowed to react in the presence of substrate for three weeks, after which reducing sugar content was measured. In all cases except for those containing crystalline style extract alone where no activity was apparent, the quantity of these substances in enzyme-substrate suspensions were higher than controls containing no substrate. Boiled controls showed that cellulose digestion was stopped by the destruction of the enzyme.

Bacterial activity in the enzyme preparations was obviated by maintaining the medium in 30 per cent alcohol. No evidence of such activity was noted at the termination of the experiments although two of the boiled preparations showed development of yeasts. Subsequent tests submitting the enzyme preparations to bacteriological culture methods also showed negative results. Whereas this demonstrates that the method of preparation of extracts was not favorable to the survival of bacteria, it still does not allow for the possible existence of cellulolytic micro-organisms in the shipworm gut under natural conditions. The authors concluded that the source of enzyme lay in the "liver" or digestive diverticula, however, no conclusion was reached as to whether it was extra- or intracellular in nature.

Recent confirmation of cellulolytic activity in other species of shipworms has been obtained by Hashimoto and Onoma (1949) in *Teredo* sp., Greenfield and Lane (1953) in *Teredo* sp. (Miami, Florida), and Deschamps (1953) in *Teredo navalis*. In *Teredo* sp. (Miami) localization of activity was more difficult to determine since the specimens were much smaller than those which had been heretofore investigated. Thus while *Bankia setacea* may reach a length of 2 or 3 feet, maximum length of *Teredo* sp. (Miami) specimens was in the order of 5 to 8 inches, most of the individuals being much smaller, and attempts at dissecting separate tissues in any quantity proved impractical. Partial localization was essayed by utilizing two tissue samples. The first, designated as "precaecal," contained all the internal organs anterior to the caecum excluding shell musculature and shells. The second, designated as "postcaecal," contained the caecum and some postcaecal gut with digestive diverticula attached. The substrate used in these experiments was a 1 per cent suspension of regenerated cellulose precipitated from a $ZnCl_2$-HCl dispersion of filter paper and washed until neutral (Parkin, 1940). Preliminary experiments involved use of homogenized tissues suspended in a NaCl medium made isotonic with sea water and buffered with M/15 phosphate buffer. Tests were run at pH's of 5.0, 5.6, 6.0, 6.7, 7.6. Analyses of reducing substances were made at various periods after first removing from sample aliquots the proteinaceous material and glycogen which was also usually present, especially at the start of the experiments. Bacterial growth was inhibited by toluene.

In the enzyme-substrate preparations, production of amounts of reducing substance above that found in controls containing no cellulose became evident within a few hours after the initiation of the procedure. Optimal pH for cellulolytic activity in postcaecal fractions appeared to lie between 5.6 and 6.0, and in subsequent experiments, the suspending media were kept buffered at pH of about 6 with M/15 phosphate buffer.

The precaecal and postcaecal fractions differed qualitatively in their response to physical conditions since the conversion of cellulose to reducing substances by precaecal material appeared to be independent of pH within the range studied. This regional difference in response suggested that more than a single enzyme might be involved in the total cellulose digesting ability of the shipworm. In comparing the action of the two fractions at pH 5.6, the postcaecal portion appeared to have approximately twice the activity of the precaecal segment. At the end of 21 hours incubation at 25°C this activity was found to be equivalent to .8 mg of reducing substance produced per mg of dry CHO-free tissue in postcaecal portions when the concentration of the tissue in the suspending medium was .48 mg CHO-free tissue per ml.

The significance of this was taken to mean that either enzyme production or production of enzyme containing bodies is mainly in the postcaecal portion of the alimentary system or that it may enter the gut in the more anterior portions and become concentrated as the intestinal contents move posteriorly.

While investigating effects of quantity of enzyme on cellulose breakdown, interference with observations was noted when tissue homogenate concentration was increased. A net increase in reducing substance content over cellulose-free controls occurred as before, however, in a homogenate concentration equivalent to the .48 mg of dry CHO-free tissue per ml previously used, this net increase after 16 hours was 10 times greater than that obtained with a concentration equivalent to 4.2 mg of dry CHO-free tissue per ml. (Although the tissue samples used were not CHO-free, this unit was used as a basis for calculations and was obtained by determining the CHO content of duplicate homogenate samples.) It has been noted that the shipworm has a high glycogen content (Lane, Posner, and Greenfield, 1952), and at higher concentrations of tissue, there is evidently either a greater concentration of inhibitory substances or an enhanced rate of glycogen breakdown as evidenced by the larger quantities of total reducing substance produced with time in preparations with and without cellulose. Either or both possibilities could effectively conceal the true rate of cellulose breakdown. Apparent attenuation of these factors was noted in the experiments with lower tissue concentrations thus revealing a more observable rate of hydrolysis.

Cell-free extracts were also prepared by centrifuging homogenates of postcaecal material, the original quantity of which was equal to that of the test media containing 4.2 mg dry CHO-free tissue per ml. The homogenates were cleared of suspended material as soon as possible and then tested for cellulolytic activity. Results showed an increase in activity above the uncentrifuged material by a factor of about 4 after 16 hours, however, activity was still not as high as in the uncentrifuged samples containing .48 mg dry Cho-free tissue per ml. These results suggested that all the enzyme was not liberated by the homogenization technique. While this experiment was repeated with similar results, the number of trials was insufficient to formulate whether autolysis of standing homogenates would have released more enzyme.

Assessment of the contribution of symbiotic microorganisms to the total cellulose digesting ability of Teredo was attempted by inoculation of sea water nutrient agar containing 1 per cent regenerated cellulose with fresh tissue homogenate and incubated at 25°C. Some colonies were found after 65 hours, but no evidence of cellulose digestion was seen at this time. Repeat cultures also yielded negative results. On the other hand, bacteria have definitely been observed in the large caecum of Xylophaga (Purchon, 1941), and Deschamps (1953) found cellulolytic microorganisms in the gut of *Teredo navalis*. This author confirmed cellulolytic activity in this species and identified one of the hydrolytic products as glucose, however, he was not able to distinguish between the activity of the shipworm itself and that of the bacteria present. It appears evident that while experimental enzyme set-ups may be treated in such a way as to prevent bacterial growth, under natural conditions the microflora present on the wood ingested or in the gut may aid in cellulose digestion.

The site of cellulase production in adult Teredo is still open to considerable investigation. Potts (1923) has shown that small wood particles are ingested by phagocytic cells in the lumen of parts of the digestive diverticula. Since such particles are necessarily quite small, he suggests that secretion of digestive juices causes some disintegration and digestion of wood. Inasmuch as an increase in production of reducing sugars was also noted in the caecum of *Bankia setacea* (Miller and Boynton, 1926), it would appear that digestion occurs to a certain extent here as well as absorption as evidenced by the presence of a typhlosole supplied with a rather large blood vessel. In the Lamellibranch Xylophaga, a marine borer sometimes found in wood (Purchon, 1941), phagocytes have also been observed in the caecum among wood particles. No mention is made, however, of the actual ingestion of wood particles. This genus does not possess phagocytes in the digestive diverticula, a situation typical of Teredinidae, and it has been suggested that it is incapable of wood digestion (Yonge, 1937). Nevertheless, in a single experiment using whole Xylophaga extract, Purchon obtained positive indication of cellulolytic activity. Since, however, bacteria are also found in the caecum, and when present in abundance appear to be accompanied by large numbers of phagocytes, the additional complicating possibility exists that the bacteria may digest wood and in turn are eaten by the phagocytes.

Reconsideration of the observation of Potts that phagocytes definitely ingest small wood particles somewhat delimits the amount of wood that can be consumed. The occurrence of particles of different size have been observed in the caecum of Teredo (Miller, 1924). Large particles may range from .30 to .40 mm in length, and from .02 to .08 mm across. Minute fragments .02 mm in length, and of approximately the same breadth also occur numerously as do those of intermediate size. Differentiation of two classes of particles, i.e., elongate fibrous ones and fine granular ones, is based on the fact that the latter are produced by action of the fine serrations of the boring valves whereas the former are rasped by the coarser denticulations. Therefore the phagocytes either consume only the smaller fibers or, in addition to this, the large fibers are reduced to appropriate size by some agency. Since fibers ejected in the feces are known to have lost most of the cellulose and yet may retain large size, it seems probable that not all of these are necessarily consumed by phagocytes. Again another cellulolytic agency is suggested. Notwithstanding the action of bacteria which has already been noted, it is not yet clear whether a cellulase is actually secreted by Teredo or whether the enzymatic tests performed thus far result from lysis of phagocytes and phagocytic cells of the digestive diverticula. It will be remembered that Boynton and Miller (1927) obtained negative results from experiments testing the digestion of cellulose in the presence of crystalline style extracts. Recent experiments of this nature utilizing certain other Lamellibranchs viz. *Mya arenaria, Mactra solidissima* (Lavine, 1946), *Ostrea edulis* (Newell, 1953) have given, on the other hand, positive evidence of the presence of a cellulase in the crystalline style. Nevertheless, such observations are further complicated by the presence of spirochaetes living in this structure. In view of these findings, it appears that further tests should be considered on the Teredo style.

It has been observed that newly attached larvae of shipworms have a well developed digestive tract (Sigerfoos, 1907; Isham and Tierney, 1953). In *Teredo bartschi* attachment occurs after about 72 hours of free-living existence, but distinctive structual changes have been noted as early as the 24-hour stage. Since the boring activity of Teredo normally follows permanent attachment to wood, and fragments of this material are found in the primitive caecum of the young organism, it is of value to ascertain whether or not the larva is adapted for wood digestion. For this purpose larvae of *Teredo bartschi* which were in the neighborhood of 24 hours in the free-living stage were collected. These were divided into lots of 100 and homogenized in glass. In five experimental set-ups the homogenized material was incubated with cellulose at 25°C and pH 6.0 along with cellulose-free controls, the physical conditions being similar to those used with the adult shipworm fractions. Parallel experiments were run using the extracts obtained by centrifugation of homogenates. After 24 hours, analyses of reducing substance showed a positive increase above cellulose-free controls in all cases. No apparent difference between whole homogenates and cell-free extracts was observed as opposed to the results of experiments using adult shipworm tissues. Average production at this time was found to be 20 μg of reducing substance per 100 larvae. Thus, pending investigations on other species, Teredo larvae apparently have the ability to digest cellulose to some extent prior to attachment.

In effect, the experimental data acquired thus far have strongly indicated both cellulolytic activity in shipworms and utilization of the end products of the process. Although it is necessary to purify the enzyme system for further definitive study of its properties, it would appear equally important to determine the site of its production. In order for this to be accomplished, separate studies on phagocytes, bacteria which may be established in the gut or present in the wood, and tissues suspect of secretory activity should be initiated. At best the studies thus far have indicated the possible regions of activity such as: caecum, gut, and hepatic diverticula. Unfortunately, the relative importance to the organism of the intracellular digestion observed in the hepatic diverticula as opposed to extracellular digestion indicated by the reducing sugar content of the caecum is somewhat masked by the utilization of generalized tissue homogenates as enzyme sources. Because of the morphological situation of Teredo tissues, dissection of specified areas suspect of cellulolytic activity have involved contamination with surrounding tissues. This is particularly emphasized in cases where removed sections of the gut contain fragments of hepatic diverticula.

Where possible, future studies may employ better isolation of such tissues. This would necessitate treatment of large numbers of specimens to obtain sufficient raw material. In some cases removal of the entire tissue is impractical due to its adherence to those surrounding it, as mentioned above. The alternative is to employ species known to acquire great size such as *Bankia setacea*. Isolation of phagocytic cells may be accomplished by Potts's (1923) method of pressure manipulation of lobules in the hepatic diverticula which he dissected for this purpose. It is suggested that histochemical assays be made of these sections of the digestive system to determine, for example, chemical changes of cells after ingestion of wood fiber as well as changes in the fiber itself. In addition microchemical determinations on the cellulolytic ability of their extracts should be checked where practicable and correlative information sought from culture reactions of the microflora in the gut.

LITERATURE CITED

Boynton, L. C., and R. C. Miller. 1927. The Occurrence of a Cellulase in the Shipworm. J. Biol. Chem., 75:613-18.

Deschamps, P. 1953. Recherche de la cellulase chez *Teredo navalis* au moyen de la chromatographie de partage. Bull. Soc. Zool. France, 78:174-77.

Dore, W. H., and R. C. Miller. 1923. The Digestion of wood by *Teredo navalis*. Univ. Calif. Publ. Zool., 22:383-400.

Greenfield, L. J., and C. E. Lane. 1953. Cellulose Digestion in Teredo. J. Biol. Chem., 204:669-72.

Harrington, C. R. 1921. A Note on the Physiology of the Shipworm (*Teredo norvegica*). Biochem. J., 15:736-41.

Hashimoto, Y., and K. Onoma. 1949. On the Digestion of Higher Carbohydrates by Mollusca *(Dolabella scapula* and *Teredo sp.)*. Bull. Soc. Sci. Fish., 15:253-58.

Isham, L. B., and J. Q. Tierney. 1953. Some Aspects of the Larval Development and Metamorphosis of *Teredo (Lyrodus) pedicellata* de Quatrefages. Bull. Mar. Sci. Gulf and Caribbean, 2:574-89.

Lane, C. E., G. S. Posner, and L. J. Greenfield. 1952. The Distribution of Glycogen in the Shipworm, *Teredo (Lyrodus) pedicellata* de Quatrefages. Bull. Mar. Sci. Gulf and Caribbean, 2:285-392.

Lavine, T. F. 1946. A Study of the Enzymatic and Other Properties of the Crystalline Style of Clams: Evidence for the Presence of a Cellulase. J. Cell. and Comp. Physiol., 28:183-95.

Miller, R.C. 1924. The Boring Mechanism of Teredo. Univ. Calif. Publ. Zool., 26:41-80.

Miller R. C., and L. C. Boynton. 1926. Digestion of Wood by the Shipworm. Science, 63:524.

Newell, B. S. 1953. Cellulolytic Activity in the Lamellibranch Crystalline style. J. Mar. Biol. Assoc. U.K., 32:491-95.

Parkin, E. A. 1940. The Digestive Enzymes of Some Wood-boring Beetle Larvae. J. Exp. Biol., 17:364-77.

Potts, F.A. 1923. The Structure and Function of the Liver of Teredo, the Shipworm. Proc. Camb. Phil. Soc., 1:1-17.

Purchon, R. D. 1941. On the Biology and Relationships of the Lamellibranch *Xylophaga dorsalis* (Thurton). J. Mar. Biol. Assoc. U.K., 25:1-39.

Sigerfoos, C.P. 1907. Natural History, Organization and Late Development of the Teredinidae, or Shipworms. Bull. U.S. Bur. Fish., 27:191-231.

Yonge, C.M. 1937. Evolution and Adaptation in the Digestion System of the Metazoa. Biol. Rev., 12:87-115.

SOME PROPERTIES OF CELLULASE FROM LIMNORIA

Dixy Lee Ray

Preliminary studies (Ray, 1951; Ray and Julian, 1952) had demonstrated that there is readily detectable cellulolytic activity in crude homogenates of the midgut caeca from Limnoria. The tests that established this were carried out by incubating homogenized tissue with finely ground filter paper. The formation of reducing sugar, detected by using Benedict's reagent, was taken as evidence of attack on cellulose; such results indicated that cellulase is produced by Limnoria. It has also been shown (Ray, 1958, 1959a) that in the case of Limnoria the animal itself, not any symbiotic microorganism, is responsible for elaboration of the effective cellulolytic enzyme. Wood is now known to be the major, if not the only, important article of food in Limnoria's diet (Ray, 1959b). These facts, coupled with the knowledge that enzyme secretion is limited to the midgut diverticula and that the tissue comprising this portion of the digestive tract, exclusive of the thin epithelium and sparse muscle elements, is made up of only two different types of cells, both secretory (Fahrenbach, 1959), make Limnoria an almost ideal source of animal cellulase.

The only drawback to the use of these experimental animals lies in their small size, and hence in the very large numbers that are needed for the somewhat tedious procedure of extracting their digestive enzymes. Nevertheless, even crude homogenates are relatively uncomplicated by the presence of large amounts of extraneous materials from cells unrelated to a digestive function, and many experiments can be done on a micro scale. Considering the scanty nature of our knowledge about cellulases in the animal kingdom, it was determined to pursue a program of study designed to elucidate as nearly as possible the nature of the enzymes responsible for the observed cellulolytic action of the homogenate. So far, purification of no single enzymatic component has been achieved, but such progress as has been made toward understanding the action of cellulase from Limnoria is here reported.

Collection of Animals for Experimental Use

For any physiological or behavioral study, it is, of course, essential that the experimental animals be in good physical condition. To assure that this is the case with Limnoria, the safest method of accumulating specimens is to remove them from their burrows by splitting the wood and gently lifting out the animals with fine forceps. Loss of animals through injury during removal runs about 5 per cent.

With collections that have been made in the north Pacific area, it is sometimes possible to force animals to leave their burrows by allowing the water surrounding the piece of wood in which they are contained slowly to become stagnant while maintaining a constant low temperature (± 10°C) over a period of 2 to 3 days. This procedure is ineffective with Neapolitan animals.* In our experience, application of none of a large number of other stimuli (addition of alcohol or other chemical agents; bubbling CH_3, CO, or CO_2 through the water; electric shock; etc.) has proved successful in forcing the animals to emerge from their burrows alive. In order, therefore, to be assured of a supply of healthy animals for extraction of the digestive enzymes, we normally remove them individually and by hand from the original wood.

Whenever possible, the animals are dissected and the crude homogenate prepared either immediately or within a few hours. Immediate dissection is not essential, and on a number of occasions we have tested enzyme preparations from animals that had been maintained for several days in dishes of clean sea water

---

*Nair and Leivestad (1958) recently reported that in Norway, if wood inhabited by Limnoria is experimentally subjected to air temperatures of -8 to -12°C for 4 or 6 hours, the animals then show a marked tendency to leave their burrows when the wood is resubmerged and the ambient temperature returns to normal. Should this reaction also occur with temperate or warm water populations, it might be possible to take advantage of such a trick to collect large numbers of animals in a healthy condition.

and in the absence of wood. No differences in cellulolytic activity between these extracts and those prepared from animals freshly removed from their burrows were detected. We assume from this that cellulase in Limnoria is not an adaptative enzyme.

The report, often repeated in the literature, that a pair of Limnoria are generally found in a single burrow, that the female occupies the blind end, and that the male does little or no burrowing is not supported by our observations. Single animals are encountered just as commonly as pairs; often there are more than two animals per burrow and they may be the same or opposite sexes; and the males burrow just as much and just as effectively as do the females.

Preparation of the Homogenate

For removal of the midgut tissue, a group of animals is first segregated into a clean watch glass, rinsed briefly with distilled water to remove excess sea water from the surface of their bodies, and drained. An individual is then oriented under a stereoscopic dissecting microscope so that the head can be held with one pair of fine watchmaker forceps, the body with another. With one quick, sharp movement, the head is pulled from the body. Nearly always the gut breaks at the posterior border of the foregut and the two pairs of midgut caeca remain attached to the head. These can then be nipped off with fine forceps, transferred to the tip of a glass rod, and ground in a small glass homogenizer kept chilled in an ice bath. The midgut tissue from approximately 600 animals makes about 1 ml of crude homogenate; this material can be frozen and stored for periods of a year or more without appreciable loss of activity.

Partial purification of the homogenate is achieved by centrifugation to remove particulate material, followed by fractionation with ammonium sulfate. The precipitate that comes down between 40 and 80 per cent saturation contains approximately 90 per cent of the activity. It is a white flocculent precipitate, readily soluble in water or in dilute buffer. After dialyzing against distilled water or dilute buffer at pH 5.4 to remove the ammonium sulfate, the partially purified extract can be used immediately, stored in the cold (4°C), frozen, or lyophilized. Following any of the latter procedures, activity is retained undiminished for very long periods of time. The preparation has no measurable reducing power, nor does any develop upon incubation in the absence of substrate. Unless otherwise indicated, all of the experiments reported have been carried out using this partially purified preparation.

Characteristics of Limnoria Cellulase

pH stability and optimum

The stability of Limnoria cellulase at different pH values has been tested by incubating the preparation, diluted 1:10 (v/v) with McIlvaine buffer solutions (citric acid-sodium phosphate), for 24 hours at 35°C. Following the incubation period, 1 ml of the enzyme-buffer solution was rapidly mixed with 40 ml of sodium carboxymethylcellulose (CMC; 1 per cent w/v) and the change in viscosity measured with a Hoeppler falling ball viscosimeter. As shown by figure 1, the greatest activity is retained in the range from pH 5.2 to pH 7.0. Therefore, for experiments of very long duration or when it was necessary to store the enzyme solution, the preparation routinely was brought to a pH of 5.4.

For experiments of short duration (one to several hours) the pH optimum lies considerably lower, at a value between pH 4.5 and pH 4.8 (figures 2 and 3). This optimum was established both by measuring the increased fluidity of a 1 per cent CMC solution made up with a series of two different buffers (McIlvaine and Sorensen) and photometrically by measuring the formation of reducing sugars after a 48-hour incubation period with CMC as substrate. For estimation of the production of reducing sugars from CMC, the Sumner (1925) DNS method was used.

Temperature optimum and thermostability

For experiments of at least 1/2 hour duration, and for those that run much longer, the temperature optimum appears to be 40°C. Results of a series of controlled temperature tests are presented graphically in figure 4. Thermo-inactivation appeared to be rapid at temperatures of 70°C and above. After only 5 minutes incubation at 60°C not more than 50 per cent activity is retained in the partially purified enzyme preparation, whereas the crude homogenate retained only 20 per cent of its activity. This suggests that extraneous proteins and other substances originally present in the homogenate provide no thermal protection. If the reduction in viscosity of 1 per cent CMC is measured following incubation of the partially purified enzyme in the absence of substrate, it can be shown that full activity is retained following exposure to a temperature of 50°C for 15 minutes. Longer exposure results in rapid loss of activity. At 35°C full activity is retained even after incubation for several hours.

Effect of substrate concentration

To determine the effect of increasing the substrate concentration, and to establish the range within which we could run micro tests on the production of reducing sugars from cellulose without wasting the small available supplies of enzyme, a series of tests was set up according to the following conditions. Partially purified enzyme preparation (0.25 ml) was added to ground filter paper suspended in 1.75 ml McIlvaine buffer at pH 4.7; the buffer also contained merthiolate as a bacteriostatic agent. The digest was covered with a layer of toluol and incubated at 35°C for 48 hours. At the end of that time the remaining fragments of filter paper were thrown down by strong centrifugation and the supernatant was analyzed for reducing sugars by the Somogyi (1945) technique. The results are shown in table 1 and presented graphically in figure 5.

TABLE 1
EFFECT OF SUBSTRATE CONCENTRATION;
μg REDUCING SUGAR PRODUCED FROM FILTER PAPER

Amount of Substrate mg	Cellulase Concentration ml	Results Calculated in μg Reducing Sugar/ml
20	0.25	48
50	0.25	68
100	0.25	104
150	0.25	112
200	0.25	148
250	0.25	164
300	0.25	192
100	0.25 (buffer only)	0
100	0.25 (boiled enzyme)	8

At the enzyme concentration used, 300 mg of ground filter paper still does not saturate the enzyme; therefore, in routine quantitative tests on production of reducing sugars from this and other cellulosic substrates, the enzyme preparation was diluted 1:4 and the amount of substrate standardized at 100 mg wherever possible.

Action of partially purified cellulase on cellulose and other substrates

In the absence of a pure cellulase preparation, it was anticipated that testing the activity of the extract on a wide variety of substrates would give information about the range of specificity and, indirectly, the probable number and kind of enzymes that might be present. These tests have been carried out by measuring the production of reducing sugars photometrically (Somogyi and Sumner techniques), measuring the loss in viscosity of solutions of substituted celluloses and other soluble polysaccharides, and by identifying the hydrolytic products by means of paper chromatography. A summary of the results appears in table 2.

There is no detectable activity on sucrose or trehalose; on the monosaccharides fructose, galactose, or glucose; or on the pentose, xylose. The data in table 2 have been verified by many repetitions, and similar results were obtained on enzyme preparations made from Limnoria of the Bay of Naples. There appear to be no significant differences, from the standpoint of their enzymatic capabilities, between the animals of the cold temperate waters around Seattle and Friday Harbor and those from the subtropical waters of the Mediterranean.

Examination of the tabulated results shows that the enzyme preparation reduces the viscosity of soluble cellulosic substrates but not of alginates, agar, dextran, levan, pectin, or inulin. It appears, therefore, that there is no action on α-linked saccharides other than starch, glycogen, and maltose. Glucose is a hydrolytic end product from wood cellulose, filter paper, cellophane, starch, glycogen, carboxymethylcellulose, hydroxyethylcellulose, and methocel. Glucose is also split from salicin, laminarin, maltose, cellobiose, and higher cellulose sugars. Galactose is split from raffinose, and both glucose and galactose appear as hydrolytic end products from lactose and melibiose.

These results may be interpreted to indicate that, in addition to cellulase and amylase, the preparation probably also contains a β-glucosidase that splits glucose from salicin, a maltase, a cellobiase, and a galactosidase that splits galactose from raffinose and is possibly responsible for the hydrolysis of lactose and melibiose. Other (or perhaps the same) carbohydrases are

TABLE 2
ACTION OF PARTIALLY PURIFIED CELLULASE ON VARIOUS SUBSTRATES

Substrate	Kind of Reaction			Reaction Products					
				From Hydrolysis:			From Synthesis:		
	Viscosity Loss	Hydrolysis	Synthesis	Glucose	Cellobiose	Galactose	Triose	Tetrose	Other
Wood		+		+					
Filter paper		+		+	(+)				
Cellophane		+		+	(+)				
Starch		+		+					
Glycogen		+		+					
Carboxymethylcellulose	+	+		+	(+)				
Hydroxyethylcellulose	+	+		+					
Methocel	+	+		+					
Alginic acid	-	-							
Na alginate	-	-							
Agar	-	-							
Dextran	-	-							
Levan	-	-							
Pectin	-	-							
Inulin	-	-							
Raffinose		+				+			
Salicin		+		+					
Laminarin		+	+	+					+
Lactose		+	+	+		+	+		
Maltose		+	+	+			+	+	+
Melibiose		+	+	+		+	+		
Cellobiose		+	+	+			+	+	+
Cellotriose		+	+	+	+				+
Cellotetrose		+	+	+	+		+		+
Cellopentose		+	+	+	+		+		+

present that are capable of reversing the hydrolytic reaction under conditions of high substrate concentration.

When disaccharides are tested, if the initial substrate concentration exceeds 2 per cent (w/v) then paper chromatographic analysis gives positive evidence of the synthesis of oligosaccharides. Cellotriose, cellotetrose, and higher cellulose sugars are produced from cellobiose. A triose and short chain oligosaccharides are formed from lactose and maltose. An as yet unidentified disaccharide, possibly gentiobiose, is also formed when any of the cellulose sugars is used as starting substrate. Similarly, a disaccharide is produced during attack on laminarin, and another, possibly isomaltose, appears during incubation with maltose. To illustrate the synthetic reactions, the results of some of these tests are reproduced in figures 6, 7, and 8.

These synthetic reactions, on the basis of present results, are believed to be transglucosylations involving only glucose transfer. Other investigators, Crook and Stone (1954) and Buston and Jabbar (1954), have obtained similar results with enzyme extracts from cellulolytic molds, and the whole question of formation of oligosaccharides by transfer reactions has been well reviewed by Edelman (1956). Evidence is accumulating that one and the same hydrolytic enzyme may catalyze reactions in both directions depending upon substrate concentration, and, if the latter is high enough, it competes with water as the receptor molecule, leading to the formation of short chain oligosaccharides. Limnoria enzymes may be of this kind, since they show a partial hydrolytic action by liberating free hexose, and a partial synthetic action in forming oligosaccharides by transglucosylation. It is planned to continue these analyses and also to investigate the effect of added hexoses that may serve as receptors for glucose. Much can be learned by pursuing these studies, but ultimate, correct interpretation of the results rests upon further purification of the enzyme preparation.

Although cellobiose is indicated in table 2 as a possible hydrolysis product from the action of cellulase on CMC, filter paper, and cellophane, it is our present opinion that this may not be the only interpretation. By spotting paper chromatograms at intervals during incubation with CMC and other insoluble cellulosic substrates, it can readily be determined that the first visible reaction product is glucose. Cellobiose appears only much later, after glucose has greatly increased in amount. Moreover, if the incubation and making of paper chromatograms is continued over many days, it can be seen that the cellobiose fluctuates widely in amount, appearing and disappearing at regular intervals. This may mean that the cellobiase in the enzyme preparation is being periodically saturated, allowing the accumulation of disaccharide, or it may mean that glucose is the only true hydrolysis product of the action of cellulase on cellulose and that the cellobiose that appears is the result of synthesis by transglucosylation once the concentration of glucose has become sufficiently high.

Action of partially purified cellulase on CMC; effect of degree of polymerization (DP) and degree of substitution (DS)

Different samples of CMC and information concerning their degree of polymerization and of substitution were obtained through the courtesy of the Hercules Powder Company. Three of these, with widely different viscosities and designated CMC 70, were selected first of all for testing the effect of the DP. These samples included CMC 70-high, lot # 5742, DS 0.72, viscosity 2100 cps; CMC 70-medium, lot # 4689, DS 0.77, viscosity 470 cps; and CMC 70-low, lot # 5008, DS 0.82, viscosity 30 cps. Solutions of each of these at a concentration of 1 per cent (w/v) were prepared in duplicate. Partially purified cellulase (0.25 ml) was added, and the experimental tubes were incubated for 1 hour at a pH of 4.8 and temperature of 40°C. Control tubes were also prepared in duplicate, with one series containing only added buffer, and another containing enzyme that had been boiled for 10 minutes. Enzyme attack was determined by estimation of reducing sugars before and after incubation using Sumner's DNS method. The results are summarized in table 3.

It is apparent that, notwithstanding the wide difference in DP, enzyme action is very similar on these substrate samples. Still, there is some difference in the amount of reducing sugar produced from the three types of CMC 70, but these samples also differed from each other in their DS, CMC 70-high having the lowest DS. A parallel experiment was therefore run, using only highly polymerized samples having widely different degrees of substitution. These included, in addition to the CMC 70-high already tested, CMC 90-high, lot # 6045, DS 0.94, viscosity 280 cps; and CMC 120-high, lot # 5920, DS 1.25, viscosity 135 cps. The results are given in table 4 and presented graphically together with those of the previous experiment in figure 9.

TABLE 3
REDUCING SUGARS PRODUCED BY CELLULASE ATTACK ON CARBOXYMETHYLCELLULOSE SOLUTIONS OF DIFFERENT DEGREES OF POLYMERIZATION

Substrate		Colorimeter Reading	Reducing Sugars, Calculated as Glucose in μg/ml
CMC 70-low;	no enzyme	1; 2	0
CMC 70-medium;	no enzyme	2; 1	0
CMC 70-high;	no enzyme	2; 3	0
CMC 70-low;	boiled enzyme	4; 2	0
CMC 70-medium;	boiled enzyme	5; 3	0
CMC 70-high;	boiled enzyme	3; 1	0
CMC 70-low;	plus enzyme	130; 132	400
CMC 70-low;	plus enzyme	132; 130	400
CMC 70-medium;	plus enzyme	156; 154	450
CMC 70-medium;	plus enzyme	152; 152	450
CMC 70-high;	plus enzyme	176; 176	510
CMC 70-high;	plus enzyme	180; 177	510

TABLE 4
REDUCING SUGARS PRODUCED BY CELLULASE ATTACK ON CARBOXYMETHYLCELLULOSE SOLUTIONS OF DIFFERENT DEGREES OF SUBSTITUTION*

Substrate	Colorimeter Reading	Reducing Sugars, Calculated as Glucose in μg/ml
CMC 70-high	168; 171	495; 500
CMC 70-high	182; 183	525; 527
CMC 90-high	81; 80	266; 265
CMC 90-high	80; 78	265; 263
CMC 120-high	6; 5	$\pm$ 75 to 80
CMC 120-high	8; 7	

*Controls were completely negative.

This series was repeated in duplicate, using amounts of CMC, calculated from the DS, to give roughly equivalent numbers of anhydroglucose units in each of the different solutions. An additional sample, CMC 50-medium, lot # 520-25-21, DS 0.54, viscosity 410 cps was also included. This sample was added because its viscosity is very close to that of CMC 70-medium, but its DS is substantially lower. Results of this experiment are presented in figure 10.

It appears from these data that the degree of polymerization of CMC has little effect upon its susceptibility to attack by Limnoria enzyme preparations, whereas the degree of substitution has a marked effect. This is similar to the results obtained by Reese (1957), using enzyme preparations from a number of different molds and microorganisms. It should be pointed out, however, that these measurements were made in 1 per cent solutions (w/v) or in solutions roughly calculated to contain the same number of anhydroglucose units, and they probably vary widely from one another in their molarity. The lack of precise information or even of reasonable estimates of the average molecular weight of different samples of CMC is a handicap to more careful testing, which should be done on a equimolar basis.

For the material at our disposal, estimates of average molecular weight are available for only two samples: CMC 70-low, MW 70,000, and CMC 70-high, MW 300,000. Using these figures, kindly supplied by the Hercules Powder Company, solutions were made at a concentration 10^{-4} molar, incubated with enzyme, and tested for the amount of reducing sugar produced. The results showed that approximately 4 times as much sugar is produced from the solution of higher molecular weight. The experimental conditions were the same as for previous tests.

Since the degree of substitution in these two samples is very close (0.72 and 0.77), this alone seems insufficient to account for the large difference in susceptibility to attack, and it may be that, when testing is done on an equimolar basis, the degree of polymerization does indeed have a large effect. Perhaps the longer chains of the more highly polymerized CMC 70-high provide more bonds available to the enzyme. Were this to be true, the implication would be that the enzyme acts by random splitting. But it must be borne in mind that the molecular weight determinations may be far from accurate; that they represent average molecular weights only; that we know nothing about the distribution curve for the population of larger and smaller molecules in this material, which is certainly heterogeneous in this respect; and that, finally, the results may reflect only the effect of substrate concentration. This last possibility is being checked, and we hope to elicit further information. For the present we are inclined to interpret these results to indicate that the DS of CMC is an important factor in determining susceptibility to enzymatic attack, as has also been found to be the case by other workers studying the action of Myrothecium enzymes on CMC (Reese, 1957). On the effect of the DP we feel that judgment should be withheld pending further experimentation.

Experiments on other soluble cellulosic substrates indicate that the kind of substituent as well as the degree of substitution may be important. For example, hydroxyethyl cellulose (Carbide and Carbon Chemicals) with a DS of 2 to 3 hydroxyethyl groups per anhydroglucose unit is readily attacked, as is also methocel (Dow Chemical Company) with a DS of 1.9. This information has come from viscosimetric studies only, and further tests are under way with these materials to learn more about the susceptibility of substituted cellulose molecules to attack by Limnoria enzymes. Cellulose acetates and nitrates are still to be analyzed.

Effect of Inhibiting Agents

Although many hydrolytic enzymes are relatively unaffected by substances that markedly reduce activity of other types of enzymes, a number of well-known inhibitors have been tested to determine their effect upon the cellulolytic activity of Limnoria enzymes. These experiments were run by determining the relative loss of viscosity in CMC 70-medium using a Hoeppler falling ball viscosimeter with 40 ml of a 1 per cent (w/v) solution of substrate, 0.1 ml enzyme preparation, pH 4.8 (citrate or phosphate buffer), and temperature 35°C, both in the presence and absence of the various inhibitors. Substances tested include formalin, phenol, copper compounds, potassium cyanide, glutathione, phenylmercuric acetate, sodium arsenite, hydroxylamine hydrochloride, sodium azide, and sodium fluoride. Controls were run with substrate and enzyme alone, substrate and inhibitor alone, and with substrate plus inhibitor and enzyme that was inactivated by boiling for 10 minutes. Wherever possible, the inhibitor was used at a concentration of $\frac{M}{100}$. Phenol, if incubated with the enzyme for 30 minutes at 35°C before the substrate is added, causes a 50 per cent inhibition at a concentration of $\frac{M}{100}$; copper carbonate at $\frac{M}{50}$ caused 50 per cent inhibition, but copper sulfate at the same concentration gave rise to only a 10 per cent inhibition; KCN resulted in no inhibition even at concentrations of $\frac{M}{10}$. None of the other inhibitors was effective in blocking the cellulolytic reaction unless used in concentrations so high that the proteins were precipitated. These results are similar to those reported by other investigators who have tested a number of mold cellulases (Basu and Whitaker, 1953; Jermyn, 1955; Reese and Mandels, 1957).

It has been reported by Reese, Gilligan, and Norkrans (1952) that the activity of Myrothecium cellulase on CMC is inhibited by cellobiose; this is not evident with Limnoria cellulase. Addition of as much as 5 per cent (w/v) of cellobiose slowed the rate of the reaction somewhat but certainly did not inhibit it, as shown by the viscosity curves reproduced in figure 11.

Further purification of the Limnoria enzyme preparation

The results of the studies of the action of Limnoria enzyme extract on a number of different substrates, of the synthetic activity, and of the effect of various inhibiting agents make it evident that for an adequate interpretation it is very important to know how many and what kind of enzymes actually are present and involved in all these reactions. Little progress has been made toward real purification of the cellulolytic extract from Limnoria. Should this provoke wonder, it has only to be recalled that the animals themselves do not exceed 3.5 mm in total body length and that the midgut tissue is only a very small part of the digestive tract. As already mentioned, approximately 600 animals must be dissected to accumulate about 1 ml of crude homogenate, and that is a very small amount on which to carry out extensive purification procedures. Nevertheless, preparations are so stable, if stored frozen, that collections can be pooled until a reasonable supply of homogenate is available.

We have reason to believe that the eventual separation of our partially purified preparation into its constituent components may not be too difficult, for from a single electrophoretic analysis on starch gel there appear to be only 5 different protein components. One of these is amylase; 3 are inactive in lowering the viscosity of CMC; and one component, not the largest, did effectively reduce the viscosity of CMC in the only test that has so far been possible to run. This active fraction appears to be an acid protein moving rapidly toward the anode. It probably contains amino acids with more than one carboxyl group, and the estimated isoelectric point would be around pH 3.5. These data must be verified by repetition. If this preliminary analysis is repeated, and the same or similar results are obtained, then it would seem that the chances for purifying Limnoria cellulase are very good, despite the obvious practical difficulties.

LITERATURE CITED

Basu, S. N., and D. R. Whitaker. 1953. Inhibition and Stimulation of the Cellulase of *Myrothecium verrucaria*. Arch. Biochem. Biophys., 42:12-24.

Buston, H. W., and A. Jabbar. 1954. Synthesis of β-linked Glucosaccharides by Extracts of *Chaetomium globosum*. Biochim. et Biophys. Acta, 15:543-48.

Crook, E. M., and B. A. Stone. 1957. The Enzymic Hydrolysis of β-glucosides. Biochem. J., 65:1-12.

Edelman, Jeffrey. 1956. The Formation of Oligosaccharides by Enzymic Transglycosylation. In F. F. Nord (ed.)., Advances in Enzymology, 17:189-232. New York, Interscience Publ. Inc.

Fahrenbach, Wolf H. 1959. Studies on the Histology and Cytology of the Limnoria Midgut Diverticula. In this volume.

Jermyn, M. A. 1955. Fungal Cellulases. VI, Substrate and Inhibitor Specificity of the β-glucosidase of *Stachybotrys atra*. Austral. J. Biol. Sci., 8:577-602.

Nair, N. Balakrishnan, and H. Leivestad. 1958. Effect of Low Temperature on the Vertical Distribution of Two Wood-boring Crustaceans. Nature, 182:814-15.

Ray, D. L. 1951. The Occurrence of Cellulase in *Limnoria lignorum*. Rept. Marine Borer Conf., U.S. Naval Civ. Eng. Res. and Eval. Lab., Port Hueneme, Calif., pp. L-1 to L-4.

-------. 1958. Recent Research on the Biology of Marine Wood Borers. Am. Wood Pres. Assoc. Proc., 54:120-28.

-------. 1959a. Marine Fungi and Wood Borer Attack. Am. Wood Pres. Assoc. Proc., 55:1-7.

-------. 1959b. Nutritional Physiology of Limnoria. In this volume.

-------, and J. R. Julian. 1952. Occurrence of Cellulase in Limnoria. Nature, 169:32.

Reese, E. T. 1957. Biological Degradation of Cellulose Derivatives. Ind. Eng. Chem., 49:89-93.

-------, and Mary Mandels. 1957. Chemical Inhibition of Cellulases and β-glucosidases. Res. Rept., Pion. Res. Div. Quartermaster Res. and Dev. Center, Natick, Mass. Microbiol. Ser., 17:1-42.-42.

-------, W. Gilligan, and B. Norkrans. 1952. Effect of Cellobiose on the Enzymatic Hydrolysis of Cellulose and its Derivatives. Physiol. Plant., 5:379-90.

Somogyi, Michael. 1945. A New Reagent for the Determination of Sugars. J. Biol. Chem., 160:61-68.

Sumner, James B. 1925. A More Specific Reagent for the Determination of Sugar in Urine. J. Biol. Chem., 65:393-95.

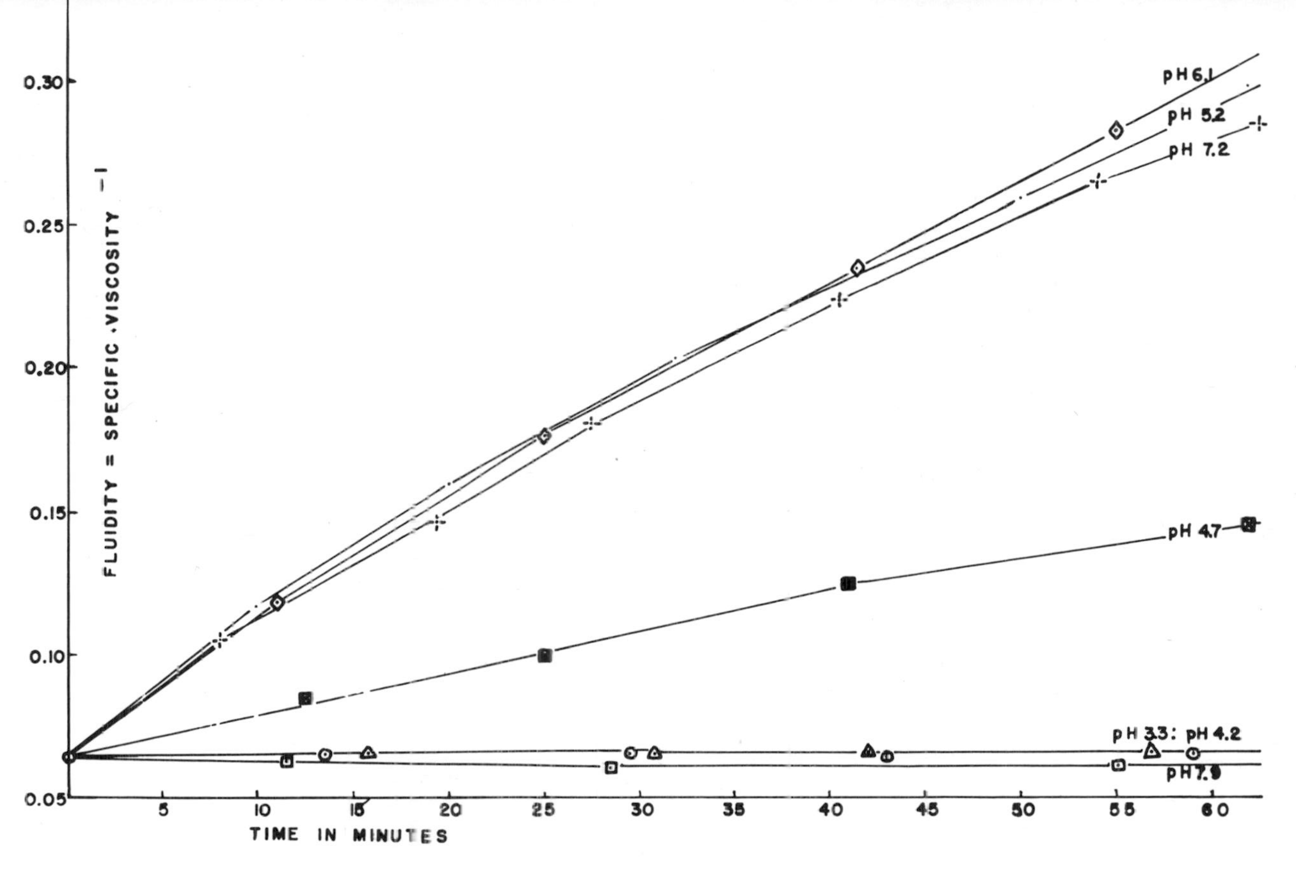

Fig. 1. pH stability. McIlvaine buffer: citric acid + Na_2HPO_4

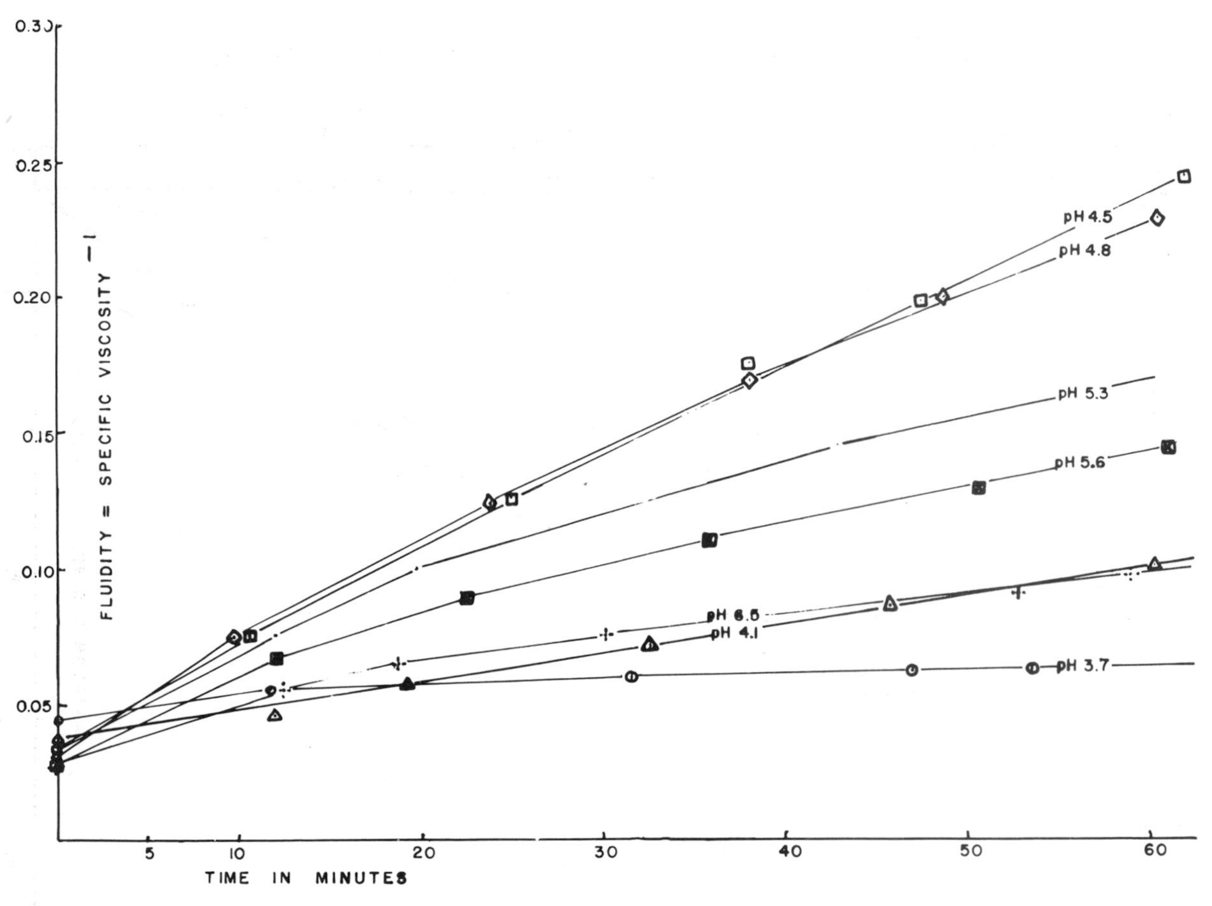

Fig. 2. pH optimum. Sorensen buffer: citric acid + NaOH

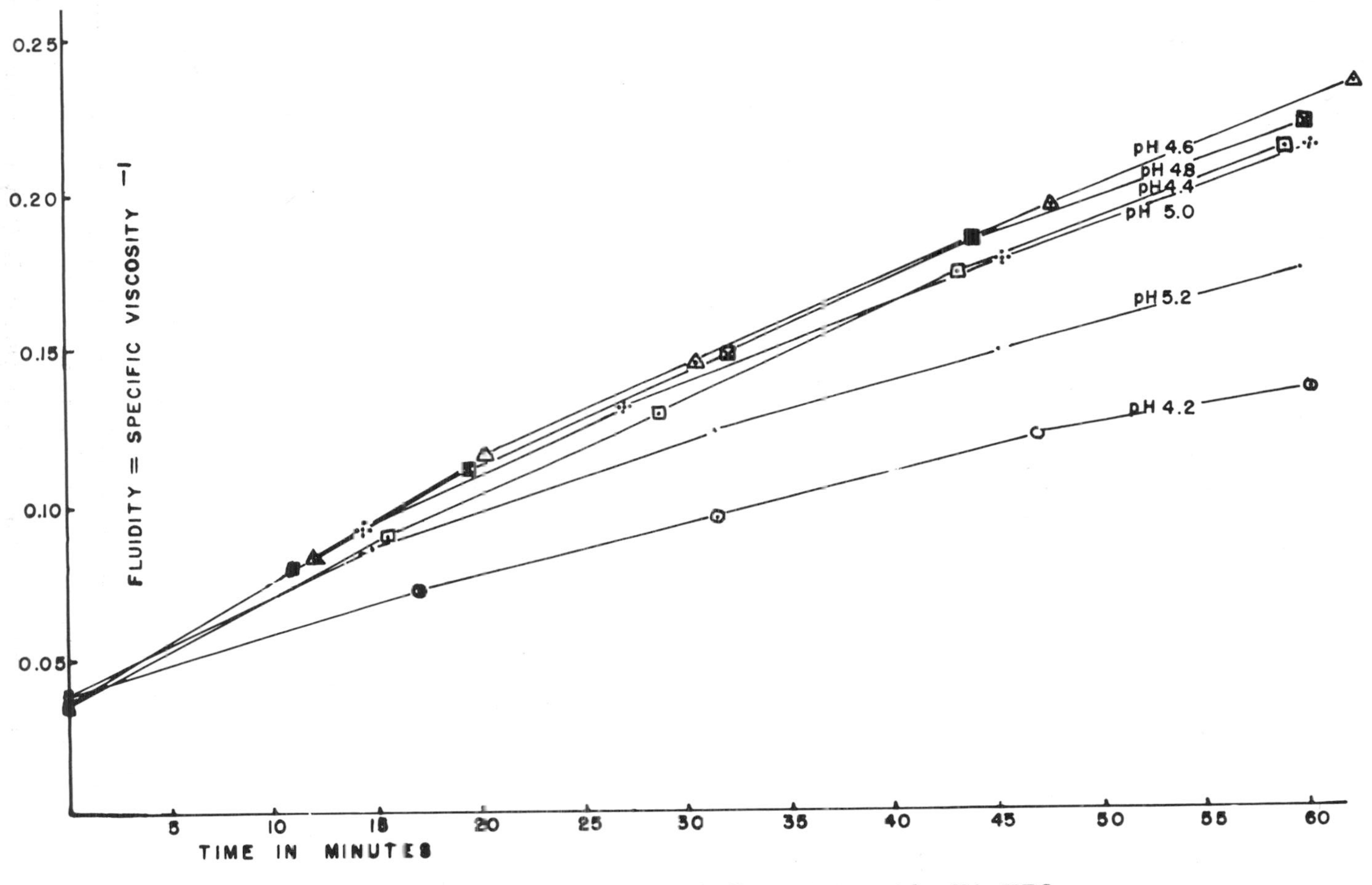

Fig. 3. pH optimum. McIlvaine buffer: citric acid + NA_2HPO_4

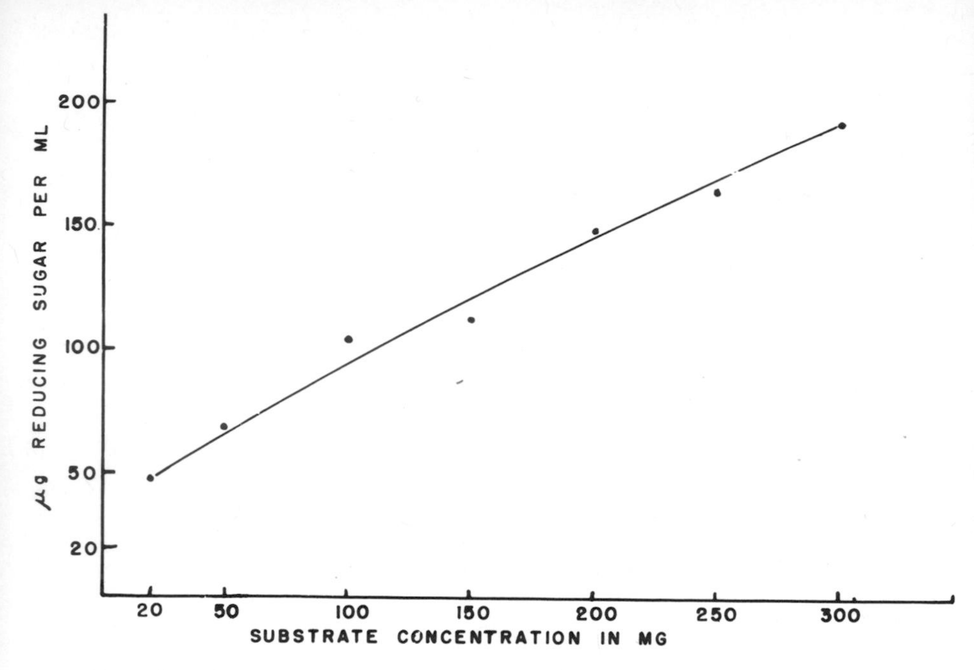

Fig. 4. Effect of substrate concentration. Production of reducing sugar from ground filter paper by 0.25 ml Limnoria cellulase, incubation at 35°C for 48 hours. Somogyi technique

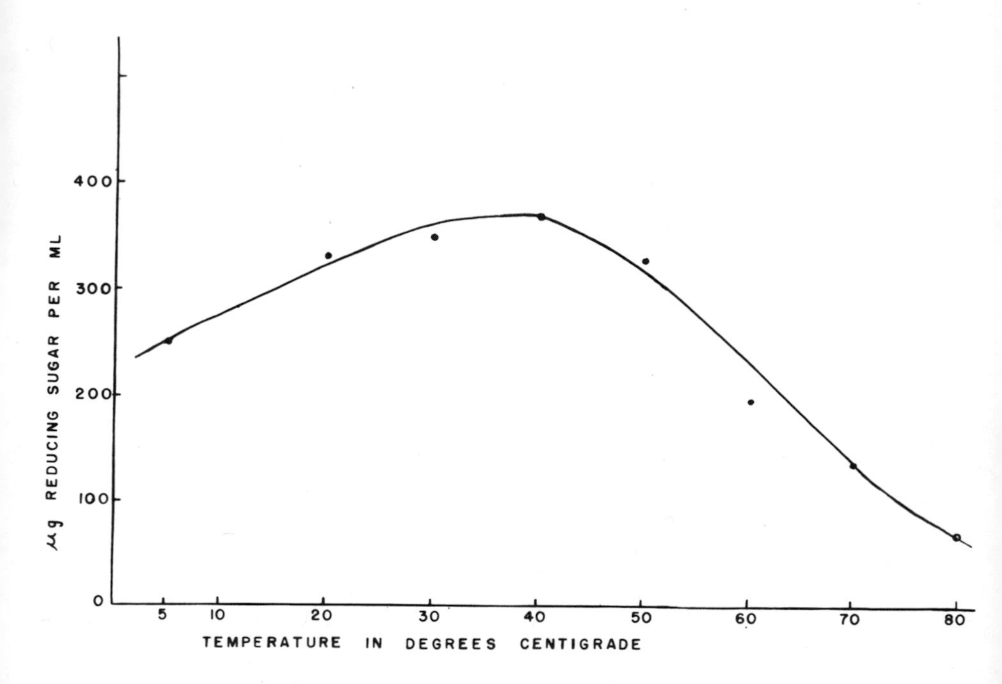

Fig. 5. Temperature optimum. Hydrolysis of CMC by Limnoria cellulase. Incubation for 30 minutes

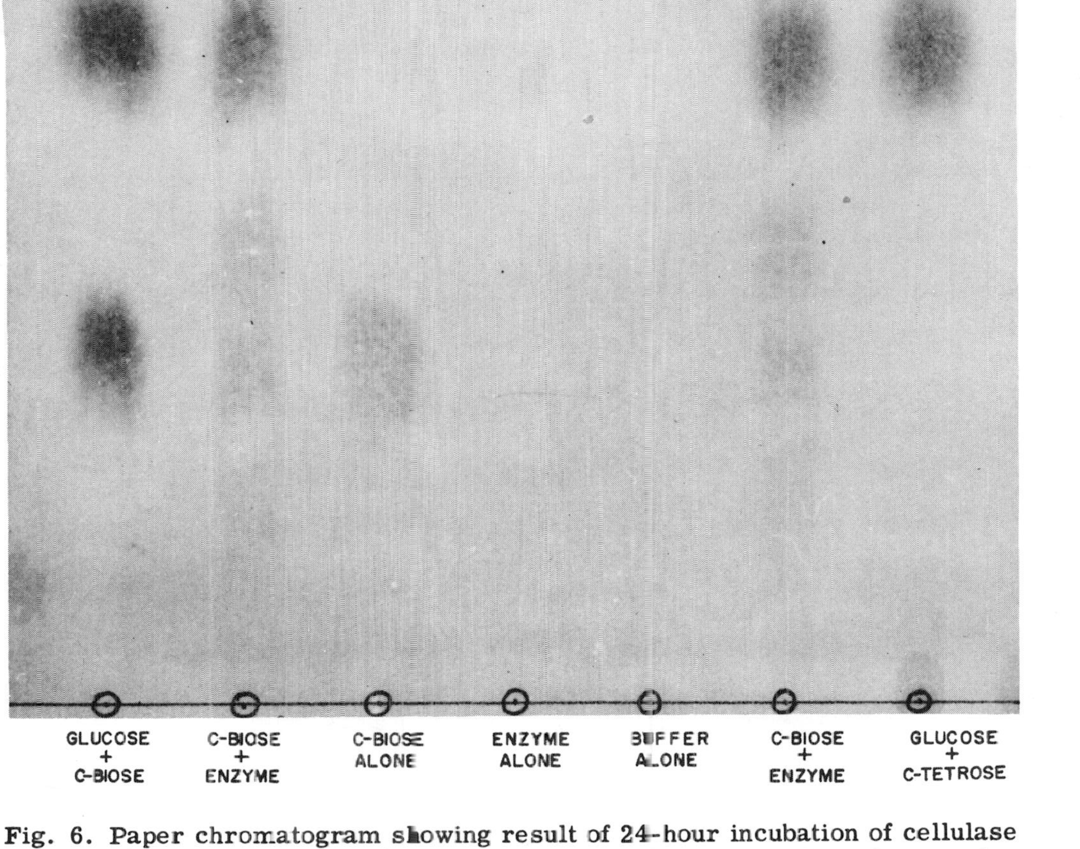

Fig. 6. Paper chromatogram showing result of 24-hour incubation of cellulase with cellobiose (5 per cent w/v). Note synthesis of disaccharide and trisaccharide in addition to hydrolytic production of glucose

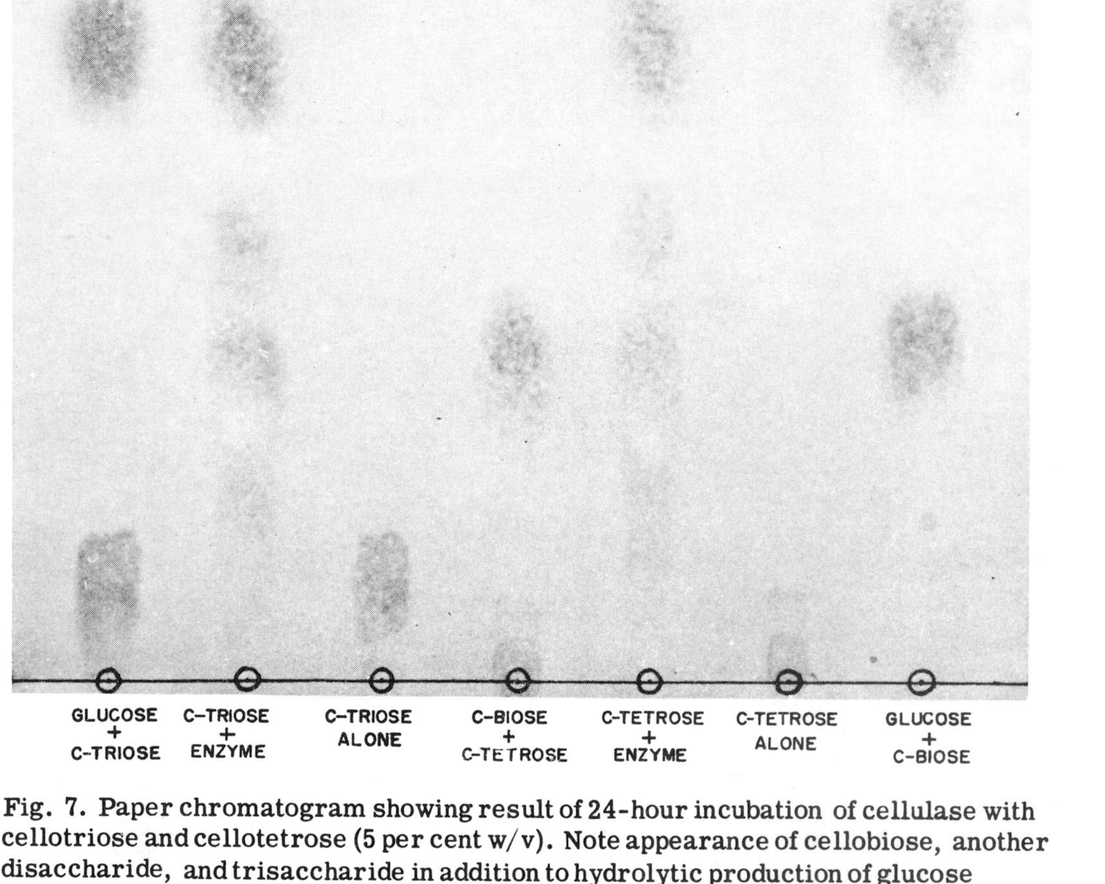

Fig. 7. Paper chromatogram showing result of 24-hour incubation of cellulase with cellotriose and cellotetrose (5 per cent w/v). Note appearance of cellobiose, another disaccharide, and trisaccharide in addition to hydrolytic production of glucose

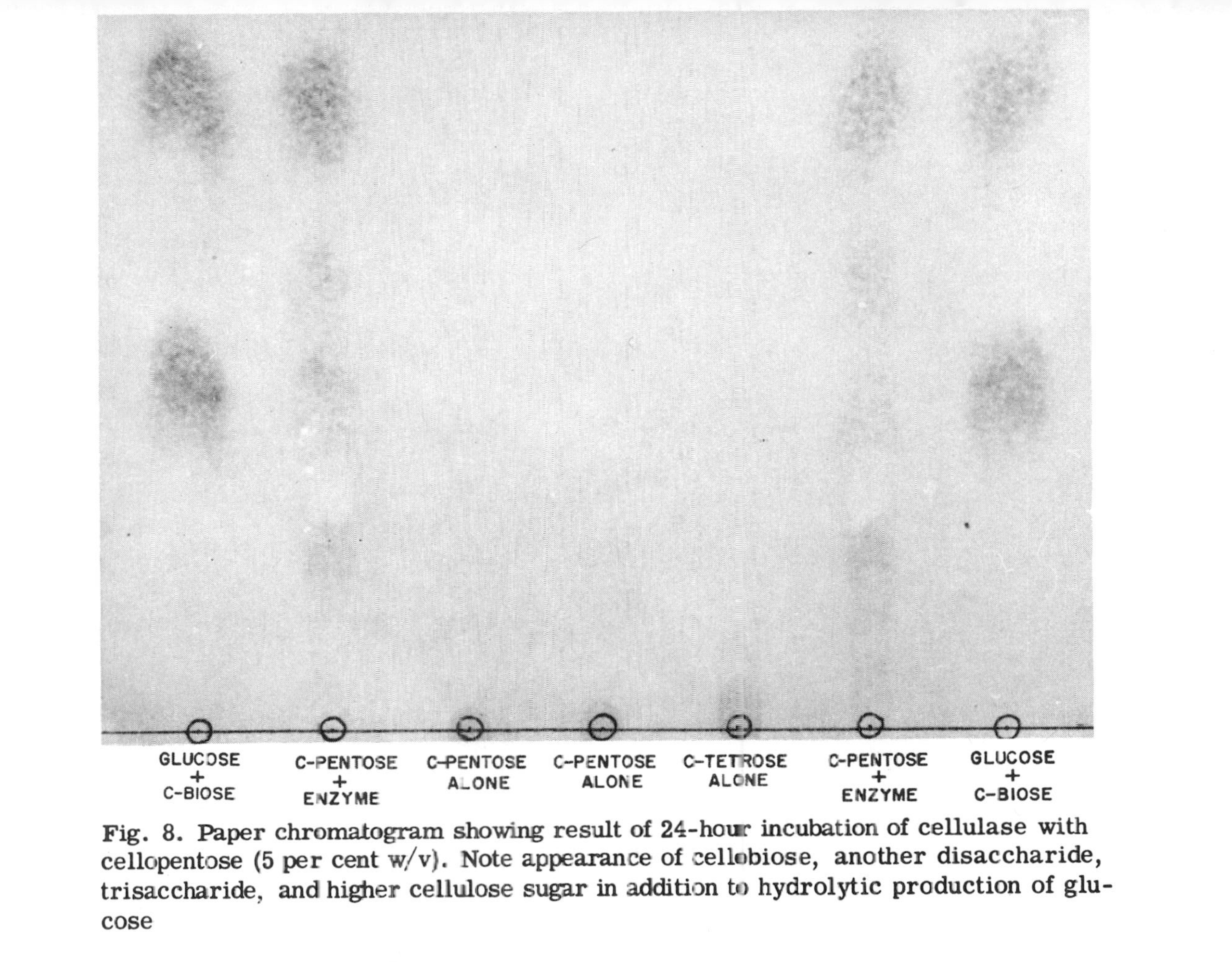

Fig. 8. Paper chromatogram showing result of 24-hour incubation of cellulase with cellopentose (5 per cent w/v). Note appearance of cellobiose, another disaccharide, trisaccharide, and higher cellulose sugar in addition to hydrolytic production of glucose

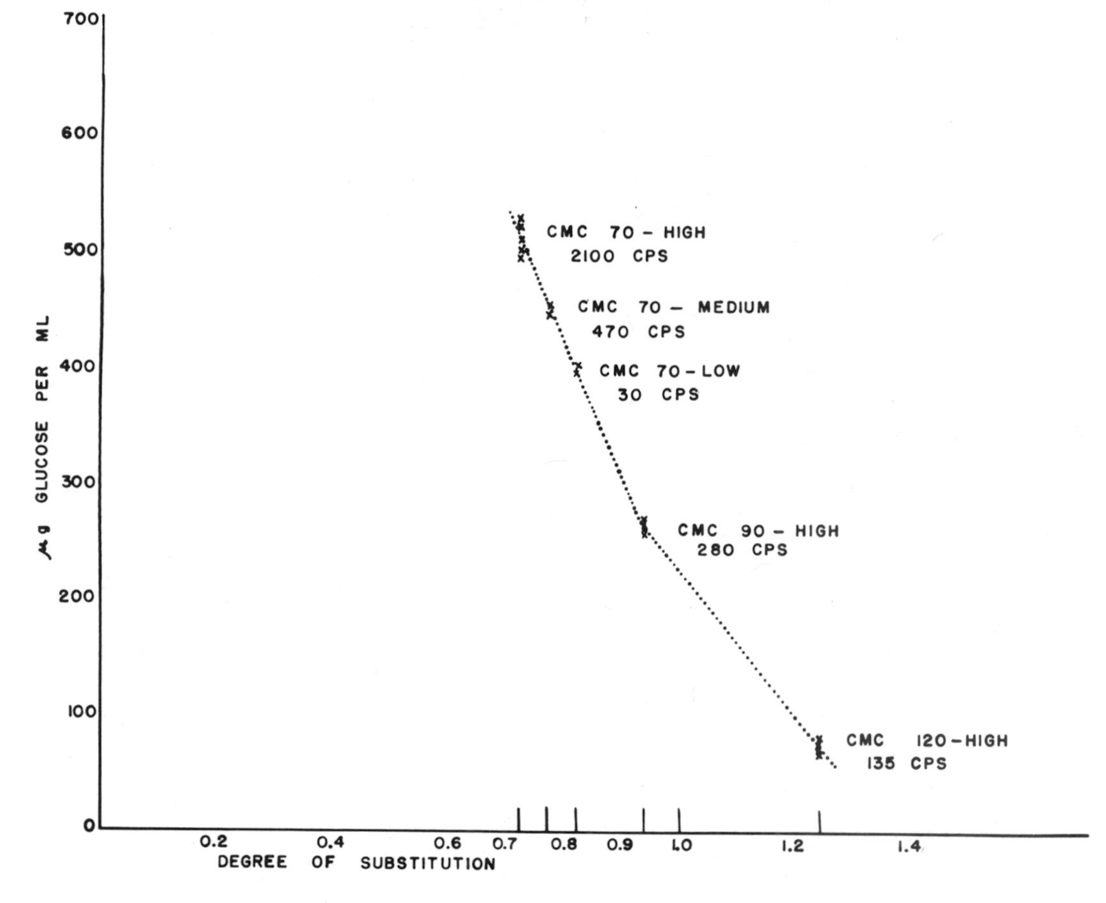

Fig. 9. Effect of cellulase on CMC solutions (1 per cent w/v) of different DP

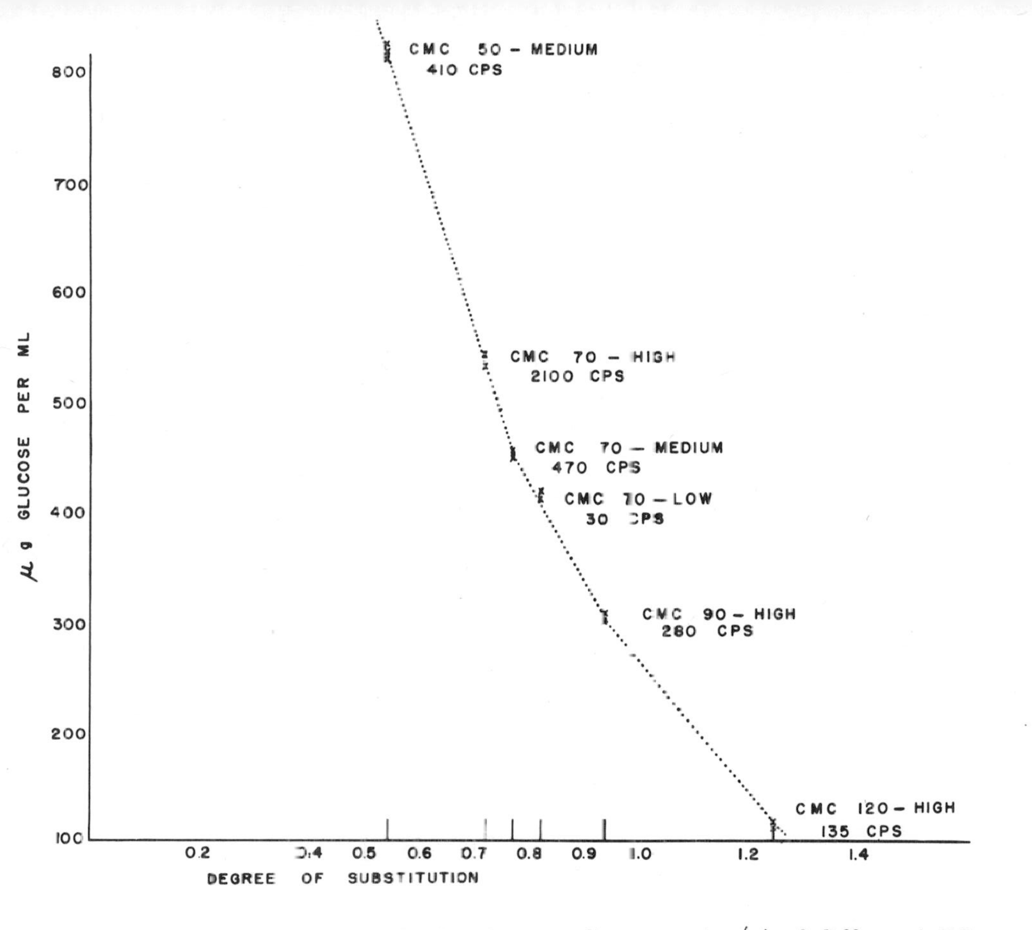

Fig. 10. Effect of cellulase on CMC solutions (1 per cent w/v) of different DS

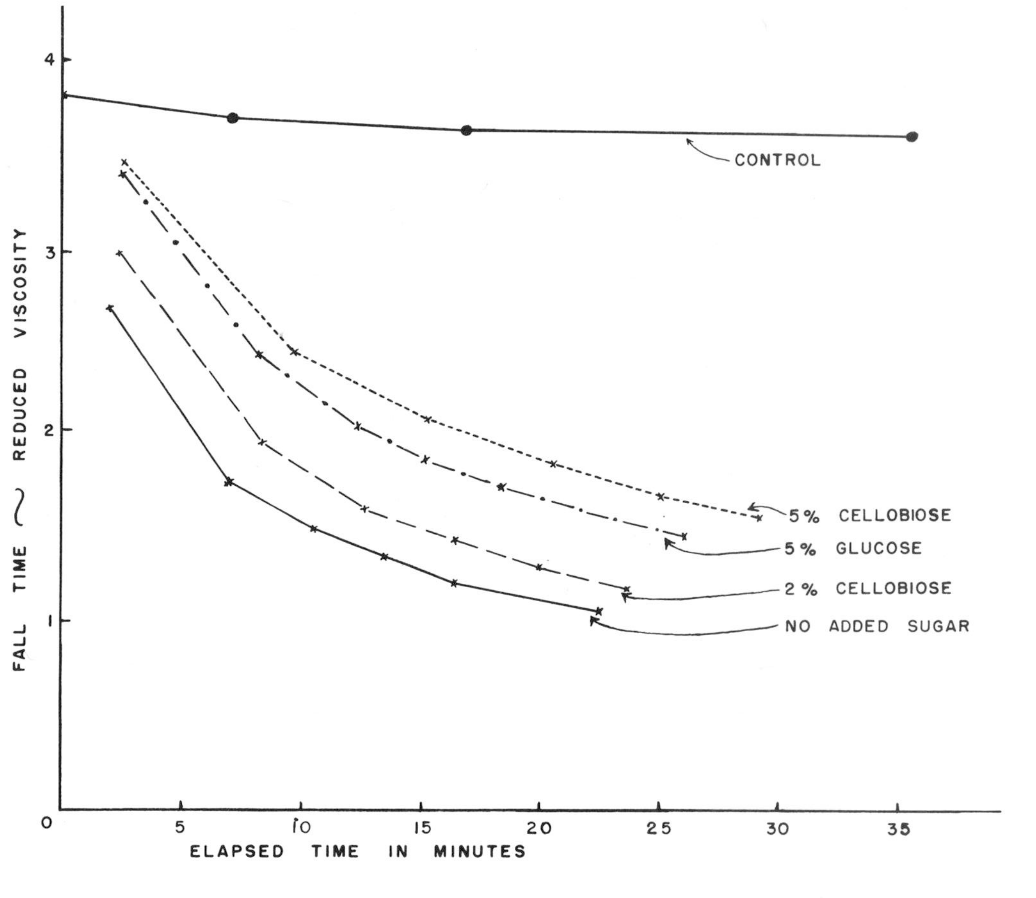

Fig. 11. Effect of added glucose or cellobiose on reduction of viscosity of 1 per cent (w/v) CMC

ENZYMES SPLITTING β-GLUCOSIDIC LINKAGES IN *STACHYBOTRYS ATRA*

G. Youatt and M. A. Jermyn

Some years ago, when work on microbiological attack on cellulose was begun in our laboratories, we faced a fundamental question of research tactics. Should we attempt to piece together a mechanism of enzymatic attack on cellulose by putting together information derived from a variety of different organisms, each of which happened to be suitable for studying a certain facet of this attack? Or should we attempt to investigate the enzymology and biochemistry of cellulose hydrolysis in a single organism? We chose the latter course because a reading of the relevant literature convinced us that "cellulase" was an effect and not an enzyme, and that there was probably no more point in talking about "the mechanism of cellulase action" than there would be in talking about "the mechanism of amylase action." This paper is an attempt to put together what we know about the enzymes of cellulose hydrolysis in the mold chosen for detailed study, *Stachybotrys atra,* and combines both published and unpublished observations and our current interpretation of these results. Like all imperfect fungi, *S. atra* is a far from perfect experimental subject, and no attempt will be made to hide the loose ends resulting from our great initial difficulties in culturing the fungus reproducibly and repeating our own results over any considerable period. Perhaps we would not again choose *S. atra* as a subject if we were starting this work today, but the very difficulties encountered emphasize, to our minds, the necessity of trying to understand the process in a single organism.

In a symposium of this kind, we feel that it is correct to place the first emphasis on the aims and ideas behind the line of investigation summarized, since these are usually regarded as mere verbiage that should not be allowed to clog up published experimental papers, and in consequence are almost never explicitly stated. Yet without some hint of these ideas the very point of half the papers may be lost. We have therefore deliberately omitted experimental detail that has been, or will be, available elsewhere.

We should also like to emphasize that, even after confining ourselves to a single organism, we have been forced to limit ourselves still more by the meager resources of time and personnel available to a small group. Thus our studies have in general been limited to shake culture at a single temperature, 28°C. This has been, in part, because the very great difficulties encountered in ensuring reproducible fungal growth have made it unprofitable to study the effects of physiological conditions on the production of, or perhaps the nature of, the enzymes. Now that these problems have been studied and solved by the persistent efforts of one member of our group, A. B. McQuade, we propose to make such experiments the next step in our program. Nevertheless, the fact that it is possible to produce the "β-glucosidase" over a very wide range of culture conditions, and the "cellulase" both in shake culture and in an aerated fermentor (Thomas, 1956), argues that there is some generality in our conclusions.

How Many Enzymes?

Table 1 summarizes the enzymes the existence of which must be invoked to account for our results to date with *S. atra*. Some of these enzymes have been isolated and purified to a degree, but the swelling factor at least has no more status than that of a useful hypothesis.

We have tried to give a general picture of the source of each of the enzymes as well as the method of assay. The latter is important not so much in itself, but because the trivial names that we have used for our enzymes have been derived from those of the assay substrates. Depending on which aspect of its behavior one was considering, it would be quite as correct to label the "cellobiase" as either a "cellulase" or a "β-glucosidase."

The only enzymes shown in the first table that we will not consider in more detail is that to which we have attached the noncommittal label of "X." The production but not the detailed properties of this enzyme were studied in the paper Fungal Cellulases III (Jermyn, 1953) under the name of "C_x." It was pro-

TABLE 1
STACHYBOTRYS ATRA ENZYMES

Enzyme	Assay Method	Inducibility	Source
β-glucosidase	Liberation of aglucones from glucosides	Nonadaptive	Mainly extracellular. Relation between concentrations inside and outside mycelium not yet investigated
Cellobiase	Glucose liberation from cellobiose and cellulose derivatives	There is some evidence that formation of the enzyme is induced by cellulose, but traces are produced on other carbon sources	Intracellular
"X"	Reducing sugar liberation from cellulose derivatives	Nonadaptive	Extracellular. Not apparently now formed; perhaps only cellobiase under different circumstances
Cellulase	Viscometry of cellulose derivatives	Minute amounts formed on noncellulosic carbon sources. Effectively induced by cellulose	Extracellular
Swelling factor	"Swelling" of cellulose	Induced by cellulose. Production very closely tied to that of cellulose	Extracellular

duced on a variety of substrates and assayed by the production of reducing sugars from carboxymethyl cellulose. The present strain of *S. atra* no longer produces this enzyme under any simple culture conditions that we have tried, and our present inclination is to rationalize the observation as the nonadaptive production of cellobiase by the original strain. But we have every hope that it will be possible to check this conclusion in the future, since it is very important for our ideas about the mechanisms that induce the formation of these enzymes. Meantime it would be unwise to take anything in the above-mentioned paper as referring to an enzyme in any way comparable with "C_X" as originally defined by Saunders, Siu, and Genest (1948). The existence of enzymes capable of liberating reducing groups from CMC had also been shown to occur nonadaptively in *Aspergillus oryzae,* and Toyama (private communication) finds a very powerful enzyme with this activity in the secretions of Trichoderma species grown on "koji," which he finds so far to be impossible to separate from the enzyme from the same source that dissolves cellulose. So the nature of "X" must at present be left *sub judice.*

β-glucosidase

During growth on a large number of carbon sources in simple media *S. atra* produces an extracellular β-glucosidase (table 2). On a glucose-salts medium that has been developed to give high and reproducible yields of the β-glucosidase of *S. atra,* we have found that a wide variety of fungi give an enzyme of very similar properties and specificity. The secretion of such an enzyme appears neither to be adaptive nor to play any large part in the metabolism of the mold; further, the secretion is controlled by physiological conditions and does not appear to take place until the most active phase of growth in shake culture has passed. Although many interesting deductions about enzyme specificity can be made by observing the pattern of action and inhibition of this enzyme, and we hope that it will be possible to use it to gain some further insight into the mechanism of enzyme synthesis, it is apparent that it has no part to play in cellulose metabolism. Since the synthesis of this enzyme, and of cellulase plus cellobiase, appear to be mutually exclusive, one speculative explanation of the facts may be that it represents the product of an enzyme-synthesizing system acting in the absence

TABLE 2
STACHYBOTRYS ATRA β-GLUCOSIDASE

In splitting Gly-O-R:

1. Gly must be glucose
   6-methylglucose
   3-methylglucose (?)
2. The configuration about -O- must be β.
3. R is preferably an aryl group
   The enzyme is less effective when R is an alkyl group, and ineffective when R is a monose residue.

Various elements of the aryl-β-glucoside structure confer affinity for the active center of the enzyme without ability to act as a substrate.

of any specific inducers. The remarkable physicochemical likeness between the β-glucosidase and the cellulase is one possible justification of this hypothesis.

Cellobiase

When the mold is grown in the presence of cellulose there can be recovered from the mycelium but not from the medium, an enzyme that, unlike the β-glucosidase discussed above, will hydrolyze cellobiose (table 3). By suitable adjustments of the conditions of growth, the amount of this second β-glucosidase can be varied, but our present finding is that its actual occurrence is induced by cellulose only. Practically every type of compound with a terminal β-glucosyl group is hydrolyzed by the enzyme, and oligosaccharides derived from cellulose are converted quantitatively to glucose, according to measurements using the glucose oxidase manometric technique. Proof that the terminal β-glucosyl linkage only is attacked by the enzyme is hard to give in spite of the fact that the results of kinetic studies and of paper chromatography are both consistent with this conclusion. This question could probably be examined by considering the nature of the transfer products from a substrate such as cellotriose if an acceptor other than water could be found. Unfortunately this has not yet been possible, a major point of difference from the first β-glucosidase which transfers the β-glucosyl group readily to a wide variety of acceptors. Since the cellobiase is not inhibited by delta-gluconolactone, a powerful inhibitor of the first enzyme, there seems little doubt that the two enzymes act by very different mechanisms in hydrolyzing the β-glucosidic linkage.

TABLE 3
STACHYBOTRYS ATRA CELLOBIASE

In splitting Gly-O-R:

1. Gly *must* be a *terminal* glucose. The degree to which glucose can be replaced by other monoses is not yet known; it certainly cannot be replaced by galactose.
2. The configuration about -O- must be β.
3. R can be almost any type of radical, even a polysaccharide chain.

Cellobiase is an exo-β-glucosidase.

There does not appear to be any true upper limit to the size of molecule that can be hydrolyzed by the cellobiase. Any sufficiently degraded cellulose is slowly broken down to glucose. We incline to the opinion that it is the architecture of cellulose aggregates rather than the chain length of the cellulose molecule that is the limiting factor in the action of this enzyme. In any case it does not seem profitable to argue about specificity until sufficient experiments have been carried out with well-defined substrates, and these experiments have simply not yet been done.

There is a partial analogue for this enzyme in the amylases of the gluc-amylase type and an almost complete analogue in the amylase of *Candida tropicalis* which is both an α-glucosidase and a maltase as well (Sawai, 1958).

Cellulase

At the same time that cellobiase is being formed in the mycelium a second enzyme is being secreted into the medium as an adaptive response to growth on cellulose (table 4). This enzyme is detected and measured by its action in reducing the viscosity of solutions of sodium carboxymethylcellulose. Thomas (1956) showed that its action was that of hydrolyzing cellulose chains at random, that it could hydrolyze cellotriose but not cellobiose, and that the end product of its action on cellulose was an approximately equimolecular mixture of glucose and cellobiose. We have subsequently shown that it is not a β-cellobiosidase. The elements of its specificity as shown in the table are really no more than a series of queries. In the case of the

TABLE 4
STACHYBOTRYS ATRA CELLULASE

In splitting Gly-O-R:

1. Gly *probably cannot* be a terminal glucose and certainly need not be.
2. R must be one or more β-linked glucose residues.
3. Alternatively the sum of the monose residues in Gly and R must be at least 3.
4. All statements about the specificity of this enzyme are likely to be tentative for a long time to come.

Cellulase is an endo-β-glucosidase.

cellulase of Irpex, which has the same requirement as that of Stachybotrys for two β-glucosidic linkages but where the third glucose residue can be replaced by an aryl group so that it is an aryl-β-cellobiosidase, Nisizawa (1955) has shown by transfer experiments that the heteroside linkage is cleaved many times as fast as the holoside linkage in *p*-nitrophenyl β-cellobioside; by analogy we would expect that the *S. atra* enzyme acting on cellotriose would preferentially hydrolyze the linkage farthest from the nonreducing end of the molecule. This point could also be decided by transfer experiments, but once more we have unfortunately been unable to establish conditions for transfer.

The enzyme is not perhaps completely adaptive since traces have been observed in noncellulosic culture media. It is rather hard to exclude all cellulose from culture vessels, and the possibility remains that these traces were an adaptive response to stray fibers.

There is an analogy between this cellulase and those α-amylases that can hydrolyze all amylose fragments down to maltotriose; an interesting partial analogy is the polygalacturonase of Ozawa (Ozawa and Okamoto, 1955) that hydrolyzes tetragalacturonic acid rapidly and trigalacturonic acid hardly at all, and many α-amylases seem to have this kind of specificity. The Irpex cellulase is, then, analogous to taka-amylase, which is also an α-maltosidase. There is no real evidence at the moment about whether these two types are fundamentally distinct.

It is an interesting further point that a cellulase inhibitor that exists in aqueous extracts of red gum *(Eucalyptus rostrata)* appears to inhibit all cellulases that we have tried from a number of sources which have, so far as we know, no more in common than the ability to reduce the viscosity of carboxymethyl cellulose solutions.

"Swelling Factor"

It should be remarked that the properties quoted above are those of a partially purified "cellulase" solution. A further property of such a purified cellulase solution is that it has little or no action on native cellulose. It is almost impossible in this context to give a watertight definition of "native cellulose" or even of "no action." It is perhaps best to be concrete and say that almost any test that can be devised for damage to the structure of such a form of cellulose as cotton linters gives very much lower results for purified cellulases than for crude culture media, and these effects show no clear relationship to "cellulase" as measured by viscosity reduction. One such test for damage to cellulose is the swelling test of Reese and Gilligan (1953, 1954), where the ability of cotton to swell in alkali after exposure to enzyme solutions is measured. They have postulated an enzyme responsible for increased ability to swell in alkali which they have called "swelling factor(s)." There is no doubt that crude *S. atra* preparations that attack native cellulose also give the swelling factor test (table 5); but there are a number of other facts that have to be taken into account in any attempt to explain the results. The swelling factor is produced, like cellulase itself, only on media that contain cellulose. This cellulose may be present as crude fiber or as purified cellulose, and growth on the hemicellulose fractions of crude fiber leads to the production of neither cellulase nor swelling factor. Swelling factor is produced during growth on all forms of cellulose, but the swelling effect is very differently manifested in different celluloses, from powerfully in cotton to negligibly in sisal. It is possible to obtain preparations of cellulase that contain no swelling factor, but not of swelling factor without cellulase. When swelling factor in a crude mixture is destroyed by heat, the affinity of the cellulase for its substrate as measured by the Michaelis constant is altered.

TABLE 5

STACHYBOTRYS ATRA SWELLING FACTOR

1. Always produced in association with cellulase.
2. It is more rapidly inactivated by heat than cellulase.
3. Conditions that inactivate "swelling factor" change the apparent K_m of cellulase toward sodium carboxymethylcellulose.
4. Is the "swelling factor" more than the synergistic action of several cellulases?

Our working hypothesis to explain these results is that swelling factor is the synergistic effect of several cellulases acting together to attack native cellulose; the swelling effect itself is in some way a consequence of the particular supermolecular architecture of a given fiber. We would emphasize that this is no more than a working hypothesis and liable to change in the light of further experimental results, e.g., the preparation of swelling factor free from cellulase.

It may be mentioned that there is nothing unique about the difficulty of attacking cellulose with purified cellulases, and the answer to this problem appears to be different in different carbohydrase systems. For starch granules, the "Z" factor of Peat (Peat, Thomas, and Whelan, 1952) appears to be a β-glucosidase; for protopectin, Kaji (1956) has shown that the "macerating enzyme" of *Clostridium felsineum* can be separated from polygalacturonase even though it is very similar in its properties. "Swelling factor" or any equivalent term may mean very different things in different organisms.

Attack on an Insoluble Substrate

The point we have been trying to make in this paper should now be fairly clear. It is that the enzymes that we know to be involved in the attack on the actual polysaccharide chains of cellulose are in no way peculiar as enzymes, and that they belong to types that can all be paralleled from other regions of carbohydrase biochemistry. The observed peculiarities in the attack on cellulose occur precisely at those points at which it becomes impossible to explain the properties of cellulose itself from simple considerations of carbohydrate chemistry, i.e., where the supermolecular architecture of the cellulose aggregates becomes more important than the properties of the individual macromolecules that make them up.

Thomas (1955) showed that, if cotton was oxidized with periodic acid and the resultant carbonyl groups reacted with phenylhydrazine to the extent that one glucose residue in a hundred was altered, the cotton then became unusable by *S. atra* as a substrate for growth. Even when a much more drastic substitution was carried out on chromatographic cellulose powder, which, unlike cotton, can be attacked by purified *S. atra* cellulase, the mold was still able to use it for growth. The conclusion we then drew was that some factor, in addition to cellulase, that the mold produced, that allowed it to act on cotton, was being inhibited by this substitution. Whether this factor in some way disrupted the highly crystalline structure of cotton cellulose or destroyed some outer layer of the fiber we were unable to say. At any rate there was some relatively small proportion of the glucose residues of cotton, presumably in a small region, whose alteration prevented the action of the factor. Alternatively, this latter must act over a length of the cellulose chain of the order of 100 glucose units. Our own preference would be for an outer fiber layer of different properties. We would base this speculation on the "swelling" behavior of cotton after the action of crude extracts, which is quite analogous to that of wool whose outer cuticle has been damaged, and the well-known fragmentation of cellulose fibers undergoing enzymatic attack, which again can be paralleled for other cuticular fibers with local weak points. The lack of "swelling" in sisal that has been exposed to crude cellulase shows, however, that it is quite illegitimate to generalize.

Too many conclusions in the field of cellulases seem to have been drawn from studies involving either cotton or chemically damaged celluloses, and it seems that to establish generally valid conclusions it will be necessary to examine the action of cellulases on a wide variety of celluloses. A pilot study of our laboratory in which *S. atra* was supplied with celluloses prepared from many sorts of green leaves and crude cellulose fibers by identical methods has shown that there is a wide difference between the availability of different celluloses to the mold. We plan to follow this up by enzymatic studies as soon as we have established consistent differences of behavior. This merely emphasizes the point that the key to explaining biological attack on cellulose lies in organization at the fiber level, or just below it.

In general the whole question of the attack on insoluble substrates by enzymes has so far not even begun to be answered; in fact it is probable that the question has not even been properly formulated. But it seems that it can only be tackled by using enzymes of clearly defined specificities on substrates of known history. It is with this idea in mind that we have preferred to move forward slowly by defining the properties of our agents exactly before attempting to find out their role in the breakdown of cellulose by *S. atra*.

Stachybotrys in Relation to Other Cellulolytic Organisms

We began by saying that we did not believe that there was a single mechanism of cellulase action, and one of us (M. A. J.) has recently had the benefit of detailed discussions with Dr. Nisizawa and his students on the action of Irpex cellulase. In table 6 are set out the differences between this cellulase and the, on the face of it, very similar enzyme from Stachybotrys. We choose these two because they are the ones with whose properties we are now best acquainted. There are very marked dif-

TABLE 6
A COMPARISON OF TWO CELLULASES

Irpex Cellulase	Stachybotrys Cellulase
Hydrolyzes cellulose, carboxymethylcellulose, cellodextrins, cello-oligosaccharides, aryl-β-cellobiosides	Hydrolyzes cellulose, carboxymethylcellulose, cellodextrins, cello-oligosaccharides including cellotriose
Does not hydrolyze alkyl -β-cellobiosides, β-glucosides, cellobiose	Does not hydrolyze aryl or alkyl cellobiosides, cellobiose
Gives cellobiose and a small amount of glucose from the hydrolysis of cellulose	Gives approximately equal amounts of glucose and cellobiose
Transient oligosaccharides apparently formed during cellulose hydrolysis by cellobiosyl transfer to cellobiose	Oligosaccharides formed at the limit of detection (?) during the hydrolysis of cellotriose
Hydrolyzes aryl -β- cellobiosides primarily to cellobiose, but also secondarily to glucose + aryl -β-glucoside	Transfer from cellotriose to alcohols not yet investigated
Transfers both cellobiosyl and glucosyl to methanol from aryl-β-cellobiosides; ratio perhaps 100:1	Point of attack on cellotriose not yet known
Appears to show extreme specificity of attack on the second bond from the nonreducing end--rather than any analogy to β-amylase	Perhaps a close analogue of α-amylase

ferences in behavior between them. Yet if the cellulases are as heterogeneous a group as the amylases, of which we set down some known types in table 7, then they must be counted rather close together. Indeed the cellulases may well be more heterogeneous than the amylases since there is no amylase with the combination of properties of the Irpex cellulase. There is no obvious reason why the cellobiase of *S. atra* should not be called a cellulase just as the analogous gluc-amylases are certainly considered as amylases, even though it yet remains to be seen if it plays anything but a subsidiary part in the breakdown of cellulose by the mold. Although it is now fashionable to consider the cellulases essentially as depolymerases, it will scarcely be surprising if other types are not very frequent. Like our work on the "swelling factor," most of the studies that have seemed to show that more than one type of cellulase is produced by a given organism are suggestive rather than conclusive, although particularly in molds--witness takadiastase--the production of multiple enzymes with a common action is frequent enough. When we have more closely defined the limits of the specificity of our cellobiase and cellulase, it will then be possible to define the action of the other enzymes that are present in active culture filtrates and to attempt to isolate them. Even this will not be a complete picture of the β-glucoside-splitting enzymes of *S. atra* until we have grown the mold under a much wider variety of conditions than we have yet surveyed. Just as the presence of cellulose apparently switches the mold from one path of enzyme synthesis to another, other growth patterns may involve the production of enzymes about which we do not yet know.

TABLE 7
A SUMMARY OF AMYLASE TYPES

Enzyme	Substrates	Transferase Action	Configuration of Liberated Reducing Group	Action on Amylose
α-amylases	poly-α-glucose chains of at least 3 to 4 residues	none ever shown	α	random splitting
Taka-amylase A	poly-α-glucose chains of at least 3 residues aryl-α-maltosides	α-maltosyl groups from maltosides	α	random splitting
Taka-amylase B	poly-α-glucose chains ? of at least 3 residues	none	β	liberate terminal non-reducing glucose
Candida amylase	amylose maltose aryl-α-glucosides, alkyl-α-glucosides	α-glucosyl groups from first three substrates	α	liberates terminal non-reducing glucose
β-amylases	poly-α-glucose chains	none	β	liberate terminal maltose

We will therefore make no apology for our opinion that the study of cellulase in the very scattered laboratories in which it is now going on will resolve itself into a study in comparative biochemistry rather than a race to determine the mechanism of action of "cellulase."

LITERATURE CITED

Jermyn, M. A. 1953. Fungal Cellulases. III, *Stachybotrys atra*. Growth and Enzyme Production on Non-cellulosic Substrates. Austral. J. Biol. Sci., 6:48-49.

Kaji, A. 1956. Macerating Enzyme Acting on Middle Lamella Pectin. I, Separation of the Macerating Enzyme Produced by *Clostridium felsineum* var. *sikokianum*. Bull. Agr. Chem. Soc. Japan, 20:8-12.

Nisizawa, K. 1955. J. Biochem. (Japan) 42:825.

Ozawa, J., and K. Okamoto. 1955. Ber. Ohara Inst. landwirtsch. Biol. Okayam Univ., 10:215.

Peat, S., G. J. Thomas, and W. J. Whelan. 1952. Enzymic Synthesis and Degradation of Starch. XVII, Z-enzyme. J. Chem. Soc., 1952:722-33.

Reese, E. T. and W. Gilligan. 1953. Separation of Components of Cellulolytic Systems by Paper Chromatography. Arch. Biochem. Biophys., 45:74-82.

-------. 1954. Swelling Factor in Cellulose Hydrolysis. Text. Res. J., 24:663-69.

Saunders, P. R., R. G. H. Siu, and R. N. Genest. 1948. A Cellulolytic Enzyme Preparation from *Myrothecium verrucaria*. J. Biol. Chem., 174:697-703.

Sawai, T. 1958. Studies on an Amylase of *Candida tropicalis* var. *japonica*. I, Maltase and Transglucosylase Activities of the Amylase. J. Biochem. (Tokyo), 45:49-56.

Thomas, R. 1955. Some Chemically Modified Celluloses and Their Resistance to Fungal Degradation. Text. Res. J., 25: 559-62.

-------. 1956. Fungal Cellulases. VII, *Stachybotrys atra*: Production and Properties of the Cellulolytic Enzyme. Austral. J. Biol. Sci., 9:159-83.

ACCEPTOR SUBSTRATE IN ENZYMATIC TRANSGLUCOSYLATION*

H. M. Tsuchiya

The mechanisms involved in certain enzymatic depolymerization and polymerization of polysaccharides are often similar. Thus, degradation consists essentially of the transfer of glycosyl residues from the donor, a polysaccharide, to a suitable acceptor. Conversely, the formation of a polysaccharide generally involves the transfer of glycosyl residues from the donor substrate to an acceptor substrate or primer capable of undergoing propagation or growth. The enzymes that mediate these transfer reactions can neither initiate nor continue the reactions unless both substrates are present. Relatively less attention has been focused upon the acceptor substrate than upon the donor substrate in studies on carbohydrases. This is unfortunate because the acceptor is as essential a substrate as the latter. Furthermore, it can profoundly modify the course of the reaction and the reaction products.

The role played by the acceptor in glycosyl transfer reactions can, in some respects, be as readily studied in a synthetic as in a degradative reaction where end products of the latter may sometimes be monomeric constituents of the polymer. Therefore, consideration of the acceptor substrate in an enzymatic synthesis of a polysaccharide may not be altogether out of order at a session devoted to cellulose depolymerizing enzymes.

This will be a discussion of the acceptor substrate in the dextransucrase reaction, an enzymatic transglucosylation that leads to the formation of dextran.

Dextran is a branched, glucose polysaccharide in which the

---

*Studies conducted at the Northern Regional Research Laboratory, U.S. Department of Agriculture, Peoria, Illinois.

principal linkage is of the α-1, 6 type (Jeanes *et al.*, 1954); the other bonds are of the 1, 2 or 1, 3 or 1, 4 type (Lohmar, 1952). Dextran, like other polymers, includes molecules that differ slightly in composition as well as molecules that display a range of molecular weights. It can be synthesized from sucrose by dextransucrase, an enzyme elaborated by *Leuconostoc mesenteroides*. This enzyme was first described by Hehre (1941).

Figure 1 shows the basic unit reaction mediated by this enzyme. It involves the transfer of glycosyl unit from the donor to the acceptor substrate. Where sucrose is the donor, fructose is liberated. Fructose is somewhat analogous to phosphate in the phosphorylase reaction (Cori, Colowick, and Cori, 1937; Hanes, 1940), or to glucose in the amylomaltase reaction (Monod and Torrani, 1948). Repetitive participation of the glucose product as the acceptor in subsequent transglucosylations leads to the polymerization of dextran. The enzyme displays greater specificity toward the donor than toward the acceptor. Although other saccharides serve as donors, sucrose appears to be the preferred substrate. On the other hand, the enzyme is less specific in its requirement for the acceptor; many sugars, low molecular weight dextran, and water function as acceptors.

The question arises as to the identity of the initial acceptor in a so-called "normal" reaction where sucrose (at relatively low concentrations), enzyme, and water are present in the reaction mixture. Presumptive evidence suggests that sucrose plays a dual role; it serves as an acceptor substrate as well as the donor. However, it is not an efficient acceptor. The formation of a dextran molecule with average molecular weight of 30, 000, 000 (Wolff *et al.*, 1954) requires the participation of only one molecule of sucrose as the initial acceptor, or primer, whereas some 200, 000 molecules of sucrose must serve as donors. This gives an approximate estimate of the relative efficiency of sucrose as an acceptor and as a donor. If sucrose were as efficient an acceptor as donor, the reaction products would probably be oligosaccharides instead of a polysaccharide. Synthesis of a high molecular weight polymer implies that a growing chain, with sucrose as the chain initiator, becomes a more efficient acceptor substrate as it grows.

High molecular weight polysaccharide is a primary reaction product in reactions with initial sucrose concentrations not exceeding 10 per cent. An entirely different situation obtains at elevated sucrose levels. Low molecular weight dextran is formed at initial sucrose levels of 70 per cent or higher; the molecular weight of dextran polymerized under such conditions does not exceed 35,000. Indeed, the reaction products include low molecular weight material extending down through the oligosaccharide range to free glucose and fructose.

Figure 2 provides evidence for the formation of lower saccharides. The left-hand side of the paper chromatogram shows sugars with reducing end groups; the right-hand side shows sugars containing fructose. The occurrence of free fructose is to be expected; for every mole of sucrose acting as glucosyl donor, one mole of fructose is liberated. The presence of glucose, formed in small amounts, implicates water as an acceptor substrate. The efficiency of water as acceptor is relatively low; nonetheless, it is very definitely a glucosyl acceptor substrate.

The formation of low molecular weight dextran at initially high sucrose concentrations can be rationalized by assuming either that the relative efficiency of sucrose as acceptor is enhanced under such reaction conditions or that fructose and glucose also serve as alternate glucosyl acceptors. Evidence presented elsewhere (Tsuchiya *et al.*, 1955) suggests that the high fructose level resulting from elevated sucrose concentration leads to formation of intermediate and low molecular weight polysaccharide. This may be analogous to the amylomaltase system. This enzyme catalyzes the transformation of maltose to amylose and glucose; presumably maltose serves as both donor and acceptor substrates. The polysaccharide formed by amylomaltase does not grow beyond the oligosaccharide stage unless glucose liberated in the transglucosylation reaction is removed. If this is done, a polysaccharide with degree of polymerization (DP) sufficiently high to react with iodine is synthesized.

Estimates of reducing values and dextran yields in reactions containing sucrose and acceptors alternate to sucrose and water indicated that isomaltose, maltose, and α-methyl glucoside increase the reaction rate and that fructose depresses the rate (Koepsell *et al.*, 1953). Another manifestation of the activity of isomaltose, maltose, and α-methyl glucoside as alternate acceptors was the decrease in yields of high DP dextran at the termination of the reaction. This means that these acceptors divert appreciable amounts of the glucosyl units to the formation of lower molecular weight material and oligosaccharides.

The formation of lower saccharides in these reactions where sucrose and the added acceptors were present in low concentrations was also examined by paper chromatography (Koepsell *et al.*, 1953. In the control reaction with sucrose alone, fructose, glucose, and leucrose (Stodola, Koepsell, and Sharpe, 1952) were the only sugars found. In the reactions with glucose or isomaltose as added acceptors, series of oligosaccharides were formed. Their mobilities coincided with those sugars obtained from the acid hydrolysis of dextran. These added acceptors divert normal dextran formation to an appreciable synthesis of sugars of the ascending α-1, 6 glucopyranosidic series. Where maltose was provided as the alternate primer, the oligosaccharides formed were panose (Pan, Andreasen, and Kolachov, 1950), a trisaccharide, and the higher sugars of this series. α-methyl glucoside also served as an efficient acceptor and led to the formation of α-methyl isomaltoside and its higher α-1, 6 linked homologs (Jones *et al.*, 1956). Fructose, melibiose, and galactose also served as primers, as evidenced by formation of oligosaccharides.

Paper chromatography evidence for the propagation of maltose to higher sugars has been presented elsewhere (Tsuchiya *et al.*, 1955). As the sucrose to maltose ratio was increased, maltose first disappeared; this was accompanied by the formation of panose. It, in turn, disappeared as the tetrasaccharide was produced. When the ratio reached 26. 5, no oligosaccharide capable of being moved on the paper by the solvent employed could be detected.

At this point, investigations were directed toward examining the acceptor efficiency of low DP dextran (Tsuchiya *et al.*, 1955; Hehre, 1953; Tsuchiya, Hellman, and Koepsell, 1953). Diversion of glucosyl units to form polymer with molecular weight lower than the high DP material was taken as an assessment of acceptor efficiency. Figure 3 demonstrates the relative effectiveness of maltose and of low molecular weight dextran as primers. The product precipitating at low alcohol concentration is high molecular weight material; the product precipitating at high alcohol concentration is lower DP material. The area under each peak gives an approximation of the yield of that polysaccharide.

The molecular weight distributions of products from three syntheses at termination of the reaction are depicted in figure 3. The first reaction contained 10 per cent sucrose and enzyme. All the product precipitated at low alcohol concentration, indicating that the dextran is of high molecular weight. The second reaction contained sucrose and maltose as substrates. The resolving of one peak into two peaks instead of a downward shift in the average DP of the polysaccharide was an unexpected finding. Presence of this acceptor, maltose, depresses the yield of high molecular weight material and induces the formation of lower DP dextran. The third reaction contained sucrose and low molecular weight dextran as substrates. Presence of this low polymeric material as primer suppresses drastically the yield of high molecular weight polymer and causes the appreciable formation of intermediate molecular weight polysaccharide. It should be noted that the two primers were compared on equivalent molar basis. These data indicate that low molecular weight dextran is superior to maltose as an acceptor substrate since it diverts more of the glucosyl units from the donor, sucrose, to the formation of lower DP product.

The synthesis of two dextran components with widely differing molecular weight deserves comment. In starch, there are two molecular species: amylose and amylopectin. However, the former is essentially a linear molecule whereas the latter is a ramified or branched molecule. The difference in molecular weight between these two starch fractions is associated with branching. No such structural difference has yet been detected between the high and lower DP dextran components. The possibility that more than one enzymatic activity may be operative in the formation of two dextran components has been considered. The two fractions may arise from the action of two different enzymes. Alternatively, high DP polysaccharide may represent condensation products of lower DP material, or the lower molecular weight fraction may represent depolymerization products derived from high DP polymer. Our efforts to demonstrate the presence of more than one enzymatic activity has thus far yielded negative results. For the present, the data can be rationalized only by assuming that a molecule of dextran which attains a certain DP becomes a vastly superior glucosyl acceptor; it grows considerably faster than do others which are of somewhat lower molecular weight. In the competition between primers for glucosyl units, the larger molecules must be markedly more efficient acceptor substrates.

However, the lower molecular weight material is capable of growth if reaction conditions are such as to permit it and if an adequate supply of the donor molecules is present (Tsuchiya *et al.*, 1955). Figure 4 provides evidence for this. If the donor to acceptor ratio is increased, molecules of intermediate molecular weight dextran grow to higher DP material. Thus, the average molecular weight of intermediate molecular weight polysaccharide rises with the donor to acceptor ratio. Indeed, increasing the ratio results in the fusion of two peaks into one.

Figure 5 shows the dependence of effectiveness of the acceptor substrate upon reaction conditions (Tsuchiya *et al.*, 1955). The effect of two enzyme concentrations were tested at two temperatures. Substrate concentrations were held constant in these four reactions. The most striking effect observed was the influence of reaction temperature. Lowering the temperature from 30° to 15° increases the yield of intermediate molecular weight fraction and suppresses the formation of high DP polymer. Presumably, the difference in effectiveness of growing chains of varying chain lengths as primer is minimized at 15° so that more of the acceptors participate in the reaction. At this temperature, enzyme concentration exerts no effect upon the distributions; the two curves are essentially superimposable. On the other hand, at 30°, increasing the enzyme level fourfold enhances the formation of intermediate molecular weight dextran and suppresses the production of high DP polysaccharide. This implies that more of the acceptors participate in the reaction at the higher enzyme concentration than at the lower level.

Data shown in figures 3, 4, and 5 indicate that a number of factors determine which acceptor candidate participates in the reaction. This depends upon both the efficiency, or quality, and the concentration, or quantity, of the prospective acceptors present. It also depends upon reaction conditions such as temperature and enzyme concentration.

The kinetic data (Tsuchiya and Stringer, 1955) also confirm the role of the acceptor in this reaction. Figure 6 shows the effect of substrate concentrations upon the initial rate of reaction. The concentrations of both the donor and the added acceptor were varied simultaneously. The alternate acceptor substrate employed in this study was α-methyl glucoside. The rates were measured by increase in reducing values calculated as fructose. Each curve represents a different level of the added acceptor. In the absence of α-methyl glucoside, the reaction is first order with respect to sucrose over the range of 0.005 to 0.1M. The Ks constant computed from points in this range lies between 0.011 and 0.013M. The appreciable velocity at zero concentration of α-methyl glucoside reflects the activity of other competing acceptors. The rate rises as expected at the lower sucrose concentrations. It then decreases as the sucrose exceeds the optimum value. This phenomenon of substrate inhibition, originally observed by Hehre (1946) for this reaction, is seen at concentrations of sucrose higher than 0.2M. As concentration of α-methyl glucoside is increased, the initial reaction rate is accelerated as would be anticipated from the basic unit reaction shown in figure 1. It is of interest to note that the highest observed rate in the absence of the added acceptor is only about 20 per cent of the potential rate predicted by the potential V_M of the system. The optimum sucrose concentration also rises. Moreover, the acceptor appears to relieve the sucrose substrate inhibition at high levels.

Figure 7 shows the effect of the added acceptor, α-methyl glucoside, upon reaction constants. The V_M and Ks values for sucrose at different levels of the added acceptor are shown. The constants vary with the concentrations of the cosubstrate, α-methyl glucoside. That this behavior is to be expected from two substrate systems was first clearly stated by Ingraham and Makower (1954).

Up to this point the discussion has been limited to consideration of the acceptor substrates in the polymerative reaction. It was stated initially that depolymerative and polymerative reactions of polysaccharides are often basically similar in nature; both involve transglycosylations. Dextransucrase appears to display slight but definite degradative activity.

Figure 8 shows the depolymerization of dextran serving as glucosyl donor (Tsuchiya and Stringer, 1954). Sucrose was not present in these reactions. It should be noted that the reaction rate with this donor is considerably slower than in reactions where sucrose serves as the donor. The presence of glucose in the reaction containing dextran as the sole saccharide substrate confirms the earlier remarks pertaining to water functioning as acceptor substrate. Where glucose was supplied as an acceptor, the anticipated series of isomaltose and higher oligosaccharides of this series are found. Where fructose was provided as acceptor, leucrose and other sugars were formed. Sucrose could not be detected in these reactions. If any was formed, it must have served as donor and thus disappeared immediately. Formation of oligosaccharides in these reactions can occur only with the concurrent depolymerization of the donor substrate, dextran.

What are the implications of these observations for investigations on the depolymerative enzymes such as the cellulases? If we may be permitted to extrapolate from these observations on this enzyme to other glucosyl transfer systems, it is suggested that more attention may with profit be directed at the acceptor substrates in the depolymerization of cellulose. Furthermore, it is also suggested that considerations arising from studies on the biological polymerization of cellulose may throw some light upon the mechanisms involved in the enzymatic degradation of cellulose.

LITERATURE CITED

Cori, C. F., S. P. Colowick, and G. T. Cori. 1937. The Isolation and Synthesis of Glucose-l-phosphoric Acid. J. Biol. Chem., 121:465-77.

Hanes, C. S. 1940. The Reversible Formation of Starch from Glucose-l-phosphate Catalysed by Potato Phosphorylase. Proc. Roy. Soc. (London), B129:174-208.

Hehre, E. J. 1941. Production from Sucrose of a Serologically Reactive Polysaccharide by a Sterile Bacteria Extract. Science, 93:237-38.

-------. 1946. Studies of the Enzymatic Synthesis of Dextran from Sucrose. J. Biol. Chem., 163:221-33.

-------. 1953. Low-molecular-weight Dextran as a Modifier of Dextran Synthesis. J. Am. Chem. Soc., 75:4866.

Ingraham, Lloyd L., and Benjamin Makower. 1954. Variation of the Michaelis Constant with the Concentrations of the Reactants in an Enzyme-catalyzed System. J. Phys. Chem., 58:266-70.

Jeanes, A., W. C. Haynes, C. A. Wilham, J. C. Rankin, E. H. Melvin, M. J. Austin, J. E. Cluskey, B. E. Fisher, H. M. Tsuchiya, and C. E. Rist. 1954. Characterization and Classification of Dextrans from Ninety-six Strains of Bacteria. J. Am. Chem. Soc., 76:5041-52.

Jones, R. W., Allene Jeanes, C. S. Stringer, and H. M. Tsuchiya. 1956. Crystalline Methyl α-Isomaltoside and Its Homologs Obtained by Synthetic Action of Dextransucrase. J. Am. Chem. Soc., 78:2499-02.

Koepsell, H. J., H. M. Tsuchiya, N. N. Hellman, A. Kazanko, C. A. Hoffman, E. S. Sharpe, and R. W. Jackson. 1953. Enzymic Synthesis of Dextran Acceptor Specificity and Chain Initiation. J. Biol. Chem., 200:793-801.

Lohmar, Rolland. 1952. Evidence of New Linkages in Dextrans. J. Am. Chem. Soc., 74:4974.

Monod, J., and A. Torriani. 1948. Synthèse d'un polysaccharide du type omidon aux dépens du maltose, en présence d'un extriot enzymatique d'origine bactérienne. Comp. Rend. Acad. Sci. (Paris), 227:240-42.

Pan, S. C., A. A. Andreasen, and Paul Kolachov. 1950. Enzymic Conversion of Maltose into Unfermentable Carbohydrate. Science, 112:115-17.

Stodola, F. H., H. J. Koepsell, E. S. Sharpe. 1952. A New Disaccharide Produced by *Leuconostoc mesenteroides*. J. Am. Chem. Soc., 74:3202.

Tsuchiya, H. M., N. N. Hellman, and H. J. Koepsell. 1953. Factors Affecting Molecular Weight of Enzymatically Synthesized Dextran. J. Am. Chem. Soc., 75:757-58.

-------, N. N. Hellman, H. J. Koepsell, J. Corman, C. S. Stringer, S. P. Rogonin, M. O. Bogard, G. Bryant, V. H. Feger, C. A. Hoffman, F. R. Senti, and R. W. Jackson. 1955. Factors Affecting Molecular Weight of Enzymatically Synthesized Dextran. J. Am. Chem. Soc., 77:2412-

-------, and C. S. Stringer. 1954. Transglycosylation Reactions of Enzyme Preparations from *Leuconostoc mesenteroides* NRRL B-512. Bact. Proc., p. 98 (P10).

-------, and C. S. Stringer. 1955. A Kinetic Study of the Dextransucrase Reaction. Bact. Proc., p. 126 (P49).

Wolff, Ivan A., C. L. Mehltretter, R. L. Mellies, P. R. Watson, B. T. Hofreiter, P. L. Patrick, and C. E. Rist. 1954. Production of Clinical Type Dextran-partial Hydrolytic Depolymerization and Fractionation of the Dextran from *Leuconostoc mesenteroides* Strain NRRL B-512. Ind. Eng. Chem., 46:271-377.

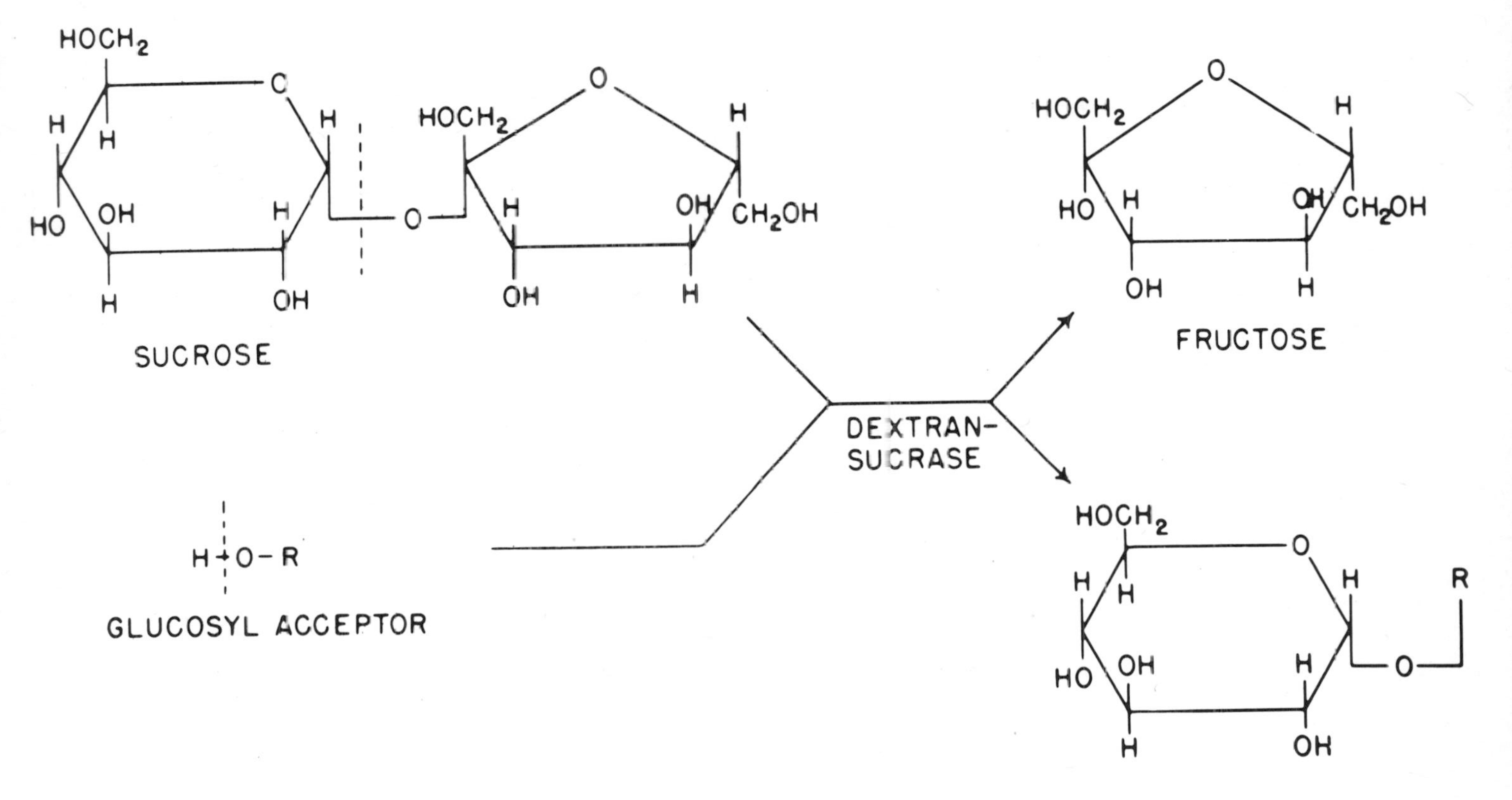

Fig. 1. Unit reaction of dextran synthesis

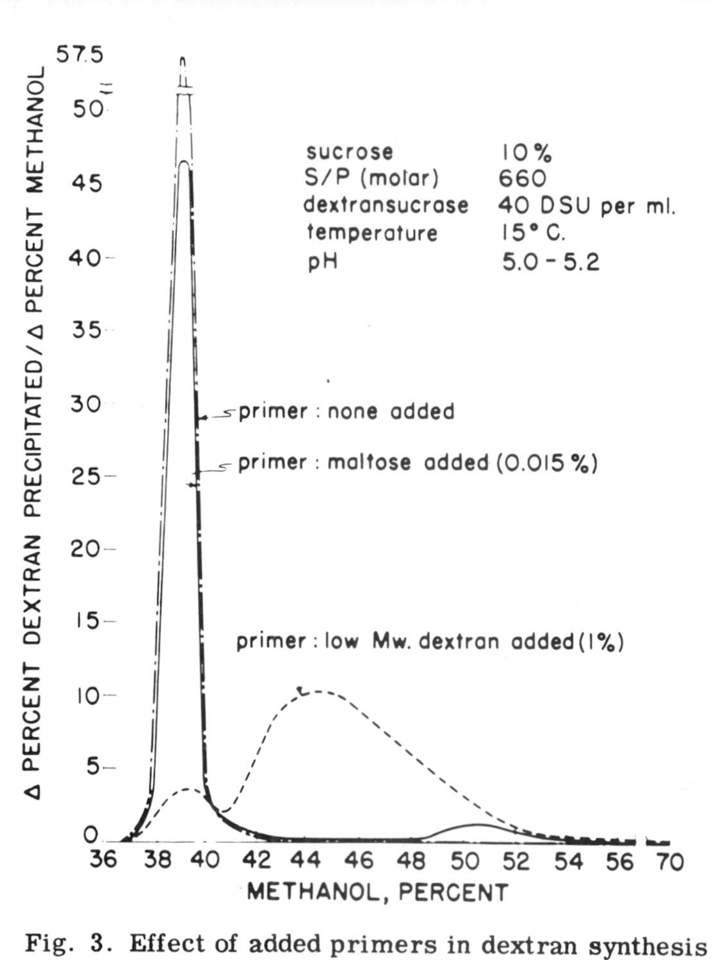

Fig. 2. Oligosaccharide synthesis in reaction with 70 per cent (w/v) sucrose

Fig. 3. Effect of added primers in dextran synthesis

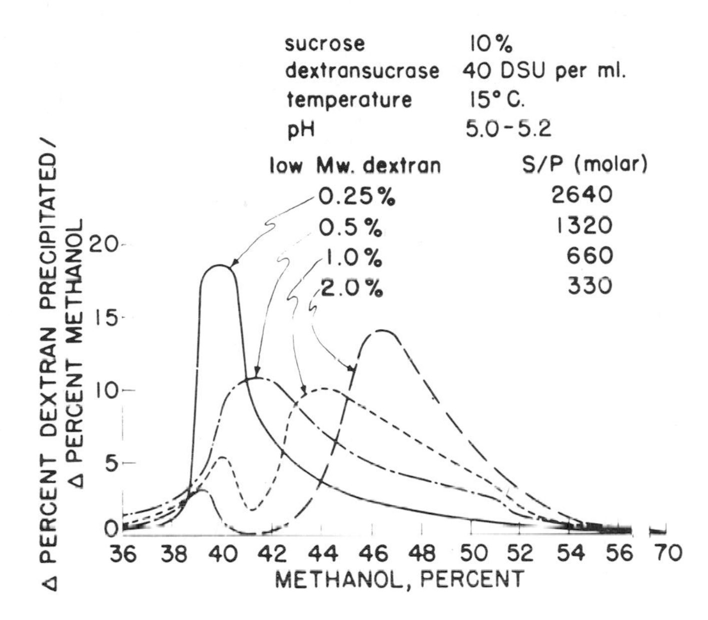

Fig. 4. Effect of primer (low Mw dextran) concentration in dextran synthesis

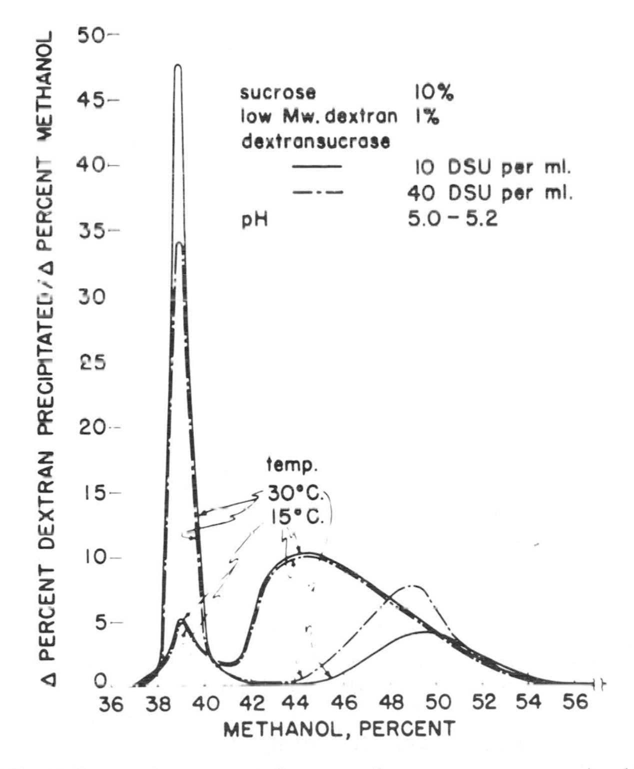

Fig. 5. Effect of temperature and enzyme concentration in primed dextran synthesis

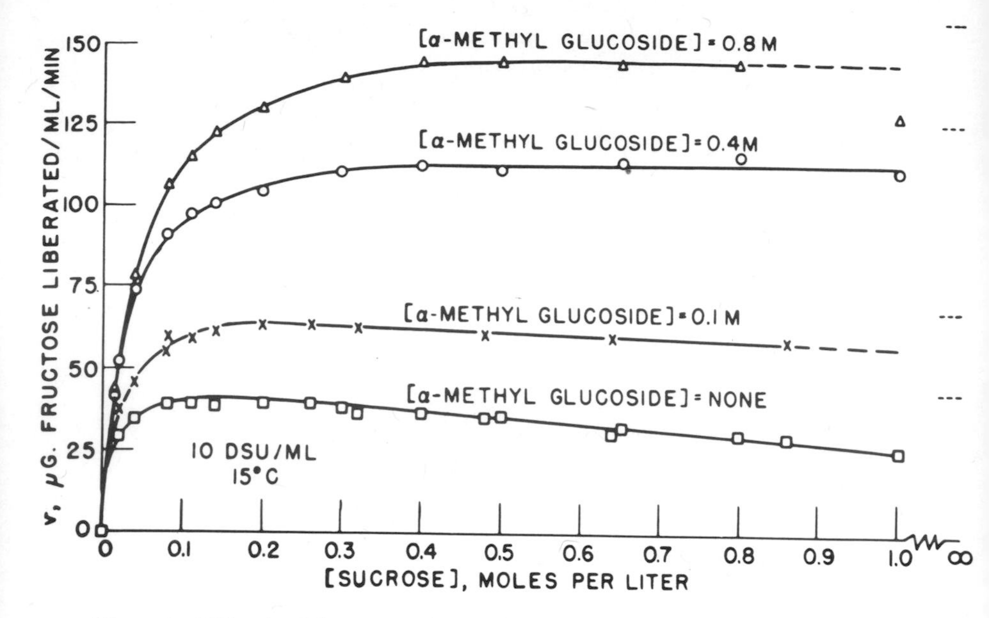

Fig. 6. Effect of [sucrose] and [α-methyl glucoside] upon initial rate of reaction (v)

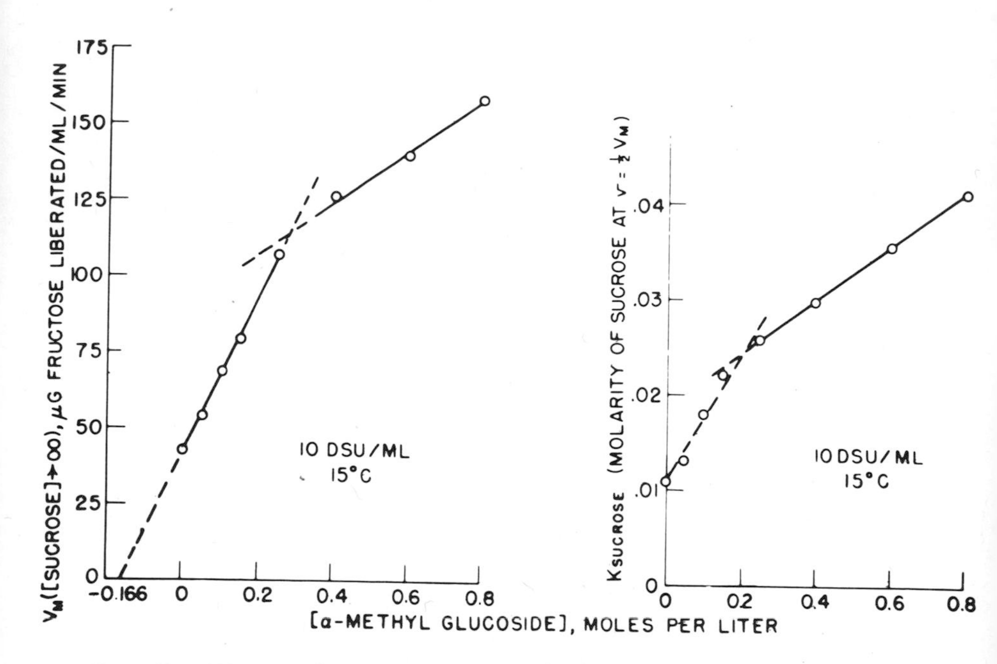

Fig. 7. Effect of [α-methyl glucoside] upon constants for the reaction step(s) involving sucrose

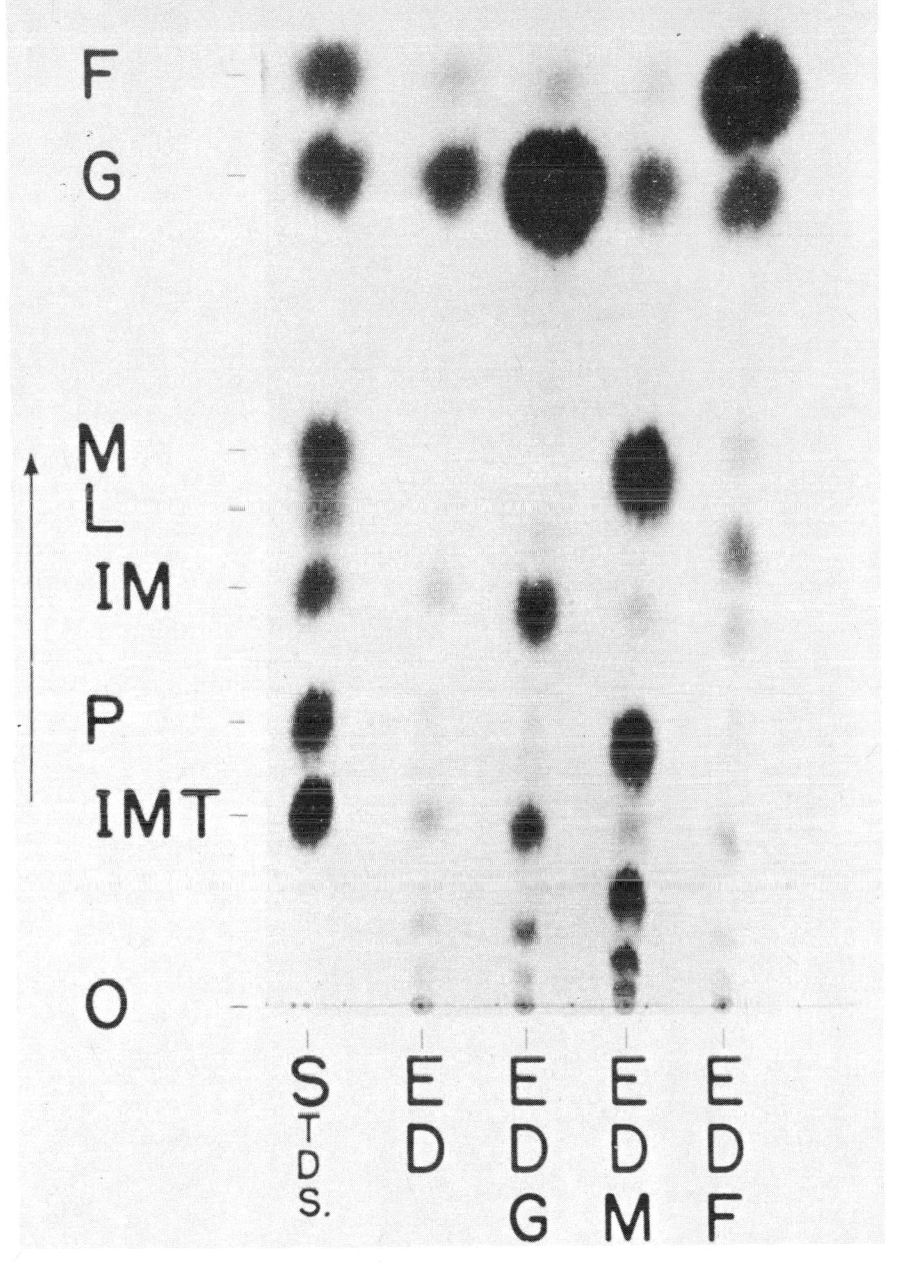

Fig. 8. Depolymerization of dextran. F (fructose), G (glucose), M (maltose), L (leucrose), IM (isomaltose), P (panose), IMT (isomaltotriose), Stds. (known standard sugars), E (enzyme), D (dextran)

Economic Implications and Evaluation

INTRODUCTION

Ralph H. Mann

At the outset I should like to express my deep sense of personal appreciation to those of you who have made this symposium in Marine Biology such a scientifically productive event as well as a very happy occasion. I desire especially to thank Dr. Ray for her long and untiring efforts together with the cooperation of the entire staff of the University's Zoology Department in organizing and planning this meeting and for their fine accomplishments in the rather intricate arrangements made for the comfort of all of us. My appreciation also is extended to Dr. Sidney Galler of the Office of Naval Research, through whose splendid cooperation and assistance this symposium was made practicable. Our thanks are rendered also to those twenty commercial interests who helped finance these meetings. I feel that it is indeed a good omen when hard-headed business people envision the economic benefits to be derived from your efforts in pure scientific research. My thanks, and I believe those of all of us here, are extended to all of these people I have mentioned who have done so much for our physical comfort and mental enjoyment here.

Today's subject is Economic Implications and Evaluations. I feel that this culminating session has been scheduled to bring out some of the real purposes of this symposium in the practical application of the results of your scientific studies. The many splendid papers presented here during the past few days are evidence that the problems of marine borers and fouling organisms are indeed joint ones for the pure scientist, for the engineer, for the producer of wood preservatives or antifouling compounds, as well as for the treaters of wood.

The proper solution to these various joint problems of ours are indeed of economic importance to our navy, which must erect, design, and maintain very large and extensive shore establishments scattered throughout the globe for the maintenance and servicing of our fleet.

This conference is unique in that it has highlighted and stressed the study of Crustacean borers of the genus Limnoria. I believe that this symposium has brought out much that is new in the marine biological field, the results of your basic research in pure science. Several of your papers have shown how marine fungi have abetted and accelerated the early attachment of Limnoria and Teredo larvae on untreated wood surfaces exposed to tidewater. Other interesting papers have shown us that Limnoria do indeed ingest the wood fibers and, through the action of enzymes within the digestive tract of the animal, enable it to digest and assimilate for body uses the wood itself as a food. This symposium, therefore, has produced a wealth of new scientific knowledge of marine wood-boring animals and fouling organisms.

Such knowledge is of interest and real importance to the wood-preserving industry, which I represent. One of the important fields of the wood-preserving industry is the treatment of piles and timbers for use in structures exposed to tidewater, in which marine environment it is usual for engineers to assume that marine boring animals of a variety of species are present.

For your general information I should like to tell you a little about the art of wood preservation and cite for you a few essential facts concerning the wood-preserving industry.

1. Wood preservation is an art. It is not a science. Your scientific help, therefore, is needed by us to provide a better treated product with which marine wooden water-front structures may be built.

2. There are in the United States (1956 statistics) 325 wood-preserving plants, of which over 250 plants are equipped to pressure treat piles and timbers.

3. The average quantity of timber products preservatively treated in the United States over the past 5 years averages 275, 000, 000 cubic feet per year. This figure is representative of a quantity approximating 10 per cent of the annual lumber cut for the whole United States. This industry does not, in fact, treat 10 per cent of all sawn lumber, but, when we include round piles, poles, posts, etc., the gross volume treated is approximately equivalent to 10 per cent of the United States annual cut of sawn lumber.

For the protection of piles or timbers exposed to marine borers in tidewater, the only preservative recommended for long periods of satisfactory service life is coal tar creosote or mixtures of creosote with coal tar. Such preservatives have been used for over 100 years. They have, up to now, proved the most effective preventatives or deterrents to attack of both molluscan and crustacean wood borers throughout the earth. For adequate treatment of wood materials for tidewater installations, pressure-impregnation processes with these preservatives are mandatory. The pressure vessels used by the wood-preserving industry for the impregnation of piles and timbers consist of heavy steel cylinders that range from 6 to 9 feet in diameter and are from 40 to 150 feet in length. The piles or timbers to be treated are placed on metal tramcars which run on standard gauge railway tracks in the bottom of the cylinder. These steel cylinders are heavy pressure vessels which must withstand pressures up to 200 psi, vacuum up to 26 inches of mercury, and temperatures as high as 260°F.

For superior pressure-treated timber materials, three basic requirements are essential. These are: (1) sound timber of a species that will "take treatment"; (2) choice of proper preservative for the exposure in which the wood is to be used; and (3) adequate absorption of preservatives to be injected. For marine use, treatments to refusal are recommended.

We have placed a section of well-treated Douglas fir pile here on the table for your inspection. It has been pressure-creosoted, and, as you see, deep and uniform penetrations of creosote have been obtained. Treated piles such as this should render long and effective service in marine-borer-infested waters.

With these introductory remarks and notes concerning the wood-preserving industry and its methods of treatment, it is now my pleasure to declare this session of the symposium open.

BIOLOGY AND COMMERCIAL TREATING PRACTICE

Reginald H. Colley

In the wood-preserving industry the words "gross absorption" are used to define the total amount of preservative that is absorbed by the wood up to the end of the pressure period, and the words "net retention" define the amount of preservative remaining in the wood at the end of the treatment process. Applying these terms figuratively, it must be admitted that over the years the gross absorption and net retention of biology by some highly important human biological factors--the suppliers and consumers of treated wood--have been way below the requirements for good performance. I shall try to show why this is so, with emphasis on the responsibility of the consumer.

I have taken a position something like that of an editorial writer in order to try to avoid entanglement in technical biology. I am leaving to others the privilege of including extensive reference citations. When I started to teach biology fifty years ago we found it convenient to divide the broad field into two parts, zoology and botany. We knew of course that there were "low" or intermediate forms, some of which now seem to be becoming more and more important in our studies of the initial stages of marine borer attack. The practical application of new biological information seems to have had more definite effect recently from a botanical point of view, i. e., on specifications for preserving utility poles, than it has on treating piles to resist marine borers. I shall begin with a revealing quotation about a most troublesome member of the latter group.

"This borer *(Limnoria lignorum)* is particularly insidious in its attacks upon creosoted piling because of its habit of creeping into small crevices. If cracks open through the saturated zone of a creosoted pile, or if bolt or dog holes or peavey holes or other breaks in the continuity of this protection occur, Limnoria is sure to find its way into the retreat thus offered and to begin its destructive work. It will reduce such a pile to a mere shell. It is sometimes found in such piles actually at work in the creosoted zone itself. Whether it does this because it slowly becomes acclimatized to the repellant substance or because of defective creosoting, low grade oil, or progressive leaching out of the toxic elements of the oil, is unknown. It is possible experimentally to acclimatize animals to live in toxic solutions by gradually increasing the dose, until they will thrive in solutions originally deadly to the species. Limnoria working near the creosoted zone might be thus acclimatized to the creosote. Since Limnoria may gnaw away the wood without eating it, it is also possible that it might for this reason be able to extend its destructive action into toxic territory for some distance. Ordinarily its digestive tract is found to be filled with the fragments of wood which seem to serve as its food."

The writer was Dr. Charles A. Kofoid. The occasion was the 1921 annual convention of the American Wood-Preservers' Association in San Francisco, at the St. Francis. The remarks were part of a report on the start of extensive studies under the direction of the San Francisco Bay Marine Piling Committee. The report and discussion run from pages 189 to 350 of the 1921 Proceedings. I suggest that it be placed on the required reading list.

Toward the end of his section on the "Biological Phase" Dr. Kofoid said (page 246): "To the end of perfecting means of protection, it is desirable therefore that the chemist, wood preserver, and biologist combine in an investigation to establish by experiment a method of treatment which shall be effectively repellant for as long a period as possible to the larvae of the marine borers."

We will come back to the longer quotation later. Dr. Kofoid prepared his manuscript almost 37 years ago. His text was good and the sermon was good, but the congregation did not have a sense of sin deep enough to compel devotion to the cooperative ideal that he suggested. I have an impression that, aside from the working group in the Marine Piling Committee, the association as a body did not appreciate the seriousness of the situation, although at that time about 8.5 million cubic feet of piles were being treated annually. This was approximately 5.8 per cent of all treated material, which is close to the present ratio. Then, as now, there was more biological information available than the producers and users of treated piles were willing or able to absorb.

In my opinion the most important human factor in the practical application of biology in the field of wood preservation is the customer. He is the one who buys the product and pays the bill. The use of the word "he" oversimplifies the situation. The customer is an economic concept--rarely a single person--and often an intricate complex, e.g., the L. and N. R. R. Company, the United States Navy, the Rural Electrification Administration, the X and Y Power and Light Company, or the Bell Telephone System.

When the customer really knows what he wants, either as a result of experience or research, or because he accepts the trusted advice of a competent consultant or of a reputable treater, he can usually get the quality of treated material he is willing to pay for. He can also take advantage of new biological information without having to wait for changes in "standard" specifications. Any successful manufacturing process implies some system of quality control to keep errors at a minimum and to prevent losses in processing, as well as to insure acceptable conformance to the manufacturing specifications. In a commercial wood-preserving plant the treater does a lot of quality control himself. At the same time he depends on additional quality control exercised by the consumer's inspector. For best results a high degree of compatibility between the producer and the customer's inspector, which really comes to mean a sort of cooperative way of life, is most desirable. Suppose we take as an example, in very condensed form, the development of new biological facts and their application to the preservative treatment of poles for Bell System use. This is obviously an example from the botanical side of biology.

It would be hard to pick a more complex customer entity than the Bell System, or one that functions more directly to use new and improved methods and products in the shortest practicable time, after they have been proved in. There is always a time lag. There are always traditions to be scrapped and personal habits and preferences to be adjusted. The adjustment period is generally longer than the enthusiastic promoter believes to be necessary. New ideas cannot be applied overnight.

We will have to go back some 33 years to 1924, to the beginning of a concerted effort to standardize the "white" poles. The first American Standard Specifications and Dimensions for Wood Poles were issued in 1933. The present standard, 05.1-

1948 was approved on April 9, 1948. The American Telephone and Telegraph Company was operating under its own material and dimension standards for poles before 1924, and so were numerous other public utilities. Operation under the new American Standards brought a majority opinion into general agreement for the first time with respect to the raw material that was to be treated. Some 125 to 140,000,000 poles have been procured under the basic principles of these pole standards.

Bell Telephone Laboratories, Inc., began paying special attention to the specifications for creosote and for creosote treatment of poles, particularly southern pine, in 1927, continuing the research and development effort that had been under way in the American Telephone and Telegraph Company. The date 1930 will serve as an approximate index of the time of the shift to the proposed new dimension classifications, to the change from the commercial 12 lb/cu ft "full cell" process of treatment to an 8 lb "empty cell" process, and to the enforcement of a quality control procedure of inspection to insure not only specified retention of a specified creosote but also a practical conformance to definite *penetration* requirements. The development work was based on fact, supported by a staggering mass of data from pole line service tests and from commercial scale experiments with full size pole charges at representative treating plants all over the southern pine area. The new A. T. and T. specifications were accepted by industry with some resistance and continuing argument, but on the whole with good grace. Actual procurement and inspection was handled by the Western Electric Company. The penetration requirements were not accepted by the American Wood-Preservers' Association until 1936.

Here then is a pattern. To work out the design took endless man hours. One must not overlook the actual power of this particular "customer" to demand and get a quality treated product, practically defined and practically produced by the treating industry. Obviously the force applied came from the consumer. Other consumer groups followed the general scheme placed into effect by the Bell System for the manufacture and for the pressure treatment of southern pine poles with creosote.

The investigation of laboratory methods for testing creosote and other preservatives at Bell Laboratories took a decisive turn in 1944, in the form of a cooperative agreement with the Forest Products Laboratory at Madison, Wisconsin, for a series of experiments with a new soil-block test. This study was supplemented by group subsidy from 1952 to 1956. The soil-block technique is now recognized as ASTM Designation D1413 56T, "Tentative Method of Testing Wood Preservatives by Laboratory Soil-Block Cultures."

Between 1944 and 1947 the work at Madison had developed the fact that certain wood-destroying fungi were more tolerant to creosote than they were to pentachlorophenol, and vice versa. Continuing culture work at the Forest Products Laboratory and at Bell Laboratories, summarized in publications by Catherine G. Duncan and L. R. Snoke, respectively, has shown that a combination of creosote and pentachlorophenol is more effective than either creosote or pentachlorophenol alone. The laboratory results were confirmed by stake and post tests and by checks on treated pole performance. Furthermore, the use of the combined preservatives offered great promise, because the required amount, 8 lb/cu ft, represented not only enhanced effectiveness but also general freedom from the bleeding that might result from higher retentions of either of the two preservatives used singly.

Commercial treatment of southern pine poles with creosote containing 2 per cent pentachlorophenol (w/v) as a substitute for straight creosote was started at one East Coast plant in September, 1953. The specifications called for 6 lb/cu ft only, on a gauge basis. The results were considered to be most satisfactory. However, since the start of this full size commercial experiment it has seemed logical and wise to increase the retention to 8 lb/cu ft. At present a limited production at this retention is going on at several eastern plants. It is understood that there are hopes of shifting all southern pine pole treatments to the creosote-pentachlorophenol solution by July 1, 1958. The story of this development is told by George Q. Lumsden in Wood Poles for Communication Lines, ASTM Bull. 222, May, 1957.

One most important feature in this example of the application of biological information about wood-destroying fungi in commercial treating practice must be emphasized at this point. The minimum penetration requirements remain as before, "not less than 2.5 inches unless at least 85 per cent of the sapwood is penetrated." The concentration of pentachlorophenol in the treating solution is to be checked periodically. However, retention is to be determined not from gauge readings but by actual toluene extraction of a specified sample of increment borer cores. The amount of *creosote* extracted from the *sample* "shall be not less than 6.0 lb/cu ft," which is considered to be the approximate equivalent of a true 8 lb/cu ft treatment by gauge. Note that this is an *assay of the treated pole* which is set up in such a way that the customer can see what he is paying for.

One more word before we leave this illustration: the picture is one of factual purpose working through a controlled and powerful system of related groups for their own best interests toward a set of practical results or end-use specifications. One can calculate a probable annual demand for the product. The poles are often made in quantity year after year by the same supplier. The supplier undertakes to furnish specified quality material. The emphasis is shifted from the age-old *caveat emptor* (let the buyer beware) to what may perhaps be considered as something very close to *caveat vendor* (let the seller beware). The poles are all branded with the supplier's name or symbol, the plant location, the pole size, the pole species, the preservative, and the year of treatment. It is easy to determine in the field whose poles show premature failure or over-all poor performance. Of course the whole scheme is designed to prevent the acceptance of substandard quality, but there is always some calculated risk.

Now I suggest a rereading of the longer quotation from Dr. Kofoid at the beginning of this article. Apparently the magnitude of the threat of destruction in his words about Limnoria was not realized to the full even by the experts--biologists, chemists, and wood preservers--on the Marine Piling Committee. Looking back on their findings as recorded in the final report published in 1927, it seems to me that rather too much emphasis was placed on the molluscan borers and not enough on the crustacean Limnoria. However, that is a matter of opinion. Dr. Kofoid certainly took pains to stress, in the discussion following the presentation of his report, "that it is a matter of considerable biological importance--to determine the repellent substance in the creosote that prevents the larvae from settling on the pile, and it will take the engineer and biologist combined to determine upon these points and bring them to a satisfactory conclusion." The only thing I can find wrong with his 1921 formula is that one most important factor, the customer, was left out.

Today we know a lot more about Limnoria. In my repeated conversations and correspondence with colleagues on both the East and West coasts I find their reactions running all the way from near panic at the thought of having to fight *Limnoria tri-*

punctata to a sort of defensive skepticism about whether three or four spots on the tail of any animal could make all that difference. I submit, without prejudice, that the fact of *L. tripunctata* is not yet quite established in enough consumer minds. However, it seems that Dr. Kofoid was more than right in saying that Limnoria could actually work in creosoted wood. He called the species *lignorum*. Did he have *tripunctata* at the time?

In spite of a lack of full appreciation of all the biology involved, it is now fairly well established that thorough treatment, i.e., maximum absorption and deep penetration with creosote, will stop Teredo, but that sometimes, and perhaps too often, creosote will not stop Limnoria. We are not yet conditioned to saying *Limnoria tripunctata* with the proper emphasis. Recent panel tests and observations on full size piles have indicated a definite superiority for solutions of creosote and coal tar, presumably against *L. tripunctata*, but the critical evidence to date is somewhat contradictory. Comprehensive experiments are under way with a view to providing a better technical base for judgments.

During the last five years our knowledge of the biology of marine boring and fouling organisms has been expanding rapidly. We are beginning to comprehend the importance of really knowing the weakest spot in the life cycle of *Limnoria tripunctata*, for example, so that effective control measures can be applied at the critical point. In this connection our improved information on its incidence, life habits, and physiology is most important. The recent work of Günther Becker and Samuel E. Meyers on marine fungi, which may serve to "condition" the outer layers of wood in sea water so that borer attack is facilitated, opens up the whole field in a relatively new direction. So also do the investigations of Luigi Provasoli on the rapidity with which marine microorganisms settle on submerged material, and on the ecological sequence of bacteria, minute algae, diatoms, hydroids, bryozoans, and mussels, for example, in connection with the food supply and possible vitamin requirements of Limnoria and of marine borer larvae.

Pin-pointing a preservative treatment for marine piling is at present an ideal, far removed from commercial practice. Still, there are degrees of application of an idea that can be realized at once, provided there is enough concentration of opinion among customer groups to permit the formulation of a rational demand and then to implement the demand with a set of practical quality control rules.

Before expanding this thesis it might be well to take a look at the similarities and differences in treating poles and in treating piles for marine use. In the last ten years the wood treated for both land and marine piles has fluctuated between a little over 10 million and about 17 million cubic feet. For the last three years, 1954, 1955, and 1956, for which data are available, the average was 14,365,900 cubic feet, or 5.69 per cent of the total wood treated. The figure for 1956 was about 16,845,000 cubic feet as a result of heavy land construction. There are no accurate estimates of the relative quantities of land and marine piles, but the latter must obviously be under the over-all total. The marine piling business is definitely limited, and one would not expect much opportunity for expansion unless the wood pile should be appreciated as a long-lasting quality unit of remarkable flexibility and general utility that can be purchased in the open market on a competitive base with steel or concrete piles.

Here is where the rub comes. The only customers who want to buy marine piles are people who must have semipermanent or permanent marine structures, groins, docks, dolphins, mooring basins, marinas, etc., either for pleasure or for business reasons. There are, of course, some steady customers for large or small lots. Some of these may be contractors.

The business of treating piling must be distributed among those wood-preserving plants that are located on or reasonably near the Atlantic, Gulf, or Pacific coasts. There are two major areas of raw timber supply, the Southeast and Gulf Coast southern pine forests, and the Douglas fir forests of the Northwest. The two types of timber are very different, not only in their characteristic sapwood depth, but also in the way they take treatment, particularly creosote and creosote-coal tar treatments. Some plants are more favorably located than others with respect to adequate timber sources for material that fits the piling customer's order.

Those last two words are key words. The treater cannot figure on recurrent demand from a large user, such as the annual pole requirements of the Bell Telephone System or the Rural Electrification Administration. He has not the remotest chance of increasing business by making up a lot of freshly treated piles and displaying them for sale in some hypothetical outdoor substitute for a display window. All he can do is maintain his reputation as a treater and wait for the customer's order. He is in effect a service man. He may rarely see the actual customer except in the form of an agent or a contractor. It must be assumed that the latter will buy the quality of material the customer has specified; but it can also be assumed that the contractor, who got his contract at least partially on a price basis, will try to get his piling from the lowest bidder. This is life, fun, business, or cut-throat competition, depending on the point of view.

Sometimes the small consumer has an advantage over the big consumer. He does not have as many people to see, and he can often act more independently. For example, the Los Angeles Harbor Board can be counseled to order special treatment for marine piling, and the order can be executed, all without agreement or consent of a corresponding authority in San Francisco. A "private" consumer can order, within reason, any treatment that the treater is equipped to perform. However, small orders that might represent much less than a full cylinder charge, and too many independent ideas and too many experimental recipes for commercial preservation create almost insurmountable difficulties for the producer. His machinery is cumbersome. Pipelines, pumps, treating cylinders, and storage tanks for preservative are massive pieces of equipment. He is generally willing to cooperate. He can modify retention of a standard preservative--creosote, for example--up or down to meet treating specifications, but he is drastically limited in the number of preservatives he can handle. The customer should understand the situation and be as sure as is humanly possible that he knows what he wants.

Going back to a question of practical biology, suppose we assume that the customer is aware of the fact that the piling he orders is to be used in southern waters that are normally heavily infested with *Limnoria tripunctata*. Perhaps that is all the general biology he needs to know. However, he may have an idea that the outer fibers of the wood should be saturated with preservative and that the surface of the treated piles should be as repugnant as possible to Limnoria and to the larvae of Teredo, Bankia, etc. At present he has his choice of two preservatives, straight coal tar creosote and creosote-coal tar solution. He decides on the latter.

He does not have to wait for the ideal perfect specification. He can use American Wood-Preservers' Association Standard P2-57 as a guide. Providing he can sell his monetary boss on his idea, which somehow is sometimes quite difficult, he starts negotiating for treatment to refusal with a creosote-coal tar solution, specifying complete sapwood penetration by a "full cell" process, final vacuum omitted so that there will be no purposeful depletion of the absorbed oil from the outer fibers of the wood. He does not care how much the piles bleed. He wants the pile surfaces to be coated with the heavier fractions of the creosote-coal tar treating solution. His contractor has "educated" the construction crew about tarry piles and why they are necessary. The net retention is specified in terms of lb/cu ft of creosote-coal tar solution per cubic foot of wood treated. A reputable supplier agrees to produce the piles in accordance with the order. If the whole transaction is not too much constrained by tradition, the customer and the treater may agree that retention shall be determined by extraction assay of the *treated sapwood,* as in the example cited above for treated southern pine poles. This last step would enable the customer to measure for himself the quality of the finished product.

The course of action suggested is open to any customer who is willing to pay the price. All the steps outlined above can be taken immediately. They would constitute practical enlightened application of modern biological information about *Limnoria tripunctata* to the treatment of piles for use in waters infested by this borer. In situations where *L. tripunctata* is not a hazard the procedures outlined could be followed with straight coal tar creosote.

There is a continuing search for possible new preservatives and new preservative processes, but until the new things are proved-in there is no reason why full advantage should not be taken of maximum penetration and retention of coal tar creosote or of creosote-coal tar solution. One cannot emphasize too strongly that a new preservative by itself will not solve the problem. It would still be necessary to define quality by strict specifications, and there would still have to be quality control procedures worked out cooperatively by customer and treater interests.

Under any reasonable system of procurement the treater's responsibility in any lot of treated piling should cease when the material becomes the property of the consumer. This of course assumes that the piles meet specifications and that the machinery of inspection and acceptance has been functioning satisfactorily. No matter what preservative is used now or in the future, the customer or his contractor must hire competent construction crews who have been trained to respect a good piece of well-treated wood. Wood butchers should be automatically excluded. The premature failure of piling that results from the mishandling of the product by the customer's agent cannot justly be blamed on the treater.

A customer is sometimes led to think in terms of a temporary structure or to underestimate the years of service that may under changed conditions be required of a given marine installation. In such circumstances he may take the risk of probable early failure, try to save money, and specify low retentions of creosote, say 10-12 lb/cu ft for southern pine piling, even where severe borer conditions prevail. Few treaters are in a position either to refuse to bid or to turn down the order if successful. The customer must bear the major share of the blame in such instances if the performance of the piling is poor. Ordering low retention is indeed retrogression from the old practice of loading longleaf and slash pine timber, particularly if it was seasoned, with 24 plus lb/cu ft. In many of these old piles the sapwood was relatively thin, and treatment resulted in essentially complete sapwood saturation. The present AWPA minimum specification of 20 lbs is a compromise, partly the result of economic forces and partly due to the difficulty of getting much more than this minimum into green timber. However, it is economic folly to try to get along with sub-substandard creosoted piles. That is almost like inviting *Limnoria tripunctata* to a submarine cocktail party.

Concerted effort is now being made to develop a definite specification for a creosote-coal tar solution for the preservation of marine piling. The aim is to define a solution that will combine the good qualities of the two components without increasing penetration difficulties. Such a solution would be recommended for universal use in treatments to refusal when the marine borer incidence indicated "severe service conditions." It should be mentioned in passing that the present AWPA Standard P2-57 carries a footnote to the effect that creosote-coal tar solutions for marine use shall be true mixtures in the approximate proportions indicated. The solutions are permitted on an either/or basis in treatment Standard C3-57 for southern pine for use in coastal waters, but they are "not recommended" for Douglas fir. The treatment of the latter with creosote-coal tar solutions involves certain processing difficulties that must be overcome by further study under commercial conditions.

It is perfectly evident that there may be a long time between the development of a "fact" and its successful commercial application. New little facts, as I like to call them, are continually bobbing up to modify our opinions. Standard specifications are generally compromises, based on weight of evidence and consensus in the points of view and several interests represented on the responsible committees. Changing a standard is far from easy. Bernard M. Baruch in his book My Own Story reviews his early experience in getting the facts before taking action. Facts meant to him those logical conclusions that he could form after a thorough study of the evidence. He points out, however, that to be evident to all a danger must be hanging right over our heads. Furthermore, he voices the opinion that, if action is delayed until the need is apparent to everyone, it will be too late. Of course he was acting on his own. He did not have to deal with a balanced membership of producers, consumers, and the general public.

Discussion of the present status of evaluation tests for screening out or proving in preservatives for marine piling has been purposely omitted from this paper. However, in my opinion we have not yet developed an accelerated laboratory test that is as natural and discriminating, with reproducible results, as the soil-block test mentioned earlier. Again expressing my personal ideas, I feel very strongly that the most constructive course we could take would be to set up, somehow, a competent small group to review, define, and evaluate the evaluation tests now being employed by various investigators.

I have referred to the work of the San Francisco Bay Marine Piling Committee. Present opinion among those who were closer to that work than I seems to be that the efforts of the committee were beginning to be effective while it was in action; that its work was purely educational; that it did an excellent job but had to quit too soon; that the associated railroad interests probably derived the greatest practical benefits, which they certainly earned and deserved; and that the over-all effect, if any, on commercial treating specifications has been very slight indeed. However, the committee did bring together the scientists and the engineers so that both groups were better informed and began to appreciate each others' problems and points of view. Half methods were shown to be economic delusions. The value of good treatment with creosote was demonstrated, and some of its limitations in controlling Limnoria were beginning to be recognized. Poor treatment and improper handling were exposed for all to see. A qualified correspondent has summed up the situation as follows: "The great shortcoming of most efforts of this kind, including symposiums, is that they usually end in talk or in a published report but do not produce much action. Intelligent discussion is good and the publication of good technical reports is valuable, but both amount to little in the long run unless progress is made in improving treatment, construction, and maintenance practice."

The Forest Products Journal for July, 1957, contains reports and discussions on the problem of decay and termites in building construction. Reading between the lines one gets the impression that nearly everyone felt that the home or building *owner* needed more education in the value of wood preservation and on the use of treated wood. In the case of marine piling treatments the treater should be in a position to supply the latest information. If he attempts to act as an instructor he must be sure of his ground, and that is no mean responsibility.

The old saying still holds. He who pays the fiddler has a right to call the tune. It is apparent that we must do all we can to be sure that the customer knows what to ask for, that means are provided for determining the true quality of the product, and that the cost of quality is justified. It is as simple as that.

TESTING IN THE FIELD OF TIMBER PRESERVATION AGAINST MARINE BORERS IN GERMANY

Günther Becker

Former Experiments and Experience in Germany

Compared with the immense testing program on marine borers which has been carried out in the United States, only a very few recent investigations can be reported from Germany. Because of the long coast lines of the States and the attendent heavy destruction of timber, especially in the southern areas, wood preservation against marine organisms is of much greater economic importance than it is in Germany.

About 20 to 50 years ago different methods of protecting timber in sea water were followed by many tests under practical conditions in Wilhelmshaven and Cuxhaven and other places along the coasts of the North Sea. The result was that creosote showed the best results when injected by pressure sufficient to yield retentions of 180...200 kg/m^3 timber. *Pinus silvestris* L., the common pine species of central, eastern, and northern Europe, proved to be the best indigenous timber species because of its growth and strength properties, available dimensions, and most especially for the good treatability of its sapwood. Several tropical timber species do well for marine structures in cooler regions, even though most of them are destroyed in their original country after a short time. Unfortunately, timber species with high natural resistance against borers are too costly for normal use. (See summarizing papers by F. Roch, 1927, and Windolf, 1936).

Heavy deterioration of marine structures due to *Teredo navalis* during the years 1933 to 1936 drew new attention to the marine borers. But some tests of newly developed preservatives carried out with panels in harbors were disappointing; storms took away many samples, as frequently happens with such tests.

Therefore laboratory tests were started in Berlin-Dahlem in 1937, following a suggestion of the late W. F. Clapp in his paper of 1937 in which he summarized the results of his long and detailed experience with marine borers.

A Laboratory Testing Method with Limnoria

We started our tests with Limnoria because breeding this organism in the laboratory seemed to be easier than breeding Teredo. Moreover we had already observed during our first inspections of attacked timber in 1937 and 1938 that *Limnoria lignorum* attacks creosoted timber sooner than *Teredo navalis* (Becker, 1944).

A Limnoria species from the Mediterranean, which later on proved to be the ubiquitous species called *L. tripunctata* by R. J. Menzies, was selected as a suitable testing organism; it could easily be bred in the laboratory at temperatures of 20°C or more (up to 27...28°C) for many generations (Becker, 1958). The tests were carried out in a 20°C air-conditioned room with artificial sea water at a salinity of 35‰ and a pH of about 7. Ten medium-sized to full-grown Limnoria specimens were transferred from infested timber to the test blocks. Careful cutting out and handling of the gribbles is essential, and the animals are used for tests only if they are quite well after a few days' observation. Of course, treated timber blocks can be exposed to attack in larger aquaria containing many Limnoria cultures. But, in addition to the endangering of numerous animals by the toxic material, it seemed better in any case to use a kind of compulsory test by transfer of counted animals, as is done in several testing methods with insects. The test has been described in detail by Becker and Schulze (1950).

The test blocks were treated with preservatives by low pressure and then kept for four weeks at 20°C and 65 per cent relative air humidity. In the case of water-soluble preservatives which are supposed to be fixed by chemical transformation, the wood samples were stored under moist conditions in glass containers. Thereafter the test blocks were leached for 4 weeks in distilled or tap water and for a further 8 weeks in artificial sea water. The water was changed daily during the first 4 weeks, and weekly later on.

At first we used test blocks of pine sapwood (*Pinus silvestris* L.) 5 cm x 2.5 cm x 1.5 cm, the normal dimensions for several German standard tests with fungi, wood-destroying beetles, and termites, and for leaching and evaportation tests. These blocks were believed to be a convenient size also for laboratory tests with Teredo. Each treated or untreated test block, with 10 Limnoria specimens, was placed in a small glass container 25 cm x 15 cm x 15 cm, containing 3,000 cm^3 water. This was aerated constantly. Recently W.-D. Kampf developed a technique with very small test pieces kept in Petri dishes in sea water without aeration. The water must be changed at least twice a week, or sometimes even more frequently. When this is used, observation is easier, there is less interference with the gribbles, and the animals are able to find the wood more readily than in a large aquarium.

The results can be compared by the degree of attack, the amount of feces, and the average and longest time of survival of the gribbles. In such a laboratory test a good preservative should prevent any deterioration of the test blocks by Limnoria.

Results of Tests in the Laboratory with Limnoria

Different fractions of creosote of the following boiling ranges were tested: 130...140°, 140...150°, 150...160°, 160...170°, 170...180°, 180...200°, 200...220°, 220...240°, 240...260°, 260...280°, 280...300°, 300...330°, 330...360°C, and the residue (Becker and Schulze, 1950). The test blocks were impregnated by low pressure with the preparations, diluted with chloroform to a 10 per cent solution (v/v). The average retention was 60 kg/m^3 timber. Limnoria tests were started immediately after the 12 weeks' leaching process.

The fractions up to 240°C failed to protect the test blocks. Probably the substances, part of which are known to be very toxic to Limnoria (Shackell, 1923; White, 1929), evaporate quickly. The preparations between 150° and 170°C seemed to be lowest in toxicity after storage and leaching. For the higher boiling fractions, lowest effectiveness toward gribbles was found with the preparation 300...330°, while the fraction 280...300° showed highest toxicity. Also the 330...360° fraction and the residue yielded promising results. It seems that α- and β-naphthol and diphenylenoxide are effective components, and acridine, anthracene, and pyrene may be worth more detailed investigations.

Probably the protective action of creosote depends not only upon a few ingredients of high toxicity, but upon the combined effect of a large number of chemicals with different properties working together and resulting in effective retention and long service time of the whole preparation (Becker, 1949).

Some preliminary tests were later carried out with inorganic salts of the following composition:

$K_2Cr_2O_7(44) + Na_2HASO\ (56)$,

$K_2Cr_2O_7(40) + Na_2HASO\ (30) + NaF\ (30)$,

$K_2Cr_2O_7(18.5) + AnCl_2(81.5)$,

$K_2Cr_2O_7(50) + CuSO_4(50)$.

After treatment, test specimens were kept under moist conditions for 4 weeks in order to provide fixation of the salts. Thereafter the blocks were dried and exposed to exhaustive leaching. The first three mixtures in concentration up to about 30 kg/m^3 timber, failed in tests with Limnoria, as could be expected from practical experience. The copper-containing mixture, however, yielded promising results with retentions of about 7 kg/m^3 timber (Becker, 1955). The copper ion seems to be effective on some enzyme system of Limnoria (Kampf, unpublished). A high efficacy of unleachable copper-containing preservatives against marine borers has recently been claimed by other investigators, testing with copper formate in the United States (Smith *et al.*, 1956), and with "Ascu" in India (Purushotham, 1955).

Since synthetic contact insecticides are known to be effective against crustaceans, the following were investigated in laboratory tests with *L. tripunctata* (Becker, 1955):

p-, p'-dichloro-diphenyl-trichlorethane (=DDT),
α-, β-, γ-, and δ-isomers of benzene hexachloride,
a Chlordane preparation,
Aldrin,
Dieldrin,
Toxaphene.

The preparations were diluted with chloroform (1.0 per cent and 0.1 per cent v/v) and injected into the test blocks by low pressure. The treated samples were stored in gradually opened containers for 4 weeks in rooms at about 20°C and thereafter for 8 weeks in sea water changed 5 times a week.

The α- and β-isomer of the benzene hexachloride totally failed even when 8 kg/m^3 timber were applied. The other insecticides showed different effectiveness toward Limnoria more or less comparable with their toxicity against wood-destroying insects (Becker, 1953). By far the most effective substance was γ-benzene hexachloride. A retention of 0.9 kg/m^3 timber killed Limnoria in the tests (after leaching as mentioned above) within 1 to 4, on an average 1.6, days. To the next groups, according to their toxicity, belonged Toxaphene, Aldrin, and Dieldrin. DDT and Chlordane affected the animals more slowly but protected the test blocks sufficiently, while δ-benzene hexachloride was effective enough to kill Limnoria with a retention of 8 kg/m^3 timber, but not with a retention of 1 kg. More tests are necessary to determine the duration of the protective action of those contact insecticides that yielded promising results.

Practical Tests of Preservatives

A Working Committee for the Control of Teredo of the Deutscher Küstenausschuss Nord- und Ostsee investigated protective measures under natural conditions. During recent years its members have carried out a series of tests of preservatives in the sea. Logs and beams 3 m long were treated under practical conditions, arranged in rafts, and exposed. These were regularly examined at several locations at the German coasts on the North Sea and the Baltic Sea, and at the southern mouth of the lagoon of Venice near Chioggia. The Mediterranean tests yielded results sooner than those in the North and Baltic seas.

Here again only creosote, applied in large quantities, rendered sufficient protection, while several recently developed preservatives failed. Preparations containing synthetic contact insecticides yielded satisfying results against Limnoria and simultaneously against Chelura but failed entirely with Teredinids. These results correspond to those of Edmondson (1953) in Hawaii. Certain water-soluble preservatives failed also even when applied together with synthetic resins, which were added to reduce the leachability.

In the future, practical tests with new preservatives shall be carried out only if they first yield promising results in laboratory tests. For open sea tests, large dimensions of timber such as are used in commercial practice seem to be advisable, so that the technical problems or preservative application will correspond as nearly as possible to the conditions in actual practice.

Possibilities for Future Development

There are a number of possibilities for improving the creosote with reference to its effectiveness against Limnoria. One may consider the addition of copper in a suitable form, bearing in mind, however, the former failures (MacLean, 1954) and the possibility of admixture of synthetic contact insecticides. According to present experience, the latter retain their effectiveness in wood relatively long and should therefore be tested for the purpose of marine timber preservation.

Perhaps in the future there will be successful development of water-soluble preservatives of such a high resistance to leaching that they can be considered for the protection of timber in sea water. Should this come about, it would be desirable to determine whether, for wood species that cannot be sufficiently treated by pressure methods, the water-soluble preservatives could be successfully applied by the aid of the diffusion process or Boucherie process.

The knowledge that Limnoria depends on the presence of certain wood-inhabiting fungi for its nutrition (Becker, Kampf, and Kohlmeyer, 1957), and that the Teredinids probably need them for their settlement and metamorphosis (Becker, 1958), also leads to new avenues for the protection of wood against marine borers. Here a new development of possibilities of wood protection should begin. In the future one will have to distinguish between wood preservatives whose toxicity prevents the settlement of marine borers and others that protect the wood against marine borers by inhibiting fungal growth.

Some of the main possibilities illustrating how wood preservatives can affect marine borers are arranged in table 1.

When testing new preparations, in Germany as elsewhere, special attention should be given to the time of effectiveness in sea water. Besides accelerated leaching tests in the case of organic preparations, special climatic testing by using high temperature in wind channels should also be included.

It goes beyond the topic of this brief survey to report on investigations concerning an improved pressure treatment of spruce (*Picea excelsa* Link) and fir (*Abies alba* Miller = *pectinata* DC.), a problem that may be compared with the impregnation of Douglas fir in North America.

TABLE 1
EXAMPLES FOR POSSIBILITIES AND PRECONDITIONS OF PROTECTIVE EFFECT OF WOOD PRESERVATIVES AGAINST MARINE BORERS

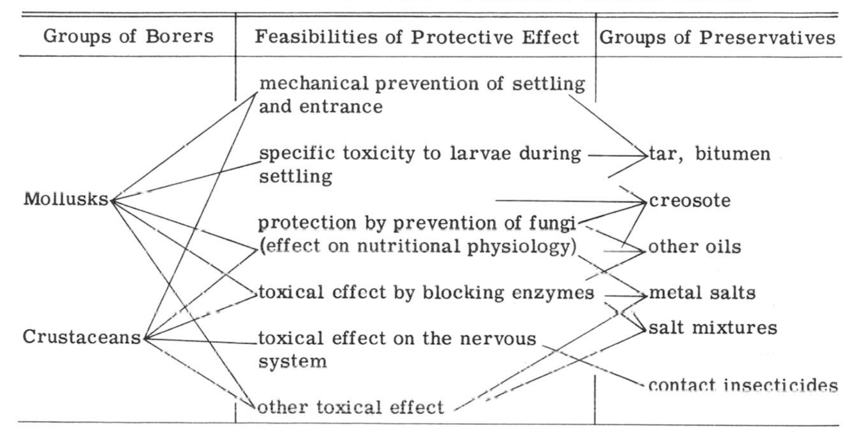

LITERATURE CITED

Becker, G. 1944. Holzschutzaufgaben gegen Meerwasser-Schädlinge. Z. hyg. Zool. u. Schädlingsbek., 36:51-67.

-------. 1949. Der Wert von Steinkohlenteeröl-Bestandteilen für den Holzschutz. Mitt. dtsch. Ges. f. Holzforschung, 37: 181-201.

-------. 1953. Untersuchungen über die Eignung von Kontaktinsektiziden für den Holzschutz. Holz als Roh- und Werkstoff, 11:45-50.

-------. 1955. Ueber die Giftwirkung von anorganischen Salzen, α-Chlornaphthalin und Kontaktinsektiziden auf die Bohrassel Limnoria. Holz als Roh- und Werkstoff, 13:457-61.

-------. 1959. Biological Investigations on Marine Borers in Berlin-Dahlem. In this volume.

-------, W. -D. Kampf, and J. Kohlmeyer. 1957. Zur Ernährung der Holzbohrasseln der Gattung Limnoria (Isopoda). Die Naturwiss., 44:473-74.

-------, and B. Schulze. 1950. Laboratoriumsprüfung von Holzschutzmitteln gegen Meerwasserschädlinge. Wiss. Abh. dtsch. Materialprüfungsanst. II, 7:76-83.

Clapp, W. F. 1937. The Activities of Marine Borers. Civ. Eng., 7:105-8.

Edmondson, C. H. 1953. Response of Marine Borers to Chemically Treated Woods and Other Products. Occas. Papers, Bernice P. Bishop Mus., Honolulu, 21:87-133.

MacLean, J. D. 1954. Results of Experiments on the Effectiveness of Various Preservatives in Protecting Wood against Marine Borer Attack. U.S. For. Prod. Lab., Madison, Rept. D1773.

Purushotham, A. 1955. A Short Account of Work Done on the Protection of Timber Against Marine Organisms at Each Centre. J. Timb. Dryers' and Pres. Assoc. (India), 1:5-8.

Roch, F. 1927. Die Holz- und Steinschädlinge der Meeresküsten und ihre Bekämpfung (Die Verhältnisse an den deutschen Küsten der Nordsee und westlich der Ostsee sowie in den früheren deutschen Kolonien). Geb. d. Medizinalverw., 24: 1-78.

Shackell, L. F. 1923. The Toxicities of Coal Tar Creosote, Creosote Distillates and Individual Constituents for the Marine Wood Borer *Limnoria lignorum*. Bull. U.S. Fish. Comm., Washington, 39:221-30.

Smith, F. G. W., R. R. Bottoms, E. Abrams, and S. M. Miller. 1956. A New Method for Long Term Preservation of Wood by Chemical Modification. For. Prod. J., 6:340-45.

Wakeman, C. M. 1947. Discussion of Recent Laboratory Biological Survey Methods for Comparing Various Toxicants Used to Combat Marine Borers. Proc. Am. Wood Pres. Assoc., 43:220-21.

White, F. D. 1929. Studies of Marine Wood-borers, I. The Toxicity of Various Substances on *Limnoria lignorum*. Contrib. Can. Biol. and Fish., 4:3-18.

Windolf. 1936. Der Bohrwurmbefall an Holzbauten in den Cuxhavener Häfen. Werft, Reed. Hafen, p. 404.

PRESERVATION OF DOUGLAS FIR MARINE PILING

R. D. Graham

A controversy has existed for some years concerning treatment of Douglas fir marine piling with creosote-coal tar solutions. Some contend that creosote-coal tar solutions are superior to straight creosote and should be used, while others question the treating results that can be obtained with creosote-coal tar solutions. These different views have resulted in conflicting treating standards. In the American Wood Preservers' Manual of Recommended Practices, for example, creosote-coal tar solutions are not recommended for treatment of Douglas fir piling for coastal waters in Standard C3, but a creosote-coal tar solution retention of 14 lb/cu ft is recommended in Standard C18. This paper will review some of the facts behind this controversy, describe tests now under way to obtain more information on the subject, and discuss important factors that must be considered in treating Douglas fir marine piling.

Marine Exposure Tests

Extensive tests of preservatives in marine waters were begun as early as 1911 by the U.S. Forest Products Laboratory, which also cooperated with the San Francisco Bay Marine Piling Committee (1927) in exposure tests. Results of these tests were summarized by MacLean (1950). He concluded that high-boiling creosotes and creosote-coal tar solutions were the most effective of the many preservatives tested, and that best treatments were obtained with heavy retentions and deep penetration. He also noted that the first point of borer attack was found commonly where penetration was shallowest. Similar conclusions had been reached by the San Francisco Bay Marine Piling Committee in 1927.

In more recent tests, Richards (1957) found a 70-30 creosote-coal tar solution more effective than the same creosote alone, although differences were small at one test site and greater at a second.

Tests by the Battelle Memorial Institute at Daytona Beach, Florida, now in their eighth year, indicate little difference in effectiveness of creosote and creosote-coal tar solutions with a tar content below 40 per cent. Close correlation is being found between behavior of full-size piling and small pine sapwood panels (3/4x3x18 inches) treated at the same time.

This evidence indicates strongly that coal tar in combination with creosote can give better protection against marine borers than can creosote alone, provided adequate quantities of tar are used. A tar content of 30 per cent can be considered the minimum necessary to effect an improvement over creosote alone in southern pine marine piling.

Treatment with Creosote and Tar Solutions

Controversies concerning use of tar in creosote are by no means of recent origin. A similar controversy occurred shortly after 1900 when addition of tar to creosote was a not uncommon practice. Von Schrenk and Kammerer (1914) discussed the propriety of this practice and voiced their opinion that tar should not be added to Grade 1 creosote because it could only detract from effectiveness of this superior creosote. On the other hand, they condoned addition of small amounts of tar to low grades of creosote because they believed that it helped reduce evaporation losses and, by so doing, enhanced their preservative value. They recommended that only low-carbon-content tars (less than 6 per cent) be used. They also noted that tar could not be separated from solution by mechanical or chemical means, and that long pressure periods were required with tar solutions to obtain penetrations similar to those with straight creosote.

Schmitz (1933) reported that coal tar did decrease toxicity of the highly toxic creosotes to fungi, but not of creosotes whose toxicities were similar to that of the tar itself. It is possible that addition of coal tar to creosote may effect an improvement in creosotes that otherwise are least effective against marine borers.

Early investigations by Teesdale and MacLean (1918) indicated that, for a given tar and creosote mixture, the three important factors influencing penetration were nature of the bitumen, percentage of free carbon in the tar, and condition and size of the free carbon particles. They also found penetration to be less uniform with those mixtures containing the higher percentages of tar and recommended that pressure periods be as long as possible and temperatures as high as possible. Subsequent studies by MacLean (1926, 1927) focused attention on the importance of the treating temperature and its influence on viscosity of the preservative. MacLean considered high temperatures of great importance to maintenance of low viscosities when refractory woods were treated.

Van Groenou, Rischen, and Van den Berge (1951) reported that the main drawback to creosote and tar mixtures was high viscosity caused by the tar and its high free carbon content.

In recent studies by Graham and Miller (1957), in which end-matched pieces of Douglas fir at similar moisture contents were treated under similar conditions, penetrations of creosote were twice as great as, and retentions 6 lb/cu ft higher than, those obtained with a 70-30 creosote-coal tar solution. Some borings of creosote-coal-tar-treated specimens had penetrations as low as 0.1 and 0.3 inch; the shallowest penetration in creosote-treated specimens was 0.9 inch.

Evidence to date shows that creosote-coal tar solutions are more difficult to force into wood than is straight creosote. This finding means that, when Douglas fir piling is treated with tar solutions, great care will be required if adequate retentions and uniform penetrations are to be obtained. Thorough inspection of every treated marine pile should be mandatory.

Exposure tests by the Forest Products Research Center

In most marine exposure tests to date, southern pine sapwood has been used for test panels. MacLean (1950) tested a few sawed and round Douglas fir test specimens on which he commented as follows: "Notwithstanding the much thinner sapwood and lower absorptions in Douglas fir specimens (10-13 pounds per cubic foot), the resistance of this wood to marine borer attack compares well with that obtained with southern pine (18-26 pounds per cubic foot)."

MacLean reasoned that, since Douglas fir sapwood was more difficult to penetrate than was pine sapwood, it might also be more resistant to leaching and washing, which would help to account for its good performance.

The present program was initiated to provide more extensive information on performance of Douglas fir sapwood and heartwood panels (3/4 x 3 x 18 inches) treated with creosote and a 70-30 creosote-coal tar solution. It was expanded to include southern pine sapwood treated with these two preservatives, and Douglas fir sapwood treated with copper formate and ammoniacal copper arsenite (Chemonite). A total of 900 panels were installed at 3 sites on both the Atlantic and Pacific coasts. Preservative retentions are shown by exposure site and exposure time in table 1. Specimens mounted on metal bars as shown in figure 1 were suspended in salt water below low tide. One complete set of replicates (5 panels in each replicate) will be removed and inspected every 2 years over a 10-year period.

TABLE 1
SUMMARY OF PRESERVATIVE RETENTIONS
BY EXPOSURE SITE

Exposure site	Type of wood and preservative*							
	DC	DT†	HC	HT	PC	PT	DA‡	DF‡
	Pounds per cubic foot							
Los Angeles, Calif.	34.4	33.9	38.0	32.8	...	...	0.39	0.27
San Francisco, Calif.	35.1	34.1	37.3	32.5	...	...	.39	.28
Friday Harbor, Wash.	34.2	33.5	37.1	33.2	...	...	.38	.29
Virginia Keys, Fla.	36.2	35.0	37.8	33.5	...	...	.38	.28
Daytona Beach, Fla.	35.8	35.4	...	...	33.9	33.9	.38	.28
Boston Harbor, Mass.	36.2	34.3	...	...	34.8	33.8	.38	.27
Average	35.3	34.4	37.5	33.0	34.3	33.8	.38	.28

*D=Douglas fir sapwood; H=Douglas fir heartwood; P=southern pine sapwood; C=creosote; T=creosote-coal tar (70/30); A=Chemonite (ammoniacal copper arsenite); F=copper formate.

†Based on weight when shipped from Forest Products Research Center. About 6 specimens in each group of 25 specimens bled after treatment. Bleeding did not occur with Douglas fir heartwood or southern pine sapwood.

‡Expressed as pounds of copper in each cubic foot of treated wood. For approximate retention on basis of volume before treatment, multiply retention by 1.08.

Some Factors Influencing Treatment Results

Although marine exposure tests are providing valuable data on behavior of preservatives, and marine organisms that attack wood are being studied intensively, comparatively little basic research is being done on the equally important task of distributing preservatives in wood to provide maximum protection. Finding the best preservative will not, by itself, solve the marine borer problem.

Need for heavy retentions and deep penetrations in marine piling has been stressed by virtually all investigators and is well recognized. Both Rhodes (1951) and MacLean (1952) have recommended that marine piling be treated to refusal. Less well recognized, perhaps, are the factors that limit retention and penetration. Effects of preservative and treating conditions have been mentioned briefly. Other important factors that influence treatment results are inherent to wood itself. They include permeability, specific gravity, moisture content, and sapwood depth.

Permeability

Permeability is determined largely by anatomical structure and by chemical substances formed within wood as its physiological functions cease. The outer band of white wood (sapwood) generally is much more permeable to liquids than is the darker-colored heartwood. Wood is more permeable up and down the stem (longitudinally) than it is from the surface toward the center. Unfortunately, influence of longitudinal penetration is negligible in the tidal area of marine piling, for one well-treated end is buried deep in the ocean floor, and the other end is cut off.

Specific gravity

Specific gravity is an index of the amount of void space present in wood at any given moisture content. There is no correlation between specific gravity and permeability. Douglas fir has an average specific gravity of 0.45 (oven-dry weight, wet-volume basis), yet Erickson (1955) reports values in the same growth ring of 0.26-0.35 for springwood and 0.51-0.77 for summerwood. This difference indicates that springwood is capable of retaining twice as much preservative as is summerwood. Most frequently, however, it is the springwood that remains untreated or poorly treated.

Moisture content

Moisture influences treating results primarily by occupying void space within the cell cavities and through its effect on development of checks and on the closure of possible avenues of penetration as the wood dries and shrinks. Freshly cut Douglas fir sapwood and heartwood have moisture contents of over 100 and about 35 per cent, respectively. As wood dries, a moisture gradient is established with low moisture content at the surface and a higher moisture content toward the center. Slope of the moisture gradient curve will vary with drying conditions, usually being steep with fast drying conditions. Moisture gradients in round Douglas fir sections from the same tree that were kiln-dried to different final moisture contents are shown in figure 2.

Preservative concentration will vary inversely as depth, not only because of this increasing moisture content, but also because of greatly increased resistance offered by the minute openings through which preservatives must pass. Checking aids, while pit closure hinders, penetration.

Mattos and Rawson (in Hill, Kofoid, *et al.*, 1927) were concerned about the effect of high wood moisture content on serviceability of marine piling in specifications they prepared for adoption by the San Francisco Bay Marine Piling Committee in 1921. This specification, which is well worth acquaintance, reads in part: "Should the inspector, upon boring the piling, find that the boring contains free moisture, he shall reject any such piling and have same re-treated under the conditions hereinbefore specified." They were attempting to prevent acceptance of piling that may have adequate penetration but low preservative retention.

Sapwood depth

Sapwood depth limits both penetration and retention in Douglas fir piling, for usually there is little or no penetration into heartwood. Although sapwood depth in this wood does increase with diameter, sapwood percentage varies inversely as the diameter. This relationship is illustrated in figure 3, compiled from several unpublished sources. Most Douglas fir piling will contain less than 50 per cent sapwood by volume, which means that concentration of preservative in the sapwood must be at least twice as great as the retention specified. An idea of theoretical *maximum* retentions possible in Douglas fir with different sapwood depths, at varying moisture contents, and with preservatives of different specific gravities can be obtained from figure 4. Retentions obtained in commercial practice will be less than these maximum values, possibly by as much as 20 or 30 per cent.

Reports that Douglas fir piling and poles now contain thicker sapwood than in past years continue to be heard in increasing number. If true, this situation would facilitate selection of thick sapwood piling for severe service locations. Thick sapwood that is deeply and heavily treated with a proved preservative will certainly help to extend the life of piling. However, to obtain improved treatment and lengthened service, we must know the material with which we are dealing. Information on distribution of preservatives in wood and effect of this distribution on wood-destroying organisms would provide a sound technical basis for treating standards. Basic research is sorely needed on permeability of wood and numerous factors influencing penetration of liquids. Such research will lead to improved treatment for all wood products.

Summary

Marine exposure tests have shown that coal tar can increase the effectiveness of creosote for protecting wood against marine borers, provided sufficient tar is added. Addition of coal tar to creosote increases the difficulty of obtaining penetrations and retentions similar to those of creosote alone. Increased care will be required in treatment of wood, particularly species difficult to treat such as Douglas fir, if uniformly deep penetrations and adequate retentions are to be obtained. Thorough inspection of all marine piling should be mandatory.

Limitations on preservative penetration and retention are imposed by differences in permeability, specific gravity, moisture content, and sapwood depth. Although exposure tests provide essential information on preservatives, basic research also is needed on factors influencing penetration of preservatives into wood.

LITERATURE CITED

Erickson, H. D. 1955. Tangential Shrinkage of Serial Sections Within Annual Rings of Douglas Fir and Western Red Cedar. For. Prod. J., 5:241-50.

Graham, R. D., and D. J. Miller. 1957. Full-Cell Treatment of Kiln-dried Coast-type Douglas Fir Piling With Creosote and Creosote-Coal Tar. Ore. For. Prod. Lab. Rept., P-2.

Groenou, H. B. van, H. W. L. Rischen, and J. van den Berge. 1951. Wood Preservation during the Past 50 Years. A. W. Sijthoff's Uitgeversmaatschappij N. V., Leiden.

Hill, C. L., C. A. Kofoid, *et al.* 1927. Marine Borers and Their Relation to Marine Construction on the Pacific Coast. Final Report, San Francisco Bay Marine Piling Committee.

MacLean, J. D. 1926. Effect of Temperature and Viscosity of Wood Preservative Oils on Penetration and Absorption. Proc. Am. Wood Pres. Assoc., 22:147-67.

-------. 1927. Relation of Treating Variables to the Penetration and Absorption of Preservatives into Wood. III, Effect of Temperature and Pressure on the Penetration and Absorption of Coal-Tar Creosote into Wood. Proc. Am. Wood Pres. Assoc., 23:57-70.

-------. 1950. Results of Experiments on the Effectiveness of Various Preservatives in Protecting Wood against Marine Borer Attack. U. S. For. Prod. Lab., Madison, Rept. D1773.

-------. 1952. Preservative Treatment of Wood by Pressure Methods. U. S. Dept. Agric. Handbook, 40.

Rhodes, E. O. 1951. History of Changes in Chemical Composition of Creosote. Proc. Am. Wood Pres. Assoc., 47:40-61.

Richards, A. P. 1957. Co-operative Creosote Program, Final Report on Marine Test Panels. Preprint, Proc. Am. Wood Pres. Assoc.

Schmitz, H. 1933. The Toxicity to Wood-destroying Fungi of Coal-Tar Creosote-Petroleum and Coal-Tar Creosote-Coal Tar Mixtures. Proc. Am. Wood Pres. Assoc., 29:125-39.

Schrenk, Hermann von, and A. L. Kammerer. 1914. The Use of Refined Coal-Tar in the Creosoting Industry. Proc. Am. Wood Pres. Assoc., 10:93.

Teesdale, C. H., and J. D. MacLean. 1918. Tests of the Absorption and Penetration of Coal-Tar and Creosote in Longleaf Pine. U. S. Dept. Agric. Bull., 607.

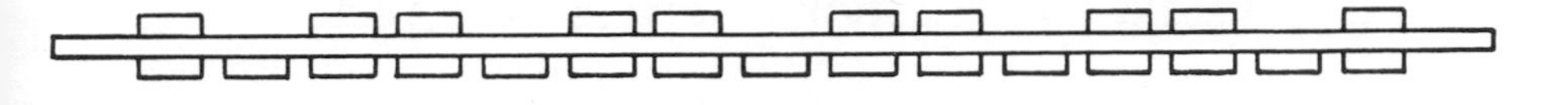

TOP VIEW

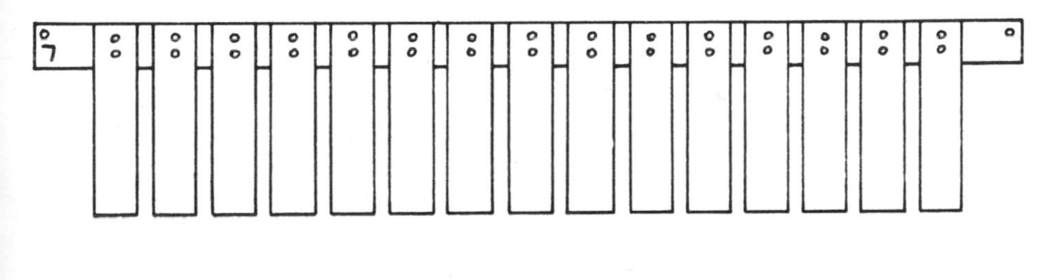

SIDE VIEW

Fig. 1. Method of mounting wood panels on metal bars for exposure tests by Forest Products Research Center

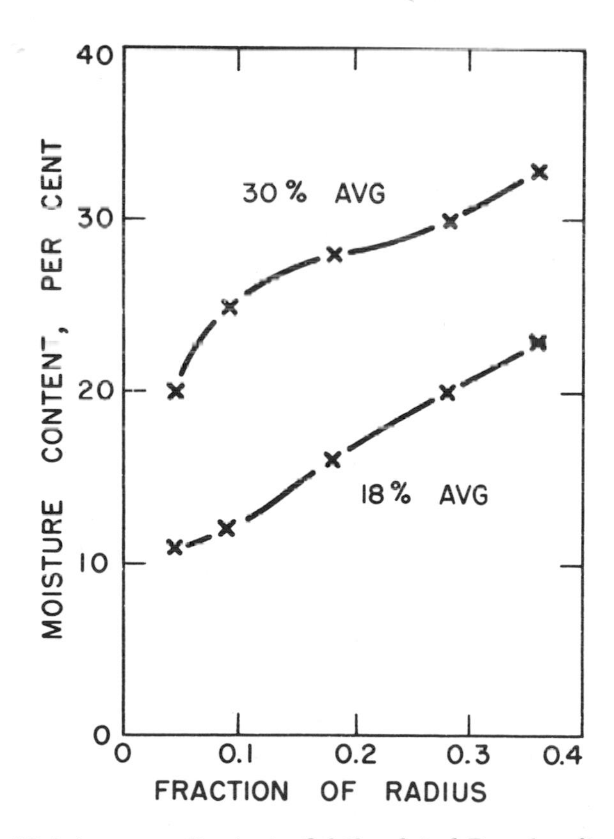

Fig. 2. Moisture gradients in 2 kiln-dried Douglas fir pole sections

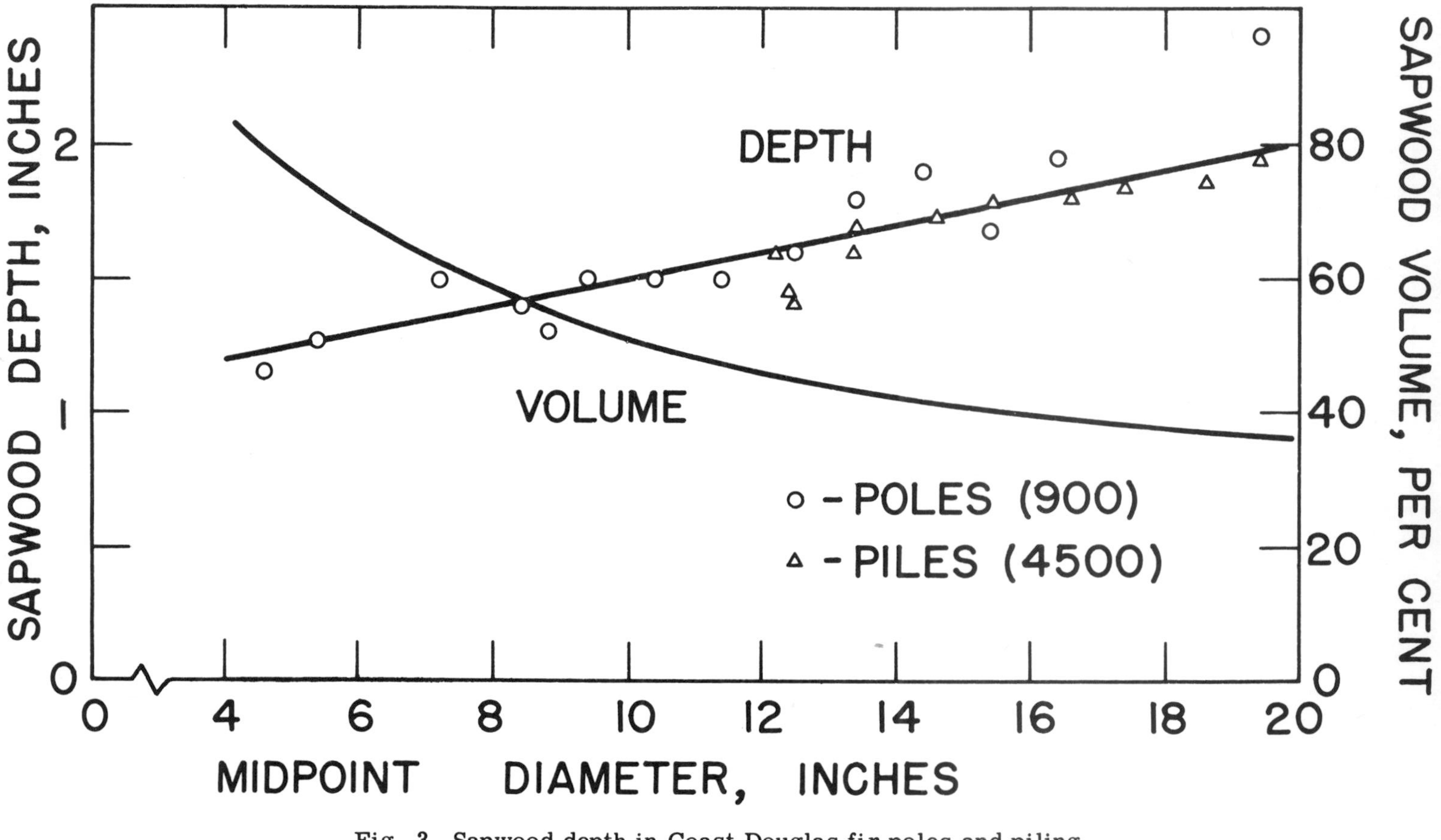

Fig. 3. Sapwood depth in Coast Douglas fir poles and piling

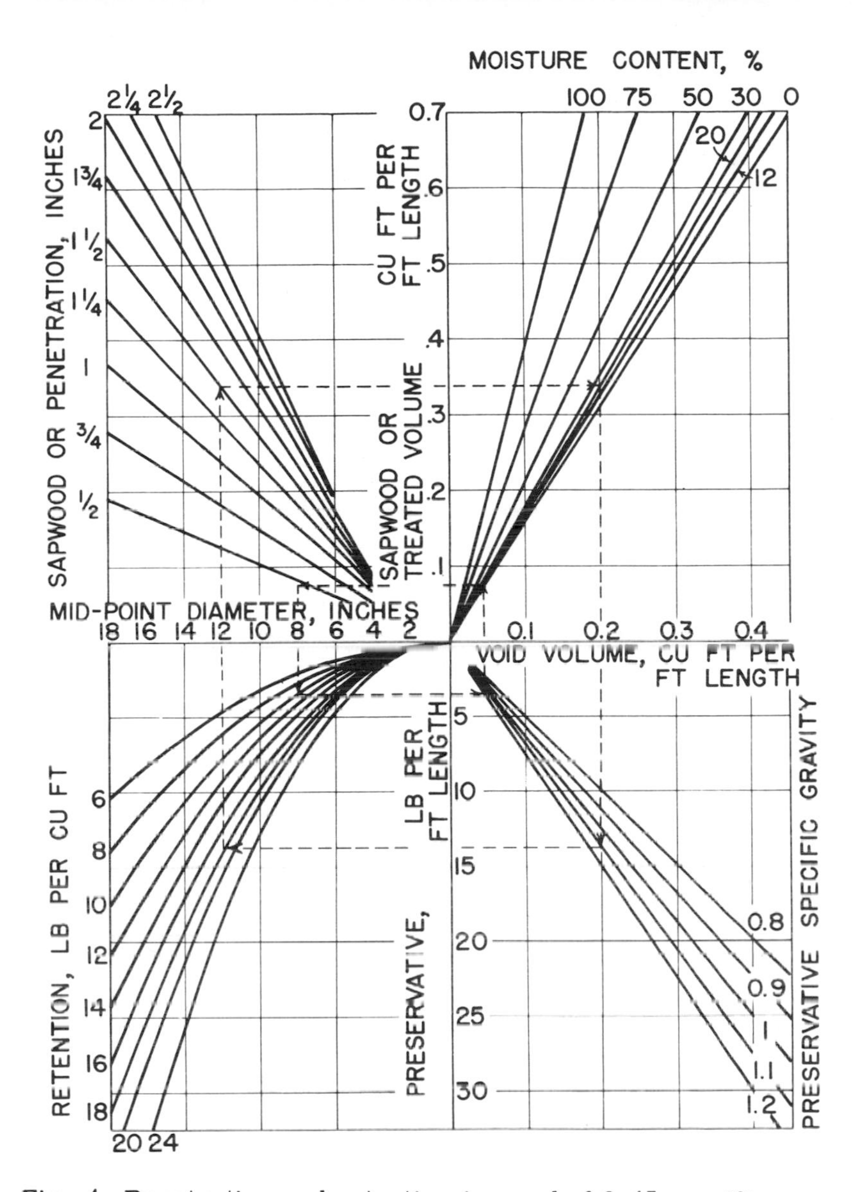

Fig. 4. Penetration and retention in wood of 0.45 specific gravity. To determine maximum theoretical retention: (1) begin at mid-point diameter; (2) proceed up to correct sapwood depth; (3) move right to assumed moisture content; (4) move down to preservative specific gravity at treating temperature; (5) proceed left to intersect vertical line through mid-point diameter; read retention at intersection. To determine penetration for a given retention: (1) begin at mid-point diameter; (2) proceed down to desired retention; (3) move right to preservative specific gravity at treating temperature; (4) proceed up to assumed moisture content; (5) move left to intersection with vertical line through mid-point diameter. Read penetration at intersection

PROTECTION OF COMMERCIAL STRUCTURES WITH SODIUM ARSENITE

Paul C. Trussell

The industrial problem that has led to the use of sodium arsenite against marine borers is one that is indigenous to the Pacific Northwest, namely, the protection of logs and temporary logging equipment which become infested as a result of transport and storage in sea water. The actual annual loss caused by *Bankia setacea* to logs along the British Columbia coast is difficult to assess but can be conservatively estimated as between $500, 000 and $1, 000, 000. In addition to this, the damage caused by this organism to standing booms and piles, A-frame floats, and floating camps brings the annual loss to the logging industry to well in excess of $1, 000, 000.

In approaching this problem the possible use of direct and alternating current and the generation of chlorine by electrolysis of sea water were tested under laboratory conditions and rejected as unpromising. The energy requirement for the application of sonic or ultrasonic waves made this approach impractical. It was finally decided that the most hopeful approach was to find a chemical toxicant for this borer.

Although the annual loss in saw logs is high, actually only about 5 to 10 per cent of the total logs passing into the salt water are involved. Thus, for a protective treatment to be economical, it had to be applied to logs *after infestation had occurred,* rather than to all the logs passing into the sea water. This need imposed a special requirement on the toxicant, that is, it had to be effective against Bankia in the adult form, rather than as the larva. This was essentially a new approach to the chemical protection of wood against teredine borers since, in the past, chemicals have always been applied to wood to prevent attack by the larvae.

In undertaking the laboratory screening of chemicals, the following requirements for the toxicant were set up: (1) it must be effective in killing adult borers present in their natural habitat (wood); (2) it must be effective in killing in excess of 99 per cent of the borers; (3) it must be active within an exposure period of 18 hours and at a concentration not in excess of 200 ppm. For commercial use, it was also desirable to have the toxicant available in large quantities and at a reasonable price.

Because of the unique sensing and protective mechanism of the test organism, the finding of an effective toxicant was much more difficult than was at first anticipated. In fact, after over a hundred chemicals selected from those available commercially, together with experimental chemicals from the research laboratories of chemical houses in the United States and Canada were tested, only one effective toxicant, sodium arsenite, was found. The various types of reactions produced by the toxicants tested are set forth with examples in table 1. In type 1, cover-

TABLE 1

TOXICITY OF CHEMICALS ON *BANKIA SETACEA* AFTER 18 HOURS' EXPOSURE

Chemical	Toxic Concentration, ppm
Type 1: Complete kill; substance not detected	
Sodium arsenite	12.5-20,000
Type 2: Complete kill, but substance detected at higher levels	
Capryldinitrophenyl acetate	6.2-12.5
Octadecyltrimethylammonium salicylate	12.5-50
Isobutyl benzoylacrylate	100
Type 3: Partial kill	
Dinitro-O-cylcohexylphenol	5-40
Triethanolamine dinitro-o-sec-butylphenate	50-100
S-benzyl-thiouronium dinitrocaprylphenate	100-200
Type 3: Nontoxic	
Sodium arsenate	
Sodium azide	
Sodium fluosilicate	
Copper sulphate	
Phenyl mercuric acetate	
Sodium pentachlorphenate	

ing those that met the full three requirements set forth above, only sodium arsenite was found; in type 2, those chemicals producing a complete kill of the test organism but detected by the animals at higher concentrations of the substance, a small number were found; in type 3, those causing a partial kill over a limited range of concentration, a larger group of chemicals were found; and in type 4, those apparently detected throughout the range tested and having no lethality, were a large number of compounds, a few of which are listed in table 1. It is interesting to note that, although Bankia did not detect sodium arsenite, it was able to detect sodium arsenate.

Field applications using sodium arsenite were first applied to logs in a flat raft formation by a tidal ponding technique. In this method, a raft of logs is moved in close to shore at high tide, and when the tide has receded about 3 feet the raft is landlocked in an artificial shallow pond in which the chemical treatment is made. Water from the pond is pumped and sprayed over the logs, and a concentrated sodium arsenite solution is introduced at the intake side of the pump. The actual addition time of the chemical was approximately 15 minutes when the pond size was 20 x 50 x 2 feet deep.

The results of a series of fourteen tests using different quantities of added sodium arsenite and different holding times between high tides are shown in table 2. The tests showed that a concentration of about 70 ppm arsenic trioxide for a 6- to 8-hour period produced complete kills of *Bankia setacea* in logs.

The second type of structure to which sodium arsenite was applied was the deep sea (Davis) raft. This raft is normally 80 to 90 feet wide, about 400 feet long, and elliptical in cross section with a maximum depth of about 20 feet. In treating such a massive structure, it was decided to apply the sodium arsenite as a fairly dilute solution by means of a fire hose over the surface of the logs and to allow the chemical to percolate down into the interstices between the logs in the raft. This series of tests was undertaken at Shannon Bay in the Queen Charlotte Islands where deep-sea rafting is normally undertaken in transporting logs to the mainland coast of British Columbia. Before applying chemicals to a raft, it was necessary to spot marked infested logs throughout the raft so that, after the raft had been treated and dissembled, the effectiveness of the chemical treatment could be assessed. These rafts are divided into ten 40-foot sections measured lengthwise along the raft, and in the first treatment it was decided to apply a different level of chemical to

TABLE 2
EFFECT OF SODIUM ARSENITE ON MARINE BORERS
(BANKIA SETACEA) IN FLAT RAFTS

Run No.	No. Logs in Raft	Arsenic Trioxide (ppm)		Hours Exposed	No. Logs with Bankia	
		Added	Found		Alive	Dead
1	6	0	0	19 1/2	6	0
2	6	1,000	865	15 1/4	0	6
3	6	1,000	830	8	0	6
4	6	400	315	17	0	6
5	6	200	206	17	0	6*
6	6	125	146	19 1/2	0	6*
7	5	125	91	9 1/2	0	5*
8	6	50	35	16 1/2	0	6
9	5	50	45	10	5	0
10	6	100	95	10	0	6
11	6	100	111	8 1/2	0	6
12	6	67	51	6 1/2	0	6
13	2	67	41	6 3/4	0	2
14	2	67	50	8 3/4	0	2

*Kill of Bankia complete except in portions of log resting in mud during treatment.

each of four alternate sections in an attempt to establish the minimum quantity of chemical necessary to kill the borers completely in all the marked logs in the section. The results of this test are shown in table 3. The highest quantity of arsenic triox-

TABLE 3
TREATMENT OF RAFT NO. 1 SECTIONS WITH SODIUM ARSENITE

Section No.	Arsenite Concentrate Applied (Imp. gal)	Arsenic Trioxide Applied (lb)	Arsenic Trioxide Concentration in Solution Applied (%)	Estimated Arsenic Trioxide in Section (ppm)
2	96	1,090	1.31	2,300
4	48	545	0.65	1,150
6	24	272	0.32	575
8	12	136	0.16	287
10	untreated	0	....	....

ide applied was 1,090 pounds per section, and the lowest, 136 pounds per section. Later, when the raft had been transported to the mainland coast of British Columbia and dissembled, it was found that all but the lowest quantity of arsenic trioxide gave a complete kill in all the marked logs in the treated sections. Therefore, about 270 pounds of arsenic trioxide, which corresponded to one drum of the particular sodium arsenite concentrate being used, was adequate for treatment of one section of a Davis raft.

A second raft had been assembled with marked infested logs placed at two levels throughout its length. The bow half of the raft was treated with the quantity of sodium arsenite found necessary by the previous large-scale experiment, and the raft was then towed to its destination. When the raft was dissembled 67 days after treatment, it was found that all Bankia in the marked infested logs in the treated half of the raft were dead, and that most of the logs in the two sections adjoining the treated half of the raft were also free of live borers.

The economic aspects of the treatment of such large-scale rafts are as follows. The value of the timber in such a raft is approximately $75, 000 to $80, 000, and a loss of up to 10 per cent of this may be sustained. The chemical cost of the treatment is approximately $400, so that the insurance cost against marine borer damage in the structures is reasonable. In actual practice, it is not necessary to treat all rafts that are constructed, as infestation occurs only at certain periods of the year and it is during these times only that treatments are necessary.

At this point, the logging industry expressed a requirement for treatment of flat rafts of logs in open water, that is, without use of a tidal pond. First attempts to hold the sodium arsenite solution at the surface by admixing it with fresh water and holding the specific gravity of the added solution to less than that of the sea water proved unsuccessful. High concentrations of sodium arsenite could be retained at the surface of the water in the raft, but the concentration at the undersides of the logs was inadequate to effect kills of Bankia. A study was then made of the use of percussion as a means of killing Bankia. Two pounds of dynamite released 8 to 10 feet below the surface of the raft was demonstrated to be the maximum the logs could withstand without shattering. Using such charges, the maximum kill that could be obtained immediately above the explosion was about 80 per cent, dropping off very rapidly with increasing distance. Attention was again turned to sodium arsenite, and this time, to compensate for the losses caused by dilution and turbulence effects, the sodium arsenite was applied in a continuous manner to the infested logs by means of a sprinkler system. In this way, although losses were occurring continuously, additions were also being made continuously with the hope that an adequate concentration could be retained during the period of treatment. After a series of experimental runs had been completed, encouraging results were forthcoming as illustrated in table 4. Using a single sprinkler over a 4-hour period to apply 15 to 20 gallons of commercial sodium arsenite to a 65x65

TABLE 4
EFFECT OF RATE OF SODIUM ARSENITE APPLICATION
ON CONCENTRATION OF ARSENIC TRIOXIDE IN RAFT SECTIONS

Run No.	Sodium Arsenite Addition			Time Sampled	Concn As_2O_3			
	Rate	Time	Volume		12 in. depth		36 in. depth	
	(gal/hr)	(hr)	(gal)	(hr)	(ppm)	(Avg)	(ppm)	(Avg)
8	6.25	4	25	1	35			
				2	36			
				3	130	69		
				4	74			
10	5.0	4	20	1	31		19	
				2	51		34	
				3	212	113	92	65
				4	156		115	
11	3.75	4	15	1	153		20	
				2	197		99	
				3	233	184	50	65
				4	154		92	

foot section of logs resulted in concentration of As_2O_3 between 100 and 200 ppm at the 1-foot depth and over 50 ppm at the 3-foot depth. During these tests it was observed that, when the chemical was sprinkled onto the logs in the raft, the chemical, in passing into the water, continued to follow the contour of the logs so that the chemical concentration reaching the infested underwater surfaces was relatively high.

Considering the economic aspects, the cost of chemicals for treating flat rafts by the tidal ponding technique was $75.00 for a raft 65 feet wide and 450 feet long. For treatment of the same size flat raft by the open water technique, the chemical cost was approximately $200. The industry has accepted the latter technique in preference to the former since rafts can be treated *in situ,* and thus towing costs are saved.

Another extensive commercial application of the use of sodium arsenite against destruction caused by both terediniens and limnoria is in the protection of floating wooden drydock pontoons. Protection of these large industrial commercial structures was initiated 5 years ago on a 13,000-ton drydock at North Vancouver. Since that time the treatment has gained rapid acceptance by the drydocking firms, and now all large commercial wooden docks in Vancouver, British Columbia; Seattle, Washington; and Galveston, Texas, are using this process. Currently over 100,000 tons of drydock capacity are being protected, and in the next few months an additional 50,000 tons capacity in the San Francisco Bay area will be coming under this system. Since drydock pontoons are self-contained units, the application of the sodium arsenite process to protect the internal portions of these structures has been a relatively simple process in theory. The chemical is introduced to the pontoon sections so as to have a uniform concentration of chemical, and the chemical is held in the pontoons for a minimum period of 48 hours in order to effect complete kills of the borers. An important aspect of the treatment is the biological phase, in which it is necessary to follow continuously the settlement of both terediniens and limnoria in the vicinity of the drydocks. All chemical treatments of drydocks are integrated with marine borer settlement studies so that the treatments are made at times when they will be most effective. Chemical costs of treating drydocks are approximately $200 to $300 per dock per treatment. Two to three treatments are required each year. It has been estimated by one drydock operator that this protective scheme applied to two docks amortized over 10 years will result in a maintenance saving of close to $1,500,000.

AN EVALUATION OF CERTAIN COAL TAR CREOSOTE FRACTIONS FOR THEIR RESISTANCE TO MARINE BORER ATTACK

T. R. Sweeney and T. R. Price

Introduction

Wooden structures, continuously or intermittently submerged in coastal waters, are subject to attack by certain crustacea and mollusca commonly referred to collectively as marine borers. These pests have plagued man since antiquity, and today the cost of their depredations runs into millions of dollars annually. Although ecological and physiological studies on marine borers may subsequently point the way to a more subtle method for their control, the two methods in general use today are armoring of the wood and impregnation of the wood with preservatives. The most widely used preservatives are coal tar creosote and solutions of creosote and coal tar.

The creosote in general use by the wood-preserving industry today is a distillate of the tar produced from the high temperature carbonization of bituminous coal with a distillation range of approximately 200°-400°C. It is a complex mixture of hydrocarbons, phenols, amines, and other classes of compounds containing sulfur and nonacidic oxygen. Why this complex mixture of chemical compounds known as creosote should function effectively as a preservative, and which constituent class of compounds or individuals make major contributions to its effectiveness, is still obscure. Ramage and Burd (1927) found that 2 x 4 x 18-inch panels that had been treated with creosote freed of low boiling tar acids (phenols) stood up well when exposed to marine borer attack. Creosotes containing high percentages of light oils or tar acids were found by MacLean (1950) to be less effective than whole creosote. He also found that the higher boiling creosotes and creosote fractions were among the most effective tested. Becker and Schulze (1950) impregnated 5 cm x

2.5 cm x 1.5 cm pine blocks with creosote fractions of various boiling ranges and, after weathering the treated blocks, hung them individually in small aquaria containing sea water. A number of Limnoria were then introduced, and from the number of surviving animals and the degree of attack on the panels the efficacy of the various fractions was determined. These workers found that fractions boiling up to 240°C were ineffective. Among the higher boiling constituents those within the boiling range 200°-300°C were most effective, and those within the range 300°-330°C least effective.

The present communication contains the results of an evaluation of several creosote fractions for their ability to resist marine borer attack.

Preparation of Materials

Creosote

The creosote used in this investigation was taken from a large stock reserved for present and future navy research. It is a medium residue creosote and meets the A.W.P.A. Standard P1-54.

Creosote fractionation

Removal of tar acids. One hundred grams of creosote were dissolved in 150 ml of benzene and the solution extracted with 4 successive portions of Claisen alkali. The Claisen alkali was prepared by dissolving 350 g of potassium hydroxide in 250 ml of water and bringing the total volume to 1000 ml with methanol. After extraction the residual benzene solution was washed well with water and the solvent removed by distillation. The quantity of tar acids removed from creosote by extraction with Claisen alkali has been shown to be twice that that can be removed by extraction with aqueous alkali (Sweeney and Walter, 1952).

Removal of tar bases. A solution of creosote in benzene as described above was extracted successively with four 100 ml portions of 10 per cent hydrochloric acid. The residual solution was washed with water until free of acid and the solvent removed by distillation.

Removal of both tar acids and tar bases. The tar acids were removed from the creosote first as described above. The residual benzene solution was then treated with hydrochloric acid as described for the removal of the tar bases.

Treatment with petroleum ether. One-third volume of petroleum ether (bp 30° to 60°C) was slowly added, with stirring, to creosote, and the precipitated solid was removed by filtration. The solvent was removed from the filtrate by distillation to yield what in this paper is termed "solids-free creosote." To obtain the so-called "petroleum ether soluble fraction," additional petroleum ether was added to the filtrate from the above filtration and the solution was allowed to stand overnight. The clear, light yellow, supernatant liquid was then decanted from the precipitated tar, and the solvent was removed by distillation. The product is an amber fluid, much less viscous than creosote, and leaves wood with a clean, attractive appearance after impregnation. To obtain the so-called "tar-free creosote," the precipitated solids were added to the petroleum ether soluble creosote.

Distillation. Creosote was distilled through a saddle-packed heated column, 25 cm long and 1.8 cm inside diameter, fitted with a heated total reflux head and fraction collector. For a distillation the stillpot was charged with about 1500 g of creosote. It was found highly desirable to seal the column to the stillpot for distillations under reduced pressure. Small leaks, which developed when ground glass joints were used, caused excessive sublimation. A summary of the distillation results is given in table 1. A summary of the creosote fractions and their percentage in whole creosote is given in table 2.

TABLE 1
BOILING RANGES OF CREOSOTE DISTILLATION FRACTIONS

Fraction	Boiling Range, °C	
	Actual	Approximate at 760 mm
1	<200/760 mm	<200
2	200-235/760 mm	200-235
3	235/760 mm-130/15 mm	235-255
4	130-160/15 mm	255-295
5	160-190/15 mm	295-330
6	190-210/15 mm	330-355
Residue	>210/15 mm	>355

Preparation of panels. Clear, edge grain, sapwood (benzidine test) southern pine panels, 5x1 1/2x1/8 inches, were used for this study. Panels were dried to constant weight at 105°C before impregnation. Weight tolerances were established at 8.25 ± 0.75 g. Panels were impregnated by submerging them in the creosote or fraction held at 80°C in an ordinary hydrogenation bomb and applying the desired pressure by the admission of nitrogen.

After impregnation the panels were allowed to bleed to constant weight in a desiccator.

TABLE 2
COMPARISON OF THE PERFORMANCES OF THE VARIOUS CREOSOTE FRACTIONS

Fraction	Per Cent of Whole Creosote	Rating*		
		Performance		Permanence
		Not Leached	Leached	
Whole creosote	100	100	100	100
Base-free creosote	96.5	141	74	53
Acid-free creosote	94.4	300	95	32
Acid- and base-free creosote	91.9	140	103	74
Solids-free creosote	92.6	121	136	112
Tar-free creosote	91.1	91	117	129
Petroleum either-soluble creosote	83.7	117	141	121
Distillation fraction 2 depleted creosote	80.1	85	68	81
Distillation fraction 3 depleted creosote	93.2	167	100	60
Distillation fraction 4 depleted creosote	92.7	105	84	81
Distillation fraction 5 depleted creosote	91.8	90	88	98
Distillation fraction 6 depleted creosote	84.5	63	60	95
Distillation residue depleted creosote	64.8	33	20	60
Distillation residue	35.2	28	66	235
Petroleum ether-soluble creosote, acid free	79.4	130	103	80

*Relative to whole creosote taken as 100.

Panel Exposure and Evaluation

The method for evaluating creosote fractions described in the following paragraphs is a precursor of the one currently in use and described in detail elsewhere (Sweeney *et al.*, in press).

Panels were prepared and exposed in 4 groups of 6 sets each. Each group contained duplicate sets of panels prepared from whole creosote and duplicate sets for each of 2 creosote fractions. Each whole creosote set consisted of 60 panels that were impregnated over a range of 5 to 20 pounds per cubic foot (pcf). Each fraction set consisted of 20 panels impregnated over a range of fraction equivalent to a range of 10 to 15 pcf of whole creosote. This was done because it was deemed desirable to compare a fraction with its equivalent weight of whole creosote rather than an equal weight. An attempt was made to vary the degree of impregnation evenly over the selected ranges.

Two procedures were used in the evaluation of each creosote fraction. One, to determine the initial borer resistance of a fraction relative to whole creosote, consisted of exposing a fraction-treated and a creosote-treated set of panels simultaneously to marine borer attack for a given time and measuring the damage sustained by each. The other procedure was similar except that before exposure the sets were subjected to leaching in hot water (80°C) for 16 days. The latter treatment was an attempt to assess the permanence, or long-term protective ability, of a fraction in a reasonable length of time without having the leaching or weathering complicated by the superimposition of marine borer attack. The leaching was accomplished in a 25-gallon stainless steel, heated tank that could accommodate about 200 panels. The tank was equipped at each end with thermoregulators capable of maintaining the continuous stream of fresh water at the desired temperature. The leached and unleached sets of panels were exposed in the sea simultaneously. During exposure the panels were scraped free of fouling biweekly and inspected and rated bimonthly. When all of the panels in a whole creosote set had sustained an attack of at least 1, that set along with its corresponding fraction sets was removed from exposure.

Creosote fractions were evaluated on the basis of their ability to impart to the wooden panels a resistance to marine borer attack. Damage, measured on a scale from 0 (no attack) to 5 (very heavy attack), was related to the actual number of borings and was expressed numerically as the sum of the Limnoria attack (L) plus twice the Teredo attack (T), i.e., L + 2T. The Teredo attack was doubled in order to compensate for the large surface-to-volume ratio of the panels. Because of the yearly and seasonal fluctuation in marine borer activity in the exposure area, absolute measurements of attack are impossible, and hence all attack ratings (L + 2T) are converted to a performance rating relative to whole creosote.

Attack ratings on individual panels for any set were converted to a performance rating for the set as follows. It was first established that, over the range of impregnation investigated, it was sufficiently accurate to express attack as a linear function of retention. On this basis, the constants m (slope) and b (intercept) of the equation $y = mx + b$, expressing a linear relationship between the intensity of marine borer attack, y, and the concentration of preservative in the panels, x, were calculated by the method of least squares from the data for each set of panels. The intensity of attack, as mentioned previously, is expressed as L + 2T; concentration of preservative is expressed in terms of whole creosote equivalent to the actual quantity of fraction impregnated. Because the damage at the mid-point of the retention range studied (12.5 pcf) is proportioned to the area under the curve, the performances of fractions and whole creosote were compared at this point. The attack rating for a fraction at this point was calculated from the appropriate derived equation, and this rating was then converted to a performance rating through the expression:

$$\frac{\text{attack on whole creosote}}{\text{attack on fraction}} \times 100 = \text{performance rating.}$$

In addition to a performance rating on each fraction, a permanence rating was obtained through the expression:

$$\frac{\text{performance rating (leached)}}{\text{performance rating (unleached)}} \times 100 = \text{permanence rating.}$$

The permanence rating indicates nothing of the extent of attack but does indicate how well a fraction stands up to the deleterious effect of leaching relative to whole creosote.

Impregnation with the distillation residue posed a special problem and emphasized the need to abandon the practice of impregnating panels with creosote fractions on the creosote equivalent basis. With fractions that constitute a large percentage of whole creosote it is not difficult to obtain a range of impregnations equivalent to 10 to 15 pcf of whole creosote; with small fractions, however, the difficulties are considerable. For example, a fraction consisting of 10 per cent of whole creosote would have a range of 1 to 1.5 pcf equivalent to 10 to 15 pcf of whole creosote. This would amount to no range at all for practical exposure work. For this reason panels were impregnated with the distillation residue on a pound for pound basis; that is, at an actual level of 10 to 15 pcf and not its equivalent in whole creosote. To obtain the rating of the distillation residue, therefore, the ratio of the attack at 12.5 pcf of creosote to the attack at 4.4 pcf of residue (12.5 pcf creosote equivalent) was determined from the appropriate equations. The attack at 4.4 pcf of residue is, of course, a value obtained by extrapolation. The performance and permanence ratings for the various fractions are shown in table 2.

Discussion of Results

Because of the large variation inherent in biological measurements, a small difference in the performance rating of a fraction relative to creosote is probably of little significance. However, some results stand out in table 2.

The removal of the tar acids gives a superior product. This

was also indicated in earlier experiments (not reported) and supports the finding of MacLean (1950) that the fortification of creosote with tar acids gave an inferior product. Even after leaching, the tar acid-free creosote appears to be as good as whole creosote. The low permanence rating is a reflection of the excellent performance of the unleached material. The beneficial effect of removing tar acids is also exhibited in the performance of the petroleum ether soluble fraction with acids removed.

The removal of the tar bases and both tar bases and acids gave products that performed somewhat better than creosote before leaching. However, like the acid-free creosote, they did not hold up to leaching very well, and the poor permanence rating of these fractions may indicate that the tar acids and bases contribute, in some way, to the ability of creosote to resist leaching from wood.

The petroleum ether soluble fraction of creosote performed somewhat better than whole creosote, especially after leaching. This creditable performance coupled with its ease of handling and its ability to leave wood clean and attractive after impregnation may have commercial significance.

It can be seen from table 2 that creosote with the distillation residue removed gave a poor performance. However, since the residue alone also gave a poor performance, it is likely that removing residue from creosote did not eliminate any highly toxic material, but rather decreased the ability of the creosote to remain in the wood. This view is supported by the very high permanence rating of the distillation residue.

Distillation fraction 6 also seems to make a distinct contribution to the effectiveness of creosote as is seen by the inferior performance of the distillation fraction 6-depleted creosote. The removal of distillation fraction 3 from creosote yields a product that is apparently superior to creosote. This is surprising and unexpected, and no explanation is offered at the present time.

On the basis of the pattern of the leaching of whole creosote from wood (Sweeney *et al.*, in press), which of course may not be exactly comparable to that for the fraction-depleted creosotes, leached panels would be expected to have lost all of distillation fractions 2 and 3, most of 4, and an appreciable percentage of 5, 6, and distillation residue. This would mean that, in the case of the leached panels, only distillation fractions 5, 6, residue, and to a lesser extent fraction 4, could be present to furnish protection. With a creosote depleted of one of these fractions by distillation, the remaining ones would be responsible for any protection. Since the residue alone offers a rather low order of protection to unleached wood but, on the other hand, is necessary to the efficient action of whole creosote as shown by the poor performance of residue-depleted creosote (table 2), there is a strong indication that distillation fractions 4, 5, and 6 may be providing a large measure of the toxicity and that the residue (and to some extent fraction 6) may be providing the means of holding these fractions in the wood. Additional experiments on panels impregnated with the individual distillation fractions have also indicated that fractions 5 and 6 make the largest contribution to the protective action afforded by creosote. It is hoped that these results can be reported in detail in the near future.

Because, as mentioned, distillation fractions 2 and 3 would be expected to be completely removed from the wooden panels by the leaching process, it might be anticipated that panel sets impregnated with whole creosote, distillation fraction 2-depleted, or distillation fraction 3-depleted creosotes would all exhibit similar resistance to marine borer attack after leaching. That this was not the case with fraction 2-depleted creosote may be a reflection of the limitations of the bioassay procedure or may be due to the possibility that fraction 2-depleted creosote does not respond to leaching in a manner similar to whole creosote.

LITERATURE CITED

Becker, G., and B. Schulze. 1950. Wiss. Abhandl. deutsch. Materialprüfungsanstalt, 2:76.

MacLean, J.D. 1950. U.S. For. Prod. Lab. Madison, Rept. D1773.

Ramage, W.D., and J. S. Burd. 1927. Ind. Eng. Chem., 19: 1234.

Sweeney, T.R., and C.R. Walter, Jr. 1952. U.S. Naval Res. Lab. Rept., 3940.

Sweeney, T. R., *et al.* Corrosion, in press.

THE EFFECTS OF ENVIRONMENT AND DECAY ON MINERAL COMPONENTS OF GRAND FIR WOOD*

Everett L. Ellis

Introduction

Large areas of the western United States support mature stands of pure coniferous forests. Past cutting practices have generally favored the more valuable species and accessible areas, often leading to a kind of unplanned, selective harvest. Naturally, the residual stands run higher in lower-valued species, creating and aggravating certain forest problems with respect to quality, growth, and composition. One of these serious problems is decay in grand fir, *Abies grandis* (Dougl.) Lindl., by the Indian paint fungus, *Echinodontium tinctorium* (Ellis) E. and E. (Hubert, 1955). Other important conifers are also affected by this organism, particularly western hemlock and white (concolor) fir. Peculiarly, however, the Indian paint fungus, as it affects grand fir, is found only east of the Cascade Mountains (Foster and Thomas, 1951), despite the fact that grand fir grows widely on many soil and site types both west and east of the Cascades.

The Indian paint fungus is considered to be the most damaging enemy of grand fir. Major forest stands of grand fir are often highly defective (up to 50 per cent of volume, or more) and of low value due solely to the frequency and severity of fungus attack. In an advanced form the rot column occupies the main portion of the stem, leaving about a 2-inch band of sound sap-

---

*This research was conducted at the University of Idaho and was supported in part by that institution and in part by an Agnes H. Anderson Research Fellowship at the University of Washington.

wood. Vertically this decayed portion may extend from 6 to 30 feet or more from the point of infection. Little lumber can be cut from such logs, and even chemical pulping is not entirely satisfactory because the decayed wood cooks differently from sound wood and is hard to bleach. Occasionally local areas of grand fir can be found in which the trees are completely sound. Mature and overmature trees most commonly show a high incidence of decay, but the author has found the incipient or early stage of decay on trees as young as 28 years.

Recent advances in the field of mineral nutrition, both in Australia where work on the effect of zinc and phosphorus deficiencies in pine plantations has been conducted (Stoate, 1950), and in North America through development of the field of biogeochemistry, i.e., prospecting for metals through the analysis of plant tissues (Warren, Delavault, and Irish, 1952), suggested that mineral analyses of grand fir wood might be a valuable approach to the decay problem of grand fir. At the outset, it was believed that differences in susceptibility of grand fir to decay in different parts of its range might parallel differences in mineral composition. Furthermore, knowledge of the mineral components of wood is meager and might be valuable to other investigators. Finally, during the course of the study, it was found that decay has a marked effect on mineral composition.

Review of Literature

The mineral analysis of wood ash has received occasional attention by investigators comparing different parts of trees (Czapek, 1925; Goodall and Gregory, 1947; Mitchell, 1948; and Riou, Delorme, and Hormisdas, 1937). Wilcox (1940) attempted to separate the wood of closely related eastern and western white pines unsuccessfully, using spectrographic methods. Wise *et al.* (1952) report qualitative spectrographic analyses for 21 elements on 6 native and 16 tropical woods, with particular reference to silica (Si) content and marine-borer resistance. Recently, Bambergs and Knavina (1950) compared mineral and micronutrient contents of the wood of several Latvian trees. Leaf and needle analyses have received much wider attention by plant physiologists and soil scientists and have often been used to explain differences in mineral nutritional behavior. Major works on foliar analysis include those of Lundegårdh (1951), Goodall and Gregory (1947), and Mitchell (1948). Other workers have investigated forest litters (Riou,

Delorme, and Hormisdas, 1937; Chandler, 1943; Scott, 1943), soils and indigenous plants (Lounamaa, 1956), and many other special interests. The application of wood and plant ash analyses to the field of prospecting for metals (biogeochemistry) has recently received widespread attention for elements such as copper, zinc, silver, iron, molybdenum, manganese, gold (Warren, Delavault, and Irish, 1952), and uranium (Cannon and Starrett, 1956).

The literature on spectrochemical methods of analysis is very extensive. Selected general texts include those of Sawyer (1944), Brode (1939), Ahrens (1950), Nachtrieb (1950), and Harrison, Lord, and Loofbourow (1948). Special mention should be made of Vanselow and Liebig (1948) and of Vanselow and Bradford (1957); the former paper was used as the principal guide for this work. Gilbert (1952) discusses the application of the Beckman Model DU Quartz Spectrophotometer for flame analyses.

The ash content and composition of plants has been investigated for a relatively long period, preceding the use of ash analyses as indicators of nutritional status. Much of the older work, as perhaps also newer work, failed to recognize problems of precision, prevention of contamination, and extreme variability encountered. Kollman (1951) reports an ash content for, presumably, *Abies pectinata* wood of 0.28 per cent; composition of the ash was reported as: CaO, 33.0 per cent; MgO, 6.2 per cent; K_2O, 22.6 per cent; Na_2O, 4.5 per cent; and SiO_2, 0.9 per cent. Similar values are reported by Czapek (1925) and Wise *et al.* (1952). Campbell (1952) refers to two investigations that report an increase of ash in rotten as compared to sound wood. Foster (1949) and Wolf and Wolf (1947) describe the nutritional function of metals in fungi and list the same essential elements as for other plants with the possible addition of gallium, perhaps cobalt and chromium, and vanadium (Bertrand, 1943). The difficulties of media purification in fungal nutritional studies add an additional problem to proof of essentiality for a mineral element. In this study the elements present or sought for are grouped according to the scheme given in table 1.

It is well to emphasize that the mineral content and composition of the wood of grand fir is of major concern in this study because of the relationship of fungus to host. Details of mineral composition in needles, bark, and other tissues, although possibly indicative of the general nutritional status of the tree,

TABLE 1

SYSTEM OF SUBDIVIDING MINERAL ELEMENTS IN SPECTROCHEMICAL ANALYSES OF GRAND FIR

Major Elements	Essential Elements	Other Common Elements	Uncommon Elements	Elements Sought but Not Found	
Ca	B	Ag	Au	As	Be
Mg	Mn	Al	Ga	Cb	Ge
K	Fe	Ba	In	Cs	Pd
Na	Mo	Co	La	Hg	Tl
(Si)	Cu	Cr	Li	Pt	
	Zn	Ni	Sn	Sb	?Cd
	(P)	Pb	V	Ta	?Bi
		Rb	Zr		
		Sr			
		Ti			

may not govern the mineral content of the wood--the eventual substrate for the Indian paint fungus, which commonly attacks the woody portion. For this and other reasons, major emphasis was placed on analyses of wood ash of grand fir.

Collection and Preparation of Samples

Grand fir grows generally in a number of forest types in western North America, as shown in figure 1. Samples of wood from 113 trees were collected, generally at breast height, and physical data, summarized in tables 2 and 3, were recorded. From these samples representative material was chosen for analyses, including wood, rotten wood, conks or fungus fruiting bodies, needles, and bark. Wood samples were dried, cross-cut to suitable lengths, and split with a soft steel hatchet into match-sized sticks for ashing. Diligent and consistent effort was made to avoid accidental contamination during all handling operations.

An electric muffle furnace was employed to ash the wood samples, using silica evaporating dishes as containers. A maximum temperature of 600°C was used. The material was ashed to virtual completion and then transferred to a desiccator for cooling. The cooled ash was transferred to new, clean polyethylene vials for weighing and storage in desiccators prior to use. Percentage of ash was based on oven-dry weight of tissue.

Analytical Procedures

For semiquantitative spectrographic work, the material to be analyzed should have known major components so that side effects will be minimized. Standards and unknowns should have a comparable composition with respect to elements occurring in large amounts, particularly if those elements have complex spectra or modify excitation characteristics. Choice of method is governed by many factors including degree of accuracy required. In this work, accuracy of the order of three (determinations within 3/2 or 2/3 of the true value) or less was desired because of the survey nature of the study. This degree of accuracy and certain other requirements can be met by slight modification of procedures developed by Vanselow and Liebig (1948) as well as generally meeting requirements for Universal Method of Spectrochemical Analysis recently reported by the American Society for Testing Materials (1956).

A base powder was compounded of specially purified spectrographic-grade salts, generally oxides and carbonates, in proportions determined from preliminary flame spectrophotometer tests. A Beckman DU instrument employing a hydrogen-oxygen flame was used to determine calcium, magnesium, sodium, and potassium, the major components of plant ash. Thirty wood ash samples were run, giving the following average values and ranges of values for these four elements:

Element	*Mean Value, %**	*Range of Values, %*
calcium	18.7	7.5-30.3
magnesium	4.1	2.4-6.7
potassium	17.2	5.0-40.3
sodium	0.6	0.1-2.3

* Means of 30 representative samples; percentages calculated as of the element.

To this base powder was added 1 per cent silicon dioxide (SiO_2), and the whole was thoroughly ground in an agate mortar and mixed.

To this base powder or matrix was then added, according to the system of Vanselow and Liebig (1948), quantities of the 33 elements thought likely to occur, in 8 concentrations ranging from 1 per cent to 0.0003 per cent, or from 10,000 to 3 ppm. Mixing of these "standard" powders was done on a dilution ba-

TABLE 2—(*First Part*)

COLLECTION DATA ON 26 TREES OF *ABIES GRANDIS*, WEST OF CASCADES

Tree No.	Sp. Gr.*	Rings Per In.	Diam. Inches	Age Years	Height Feet	Elev. Feet	Ann. Ins. Rainfall†	Vigor
1	0.416	7	9.0	x	40	50	46.3	Healthy
2	0.421	8	14.2	x	60	150	41.5	Partly dead
4	0.445	18	9.3	102	45	1200	45.7	Badly suppressed
7	0.492	3	18.2	75	75	400	52.4	Very vigorous
10	0.501	6	25.4	54	124	700	39.5	Dead, logged
13	0.453	2	19.0	40	70	300	40.4	Healthy
14	0.499	11	8.9	61	78	1890	51.3	Vigorous
15	0.472	26	55.3	>300	185	1890	51.3	Old, decadent
18	0.370	5	5.4	18	22	400	45.6	Young reproduction
19	0.513	6	11.4	>38	58	470	45.6	Vigorous
21	0.542	13	10.8	65	70	120	51.4	Vigorous
22	0.569	6	17.8	35	>50	130	83.7	Almost dead
23	0.417	3	33.5	To rot 80	95	280	84.6	Healthy, vigorous
25	0.431	20	9.2	60	52	600	70.3	Nearly dead
28	0.529	5	17.9	106	72	800	46.0	Vigorous
31	0.586	10	15.5	90	63	350	40.0	Vigorous
32	0.577	4	25.2	x	130	170	39.5	Very vigorous
37	0.452	6	10.4	34	92	1350	68.5	Vigorous
39	0.482	5	10.4	31	68	550	43.1	Vigorous
40	0.336	3	8.0	17	36	100	78.0	Very healthy
41	0.485	7	22.2	>80	138	35	55.0	Healthy
42	0.410	22	54.3	>250	x	50	68.0	Dead blowdown
44	0.403	7	29.8	92	150	60	75.9	Very healthy
54	0.450	15	5.1	28	30	1200	55.0	Young, vigorous
60	0.380	8	6.5	36	36	35	26.5	Young, vigorous
62	0.506	28	13.5	98	68	50	18.8	Healthy
Mean	0.467	9.8	17.93	77.8	76.3	513	52.4	
σ	0.070	7.2	12.96	66.9	39.8	545	16.5	
σ_x	0.001	1.4	2.59	13.4	8.0	109	3.3	

*Specific gravity based on oven-dry weight and oven-dry volume.

†Rainfall data from U.S.D.A., Climate and Man; average value reported by closest weather station up to 1940.

xSignifies not determined.

>Signifies value "greater than" reported.

TABLE 2—(*Second Part*)

COLLECTION DATA ON 26 TREES OF *ABIES GRANDIS*, WEST OF CASCADES

Tree No.	Decay‡	Soil	Collection Location
1	Sound	Loam, alluvial, granite, deep	Skagit Co., Wash. 1½ mi. s. Mt. Vernon
2	Sound	Clay, glacial, bluff above Sound	British Columbia, Canada s. U.B.C. campus
4	Sound	Granitic sandy loam, glacial	Pierce Co., Wash. Pack Forest U. of W.
7	Sound	Granitic sandy loam, glacial	Mason Co., Wash. 2 mi. s. Shelton
10	Sound	Rich sandy granitic loam	Jefferson Co., Wash. Duckabush R.
13	Sound	Alluvial, sandy clay agglomerate	Clallam Co., Wash. Elwha R.
14	Sound	Granitic alluvial loam over gravel	Pierce Co., Wash. White R.
15	Partly butt-rotted	Granitic alluvial loam over gravel	Pierce Co., Wash. White R.
18	Sound	Glacial, gravelly	Thurston Co., Wash. Near Vail
19	Butt rotted, catface	Alluvial sandy loam, granitic	Lewis Co., Wash. 8 mi. e. Forest
21	Sound	Brown alluvial sandy loam	Grays Harbor Co., Wash. 8½ mi. s.e. Elma
22	Center butt rotted	Coarse clayey loam	Pacific Co., Wash. 2½ mi. e. Menlo
23	Sound	Thin clay loam over basalt	Wahkiakum Co., Wash. Cathlamet
25	Sound	Alluvial silty clay loam	Columbia Co., Ore. 2 mi. s. Mist
28	Sound	Alluvial gray clay	Washington Co., Ore. 3½ mi. n. Timber
31	Sound (wet core)	Clay over basalt	Benton Co., Ore. 4 mi. w. Corvallis
32	Sound	Deep alluvial loam	Linn Co., Ore. 5 mi. s. Albany
37	Sound	Alluvial sand and clay over basalt	Lane Co., Ore. 2.4 mi. e. Blue R.–MacKenzie R.
39	Sound	Shallow reddish loam over basalt	Douglas Co., Ore. 10 mi. w. Drain
40	Sound	Alluvial sand	Douglas Co., Ore. 3 mi. e. Stottsburg, Umpqua R.
41	Sound	Alluvial clay and sand	Coos Co., Ore. 2½ mi. n. North Bend, Coos Bay
42	Heavily rotted butt	Alluvial sand and clay	Coos Co., Ore. 16 mi. n. Port Orford
44	Sound	Deep clay	Del Norte Co., Calif. 8 mi. n. Crescent City
54	Sound	Deep alluvial sandy loam	Clackamas Co., Ore. ½ mi. e. Brightwood
60	Sound	Deep granitic sand	Skagit Co., Wash. Deception Pass State Park
62	Slight center rot	Glacial clay and gravel	Island Co., Wash. 7 mi. s. Coupeville

‡No decay from *Echinodontium tinctorium;* generally white saprots.

TABLE 3—(*First Part*)
COLLECTION DATA ON 34 TREES OF *ABIES GRANDIS*
EAST OF CASCADES

Tree No.	Sp. Gr.*	Rings Per In.	Diam. Inches	Age Years	Height Feet	Elev. Feet	Ann. Ins. Rainfall†	Vigor
34	0.456	18	13.2	103	87	3600	55.0	Healthy
45	0.659	10	20.9	95	70	3240	32.5	Healthy
49	0.515	32	4.0	46	26	4350	32.5	Suppressed
51	0.465	28	10.5	90	71	3750	38.6	Healthy
52	0.396	19	13.0	132	74	2900	84.0	Healthy
53	0.468	22	13.5	157	88	2900	84.0	Healthy
57	0.465	52	11.5	101	85	2750	38.0	Healthy
58	0.463	20	6.6	67	71	2850	16.3	Healthy
64	0.434	14	7.0	107	56	2700	51.4	Suppressed
65	0.532	15	6.5	86	17	2250	23.0	Badly suppressed
70	0.365	7	14.2	80	36	2160	22.6	Very vigorous
72	0.506	20	7.3	55	33	2240	21.5	Vigorous understory
74	0.537	22	5.8	78	31	2050	18.8	Badly suppressed
75	0.399	14	7.9	36	53	3250	28.8	Vigorous understory
78	0.415	16	7.1	60	64	2850	38.7	Vigorous
79	0.449	20	6.0	52	32	2850	38.7	Partly suppressed
80	0.505	16	7.5	49	55	2270	22.8	Vigorous
82	0.431	21	13.4	177	72	3900	22.5	Vigorous
83	0.491	9	5.9	31	65	4000	23.6	Vigorous
85	0.373	9	8.7	45	33	4750	24.7	Vigorous
88	0.385	5	16.3	155	64	5020	24.7	Vigorous
90	0.386	10	8.0	To rot 40	53	5350	21.3	Overtopped
92	0.519	13	5.4	52	41	4000	25.2	Vigorous
95	0.530	16	6.2	121	34	4100	23.0	Poor residual
96	0.492	15	10.9	108	64	2820	21.8	Poor codominant
97	0.511	6	10.9	106	63	2820	21.8	Vigorous codominant
99	0.428	15	7.0	70	65	3250	25.2	Vigorous codominant
103	0.530	29	12.0	131	90	3550	38.7	Healthy
104	0.399	20	10.5	132	90	3550	38.7	Poor, suppressed
105	0.468	15	31.4	x	120	3550	38.7	Healthy residual
107	0.385	11	18.0	166	83	4050	38.7	Poor, residual
109	0.434	11	5.0	21	16	4750	21.8	Healthy
111	0.463	20	23.9	192	112	3100	23.0	Poor, residual
112	0.464	52	7.4	111	54	3300	23.0	Badly suppressed
Mean	0.462	18.3	10.69	92.5	60.8	3377	32.5	
σ	0.061	10.5	5.82	44.5	25.1	770	15.8	
σ_x	0.011	1.8	1.01	7.9	4.4	134	2.75	

*Specific gravity based on oven-dry weight and oven-dry volume.
†Rainfall data from U.S.D.A., Climate and Man; average value reported by closest weather station up to 1940.

TABLE 3—(*Second Part*)

COLLECTION DATA ON 34 TREES OF *ABIES GRANDIS* EAST OF CASCADES

Tree No.	Decay‡	Soil	Collection Location
34	Sound	Lava beds, very little soil, pumice	Jefferson Co., Ore. 27 mi. w. Sisters
45	Sound (E.t. area)	Sandy, volcanic, basalt	Jackson Co., Ore. 4 mi. s. Ashland
49	Sound (E.t. area)	Reddish clay, volcanic	Klamath Co., Ore. Parker Mountain Pass
51	Sound (E.t. area)	Deep fine volcanic ash	Jackson Co., Ore. 2 mi. s. Union Creek R.S.
52	Badly rotted E.t.	Deep fine volcanic ash, basalt	Wasco Co., Ore. 21 mi. s.e. Timberline on U.S. 26
53	Sound (E.t. area)	Deep fine volcanic ash, basalt	Wasco Co., Ore. 21 mi. s.e. Timberline on U.S. 26
57	Sound (E.t. area)	Sandy, basaltic	Klickitat Co., Wash. 6 mi. w. Troutlake
58	Sound (E.t. area)	Alluvial over basalt	Klickitat Co., Wash. 3 mi. s. Satus Pass
64	Sound	Stony, sandy loam, granitic	Kittitas Co., Wash. 3½ mi. n. Easton on U.S. 10
65	Rotten (not E.t.)	Alluvial sand and gravel, granitic	Kittitas Co., Wash. 1½ mi. w. Cle Elum
70	Early decay in butt E.t.	Alluvial sandy loam	Pend Oreille Co., Wash. 1½ mi. n. Ruby below Newport
72	Sound	Alluvial sandy loam, granitic	Boundary Co., Ida. 11 mi. s. Bonners Ferry
74	Badly rotted E.t.	Alluvial, sandy, granitic	Lincoln Co., Mont. 3½ mi. e. Troy on U.S. 2
75	Sound	Alluvial, sandy, over gray clay	Sanders Co., Mont. 4 mi. s. Cabinet Divide
78	Sound	Alluvial sandy loam, mine tailings	Shoshone Co., Ida. ½ mi. w. Wallace
79	Sound	Alluvial sandy loam, mine tailings	Shoshone Co., Ida. ½ mi. w. Wallace
80	Sound	Alluvial sandy clay	Kootenai Co., Ida. 1 mi. w. Dudley, Rose Lake Cutoff
82	Badly rotted E.t.	Loessial, deep	Lewis Co., Ida. 1½ mi. w. Winchester
83	Sound (E.t. area)	Sandy clay loam over basalt	Idaho Co., Ida. 2 mi. s. Grangeville
85	Sound (E.t. area)	Sandy, granitic	Valley Co., Ida. 4 mi. w. Donnelley (w. side valley)
88	Badly rotted E.t.	Reddish volcanic ash, basalt	Adams Co., Ida. 4 mi. w. McCall, Red Ridge
90	Sound (E.t. area)	Sandy, granitic	Idaho Co., Ida. 13 mi. n.w. Burgdorf
9[illegible]	Sound (E.t. area)	Volcanic ash and loess	Benewah Co., Ida. 1 mi. s. Emida
95	Badly rotted E.t.	Volcanic ash and loess	Benewah Co., Ida. 9 mi. n. Potlatch, Benewah Summit
96	Badly rotted E.t.	Volcanic ash and loess, some clay	Latah Co., Ida. 1 mi. e. Deary
97	Small core rot E.t.	Volcanic ash and loess, some clay	Latah Co., Ida. 1 mi. e. Deary
99	Sound (E.t. area)	Volcanic ash and loess, deep	Shoshone Co., Ida. 5 mi. s. Clarkia
103	Sound (E.t. area)	Decomp. granite topped with loess	Clearwater Co., Ida. PFI Camp T Elkberry Creek
104	Sound (E.t. area)	Decomp. granite, some clay and loess	Clearwater Co., Ida. PFI Camp T Elkberry Creek
105	Center rot E.t.	Decomp. granite, topped with loess	Clearwater Co., Ida. PFI Camp T Elkberry Creek
107	Badly rotted E.t.	Sandy, granitic	Clearwater Co., Ida. 18 mi. e. Bovill, Robinson's Ridge
109	Sound	Granite topped with loess	Latah Co., Ida. 10 mi. n.e. Moscow
111	Badly rotted E.t.	Sandy clay loam, decomposed granite	Latah Co., Ida. 8 mi. s. Harvard, U. of I. Forest
112	Sound (E.t. area)	Reddish sandy clay loam, decomposed granite	Latah Co., Ida. 8 mi. s. Harvard, U. of I. Forest

‡Initials E.t. refer to the Indian paint fungus, *Echinodontium tinctorium*.

sis, using the base powder (matrix) as a diluent. Extreme care was taken to prevent contamination and moisture pickup throughout all operations. "Standard" spectrograms were prepared from the base powders after dilution with 1 part of internal standard with 4 parts of "standard" powder. Line intensities were estimated from these "standard" plates for all detectable concentrations of the 33 elements included. Reference spectrograms were prepared in a similar manner using iron rods and RU powder for wave length and element detection line determinations.

Both "standards" and unknowns alike were packed into specially purified graphite electrodes. A ratio of 4 parts powder ("standard" or unknown ash) to 1 part internal standard containing sodium to repress cyanogen bands was used for filling each electrode crater. Arcing was for 20 seconds, 1 minute, and 3 minutes for each "standard" or unknown, under the following conditions: electrode spacing 4 mm; lower electrode positive; DC current of 240 volts and 8 amps; slit 0.02 mm; range 2400-4800 Å on two 4" x 10" EK SA #1 plates; 16 spectra (including Fe's) $1\frac{1}{2}$ mm tall taken on each pair of spectrogram plates. Plates were developed in pairs under rigidly controlled time and temperature conditions. Intensities of elemental lines were estimated from the standard (reference) plates, and results were recorded in terms of ppm of ash for the elements concerned. In most cases 3 selected spectrum lines were estimated for each element; for a few elements only 2 lines were available in the spectral range employed. From the recorded ppm of ash data it was possible to compute ppm of dry weight using ash percentages.

Results

Reference to tables 2 and 3, which detail essential collection data for trees selected for spectrographic analyses, indicates the possibilities of numerous comparisons of mineral content of grand fir wood. At the outset it was believed that differences might exist in mineral content and composition of trees grown in the two major portions of the native range of this tree divided by the Cascade Mountain Range. Accordingly, analytical results are separated in tables 4 and 5 according to geographical location, that is, for trees grown west and east of the Cascades, respectively. Table 6 shows inherent variations within

TABLE 4—(*First Part*)

MINERAL COMPOSITION OF SOUND GRAND FIR WOOD FROM 25 TREES GROWN WEST OF CASCADES

	Major Constituents										
	Ash	Ca		Mg		K		Na		SiO_2*	
Tree No.	% of Dry Wt.	% of Ash	Ppm Wood	% of Ash	Ppm Wood	% of Ash	Ppm Wood	% of Ash	Ppm Wood	% of Ash	Ppm Wood
1	0.52	x	x	x	x	x	x	x	x	x	x
2	0.42	15.5	650	6.7	280	21.3	890	1.8	76	x	x
4	0.42	20.5	860	5.4	230	13.0	550	1.2	50	x	x
7	0.27	x	x	x	x	x	x	x	x	x	x
10	0.22	15.2	330	4.5	90	21.3	470	0.4	9	x	x
13	0.56	x	x	x	x	x	x	x	x	x	x
14	0.27	23.7	640	6.0	160	8.0	220	0.2	5	1.5	40
15	0.35	9.8	340	3.6	130	33.5	1170	0.5	18	x	x
18	0.29	x	x	x	x	x	x	x	x	x	x
19	0.24	x	x	x	x	x	x	x	x	x	x
21	0.29	13.5	390	4.8	140	26.1	760	1.8	52	x	x
22	0.49	x	x	x	x	x	x	x	x	x	x
23A	0.29	x	x	x	x	x	x	x	x	x	x
25	0.68	9.3	630	4.3	290	39.1	2660	0.5	34	x	x
28	0.28	19.8	550	3.7	100	12.5	350	0.9	25	x	x
31	0.32	x	x	x	x	x	x	x	x	x	x
32	0.37	x	x	x	x	x	x	x	x	x	x
37	0.27	x	x	x	x	x	x	x	x	x	x
39	0.23	x	x	x	x	x	x	x	x	x	x
40	0.32	13.8	440	3.3	110	24.3	780	0.3	10	x	x
41	0.19	x	x	x	x	x	x	x	x	x	x
42	0.43	x	x	x	x	x	x	x	x	x	x
44	0.28	x	x	x	x	x	x	x	x	x	x
54	0.43	17.8	770	6.0	260	22.0	950	0.6	26	x	x
60	0.79	7.5	590	5.4	430	40.3	3180	1.2	95	x	x
62	0.37	10.8	400	4.9	130	30.5	1130	2.3	85	x	x
Mean	0.331	14.8	549	4.9	200	24.3	1092	1.0	40	1.5	40
σ	0.215	4.8	164	1.0	97	9.8	870	0.6	30	x	x
σ_x	0.043	1.4	50	0.1	29	3.0	72	0.2	9	x	x

*Determined gravimetrically.

xSignifies not determined.

TABLE 4—(*Second Part*)

MINERAL COMPOSITION OF SOUND GRAND FIR WOOD FROM 26 TREES GROWN WEST OF CASCADES

	Essential Elements													
	B		Mn		Fe		Mo		Cu		Zn		P†
Tree No.	Ppm Ash	Ppm Wood	Ppm Ash	Ppm Wood	Ppm Ash	Ppm Wood	Ppm Ash	Ppm Wood	Ppm Ash	Ppm Wood	Ppm Ash	Ppm Wood	Rel. Line Intens.
1	77	0.4	7720	40.1	316	1.6	8	0.04	1000	5.2	77	0.4	MS
2	316	1.3	10000	42.0	1000	4.2	2	0.01	1000	4.2	172	0.7	W
4	316	1.3	10000	42.0	1000	4.2	2	0.01	1000	4.2	77	0.3	W
7	77	0.2	3160	8.5	544	1.5	2	0.01	1000	2.7	77	0.2	MW
10	254	0.6	3160	7.0	544	1.2	—	—	544	1.5	—	—	MW
13	316	1.8	3160	17.7	544	3.0	—	—	544	3.0	—	—	MW
14	172	0.5	7720	20.8	544	1.5	—	—	1000	2.7	316	0.9	W
15	316	1.1	7720	27.0	772	2.7	—	—	1000	3.5	—	—	W
18	1000	2.9	5440	15.8	772	2.2	—	—	1000	2.9	77	0.2	MW
19	77	0.2	3160	7.6	544	1.3	—	—	544	1.3	172	0.4	MW
21	77	0.2	7720	22.4	772	2.2	—	—	544	1.6	—	—	MW
22	77	0.4	3160	15.5	316	1.6	—	—	544	2.7	—	—	W
23A	1000	2.9	7720	22.4	1000	2.9	—	—	1000	2.9	—	—	W
25	172	1.2	2540	17.0	544	3.7	—	—	1000	6.8	—	—	—
28	77	0.2	3160	8.8	772	2.2	—	—	544	1.5	—	—	MW
31	77	0.2	5440	17.4	544	1.7	—	—	772	2.5	77	0.2	WT
32	100	0.4	7720	28.6	544	2.0	—	—	1000	3.7	—	—	MW
37	172	0.5	10000	27.0	772	2.8	—	—	1000	2.7	172	0.5	M
39	316	0.7	10000	23.0	1000	2.3	—	—	772	1.8	172	0.4	MW
40	772	2.5	10000	32.0	1000	3.2	—	—	1000	3.2	77	0.3	MW
41	316	0.6	10000	19.0	772	1.5	—	—	772	1.5	77	0.2	MW
42	—	—	5440	23.4	316	1.4	—	—	54	0.2	—	—	—
44	77	0.2	5440	15.2	772	2.2	—	—	1000	2.8	—	—	MW
54	254	1.1	7720	33.2	772	3.3	—	—	1000	4.3	172	0.7	MW
60	—	—	1720	13.5	316	2.5	—	—	316	2.5	77	0.6	T
62	254	0.9	772	2.9	254	0.9	—	—	254	0.9	77	0.3	W
Mean	256	0.9	6144	21.1	656	2.3	0.50	0.002	777	2.8	71.9	0.24	x
σ	265	0.8	2933	10.6	234	0.9	1.60	0.008	279	1.4	80.0	0.26	x
σ_x	53	0.16	587	2.1	47	0.2	0.32	0.002	56	0.3	16.0	0.05	x

†Relative intensity of lines only.
xSignifies not determined.

--Signifies not found or too little to detect.

TABLE 4—(*Third Part*)

MINERAL COMPOSITION OF SOUND GRAND FIR WOOD FROM 25 TREES GROWN WEST OF CASCADES

OTHER COMMON ELEMENTS

Tree No.	Ag Ppm Ash	Ag Ppm Wood	Al Ppm Ash	Al Ppm Wood	Ba Ppm Ash	Ba Ppm Wood	Co Ppm Ash	Co Ppm Wood	Cr Ppm Ash	Cr Ppm Wood	Ni Ppm Ash	Ni Ppm Wood	Pb Ppm Ash	Pb Ppm Wood	Rb Ppm Ash	Rb Ppm Wood	Sr Ppm Ash	Sr Ppm Wood	Ti Ppm Ash	Ti Ppm Wood
1	17	0.09	1720	8.8	772	4.0	5	0.03	2	0.01	54	0.30	25	0.13	772	4.0	1000	5.2	54	0.28
2	25	0.11	3160	13.3	3160	13.3	3	0.01	5	0.02	54	0.20	54	0.23	772	3.2	1000	4.2	100	0.42
4	25	0.11	1000	4.2	10000	42.0	5	0.02	5	0.02	32	0.13	77	0.32	254	1.1	1000	4.2	32	0.13
7	8	0.02	544	1.5	2540	6.9	3	0.01	2	< 0.01	100	0.27	—	—	100	0.3	254	0.7	2	< 0.01
10	3	0.01	316	0.7	544	1.2	2	< 0.01	8	0.02	77	0.17	32	0.07	254	0.6	254	0.6	32	0.07
13	3	0.02	1000	5.6	1000	5.6	2	0.01	5	0.03	8	0.04	3	0.02	172	1.0	544	3.0	77	0.43
14	5	0.01	316	0.9	1720	4.6	3	0.01	8	0.02	10	0.03	25	0.07	316	0.9	1000	2.7	2	< 0.01
15	2	0.01	3160	11.1	772	2.7	5	0.02	8	0.03	10	0.04	25	0.09	544	1.9	772	2.7	100	0.35
18	10	0.03	1720	4.9	2540	7.4	5	0.01	17	0.05	172	0.50	25	0.07	172	0.5	1000	2.9	54	0.16
19	17	0.04	544	1.6	2540	6.1	5	0.01	8	0.02	100	0.24	17	0.04	100	0.2	544	1.3	3	0.01
21	5	0.01	1720	4.9	2540	7.4	3	0.01	3	0.01	100	0.29	10	0.03	316	0.9	772	2.2	54	0.16
22	8	0.04	1000	4.9	3160	15.5	3	0.01	8	0.04	32	0.16	10	0.05	254	1.2	544	2.7	54	0.26
23A	25	0.07	1000	2.9	5440	15.8	3	0.01	77	0.22	172	0.50	17	0.05	544	1.6	1720	5.0	32	0.09
25	8	0.05	316	2.1	1720	11.7	2	0.01	2	0.01	8	0.05	—	—	1720	11.7	1000	6.8	8	0.05
28	3	0.01	544	1.5	3160	3.8	—	—	3	0.01	77	0.22	8	0.02	77	0.2	316	0.9	25	0.07
31	2	0.01	316	1.0	2540	3.1	2	0.01	5	0.02	32	0.10	—	—	172	0.6	1720	5.5	—	—
32	3	0.01	316	1.2	3160	11.7	—	—	8	0.03	100	0.40	8	0.03	772	2.9	772	2.9	8	0.03
37	3	0.01	1000	2.7	254	0.7	2	< 0.01	5	0.01	17	0.05	—	—	254	0.7	772	2.1	2	< 0.01
39	17	0.04	544	1.3	5440	12.5	3	0.01	77	0.18	100	0.20	8	0.02	100	0.2	1720	4.0	32	0.07
40	5	0.02	2540	8.1	5440	17.4	2	0.01	25	0.08	100	0.30	32	0.10	544	1.7	772	2.5	100	0.32
41	2	< 0.01	1720	3.2	3160	5.0	3	0.01	17	0.03	100	0.20	8	0.02	100	0.2	1720	3.3	32	0.06
42	3	0.01	772	3.3	772	3.3	—	—	3	0.01	8	0.03	2	0.01	100	0.4	316	1.4	8	0.03
44	5	0.01	2540	7.0	3160	3.8	2	0.01	10	0.03	254	0.70	10	0.03	544	1.5	772	2.2	25	0.07
54	8	0.03	1720	7.3	7720	33.2	2	0.01	5	0.02	25	0.10	25	0.10	1000	4.3	1000	4.3	77	0.33
60	3	0.02	1720	13.4	254	2.0	—	—	2	0.01	17	0.13	3	0.02	254	2.0	1000	7.9	25	0.20
62	2	0.01	3160	11.7	2540	9.4	—	—	3	0.01	8	0.03	3	0.01	172	0.6	316	1.2	8	0.03
Mean	8.3	0.03	1323	5.0	2924	10.2	2.4	0.01	12.3	0.035	58	0.15	16.4	0.06	399	1.7	869	3.2	36	0.14
σ	6.1	0.03	927	3.9	2269	9.2	1.6	0.01	19.4	0.05	51	0.17	17.7	0.07	366	2.3	442	1.8	32	0.14
σ_x	1.2	0.006	186	0.8	454	1.8	0.3	0.0001	3.9	0.01	12	0.035	3.6	0.001	73	0.5	88	0.4	6.4	0.03

—Signifies not found or too little to detect.

< Signifies "less than" reported.

TABLE 4—(*Fourth Part*)

MINERAL COMPOSITION OF SOUND GRAND FIR WOOD FROM 26 TREES GROWN WEST OF CASCADES

	Uncommon Elements									
	Au		In		La		Li		Sn	
Tree No.	Ppm Ash	Ppm Wood	Ppm Ash	Ppm Wood	Ppm Ash	Ppm Wood	Ppm Ash	Ppm Wood	Ppm Ash	Ppm Wood
1	—	—	54	0.28	8	0.04	—	—	—	—
2	—	—	254	1.07	8	0.03	—	—	8	0.03
4	—	—	—	—	5	0.02	—	—	8	0.03
7	—	—	—	—	5	0.01	—	—	—	—
10	—	—	—	—	3	0.01	—	—	—	—
13	—	—	—	—	5	0.03	—	—	—	—
14	—	—	—	—	2	< 0.01	—	—	—	—
15	—	—	—	—	—	—	316	1.11	—	—
18	—	—	—	—	8	0.02	—	—	—	—
19	—	—	—	—	5	0.01	—	—	—	—
21	—	—	—	—	3	0.01	—	—	—	—
22	—	—	—	—	5	0.02	—	—	—	—
23A	—	—	—	—	2	0.01	—	—	—	—
25	—	—	—	—	—	—	77	0.52	—	—
28	—	—	—	—	—	—	—	—	—	—
31	—	—	—	—	—	—	—	—	—	—
32	—	—	—	—	3	0.01	—	—	—	—
37	—	—	—	—	—	—	—	—	—	—
39	—	—	—	—	—	—	—	—	—	—
40	25	0.08	—	—	5	0.02	—	—	—	—
41	—	—	—	—	10	0.02	—	—	—	—
42	—	—	—	—	5	0.02	—	—	—	—
44	—	—	—	—	8	0.02	—	—	—	—
54	—	—	—	—	—	—	—	—	—	—
60	—	—	—	—	8	0.06	—	—	—	—
62	—	—	—	—	8	0.03	—	—	—	—
Mean	1.0	0.003	11.8	0.05	4.0	0.02	15.0	0.006	0.6	0.002
σ	4.8	0.015	15.7	0.21	3.2	0.02	62.0	0.240	2.1	0.005
σ_x	1.0	0.003	3.1	0.04	0.6	0.003	12.4	0.048	0.4	0.001

—Signifies not found or too little to detect.
No Ga, V, or Zr found.

< Signifies "less than" reported.

TABLE 5—(*First Part*)

MINERAL COMPOSITION OF SOUND GRAND FIR WOOD FROM 34 TREES GROWN EAST OF CASCADES

	Major Constituents											
	Ash	Ca		Mg		K		Na		SiO_2*	
Tree No.	% of Dry Wt.	% of Ash	Ppm Wood	% of Ash	Ppm Wood	% of Ash	Ppm Wood	% of Ash	Ppm Wood	% of Ash	Ppm Wood
34	0.34	19.3	560	4.5	150	13.5	[illegible]60	0.2	7	x	x
45	0.70	30.3	2120	2.3	160	5.0	[illegible]50	0.4	28	x	x
49	0.45	24.0	1080	5.4	240	8.5	[illegible]80	0.1	5	x	x
51	0.52	x	x	x	x	x	x	x	x	x	x
52	0.50	22.5	1130	4.9	250	7.8	[illegible]90	0.1	7	0.8	40
53	0.33	21.5	710	5.6	180	11.3	[illegible]70	0.4	13	x	x
57	0.34	x	x	x	x	x	x	x	x	x	x
58	0.31	22.8	710	4.9	150	12.0	[illegible]70	0.2	6	x	x
64	0.34	x	x	x	x	x	x	x	x	x	x
65	0.28	21.0	590	3.9	110	13.0	[illegible]60	0.4	11	x	x
70	0.43	14.3	510	3.3	140	19.0	[illegible]20	2.1	90	x	x
72	0.40	x	x	x	x	x	x	x	x	x	x
74	1.90	x	x	x	x	x	x	x	x	x	x
75	0.31	19.8	510	3.9	120	16.3	[illegible]10	0.2	6	x	x
78	0.48	x	x	x	x	x	x	x	x	x	x
79	0.39	25.0	980	5.0	200	8.8	[illegible]40	0.2	8	x	x
80	0.65	x	x	x	x	x	x	x	x	x	x
82	0.63	17.8	1120	5.0	320	10.8	[illegible]80	0.2	12	0.1	6
83	0.32	15.8	510	3.3	110	12.8	[illegible]10	0.1	3	x	x
85	0.33	x	x	x	x	x	x	x	x	x	x
88	0.36	21.5	770	3.1	110	14.0	[illegible]00	0.4	14	x	x
90	0.36	24.5	910	3.1	110	8.5	[illegible]10	0.1	4	x	x
92	0.31	x	x	x	x	x	x	x	x	x	x
95	0.38	x	x	x	x	x	x	x	x	x	x
96	0.38	18.0	580	3.4	130	16.3	[illegible]20	0.3	11	x	x
97	0.35	x	x	x	x	x	x	x	x	x	x
99	0.38	22.0	840	3.3	130	10.0	[illegible]80	0.8	30	x	x
103	0.38	26.3	1000	3.3	130	8.3	[illegible]20	0.4	15	x	x
104	0.45	x	x	x	x	x	x	x	x	x	x
105	0.28	x	x	x	x	x	x	x	x	x	x
107	0.26	19.6	510	3.0	78	11.9	[illegible]10	0.2	5	x	x
109	0.33	18.5	610	4.3	140	15.3	[illegible]00	0.1	3	x	x
111	0.51	17.0	870	2.4	120	23.3	1[illegible]90	0.2	10	x	x
112	0.48	26.8	1290	3.3	160	5.5	[illegible]60	0.3	14	x	x
Mean	0.446	21.4	872	3.9	154	12.0	[illegible]68	0.35	14	0.4	23
σ	0.275	3.8	354	0.9	56	4.3	[illegible]09	0.42	15	—	—
σ_x	0.048	0.9	79	0.2	12	1.0	47	0.09	3	—	—

xSignifies not determined. *Determined gravimetrically. —Signifies not found or too little to detect.

TABLE 5—(*Second Part*)

MINERAL COMPOSITION OF SOUND GRAND FIR WOOD FROM 34 TREES GROWN EAST OF CASCADES

	Essential Elements												
	B		Mn		Mo		Fe		Cu		Zn		P†
Tree No.	Ppm Ash	Ppm Wood	Ppm Ash	Ppm Wood	Ppm Ash	Ppm Wood	Ppm Ash	Ppm Wood	Ppm Ash	Ppm Wood	Ppm Ash	Ppm Wood	Rel. Line Intens.
34	254	0.9	5440	18.5	1000	3.4	—	—	544	1.8	1000	3.4	MW
45	77	0.5	1720	12.0	544	3.8	—	—	316	2.2	316	2.2	W
49	316	1.4	5440	24.5	544	2.4	8	0.04	316	1.4	316	1.4	W
51	172	0.9	5440	28.3	316	1.6	8	0.04	544	2.8	1000	5.2	W
52	77	0.4	1720	8.5	254	1.3	—	—	316	1.6	—	—	T
53	172	0.6	5440	18.0	544	1.8	—	—	544	1.8	1000	3.3	WT
57	172	0.6	3160	10.7	316	1.1	—	—	544	1.8	316	1.1	WT
58	100	0.3	5440	16.9	772	2.4	—	—	316	1.0	316	1.0	W
64	100	0.3	5440	18.5	316	1.1	8	0.03	172	0.6	316	1.1	T
65	254	0.7	5440	15.2	1000	2.8	—	—	316	0.9	—	—	W
70	77	0.3	1720	7.4	772	3.3	—	—	1000	4.3	—	—	WT
72	254	1.0	3160	12.6	772	3.1	—	—	1000	4.0	1000	4.0	T
74	77	1.5	5440	103.4	316	6.0	—	—	316	6.0	—	—	WT
75	254	0.8	3160	9.8	1000	3.0	—	—	1000	3.0	1720	5.4	WT
78	172	0.8	5440	26.1	772	3.7	—	—	254	1.2	254	1.2	T
79	77	0.3	1720	6.7	772	3.0	—	—	544	2.1	—	—	T
80	77	0.5	1720	11.2	316	2.0	3	0.02	544	3.5	—	—	—
82	172	1.1	3160	19.9	772	4.9	3	0.02	316	2.0	—	—	—
83	316	1.0	5440	17.4	1000	3.2	—	—	1000	3.2	254	0.8	WT
85	254	0.8	5440	18.0	772	2.6	8	0.03	544	1.8	1000	3.3	WT
88	172	0.6	5440	19.6	772	2.8	—	—	544	2.0	1720	6.2	W
90	316	1.1	3160	11.4	772	2.8	—	—	316	1.1	—	—	MW
92	254	0.8	3160	9.8	1000	3.1	3	0.01	544	1.7	254	0.8	WT
95	x	x	3160	12.0	544	2.1	—	—	544	2.1	254	1.0	x
96	54	0.2	1720	6.5	772	2.9	—	—	316	1.2	—	—	—
97	77	0.3	1720	6.0	772	2.7	3	0.01	544	1.9	—	—	—
99	544	2.1	5440	20.7	1000	3.8	—	—	544	2.1	254	1.0	T
103	544	2.1	5440	20.7	1000	3.8	—	—	544	2.1	254	1.0	T
104	316	1.4	5440	24.5	544	2.4	—	—	1000	4.5	254	1.1	T
105	100	0.3	1720	4.8	772	2.2	—	—	544	1.5	—	—	WT
107	544	1.4	3160	8.2	772	2.0	—	—	1000	2.6	254	0.7	W
109	772	2.6	5440	18.0	1000	3.3	3	0.01	1000	3.3	772	2.6	WT
111	100	0.5	3160	16.1	772	3.9	—	—	316	1.6	—	—	T
112	544	2.6	5440	26.1	1000	4.8	8	0.04	1000	4.8	254	1.2	T
Mean	235	0.9	3961	17.9	707	2.9	1.6	0.007	565	2.3	385	1.4	—
σ	174	0.6	1568	16.2	251	1.1	2.8	0.013	265	1.2	472	1.7	—
σ_x	31	0.1	273	2.8	44	0.2	0.5	0.002	46	0.2	82	0.3	—

†Relative intensity of lines only. xSignifies not determined. —Signifies not found or too little to detect.

TABLE 5—(*Third Part*)

MINERAL COMPOSITION OF SOUND GRAND FIR WOOD FROM 34 TREES GROWN EAST OF CASCADES

OTHER COMMON ELEMENTS

Tree No.	Ag		Al		Ba		Co		Cr		Ni		Pb		Rb		Sr		Ti	
	Ppm Ash	Ppm Wood	Ppm Ash	Ppm Wood	Ppm Ash	Ppm Wood	Ppm Ash	Ppm Wood	Ppm Ash	Ppm Wood	Ppm Ash	Ppm Wood	Ppm Ash	Ppm Wood	Ppm Ash	Ppm Wood	Ppm Ash	Ppm Wood	Ppm Ash	Ppm Wood
34	32	0.11	5440	18.5	3160	10.7	100	0.34	17	0.06	100	0.34	25	0.09	1720	5.8	3160	10.7	17	0.06
45	10	0.07	772	5.4	1720	12.0	2	0.01	3	0.02	10	0.07	3	0.02	—	—	2540	17.8	8	0.06
49	17	0.08	1720	7.7	3160	14.2	2	0.01	5	0.02	25	0.11	10	0.04	254	1.1	5440	24.5	8	0.04
51	32	0.17	2540	13.2	7720	40.1	2	0.01	8	0.04	17	0.09	17	0.09	1720	8.9	10000	52.0	3	0.02
52	8	0.04	1720	8.5	1720	8.5	3	0.02	8	0.04	5	0.02	—	—	172	0.8	1000	5.0	5	0.02
53	772	2.55	1720	5.6	7720	25.5	2	0.06	8	0.03	10	0.03	54	0.18	544	1.8	5440	18.0	8	0.03
57	5	0.02	772	2.6	7720	26.2	2	0.01	17	0.06	8	0.03	3	0.01	1720	5.8	3160	10.7	10	0.03
58	10	0.03	1000	3.1	5440	16.0	3	0.01	5	0.02	10	0.03	10	0.03	172	0.5	3160	9.8	17	0.05
64	17	0.06	1000	3.4	5440	18.5	3	0.01	17	0.06	17	0.06	3	0.01	172	0.6	1000	3.4	54	0.18
65	32	0.09	1000	2.8	2540	7.1	2	0.05	5	0.01	25	0.07	3	0.01	1000	2.8	3160	8.8	32	0.09
70	3	0.01	1000	4.3	1000	4.3	2	0.01	8	0.03	10	0.04	17	0.07	544	2.3	5440	23.4	32	0.14
72	54	0.22	1000	4.0	3160	12.5	2	0.01	100	0.40	10	0.04	25	0.10	254	1.0	5440	21.8	17	0.07
74	254	4.83	544	10 3	>10000	>190.0	2	0.03	5	0.10	3	0.15	17	0.32	254	4.8	1000	19.0	5	0.10
75	17	0.05	772	2.4	10000	31.0	2	0.01	10	0.03	54	0.17	77	0.24	772	2.4	3160	9.8	32	0.10
78	17	0.08	1000	4.8	2540	12.2	2	0.01	2	0.01	5	0.02	8	0.04	77	0.4	1720	8.3	17	0.08
79	10	0.04	1000	3.9	5440	21.2	2	0.01	8	0.03	3	0.03	772	3.01	172	0.7	772	3.0	5	0.02
80	10	0.06	1000	6.5	3160	20.5	—	—	3	0.02	5	0.03	3	0.02	1000	6.5	3160	20.5	32	0.20
82	3	0.02	1000	6.3	>10000	>63.0	2	0.01	25	0.16	17	0.11	—	—	254	1.6	5440	34.3	5	0.03
83	25	0.08	1720	5.5	5440	17.4	32	0.10	10	0.03	172	0.55	54	0.17	544	1.7	1720	5.5	32	0.10
85	8	0.03	3160	10.4	5440	18.0	3	0.01	32	0.10	17	0.06	10	0.03	544	1.8	10000	33.0	17	0.06
88	10	0.04	1720	6.2	10000	36.0	5	0.02	17	0.06	8	0.03	32	0.11	772	2.8	1720	6.2	25	0.09
90	3	0.01	772	2.8	3160	11.4	2	0.01	5	0.02	10	0.04	3	0.01	25	0.1	772	2.8	25	0.09
92	10	0.03	772	2.4	2540	7.3	5	0.02	77	0.24	5	0.06	10	0.03	772	2.4	5440	16.9	25	0.08
95	17	0.06	772	2.9	5440	20.7	2	0.01	8	0.03	17	0.02	2	0.01	544	2.1	772	2.9	17	0.06
96	32	0.12	1000	3.8	2540	9.5	3	0.01	3	0.01	3	0.01	10	0.04	544	2.1	2540	9.6	17	0.06
97	17	0.06	2540	8.9	7720	27.0	2	0.01	5	0.02	17	0.06	8	0.03	544	1.9	3160	11.1	5	0.02
99	17	0.06	1000	3.8	10000	38.0	2	0.01	5	0.02	25	0.10	25	0.10	544	2.1	3160	12.0	17	0.06
103	32	0.12	1720	6.5	10000	38.0	3	0.01	3	0.01	10	0.04	54	0.21	316	1.2	2540	9.6	17	0.06
104	54	0.24	772	3.5	10000	45.0	2	0.01	10	0.04	17	0.08	2	0.01	316	1.4	5440	24.5	17	0.08
105	54	0.15	772	2.2	3160	8.8	3	0.01	10	0.03	17	0.05	2	0.01	54	0.2	772	2.2	54	0.15
107	172	0.44	772	2.0	10000	26.0	5	0.01	10	0.03	17	0.04	100	0.30	772	2.0	10000	26.0	32	0.08
109	254	0.80	772	2.6	10000	33.0	2	0.01	25	0.08	17	0.06	54	0.18	772	2.6	10000	33.0	32	0.10
111	100	0.51	1000	5.1	5440	27.7	2	0.01	3	0.02	5	0.02	17	0.09	316	1.6	1720	8.8	32	0.16
112	316	1.52	2540	12.2	10000	48.0	5	0.02	8	0.04	25	0.12	32	0.15	544	2.6	5440	26.1	17	0.08
Mean	71	0.38	1377	5.7	5956	27.8	6.1	0.02	14	0.06	21.4	0.08	43.0	0.17	551	2.25	3806	15.6	20.2	0.08
σ	144	0.92	947	3.7	3138	31.2	19.8	0.06	20	0.08	31.4	0.10	132.5	0.50	450	1.94	2770	10.8	12.9	0.045
σ_x	25	0.16	165	0.6	546	5.4	3.4	0.001	3.5	0.013	5.5	0.02	23.1	0.09	78	0.34	482	1.9	2.2	0.008

—Signifies not found or too little to detect.

>Signifies value "greater than" reported

TABLE 5—(*Fourth Part*)

MINERAL COMPOSITION OF SOUND GRAND FIR WOOD FROM 34 TREES GROWN EAST OF CASCADES

	Uncommon Elements													
	Au		Ga		In		La		Li		Sn		Zr	
Tree No.	Ppm Ash	Ppm Wood	Ppm Ash	Ppm Wood	Ppm Ash	Ppm Wood	Ppm Ash	Ppm Wood	Ppm Ash	Ppm Wood	Ppm Ash	Ppm Wood	Ppm Ash	Ppm Wood
34	—	—	—	—	8	0.03	10	0.03	77	0.26	—	—	—	—
45	—	—	—	—	—	—	77	0.54	—	—	—	—	—	—
49	77	0.34	—	—	—	—	8	0.04	—	—	—	—	—	—
51	77	0.40	—	—	8	0.04	25	0.13	—	—	—	—	—	—
52	—	—	—	—	—	—	—	—	32	0.16	—	—	—	—
53	—	—	—	—	8	0.03	—	—	—	—	—	—	—	—
57	—	—	—	—	25	0.08	—	—	—	—	—	—	—	—
58	—	—	—	—	8	0.02	10	0.03	—	—	—	—	—	—
64	77	0.26	—	—	8	0.03	8	0.03	—	—	—	—	—	—
65	—	—	—	—	—	—	—	—	—	—	—	—	—	—
70	—	—	—	—	—	—	25	0.11	100	0.43	—	—	—	—
72	—	—	—	—	—	—	8	0.03	54	0.22	—	—	8	0.03
74	—	—	—	—	—	—	—	—	54	1.03	—	—	—	—
75	—	—	—	—	—	—	32	0.10	54	0.17	10	0.03	—	—
78	—	—	—	—	—	—	77	0.37	54	0.26	—	—	—	—
79	—	—	—	—	—	—	8	0.03	32	0.12	—	—	10	0.04
80	—	—	—	—	—	—	25	0.16	77	0.50	—	—	—	—
82	—	—	—	—	—	—	—	—	54	0.34	—	—	—	—
83	—	—	—	—	—	—	8	0.03	—	—	—	—	—	—
85	—	—	—	—	—	—	—	—	54	0.18	—	—	—	—
88	—	—	—	—	—	—	—	—	54	0.20	—	—	—	—
90	—	—	—	—	—	—	8	0.03	32	0.11	—	—	—	—
92	—	—	—	—	—	—	—	—	100	0.31	—	—	8	0.02
95	—	—	—	—	32	0.12	—	—	54	0.21	—	—	—	—
96	—	—	—	—	—	—	25	0.10	77	0.29	—	—	—	—
97	—	—	—	—	—	—	—	—	77	0.27	—	—	—	—
99	—	—	—	—	—	—	8	0.03	100	0.38	—	—	—	—
103	—	—	—	—	—	—	10	0.04	77	0.29	—	—	—	—
104	—	—	—	—	8	0.03	10	0.04	54	0.24	—	—	—	—
105	—	—	—	—	10	0.03	8	0.02	—	—	10	0.03	—	—
107	—	—	25	0.06	—	—	25	0.06	77	0.20	—	—	—	—
109	—	—	—	—	—	—	25	0.08	54	0.18	—	—	—	—
111	—	—	—	—	—	—	10	0.05	54	0.28	—	—	—	—
112	100	0.48	—	—	10	0.05	8	0.04	54	0.26	25	0.12	—	—
Mean	9.7	0.04	0.7	0.002	3.7	0.014	13.5	0.06	44.3	0.20	1.3	0.005	0.8	0.003
σ	26.9	0.12	—	—	7.2	0.026	18.4	0.11	32.9	0.20	4.7	0.021	2.2	0.009
σ_x	4.7	0.02	—	—	1.3	0.004	3.2	0.02	5.7	0.04	0.8	0.004	0.4	0.002

—Signifies not found or too little to detect.

No V found.

TABLE 6—(*First Part*)

MINERAL COMPOSITION OF SOUND GRAND FIR WOOD AT DIFFERENT HEIGHTS IN THE SAME TREE
(EXPRESSED AS PPM ELEMENT, BASED ON OVEN-DRY WEIGHT OF WOOD)

Tree No.	Ash % of Dry Wt.	Ca	Mg	K	Na	SiO_2*	B	Mn	Fe	Mo	Cu	Zn
62	0.37	400	180	1130	85	x	0.9	2.9	0.9	—	0.9	0.3
62 (15′)	0.43	730	270	630	20	33	1.1	23.4	4.3	—	2.3	2.3
82	0.63	1120	320	680	12	6	1.1	19.9	4.9	0.02	2.0	—
82 (32′)	0.79	1090	350	1190	12	8	0.2	7.9	2.5	—	4.3	—
107	0.26	510	78	310	5	x	1.4	8.2	2.0	—	2.6	0.7
107 (28′)	0.52	550	220	1070	8	31	x	13.2	4.0	—	1.3	—
111	0.51	870	120	1190	10	x	0.5	16.1	3.9	—	1.6	—
111 (70′)	1.06	x	x	x	x	x	—	8.2	2.7	0.03	1.1	—

*Determined gravimetrically.
xSignifies not determined.

—Indicates not found or too little to detect.
No Au, In, or V found.

TABLE 6—(*Second Part*)

Tree No.	Ash % of Dry Wt.	Ag	Al	Ba	Co	Cr	Ni	Pb	Rb	Sr	Ti	Ga	La	Li	Sn	Zr
62	0.37	0.01	11.7	9.4	—	0.01	0.03	0.01	0.6	1.2	0.03	—	0.03	—	—	—
62 (15′)	0.43	0.33	13.6	23.4	0.02	0.33	0.04	0.23	1.3	13.6	0.11	—	0.02	0.33	—	—
82	0.63	0.02	6.3	>63.0	0.01	0.16	0.11	—	1.6	34.3	0.03	—	—	0.34	—	—
82 (32′)	0.79	0.06	7.9	25.0	0.01	0.06	0.04	—	0.2	13.6	0.06	—	0.13	0.25	—	—
107	0.26	0.44	2.0	26.0	0.01	0.03	0.04	0.30	2.0	26.0	0.08	0.06	0.06	0.20	—	—
107 (28′)	0.52	0.04	5.2	16.4	0.01	0.04	0.03	0.01	2.8	13.2	0.17	—	—	0.17	—	0.09
111	0.51	0.51	5.1	27.7	0.01	0.02	0.02	0.09	1.6	8.8	0.16	—	0.05	0.28	—	—
111 (70′)	1.06	0.34	57.7	33.5	0.03	0.11	0.05	0.08	5.8	106.0	0.57	—	0.11	—	0.08	0.11

—Indicates not found or too little to detect.
No Au, In, or V found.

>Signifies value "greater than" reported.

selected individual trees for wood taken from different heights in the tree.

The influence of two parent soil materials on mineral composition of grand fir wood is illustrated in table 7, which compares average analyses for samples grown on basaltic and granitic soil types for areas west and east of the Cascades. Tables 8 and 9 present statistical data, including significance of differences in means, for comparing analyses of mineral content as follows: wood from trees grown in the two major areas mentioned; wood from trees grown on basaltic and granitic soils east of the Cascades; and wood from sound and rotten portions of selected trees. Table 10 gives data on ash content and specific gravity of matched sound and rotten wood specimens.

Results are expressed generally both as ppm of the element for ash (ppm ash), and ppm of the element for oven-dry weight (ppm dry wt.). Mineral elements are classified as major elements, essential elements, other common elements, and uncommon elements, as indicated in table 1. Elements of doubtful occurrence and those sought but not found are also listed in table 1. All spectrographic results, as previously explained, should be considered as semiquantitative since they are based on comparisons with a limited concentration range of standards of similar composition; they are comparative, however, and follow published data very closely. Failure to report an element does not necessarily mean that the element was missing, but, more likely, that it occurred below detection limits (generally from about 3 to 32 ppm ash). Spectrophotometric results for the major elements Ca, Mg, K, and Na should be considered accurate values for these elements.

Discussion

For convenience this discussion will be divided into a series of short statements with reference to specific observations and findings.

1. *Mineral composition of grand fir wood ash (ppm dry weight only) for different heights in the same tree* (table 6)

No consistent pattern is apparent except that the wood of higher portions of the tree contains more ash and higher values for Al and Ti. It is believed that these results illustrate normal variations within sound portions of individual trees.

TABLE 7—(*First Part*)

MINERAL COMPOSITION OF SOUND GRAND FIR WOOD FROM TREES GROWN ON BASALTIC AND GRANITIC SOILS, WEST AND EAST OF CASCADES

					Major Constituents										
					Ash	Ca		Mg		K		Na		SiO_2†	
No. Samples	Tree Nos.	Area	Soil Type*		% of Dry Wt.	% of Ash	Ppm Dry Wt.	% of Ash	Ppm Dry Wt.	% of Ash	Ppm Dry Wt.	% of Ash	Ppm Dry Wt.	% of Ash	Ppm Dry Wt.
4	23A, 31, 37, 39	West	Basaltic	Mean	0.28	x	x	x	x	x	x	x	x	x	x
				σ	0.03	x	x	x	x	x	x	x	x	x	x
				σ_x	0.02	x	x	x	x	x	x	x	x	x	x
4	4, 10, 14, 60	West	Granitic	Mean	0.42	16.7	605	5.3	228	20.6	1105	0.8	39.8	1.5	40
				σ	0.22	6.1	188	0.5	129	7.2	671	0.5	36.4	x	x
				σ_x	0.13	3.5	109	0.3	75	4.1	387	0.3	21.0	x	x
8	34, 45, 52, 57, 83, 88, 92, 96	East	Basaltic (& loess)	Mean	0.41	21.2	979	3.6	152	11.6	455	0.2	11.9	0.8	40
				σ	0.12	4.6	544	0.9	48	3.9	88	0.1	7.8	x	x
				σ_x	0.05	2.1	243	0.4	21	1.7	39	0.06	3.5	x	x
8	64, 72, 74, 85, 90, 103, 107, 109	East	Granitic	Mean	0.54	22.2	758	3.4	114	11.0	360	0.2	6.8	x	x
				σ	0.52	3.3	203	0.5	24	2.9	81	0.1	4.7	x	x
				σ_x	0.20	1.9	117	0.3	14	1.7	47	0.07	2.7	x	x

*Soils from basaltic or granitic parent materials, often topped with glacial drift west of Cascades and with loess east of Cascades.

†Determined gravimetrically.

‡Signifies not determined.

TABLE 7—(*Second Part*)

MINERAL COMPOSITION OF SOUND GRAND FIR WOOD FROM TREES GROWN ON BASALTIC AND GRANITIC SOILS, WEST AND EAST OF CASCADES

					Essential Elements													
					Ash	B		Mn		Fe		Mo		Cu		Zn		
No. Samples	Tree Nos.	Area	Soil Type*		% of Dry Wt.	Ppm Ash	Ppm Wood	Ppm Ash	Ppm Wood	Ppm Ash	Ppm Wood	Ppm Ash	Ppm Wood	Ppm Ash	Ppm Wood	Ppm Ash	Ppm Wood	
4	23A, 31, 37, 39	West	Basaltic	Mean	0.28	391	1.1	8290	22.4	829	2.4	—	—	886	2.5	105	0.3	
				σ	0.03	362	1.1	1917	3.4	192	0.5	x	x	114	0.5	72	0.2	
				σ_x	0.02	209	0.6	1107	2.0	111	0.3	x	x	66	0.3	42	0.1	
4	4, 10, 14, 60	West	Granitic	Mean	0.42	186	0.6	5650	20.8	601	2.4	0.4	< 0.01	715	2.7	118	0.4	
				σ	0.22	119	0.5	3349	13.2	248	1.2	x	x	296	1.0	119	0.3	
				σ_x	0.13	68	0.3	1933	7.6	143	0.7	x	x	171	0.6	69	0.2	
8	34, 45, 52, 57, 83, 88, 92, 96	East	Basaltic (& loess)	Mean	0.41	172	0.6	3475	12.9	707	2.7	0.4	< 0.01	516	1.9	482	1.8	
				σ	0.12	91	0.2	1621	4.6	285	0.9	x	x	211	0.6	550	2.0	
				σ_x	0.05	34	0.1	613	1.7	108	0.3	x	x	80	0.2	208	0.7	
8	64, 72, 74, 85, 90, 103, 107, 109	East	Granitic	Mean	0.54	358	1.4	4585	26.4	715	3.1	2.4	0.01	612	2.7	450	1.6	
				σ	0.52	226	0.7	1104	29.4	248	1.3	3.4	0.01	322	1.6	389	1.4	
				σ_x	0.20	86	0.3	417	11.1	94	0.5	1.3	0.01	122	0.6	147	0.5	

*Soils from basaltic or granitic parent materials, often topped with glacial drift west of Cascades and with loess east of Cascades.
xSignifies not determined.

< Signifies "less than" reported.
—Indicates not found or too little to detect.

TABLE 7—(*Third Part*)

MINERAL COMPOSITION OF SOUND GRAND FIR WOOD FROM TREES GROWN ON BASALTIC AND GRANITIC SOILS, WEST AND EAST OF CASCADES

OTHER COMMON ELEMENTS

No. Samples	Tree Nos.	Area	Soil Type*		Ash	Ag		Al		Ba		Co		Cr	
					% of Dry Wt.	Ppm Ash	Ppm Wood	Ppm Ash	Ppm Wood	Ppm Ash	Ppm Wood	Ppm Ash	Ppm Wood	Ppm Ash	Ppm Wood
4	23A, 31, 37, 39	West	Basaltic	Mean	0.28	11.7	0.03	715	2.0	3419	9.3	2.4	0.01	41.0	0.11
				σ	0.03	9.7	0.03	296	0.8	2177	5.7	0.6	<0.01	36.0	0.09
				σ_x	0.02	5.6	0.02	171	0.5	1257	3.3	0.4	<0.01	20.8	0.05
4	4, 10 14, 60	West	Granitic	Mean	0.42	9.0	0.04	838	4.8	3130	12.4	2.4	0.01	5.7	0.02
				σ	0.22	9.3	0.04	581	5.2	4004	17.1	1.8	0.01	2.6	<0.01
				σ_x	0.13	5.4	0.02	335	3.0	2312	9.9	1.1	<0.01	1.5	<0.01
8	34, 45, 52, 57, 83, 88, 92, 96	East	Basaltic (& loess)	Mean	0.41	16.5	0.06	1740	6.6	4355	16.0	19.0	0.07	19.0	0.06
				σ	0.12	10.5	0.03	1460	4.9	2878	9.4	32.0	0.11	22.6	0.07
				σ_x	0.05	4.0	0.01	552	1.8	1088	3.6	12.1	0.04	8.5	0.03
8	64, 72, 74, 85, 90, 103, 107, 109	East	Granitic	Mean	0.54	99.2	0.81	1218	5.2	7150	43.4	2.6	0.01	24.6	0.10
				σ	0.52	102.5	1.54	803	3.2	2962	56.1	1.1	0.01	30.1	0.12
				σ_x	0.20	38.7	0.58	303	1.2	1119	21.2	0.4	<0.01	11.4	0.04

*Soils from basaltic or granitic parent materials, often topped with glacial drift west of Cascades and with loess east of Cascades.

< Signifies "less than" reported.

TABLE 7—(*Fourth Part*)

MINERAL COMPOSITION OF SOUND GRAND FIR WOOD FROM TREES GROWN ON BASALTIC AND GRANITIC SOILS, WEST AND EAST OF CASCADES

OTHER COMMON ELEMENTS

No. Samples	Tree Nos.	Area	Soil Type*		Ash	Ni		Pb		Rb		Sr		Ti	
					% of Dry Wt.	Ppm Ash	Ppm Wood	Ppm Ash	Ppm Wood	Ppm Ash	Ppm Wood	Ppm Ash	Ppm Wood	Ppm Ash	Ppm Wood
4	23A, 31, 37, 39	West	Basaltic	Mean	0.28	80.2	0.17	6.2	0.02	268	0.8	1483	4.1	16.4	0.04
				σ	0.03	61.5	0.19	7.0	0.02	169	0.5	410	1.3	15.6	0.04
				σ_x	0.02	35.5	0.11	4.0	0.01	97	0.3	237	0.8	9.0	0.02
4	4, 10, 14, 60	West	Granitic	Mean	0.42	34.0	0.12	34.2	0.12	270	1.1	814	3.8	22.7	0.10
				σ	0.22	26.1	0.05	26.9	0.12	26	0.5	323	2.7	12.4	0.07
				σ_x	0.13	15.1	0.03	5.5	0.07	15	0.3	186	1.6	7.2	0.04
8	34, 45, 52, 57, 83, 88, 92, 96	East	Basaltic (& loess)	Mean	0.41	38.9	0.13	17.1	0.06	780	2.7	2660	10.3	17.4	0.06
				σ	0.12	58.9	0.21	17.5	0.06	598	2.0	1264	4.6	8.8	0.03
				σ_x	0.05	22.3	0.08	6.6	0.02	226	0.8	478	1.7	3.3	0.01
8	64, 72, 74, 85, 90, 103, 107, 109	East	Granitic	Mean	0.54	13.2	0.06	33.2	0.14	389	1.8	5094	18.6	24.9	0.09
				σ	0.52	3.8	0.03	33.2	0.12	260	1.4	4048	11.4	13.8	0.04
				σ_x	0.20	1.4	0.01	12.5	0.04	98	0.5	1530	4.3	5.2	0.01

*Soils from basaltic or granitic parent materials, often topped with glacial drift west of Cascades and with loess east of Cascades.

TABLE 7—(*Fifth Part*)

MINERAL COMPOSITION OF SOUND GRAND FIR WOOD FROM TREES GROWN ON BASALTIC AND GRANITIC SOILS, WEST AND EAST OF CASCADES

UNCOMMON ELEMENTS

No. Samples	Tree Nos.	Area	Soil Type*		Ash	Au		Ga		In		La		Li		Sn		Zr	
					% of Dry Wt.	Ppm Ash	Ppm Wood	Ppm Ash	Ppm Wood	Ppm Ash	Ppm Wood	Ppm Ash	Ppm Wood	Ppm Ash	Ppm Wood	Ppm Ash	Ppm Wood	Ppm Ash	Ppm Wood
4	23A, 31, 37, 39	West	Basaltic	Mean	0.28	—	—	—	—	—	—	0.4	< 0.01	—	—	—	—	—	—
				σ	0.03	x	x	x	x	x	x	x	x	x	x	x	x	x	x
				σ_x	0.02	x	x	x	x	x	x	x	x	x	x	x	x	x	x
4	4, 10, 14, 60	West	Granitic	Mean	0.42	—	—	—	—	—	—	4.4	0.02	—	—	2.0	0.01	—	—
				σ	0.22	x	x	x	x	x	x	2.4	0.02	x	x	x	x	x	x
				σ_x	0.13	x	x	x	x	x	x	1.4	0.01	x	x	x	x	x	x
8	34, 45, 52, 57, 83, 88, 92, 96	East	Basaltic (& loess)	Mean	0.41	—	—	—	—	4.1	0.01	15.0	0.09	42.5	0.15	—	—	1.0	< 0.01
				σ	0.12	x	x	x	x	8.3	0.03	24.8	0.17	37.7	0.13	x	x	x	x
				σ_x	0.05	x	x	x	x	3.1	0.01	9.4	0.06	14.2	0.15	x	x	x	x
8	64, 72, 74, 85, 90, 103, 107, 109	East	Granitic	Mean	0.54	9.6	0.03	3.1	0.01	1.0	< 0.01	10.5	0.03	50.2	0.28	—	—	1.0	< 0.01
				σ	0.52	x	x	x	x	x	x	9.1	0.02	23.3	0.09	x	x	x	x
				σ_x	0.20	x	x	x	x	x	x	3.4	0.01	8.8	< 0.01	x	x	x	x

*Soils from basaltic or granitic parent materials, often topped with glacial drift west of Cascades and with loess east of Cascades.

xSignifies not determined.

—Indicates not found or too little to detect.

< Signifies "less than" reported.

TABLE 8

GRAND FIR ASH AND PHYSICAL FEATURES: SIGNIFICANCE OF DIFFERENCES BETWEEN MEANS

I. MAJOR CONSTITUENTS AND ESSENTIAL ELEMENTS*

		Buttwood West and East of Cascades					Wood Grown on Basaltic and Granitic Soils, East of Cascades					Sound and Rotten Wood, East of Cascades†				
		Mean West	Mean East	τ	LDF	Signif. Level %	Mean Basalt	Mean Granite	τ	LDF	Signif. Level %	Mean Sound	Mean Rotten	τ	LDF	Signif. Level %
	No. samples	26	34				8	8				8	9			
Major constituents	Ash % of dry wt.	0.40	0.40	x			0.41	0.54	0.63	7	30	0.77	3.40	3.73	7	99
	Ca % of ash	14.8	21.4	3.92	11	99	21.2	22.2	0.35	7	20	16.5	18.1	0.37	4	20
	Ca ppm dry wt.	549	872	3.45	11	99	979	758	0.82	7	50	914	7747	3.38	4	95
	Mg % of ash	4.9	3.9	4.59	11	99	3.6	3.4	0.40	7	30	4.4	5.0	0.22	4	10
	Mg ppm dry wt.	200	154	1.44	11	80	152	114	1.51	7	80	254	1704	5.10	4	99
	K % of ash	24.3	12.0	3.96	11	99	11.6	11.0	0.25	7	10	14.1	11.5	0.77	4	50
	K ppm dry wt.	1092	468	7.23	11	99	455	360	1.55	7	80	790	3990	3.71	4	95
	Na % of ash	1.0	0.4	2.86	11	95	0.2	0.2	x			0.2	0.1	1.00	4	50
	Na ppm dry wt.	40	14	2.70	11	95	12	7	1.15	7	70	10	51	3.73	4	95
Essential elements	Ash % of dry wt.	0.33	0.45	1.80	25	90	0.41	0.54	0.63	7	30	0.77	3.40	3.73	7	99
	B ppm ash	256	235	0.34	25	20	172	358	2.01	7	90	68	—	x		
	B ppm dry wt.	0.9	0.9	0.35	25	20	0.6	1.4	2.53	7	95	0.6	—	x		
	Mn ppm ash	6144	3961	3.37	25	99	3475	4585	1.50	7	80	2440	2880	0.51	7	30
	Mn ppm dry wt.	21.1	17.9	0.97	25	50	12.9	26.4	1.20	7	70	22.4	100.5	2.81	7	95
	Fe ppm ash	656	707	0.80	25	50	707	715	0.05	7	0	500	375	1.03	7	50
	Fe ppm dry wt.	2.3	2.9	2.40	25	95	2.7	3.1	0.69	7	30	3.3	11.0	3.72	7	99
	Mo ppm ash	0.5	1.6	1.88	25	90	0.4	2.4	1.47	7	80	0.8	0.01	x		
	Mo ppm dry wt.	< 0.01	0.01	17.80	25	99	< 0.01	0.01	x			0.33	0.01	x		
	Cu ppm ash	777	565	2.93	25	99	516	612	0.66	7	30	338	186	2.03	7	90
	Cu ppm dry wt.	2.8	2.3	1.60	25	80	1.9	2.7	1.27	7	70	2.4	4.9	1.81	7	50
	Zn ppm ash	72	385	3.81	25	99	482	450	0.12	7	0	32	197	x		
	Zn ppm dry wt.	0.24	1.44	3.94	25	99	1.8	1.6	0.23	7	10	0.12	4.70	x		
Physical features	Specific gravity	0.467	0.462	0.42	25	30	0.483	0.448	0.92	7	50	0.444	0.415	0.66	7	30
	Rings per inch	9.8	18.3	3.67	25	99	17.6	15.8	0.31	7	20	x	x			
	Diam., inches	17.9	10.7	2.61	25	95	12.1	9.0	1.32	7	70	x	x			
	Age, years	78	92	0.94	25	50	97	80	0.91	7	50	x	x			
	Height, feet	76	61	1.71	25	80	69	49	1.83	7	80	x	x			

*Statistical data: $\tau = \frac{\text{diff. in means}}{SE_{\text{diff.}}}$.

LDF is lesser degrees of freedom.

Significance level % indicates the probability that a difference in means is real and not due alone to chance (many authors list values below 50% as "no difference").

†All decay by *Echinodontium tinctorium*.

—Signifies not found or too little to detect.

xSignifies value not calculated because of too few samples, or that the means are identical.

< Signifies value "less than" reported.

GRAND FIR ASH AND PHYSICAL FEATURES: SIGNIFICANCE OF DIFFERENCES BETWEEN MEANS
II. OTHER COMMON ELEMENTS AND UNCOMMON ELEMENTS*

		BUTTWOOD WEST AND EAST OF CASCADES					WOOD GROWN ON BASALTIC AND GRANITIC SOILS, EAST OF CASCADES					SOUND AND ROTTEN WOOD, EAST OF CASCADES†				
		Mean West	Mean East	τ	LDF	Signif. Level %	Mean Basalt	Mean Granite	τ	LDF	Signif. Level %	Mean Sound	Mean Rotten	τ	LDF	Signif. Level %
	No samples	26	34				8	8				8	9			
Other common elements	Ash % of dry wt.	0.33	0.45	1.80	25	90	0.41	0.54	0.63	7	30	0.77	3.40	3.73	7	99
	Ag ppm ash	8	71	2.51	25	95	16	99	2.12	7	90	45	59	1.13	7	70
	Ag ppm dry wt.	0.03	0.38	2.02	25	90	0.06	0.81	1.29	7	70	0.7	1.7	1.28	7	70
	Al ppm ash	1316	1377	0.24	25	10	1740	1218	0.83	7	50	1560	1604	0.06	7	0
	Al ppm dry wt.	5.0	5.7	0.74	25	30	6.6	5.2	0.65	7	30	12.8	43.2	3.00	7	95
	Ba ppm ash	2924	5956	4.27	25	99	4355	7150	1.79	7	80	4898	7493	1.47	7	80
	Ba ppm dry wt.	10.2	27.8	3.07	25	99	16.0	43.4	1.27	7	70	45.8	254.3	2.94	7	95
	Co ppm ash	2	6	1.06	25	99	19	3	1.36	7	70	2	2	x		
	Co ppm dry wt.	0.01	0.02	15.00	25	99	0.07	0.01	1.40	7	70	0.02	0.07	1.60	7	80
	Cr ppm ash	12	14	0.36	25	20	19	25	0.39	7	20	9	3	2.34	7	90
	Cr ppm dry wt.	0.04	0.06	1.32	25	80	0.06	0.10	0.80	7	50	0.07	0.14	1.44	7	80
	Ni ppm ash	68	21	3.50	25	99	39	13	1.15	7	70	9	7	0.76	7	50
	Ni ppm dry wt.	0.15	0.08	1.72	25	90	0.13	0.06	0.86	7	50	0.06	0.18	2.48	7	95
	Pb ppm ash	16	43	1.14	25	70	17	33	1.14	7	70	3	5	0.46	7	30
	Pb ppm dry wt.	0.05	0.17	1.26	25	70	0.06	0.14	1.81	7	80	0.02	0.08	1.20	7	70
	Rb ppm ash	399	551	1.42	25	80	780	389	1.59	7	80	361	332	0.27	7	20
	Rb ppm dry wt.	1.7	2.2	0.87	25	50	2.7	1.3	0.96	7	50	2.5	9.2	3.81	7	99
	Sr ppm ash	869	3806	5.99	25	99	2660	5094	1.52	7	80	3126	3150	0.02	7	0
	Sr ppm dry wt.	3.2	15.6	6.48	25	99	10.3	18.6	1.80	7	80	25.4	101.2	2.59	7	95
	Ti ppm ash	36	20	2.37	25	95	17	25	1.22	7	70	18	123	1.22	7	70
	Ti ppm dry wt.	0.14	0.08	2.21	25	95	0.06	0.09	2.14	7	90	0.13	2.82	1.29	7	70
Uncommon elements‡	Ash % of dry wt.	0.33	0.45	1.80	25	90	0.41	0.54	0.63	7	30	0.77	3.40	3.73	7	99
	Au ppm ash	1	10	1.68	25	80	—	9.6	x			—	63	x		
	Au ppm dry wt.	< 0.01	0.04	0.92	25	50	—	0.03	x			—	2.0	x		
	Ga ppm ash	—	1	x			—	3	x			—	—	x		
	Ga ppm dry wt.	—	< 0.01	x			—	0.01	x			—	—	x		
	In ppm ash	12	4	2.40	25	95	4	1	1.00	7	50	4	—	x		
	In ppm dry wt.	0.05	0.01	1.00	25	50	0.01	< 0.01	0.50	7	30	0.02	—	x		
	La ppm ash	4	14	2.89	25	99	15	10	0.45	7	30	6	5	0.47	7	30
	La ppm dry wt.	0.02	0.06	2.45	25	95	0.09	0.03	1.00	7	50	0.04	0.13	0.96	7	50
	Li ppm ash	2	44	2.13	25	95	42	50	0.46	7	30	42	41	0.06	7	0
	Li ppm dry wt.	0.01	0.20	3.31	25	99	0.15	0.28	2.55	7	95	0.31	1.32	1.89	7	90
	Sn ppm ash	1	1	0.75	25	50	—	—	x			1	—	x		
	Sn ppm dry wt.	< 0.01	0.01	0.79	25	50	—	—	x			0.01	—	x		
	Zr ppm ash	—	1	x			1	1	x			3	6	0.59	7	30
	Zr ppm dry wt.	—	< 0.01	x			< 0.01	< 0.01	x			0.02	0.13	1.38	7	70

*Statistical data: $\tau = \frac{\text{diff. in means}}{SE_{\text{diff.}}}$.

LDF is lesser degrees of freedom.

Significance level % indicates the probability that a difference in means is real and not due alone to chance (many authors list values below 50% as "no difference").

†All decay by *Echinodontium tinctorium*.

—Signifies not found or too little to detect.

xSignifies value not calculated because of too few samples, or that the means are identical.

‡No V found.

< Signifies value "less than" reported.

TABLE 10

RATIOS OF SPECIFIC GRAVITIES AND ASH CONTENTS OF SOUND AND ROTTEN WOOD OF GRAND FIR*

Area	Type of Rot†	Tree No.	Sp. Gr.‡	Ash	Ratio	Ratio
				% of Dry Wt.	Sp. Gr. Rotten to Sp. Gr. Sound	Ash % Dry Wt. Rotten to Ash % Dry Wt. Sound
West	Sound	3	0.441	0.311	1.00	3.21
	Saprot, firm	3	0.442	0.997		
	Sound	22	0.537	0.486	0.79	1.92
	Saprot, firm	22	0.426	0.934		
	Sound	42	0.551	0.427	0.28	2.96
	Saprot, soft, spongy	42	0.152	1.259		
	Sound	62-52′	0.494	0.330	0.92	7.82
	White rot, firm	62-52′	0.455	2.579		
					Mean 0.75	Mean 3.63
East	Sound	52-15′	0.452	0.361	0.71	6.84
	E.t. soft, advanced	52-15′	0.322	2.471		
	Sound	74	0.579	0.734	0.71	4.93
	E.t. soft to med., adv.	74	0.410	3.620		
	Sound	82	0.410	0.630	1.09	3.98
	E.t. soft, advanced	82	0.447	2.513		
	Sound	82-32′	0.357	0.794	1.25	8.57
	E.t. soft, advanced	82-32′	0.437	6.768		
	Sound	82-55′	0.362	0.585	1.27	11.93
	E.t. medium, adv.	82-55′	0.461	6.981		
	Sound	95	0.534	0.382	0.60	5.97
	E.t. very soft, adv.	95	0.323	2.284		
	Sound	96	0.489	0.379	1.06	5.25
	E.t. firm, advanced	96	0.518	1.993		
	Sound	107-28′	0.414	0.524	1.02	4.75
	E.t. soft, advanced	107-28′	0.422	2.470		
	Sound	111-70′	0.372	1.058	1.06	1.35
	E.t. firm, advanced	111-70′	0.374	1.434		
					Mean 0.98	Mean 5.95

*Sound and rotten wood samples chosen as matched radial pairs.
†E.t. signifies decay by *Echinodontium tinctorium*.
‡Specific gravity based on oven-dry weight and oven-dry volume.

2. *Mineral composition of sound-wood ash for 26 grand fir trees grown west of the Cascades* (table 4)

Mean values, ranges of values, standard deviations, and standard errors of means are given for 27 elements classified as major constituents, essential elements, other common elements, and uncommon elements (see table 1 for classification). These results, with their statistical measures, show inherent ranges of variability and reliability of means. The higher Na values for trees 2, 21, 60, and 62 may reflect the fact that these trees all grew close to salt water. Numerous examples of extreme variability are found, as illustrated by σ (standard deviation) values of magnitudes which approach or exceed mean values.

3. *Mineral composition of sound-wood ash for 34 grand fir trees grown east of the Cascades* (table 5)

Inherent variability and reliability of mean values compare closely with values of table 4 for trees grown west of the Cascades. Differences between the two areas will be pointed out subsequently.

4. *Mineral composition of sound-wood ash for grand fir trees grown on basaltic and granitic soils* (table 7)

Major differences in many elements, as a function of parent soil material upon which the trees grew, are apparent. Other than for percentage of ash, few consistent differences are noticeable. The frequent occurrence of glacial overburden west of the Cascades and wind-blown loess east of the Cascades may obscure greater and/or more consistent differences.

5. *Significance of differences between means of grand fir ash: major constituents, essential elements, and physical features* (table 8); *other common elements and uncommon elements* (table 9)

Computed mean values and statistical measures of the significance (reliability) of differences between means are summarized for three major groups of grand fir wood samples, as follows: (1) from trees grown west and east of the Cascades; (2) from trees grown on basaltic and granitic parent soils east of the Cascades; (3) from sound and decayed wood grown east of the Cascades (all decay caused by *Echinodontium tinctorium*, the Indian paint fungus).

a. Major constituents (table 8)

(1) Contents of these elements were found to be significantly different except perhaps in the case of Mg. Wood of trees grown west of the Cascades contained more Ca but less Mg, K, and Na. There was no difference in ash content.

(2) Trees grown on granitic parent soil materials showed slightly more ash, but no differences in major constituents were found to be highly significantly different between the two soil types for trees grown east of the Cascades. Differences in ash content are reflected by greater differences for the major elements, expressed as ppm of dry weight.

(3) Samples of sound and decayed wood were matched radially and were not separated until just before ashing. The increased ash content of grand fir wood decayed by the Indian paint fungus seems quite remarkable--an increase from 0.77 per cent to 3.40 per cent in 8 matched samples. Naturally, the increased ash is reflected in highly significant differences in all cases for major constituent elements when expressed as ppm of dry weight. The changes in ash content as a function of decay will be discussed more fully later.

b. Essential elements (table 8)

(1) In most cases significant differences were found between essential element contents of wood from trees grown west and east of the Cascades. For the elements B and Mn no difference was found when expressed as ppm of dry wood. Total amounts present are quite similar except that more Mn occurred in ash of trees west of the Cascades and more Zn east. The ash content was significantly different and higher in the wood from trees grown east of the Cascades.

(2) In wood from trees grown on granitic versus basaltic parent soil materials (all east of the Cascades), differences in essential element content were significant only in the case of the element B; they are probably also different for Mn, Mo, and Cu. With the exception of Zn, granitic soils grew trees with higher essential element content.

(3) The matched sound and rotten wood ashes showed significant differences in content of Mn and Fe, and are probably different in B, Mo, Zn, and Cu. In connection with the last element, Cu, it should be mentioned that the fruiting body of *Echinodontium tinctorium* is abnormally high in Cu, and this may account for the low Cu content in rotten wood owing to translocation of this element by the fungal hyphae. Surprisingly, No B was found in rotten wood ash, but the conks (fruiting bodies) showed approximately 10 times more B than sound wood; Zn

presents a similar situation. Rotten wood seems relatively high in its Fe and Mn content as compared to sound wood, and these metals may adversely affect certain properties of highly purified pulp produced from decayed wood.

c. Physical features (table 8). Table 8 shows very few significant differences between physical features of trees collected west and east of the Cascades except in rate of growth and size of the trees from which the wood samples were chosen. These differences emphasize the fact that grand fir west of the Cascades grows more rapidly and generally somewhat larger than trees of similar age growing east of these mountains. No major differences were found among 8 trees, each growing on basaltic and granitic parent soils east of the Cascades. Comment 6, below, will cover in detail differences between certain properties of matched sound and rotten wood samples.

d. Other common elements (table 9). The common occurrence of the elements considered in this category suggests their inclusion, despite the fact that the nutritional status of most of them remains to be established. It is generally believed that these elements occur widely in soils and plants but are not required or essential to plant growth.

(1) Comparison between wood ash from trees grown west and east of the Cascades reveals apparently significant differences for the elements Ag, Ba, Co, Ni, Sr, and Ti. Actual amounts of commonly occurring elements are generally higher east of the Cascades except for Ni and Ti; ash content is also greater on the east side of the mountains. Certain of these metallic elements are potentially "essential" or may substitute for proven essential elements.

(2) Few significant differences in other common elements were found between wood grown on basaltic and granitic soil types east of the mountains. Actual amounts are very variable, leading to greater difficulty in establishing true differences for the 8 wood samples in each group. The common occurrence of loessial soil topping in this region probably masks differences otherwise attributable to parent soil materials. Granitic soils seem to favor higher levels for Ag, Ba, Sr, and Ti; more Co, Ni, and Rb were found in wood grown on basaltic soils.

(3) Sound and rotten wood samples, matched as described, present a seemingly undefinitive picture with respect to other common element content. The ppm ash figures are generally comparable except for the elements Ba, Cr, and perhaps Ag and Ti. Analytical results expressed as ppm dry wt., however,

are significantly different (and higher for rotten wood) for the elements Al, Ba, Ni, Rb, Sr, and perhaps for Ag, Co, Cr, Pb, and Ti. These results imply a similar composition of these mineral elements in rotten wood but in an increasing concentration. Some selectivity may be exhibited by the fungus in accumulating certain elements, as previously discussed under major constituents and essential elements.

e. Uncommon elements (table 9). The occasional occurrence of elements in this classification illustrates the fact that plants will absorb mineral elements from the soils on which they are grown. The small number of reported analyses do not offer a highly valid basis for comparisons of wood west and east of the Cascades; of wood grown on different soil types; or for sound and rotten matched wood samples. It must be reported that vanadium (V) was consistently found in four fungus fruiting bodies (conks), in one sap-rotted wood specimen, and one sample of bark. This element has been suggested (as has the element gallium) as being essential for the growth of certain higher fungi (Bertrand, 1943). The occurrence of V in the conks and not in rotten wood (wood acted upon by the fungus mycelium forming the conk) may also indicate an accumulation by the fungus of V occurring in wood below the limits of detection (10 ppm ash).

6. *Specific gravities and ash contents of sound and rotten wood of grand fir* (table 10)

Previous reference has been made to an increase in ash in rotten (decayed) wood. One potential explanation of this increase is that loss of metabolized carbohydrates (cellulose and hemicelluloses) and polyuronides through the action of the fungus leads to a concentration of minerals that are not volatilized. To check this possibility, a series of specific gravities and ash contents of matched sound and rotten wood were made. These results, for two types of decay, are presented in table 10. It is readily apparent that decay of wood by these fungi is accompanied by an increase in ash content, and that the early stages (incipient) of decay may actually result in slightly increased specific gravity values. Hence the use of density measurements as an indicator of strength properties is rendered less useful and possibly actually misleading in cases of incipient decay from the Indian paint fungus, *Echinodontium tinctorium*.

Summary

Decay of grand fir by the Indian paint fungus, virtually unknown west of the Cascade Mountains, is the most serious cause of loss in this species growing east of these mountains. Large areas and volumes of grand fir wood are rendered almost worthless from decay, which is apparently more critical in localized areas and on basaltic soils. Recent advances in mineral nutrition and biogeochemistry suggested the desirability of more knowledge of the composition and content of mineral elements in grand fir. Collections were made from over 100 trees in the major range of grand fir. Selected samples of wood were analyzed for the elements Ca, Mg, K, and Na by flame spectrophotometry and for 33 other elements spectrographically. More than 100 spectrographic analyses were made, including analyses of buttwood, rotten wood, fruiting bodies of the fungus, bark, and other tissues.

Analytical procedure involved the direct-current arcing of ash in purified graphite electrodes on a Jarrell-Ash 21-foot grating spectrograph. Three arcings--20 seconds, 1 minute, and 3 minutes--were made on each sample, photographing the range of about 2400 to 4800 Å on Spectrum Analysis No. 1 plates. After development, spectra were examined and line intensities estimated and compared with intensities of standards prepared from highly purified "Specpure" reagent salts to provide semi-quantitative data for the elements detected.

Results of spectrographic analyses are presented in a series of tables giving ppm of elements in terms of ash and of oven-dry weight. Elements were classified as major: Ca, Mg, K, Na, determined by flame spectrophotometry, and Si by gravimetric means; essential: B, Mn, Fe, Mo, Cu, Zn; other common elements: Ag, Al, Ba, Co, Cr, Ni, Pb, Rb, Sr, Ti; and uncommon: Au, Ga, La, Li, Sn, V, and Zr. The elements C, H, O, N, S, Cl, Se, Br, I, and F were not determined; elements sought for but not found included As, Cb, Cs, Hg, Pt, Sb, Ta, Be, Ge, Pd, Tl, and perhaps Cd and Bi. Where an element is reported as missing, it may occur but be below detection limits for that element, generally less than 30 ppm of ash.

Mean values, selected statistical values, and significance of difference in means are presented for 4 selected groups of data: buttwood ash components for 26 trees west of the Cascades and 34 trees east of the Cascades; buttwood ash components for 16 trees east of the Cascades, 8 each on basaltic and granitic parent soils (often topped with loess); 8 matched sound and rottenwood ash samples from trees east of the Cascades infected with the Indian paint fungus; and wood from different heights in the same tree. Additional data are presented concerning the increase of ash and changes in specific gravity of grand fir wood infected with the Indian paint fungus and certain unidentified saprot fungi of grand fir. Results generally agree closely with published data.

Major conclusions are as follows:

1. Mineral composition at different heights in an individual grand fir tree does not vary appreciably.

2. Significant differences are evident in mean values of most elements in buttwood ash of trees grown west and east of the Cascades.

3. A few differences were established for elements detected in buttwood ash of trees grown on different parent soils within the area east of the Cascades.

4. Wood decayed by the fungus *Echinodontium tinctorium* shows an increased ash content and an apparent accumulation of certain elements; similar accumulations of certain elements were found in the fungus fruiting bodies (conks).

5. Increased ash content of grand fir wood decayed by the Indian paint fungus is not accompanied, in the early stages of decay, by corresponding decreases in specific gravity of the wood.

Methods employed in this study and results presented should prove valuable to investigations of mineral nutrition of fungi during decay of wood, to possible correlations of growth characteristics and mineral components, to measurements of fertilization effectiveness, to studies of physical and mechanical properties as a function of region of origin of wood products, and to the establishment of normal levels of occurrence of given mineral elements. The resistance of wood to attacks of Teredo and Limnoria as a function of its silica content has been established for certain species of trees. The methods of study and results outlined in this paper, if applied to species of value as piling, might reveal comparable relationships between properties of the wood, including mineral composition, and resistance to attacks of boring organisms.

LITERATURE CITED

Ahrens, L. H. 1950. Spectrochemical Analysis. Cambridge, Mass., Addison-Wesley Press.

A.S.T.M. Committee E-2 on Emission Spectroscopy. 1956. A Universal Method of Spectrochemical Analysis. ASTM Bull., 216:29-32.

Bambergs, K., and M. Knavina. 1950. Content of Mineral Substances and Micro-elements in Healthy and Inner-rotten Trees. Latvijas PSR Zinatnu Akad, Vestis, 8:37:21-30.

Bertrand, D. 1943. V. in Fungi, Especially *Amanita*. Bull. Soc. Clim. Biol., 25:194-97.

Brode, W. R. 1939. Chemical Spectroscopy. New York, John Wiley and Sons.

Campbell, W. G. 1952. The Biological Decomposition of Wood. In L. C. Wise and E. C. Jahn, eds., Wood Chemistry. 2nd ed. Vol. II, pp. 1061-1116. New York, Reinhold Publ. Corp.

Cannon, H. L., and W. H. Starrett. 1956. Botanical Prospecting for Uranium on La Ventana Mesa, Sandoval County, New Mexico. Geol. Sur. Bull., 1009-M:391-407.

Chandler, R. F., Jr. 1943. Amount and Mineral Nutrient Content of Freshly Fallen Needle Litter of Some Northeastern Conifers. Soil Sci. Soc. Am. Proc., 8:409-11.

Czapek, F. (1925). Biochemic der Pflanzen. 3 vols. Jena, G. Fischer.

Foster, J. W. 1949. Chemical Activities of Fungi. New York, Academic Press.

Foster, R. E., and R. W. Thomas. 1951. A Record of the Indian Paint Fungus on Vancouver Island. For. Path. Note 5, Dom. Dept. of Agric., Lab. For. Path., Victoria.

Gilbert, P. T., Jr. 1952. Flame Photometry--New Precision in Elemental Analysis. Indust. Lab. Beckman Inst., South Pasadena.

Goodall, D. W., and F. G. Gregory. 1947. Chemical Composition of Plants as an Index of their Nutritional Status. Imp. But. of Hort. and Plant. Crops. Tech. Comm. 17. Penglais.

Harrison, G. R., R. C. Lord, and J. R. Loofbourow. 1948. Practical Spectroscopy. New York, Prentice-Hall.

Hubert, E. E. 1955. Decay--a Problem in the Future Management of Grand Fir. Jour. For., 53:6, 409-11.

Kollman, F. 1951. Technologie des Holzes und der Holzwerkstoffe. 2nd ed. Vol. I. Berlin, J. Springer.

Lounamaa, J. 1956. Trace Elements in Plants Growing Wild on Different Rocks in Finland. A Semi-quantitative Spectrographic Survey. Ann. Bot. Soc. Zool. Bot. Fennicae "Vanamo," 29.

Lundegårdh, H. 1951. Trans. by R. L. Mitchell. Leaf Analysis. London, Hilger and Watts.

Mitchell, R. L. 1948. The Spectrographic Analysis of Soils, Plants and Related Materials. Comm. Bur. of Soil Sci., Tech. Comm. 44. Harpenden.

Nachtrieb, N. W. 1950. Principles and Practice of Spectrochemical Analysis. New York, McGraw-Hill.

Riou, P., G. Delorme, and Hormisdas. 1937. Distribution of Manganese and Iron in the Coniferae of the Province of Quebec. Compt. rend., 205:743-5.

Sawyer, R. A. 1944. Experimental Spectroscopy. New York, Prentice-Hall.

Scott, D. R. M. 1955. Amount and Chemical Composition of the Organic Matter Contributed by Overstory and Understory Vegetation to Forest Soil. Yale Univ. Sch. For. Bull., 62.

Stoate, T. N. 1950. Nutrition of the Pine. Austral. For. and Tmbr. Bur. Bull., 30. Canberra.

Vanselow, A. P., and G. R. Bradford. 1957. Techniques and Applications of Spectroscopy in Plant Nutrition Studies. Soil Sci., 75-83.

-------, and G. F. Liebig, Jr. 1948. Spectrochemical methods for the Determination of Minor Elements in Plants, Waters, Chemicals, and Culture Media. Univ. of Calif. Agr. Expt. Sta. Berkeley. (Mimeographed.)

Warren, H. V., R. E. Delavault, and Ruth Irish. 1952. Biogeochemical Investigations in the Pacific Northwest. Bull. Geol. Soc. Am., 63:435-84.

Wilcox, H. E. 1940. The Spectrographic Analysis of White Pine. Unpubl. M.S. thesis, New York State College of Forestry.

Wise, L. E. 1952. Miscellaneous Extraneous Components of Wood. In L. E. Wise and E. C. Jahn, eds., Wood Chemistry. 2nd ed. Vol. I, pp. 641-65. New York, Reinhold Publ. Corp.

-------, Ruth C. Rittenhouse, E. E. Dickey, O. H. Olson, and C. Garcia. 1952. The Chemical Composition of Tropical Woods. J. For. Prod. Res. Soc., II (5): 227-49.

Wolf, F. A., and F. T. Wolf. 1947. The Fungi. Vol. II. New York, John Wiley and Sons.

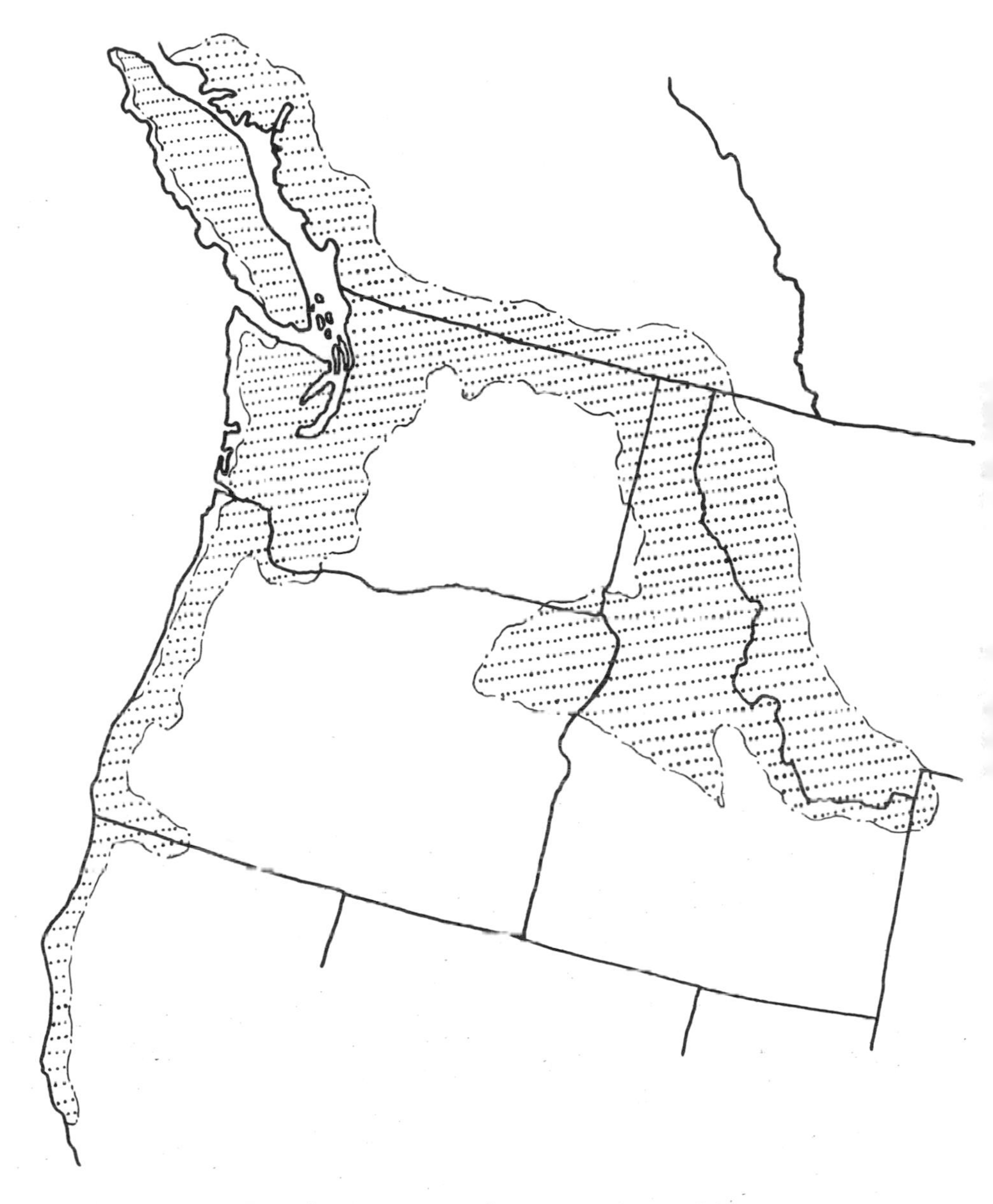

Fig. 1. Approximate range of grand fir

MINERAL ABSORPTION, TRANSPORT, AND DISTRIBUTION IN THE TISSUES OF HIGHER PLANTS

Richard B. Walker

Data presented in the preceding paper by Everett Ellis illustrate the differences that may exist in the mineral composition of wood samples collected from a particular species in a variety of habitats. Dr. Ellis offered the hypothesis that resistance of woody tissues to the attacks of organisms may be influenced by their mineral composition. If this hypothesis is valid, then the mechanisms of mineral absorption and translocation, and other factors that influence mineral composition of wood, become of importance. In this paper, I shall attempt to outline the general nature of ion uptake by higher plants, with emphasis upon woody plants and upon specific ions, depending upon the information available. It must be admitted, however, that most of the information available in this field has been derived from experiments or observations upon crop plants (mostly annuals), and that the emphasis in such experimentation generally has been placed upon elements which are essential to the growth of plants.

Mineral Absorption

The usual medium for the growth of higher plants, the soil, is a complex substrate from which some 30 to 40 elements may be absorbed by plants in amounts which are readily detectable. Of these, only 13 are known to be essential, in addition to C, H, and and O (table 1), but a number of others are often absorbed in relatively large amounts (Al, Si, Na), and others absorbed in lesser quantities may readily influence either the metabolism of the plant or the establishment of parasites or saprophytes in its tissues.

This complex of elements may exist in a number of forms in the soil, as shown in figure 1. The reactions and equilibria in-

TABLE 1
ELEMENTS ABSORBED BY PLANTS FROM SOILS

1. Essential elements required in relatively large amounts (macronutrients)	N, P, K, Ca, Mg, S.
2. Essential elements required in trace amounts (micronutrients)	Fe, B, Mn, Zn, Cu, Mo, Cl.
3. Other elements usually present in measurable amounts in plant tissues*	Al, As, Ba, I, Na, Ni, Pb, Si, Ti.

*Many other elements may often be present, especially in plants growing on soils which are peculiarly high in particular elements, e.g., Se, Cr, Ag, F, etc.

volved may be affected by pH, by additions of fertilizers or other other amendments, and by the excretion from roots of CO_2 and perhaps organic acids. Traditionally, it has been considered that absorption by roots occurs principally from the so-called soil solution, in which the concentration of solutes is relatively low (usually equivalent to the order of 0.5 atmospheres or less in osmotic pressure). However, there is much evidence that plants may absorb as well by direct exchange of ions adsorbed on colloidal surfaces of the soil for H^+ or anions (perhaps HCO_3^-) attached to the root surface. The pH of the soil is governed primarily by the proportion of H^+ to other adsorbed cations on the soil colloids, and by the nature of the colloids themselves. The solubility and availability for absorption of a number of ions is markedly affected by pH. In particular, at pHs below 5, the availability of Al, Fe, Mn, Cu, Ni, Cr, and certain other ions is much higher than in the neutral or alkaline range. On the other other hand, Mo is more available in the alkaline range.

Considerable mineral absorption may occur by diffusion if the concentration of an absorbable ion remains lower in the root cells than in the soil. Precipitation or binding of an ion in the cells, or its rapid translocation from the root into the shoot, can serve to maintain such a gradient in concentrations. In many cases, however, the concentrations of absorbable ions are found to be *higher in the roots* than in the soil. The development of such concentrations internally and continued absorption in the face of adverse diffusion gradients is possible only through expenditure of energy by the root cells. Although the general nature of such a process of ion accumulation by means of energy use is well known, the manner in which the energy is applied and the ions absorbed has not been elucidated in spite of intensive research on the subject (see Epstein, 1956, for progress in this aspect).

A number of external and internal factors influence the absorption process, the principal ones of which are (Hoagland, 1944; Broyer, 1951): (1) temperature, (2) oxygen tension in the soil, (3) carbohydrate supply in the roots, (4) internal and external concentrations of the ion in question, (5) the nature of the specific ion being absorbed, and (6) the species (and often the variety) of the plant. The first three of these affect primarily the amount of energy available through cellular respiration. The fourth reflects the previous absorption by the plant and, as well, the nature of and any manipulation (e.g., fertilization) of the medium. An example of the effect on absorption of the levels of a nonnutrient ion in the medium is seen in unpublished data of our laboratory (table 2).

TABLE 2
THE EFFECT OF THE NICKEL CONCENTRATION IN THE CULTURE SOLUTION ON THE CONTENT OF NICKEL IN TOMATO LEAVES
(Walker, unpublished)

Ni^{++} in Culture Solution (ppm)	Ni in dry leaf tissue (ppm)
none added	2.0
0.05	4.1
1.0	89.

Some ions enter plants much more readily than others. Among those known to enter readily are K^+, NH_4^+, NO_3^-, and Cl^-, whereas Ca^{++} and $SO_4^=$ generally penetrate much more slowly. Although permeability probably is the principal factor controlling ease of penetration, ion competition is often involved as well. An example of the latter is seen in the absorption of the nutrient cations. As a general rule, the sum of Ca^+ + Mg^+ + K^+ (+Na in some species) is nearly constant in leaf tissue, regardless of the levels of the individual cations. This tendency for the total cation content to be similar regardless of composition has been called "cation-equivalent constancy" by Bear and Prince (1945). The data in table 3 illustrate this principle, as wide fluctuations in calcium and magnesium contents have little effect on the total cation contents of the leaves. Many other examples are known of the level in the medium of one ion affecting the absorption of another.

TABLE 3

EFFECT OF THE Ca/Mg RATIO IN THE CULTURE SOLUTION ON THE CATION CONTENT AND THE SUM OF CATIONS IN THE LEAF TISSUE OF SUNFLOWER, *HELIANTHUS ANNUUS* (Walker, Walker, and Ashworth, 1955)

Ca/Mg Ratio	Cations in dry leaf tissue (meq/100 gm)			
	Ca	Mg	K	Sum
2:1	56	18	139	213
1:11	16	48	126	190
1:19	13	65	141	219

Also, plant species and varieties may vary greatly in their abilities to absorb ions even from the same or comparable substrates. For example, Collander (1941) found in a classical experiment that ability to accumulate sodium from the same culture solution varied greatly among the 21 species tested, whereas accumulations of potassium and rubidium were roughly parallel to each other but neither differed by a factor of more than 2 or 3 among the species tested (table 4). He similarly found no striking over-all trends in calcium or magnesium absorption, although he made the interesting observation that all members of the family Chenopodiaceae (Atriplex, Salicornia, Salsola, and Spinacia) accumulated particularly large amounts of magnesium. Similar differences between species have been observed in absorption of ions from soils. For example, from soils in which exchangeable calcium is low, some species may absorb little calcium and become calcium-deficient, whereas other species may absorb more calcium and grow well at the same low calcium

TABLE 4
ABSORPTION OF SODIUM, POTASSIUM, AND RUBIDIUM FROM THE SAME CULTURE SOLUTION* BY DIFFERENT SPECIES LISTED IN ORDER OF INCREASING SODIUM ABSORPTION
(Collander, 1941)

Plant Species	% of Total Cations Absorbed		
	Na	K	Rb
Buckwheat *(Fagopyrum esculentum)*	0.5	19	18
Corn *(Zea mays)*	0.7	31	28
Sunflower *(Helianthus annuus)*	0.7	24	29
Goosefoot *(Chenopodium bonus-Henricus)*	0.9	23	27
Russian thistle *(Salsola kali)*	1.3	16	15
Pea *(Pisum sativa)*	1.4	23	27
Tobacco *(Nicotiana tabacum)*	3.5	28	28
Tomato *(Solanum lycopersicum)*	3.5	15	22
Spinach *(Spinacia oleracea)*	4.6	23	24
Oats *(Avena sativa)*	4.7	25	41
Aster *(Aster tripolium)*	5.2	27	30
Poppy *(Papaver somniferum)*	5.4	25	37
Lettuce *(Lactuca sativa)*	5.7	24	34
English plantain *(Plantago lanceolata)*	6.9	22	30
Sweet clover *(Melilotus albus)*	8.4	22	25
Vetch *(Vicia sativa)*	8.6	22	28
Saltbush *(Atriplex litorale)*	10.3	11	20
Mustard *(Sinapsis alba)*	10.9	18	26
Glasswort *(Salicornia herbacea)*	12.5	11	13
Plantain *(Plantago maritima)*	16.3	12	12
Saltbush *(Atriplex hortense)*	27.9	12	14

*The solution contained 2 mgm equivalents per liter of each of the following: Na, K, Rb, Mg, and Ca. Traces of Li, Mn, and Sr were also present. Analyses were made on the tops of the plants only.

level. The data of table 5 illustrate such a case in which a native sunflower of California (*Helianthus bolanderi*) absorbed more calcium and yielded better than the common sunflower (*H. annuus*) at the lower calcium levels. Although these examples have emphasized the absorption of cations, differences in the tendencies of plants to accumulate anions have been demon-

TABLE 5
YIELD AND Ca ABSORPTION OF TWO SUNFLOWER SPECIES AS AFFECTED BY % EXCHANGEABLE Ca IN THE SOIL
(Walker, Walker, and Ashworth, 1955)

	Helianthus annuus		*Helianthus bolanderi*	
% Ca in Soil Cations	Average Yield (gm dry wt)	Ca in dry leaf tissue (meq/100 gm)	Average Yield (gm dry wt)	Ca in dry leaf tissue (meq/100 gm)
6.1	0.31	22	2.75	32
7.3	0.47	18	3.30	40
10.9	2.00	17	2.44	54
13.5	3.26	24	2.80	65
18.4	4.51	38	3.07	78
25	5.46	58	2.50	95

strated involving diverse ions such as nitrate, molybdate, chloride, fluoride, silicate, phosphate, and selenate. Such variations in the ability to absorb cations or anions may have marked effects on the geographic distribution of different species, and influence as well the resistance of wood or other plant tissues to the attacks of destructive organisms.

Although we have been dealing with mineral absorption in general, perhaps a more specific consideration of some anatomical aspects of such absorption would be useful. Most mineral absorption occurs through the terminal portions of roots, usually involving a few centimeters of the root tip at the most. Absorption in somewhat older regions of roots is probably inhibited by thickenings on the cell walls of the endodermis. At the extreme tip of the root, the tissues are not differentiated, but usually within a few millimeters of the tip there is maturation into epidermis, cortex, endodermis, and a central core of conducting tissue (fig. 2). In this maturation region, some epidermal cells may have long protuberances called root hairs. The root hairs probably have more influence on water uptake than in mineral absorption. The xylem of the small roots is part of a continuous system that extends upward through the larger roots, the sapwood of the stem, and out into the branches and leaves. It is this system through which minerals and water are conducted throughout the plant. In brief, mineral absorption is the movement of mineral ions from the soil into the xylem of the small rootlets. Classically, plant physiologists have thought of this process in terms of uptake by the epidermal cells followed by transfer, cell by cell, across the cortex region and endodermis, and then into the xylem. Although this mode of transfer must occur, it probably cannot account for all absorption. In recent years evidence has accumulated that ions of the external medium may penetrate freely and rather quickly via the cell walls or even via the cytoplasm and plasmodesmata through the epidermis and cortex, and perhaps into the xylem itself. Prerequisites for this sort of passage through the cytoplasm are the presence of cytoplasmic connections between cells (plasmodesmata), and a relatively free permeation of ions through the outer cytoplasmic membrane. It is quite possible that the absorption of elements such as silicon, aluminum, and others which do not penetrate readily into vacuoles may occur via the so-called "symplast" of plasmodesmata-connected cytoplasm. Broyer (1950) has discussed this possible mechanism in some detail.

Transport and Distribution in the Plant

Once into the xylem at the root level, ions are carried upward in the xylem system and distributed into the stem, branches, and leaves. This movement may be by diffusion but is often more rapid than diffusion as the ions are carried along in the conducting cells of the xylem by the "transpiration stream," i.e., the water moving upward in replacement of water lost by evaporation (transpiration) from leaves and small branches. All tissues of the shoot are supplied to some extent with minerals, but the minerals have a strong tendency to be concentrated in the young, actively growing tissue of the buds and immature leaves, presumably because of the high respiratory rate and ability to absorb ions actively in these young tissues. Even in the mature leaves, there is a tendency for ions to collect as they reach the upper terminus of the xylem system. There is a paucity of information on the mineral composition of different portions of large trees, but the data from Gäumann (1935) listed in table 6 indicates the usual pattern. Mineral content of a tissue region can be roughly associated with the number and activity of the living cells present. The heartwood consists almost entirely of dead cells and is the lowest in mineral content. The sapwood, active in mineral and water conduction, contains a small but significant number of living cells in the rays and is usually somewhat higher in minerals than the heartwood. The bark contains larger amounts of minerals than wood because the inner bark consists

TABLE 6
MINERAL CONTENTS OF VARIOUS PARTS OF 110-YEAR-OLD BEECH TREE, *FAGUS SYLVATICA*; AVERAGE VALUES FOR SEVERAL TREES CUT IN DIFFERENT MONTHS
(Gäumann, 1935)

Part of Tree	Estimated Dry Weight (Kg)	Contents of Elements (% of dry weight)					
		N Total	P	K	Ca	Mg	SiO_2
Leaves	28	2.6	0.13	0.95	0.83	0.17	1.52
Twigs	10	0.88	0.070	0.31	1.76	0.096	0.40
Branches (< 7 cm diam)							
wood	130	0.22	0.02	0.13	0.12	0.066	0.03
bark	10	0.01	0.070	0.38	2.47	0.11	0.60
Stems (> 7 cm diam)							
sapwood	300	0.15	0.013	0.091	0.11	0.025	0.03
heartwood	1100	0.13	0.009	0.083	0.12	0.025	0.05
bark	75	0.86	0.065	0.41	2.82	0.097	0.38
Roots							
wood	130	0.20	0.017	0.15	0.15	0.036	0.03
bark	20	0.82	0.022	0.14	2.74	0.078	1.66

mostly of the living cells of the phloem. Leaves and buds are made up almost entirely of active living cells and thus are the highest in mineral content. If an element is higher or lower than normal in one tissue or organ of a plant, it will tend to be similarly higher or lower than normal in the other tissues. Thus leaf tissue analyses (the most commonly made) can often be used to predict the relative levels to be found in the wood or other tissue.

There is a marked difference between elements in regard to their tendency to be relocated in the plant once they have been transported into an organ or tissue (table 7). Nitrogen, potassium, phosphorus, and magnesium (mobile elements) move freely through the plant, but all other elements (except sometimes sulfur) tend to remain in the location where the atoms are first incorporated into the plant (immobile elements). Often the concentrations of the immobile elements increase with age of a tissue or organ--this is especially true of calcium and silicon.

TABLE 7
ELEMENTS CLASSIFIED ACCORDING TO THEIR TENDENCIES TO BE REDISTRIBUTED IN THE PLANT

Mobile elements	N, P, K, Mg
Immobile elements	Ca, Fe, B, Mn, Cu, Zn, Si, Ni, etc.

General Considerations

It is quite possible to find natural differences in the mineral composition of different species of plants and also in the individuals of the same species growing under different conditions (Ellis, 1958; Kruckeberg, 1958). There appears to be a reasonably good possibility that the content of one or more elements in wood affects its resistance to biological deterioration. It may be unduly restrictive to limit our consideration of the effects of mineral nutrition to the mere presence of the element itself. There is some evidence that mineral nutrition can markedly affect the organic constituents of wood. In our department, Blaser and Takahashi have reported a striking effect of boron deficiency on the strength and cell wall structure of the wood of the western red cedar (*Thuja plicata*). In connection with the paper of Robert Graham (1958), the mineral nutrition may affect growth rate and percentage of sapwood and thus affect the quality of the wood with regard to penetration and retention of creosote.

Unfortunately, these possible effects of mineral nutrition have not been isolated in terms of specific elements. Perhaps extensive and carefully coordinated analyses of the spectrographic type on samples of susceptible and resistant wood could suggest which elements are of importance. Once this is established, it might be feasible, as suggested by Kruckeberg (1958) to produce the desired mineral composition by fertilization of the soil. Success in this approach would necessitate some attention to the principles involved in the absorption and redistribution of ions in the plant as outlined in the early part of this paper. Some obvious difficulties would be met in certain cases, including the slight solubility of some fertilizer materials, possible toxicities to the trees, and the cost of such a program over the long periods required for the growth of woody plants. These difficulties do not appear to be insurmountable, however.

Summary

The major principles governing the absorption and distribution of mineral elements in plants have been outlined. In general, plants absorb appreciable amounts of any element that is present in the soil in soluble form, and the amount absorbed is roughly dependent upon the concentration in the soil. Even elements such as the heavy metals that are rather toxic to plants may be absorbed in appreciable amounts. Plant species vary considerably in the amounts of different elements that are absorbed and in their sensitivity to toxic elements. Absorption of a particular element may be much affected in addition by the concentrations of other ions in the medium.

The mineral elements absorbed by plants are distributed throughout the plant body but tend to concentrate in the regions where metabolism is most active, such as the buds and young leaves. Concentrations in the wood are usually low in comparison to leaves but are nevertheless of appreciable magnitude.

The principles of mineral absorption and translocation need to be kept in mind in any attempts to produce timber of particular mineral composition.

LITERATURE CITED

Bear, F. E., and A. L. Prince. 1945. Cation-equivalent Constancy in Alfalfa. J. Am. Soc. Agron., 37:219-22.

Broyer, T. C. 1950. Further Observations on the Absorption and Translocation of Inorganic Solutes Using Radioactive Isotopes with Plants. Plant Physiol., 25:367-76.

Broyer, T. C. 1951. The Nature of the Process of Inorganic Solute Accumulation in Roots. In Truog, E. (ed.). Mineral Nutrition of Plants, pp. 187-249. Univ. of Wisconsin Press.

Collander, R. 1941. Selective Absorption of Cations by Higher Plants. Plant Physiol., 16:691-720.

Ellis, E. L. 1959. The Effects of Environment and Decay on Mineral Components of Grand Fir Wood. In this volume.

Epstein, E. 1956. Mineral Nutrition of Plants: Mechanisms of Uptake and Transport. Ann. Rev. Plant Physiol., 7:1-24.

Gäumann, E. 1935. Der Stoffhaushalt der Buche im Laufe eines Jahres. Ber. Schweiz. bot. Ges., 44:157-334.

Graham, R. D. 1959. Preservation of Douglas Fir Marine Piling. In this volume.

Hoagland, D. R. 1944. Lectures on the Inorganic Nutrition of Plants. Chronica Botanica. Lectures 3 and 4.

Kruckeberg, A. R. 1959. Ecological and Genetic Aspects of Metallic Ion Uptake by Plants and Their Possible Relation to Wood Preservation. In this volume.

Walker, R. B., Helen M. Walker, and P. R. Ashworth. 1955. Calcium-Magnesium Nutrition with Special Reference to Serpentine Soils. Plant Physiol., 30:214-21.

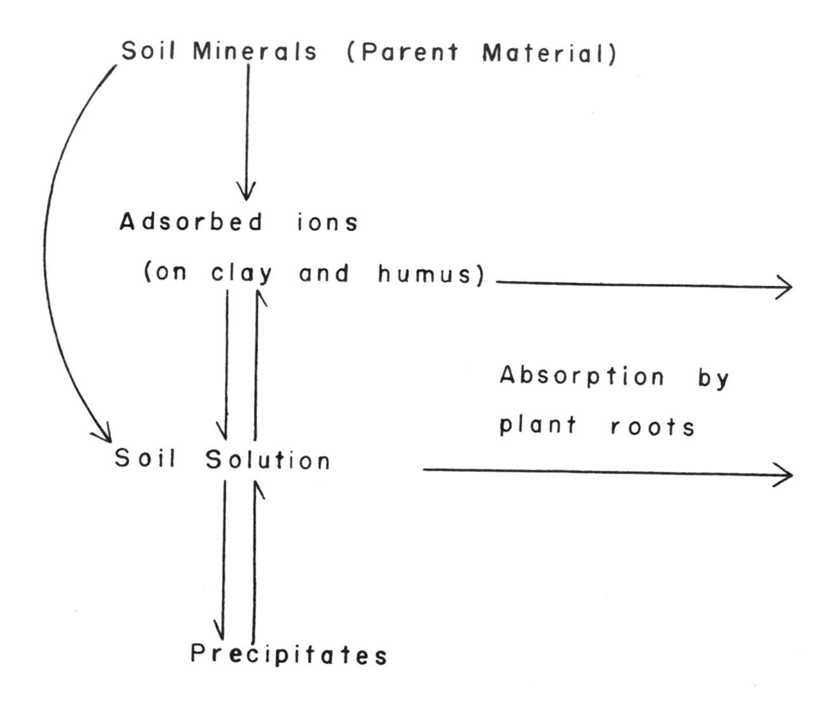

Fig. 1. Mineral element equilibria in soils

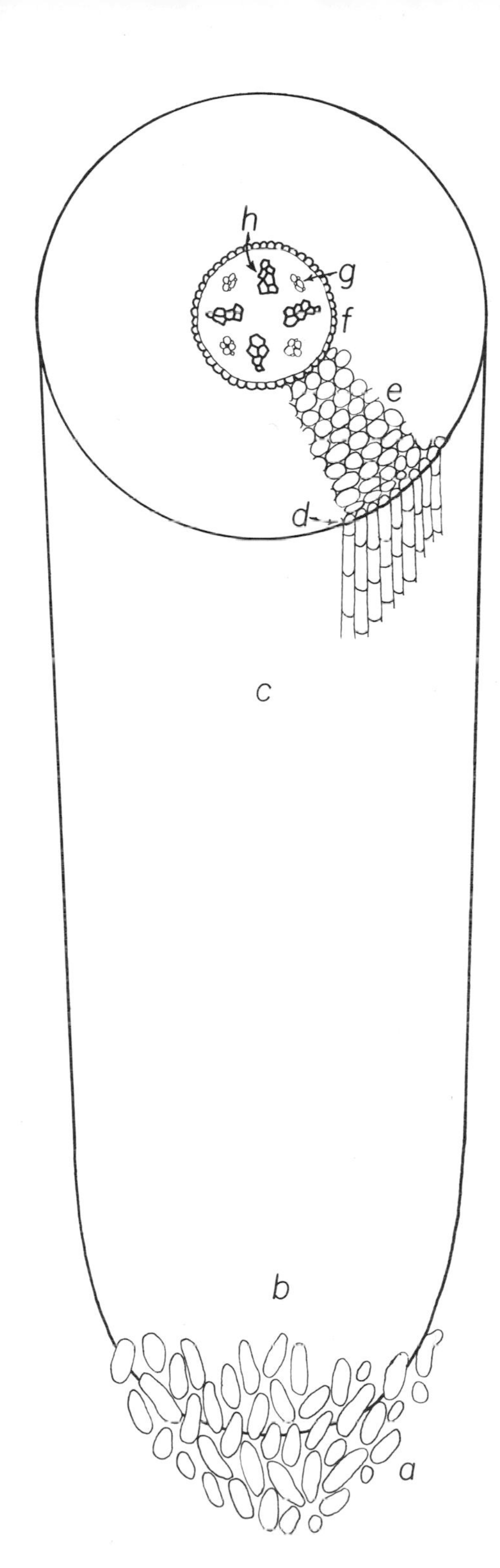

Fig. 2. Schematic diagram of the terminal few mm of a young root: a, root cap; b, meristematic region; c, elongation region; d, epidermis; e, cortex; f, endodermis; g, phloem (active in organic food conduction); h, xylem or "wood" (active in conduction of water and minerals)

ECOLOGICAL AND GENETIC ASPECTS OF METALLIC ION UPTAKE BY PLANTS AND THEIR POSSIBLE RELATION TO WOOD PRESERVATION

A. R. Kruckeberg

Man's search for methods of controlling the destruction of useful biological productivity has often been wastefully empirical. Suppression of pests by various toxic or repellent agents has usually involved a trial-and-error approach. Only in recent years has biological control--pitting one organism against another--been exploited successfully. In the field of wood preservation against marine organisms, biological control could take two mutually exclusive pathways. The first approach might involve the discovery of parasites that could eradicate the wood-destroying pests. To my knowledge, Mohr's (1958) study of commensals and parasites of Limnoria is the first attempt to explore the potentialities of this type of biological control. The other approach is that of finding, or producing by biological means, trees that are naturally resistant to marine wood-destroying organisms.

Scattered through the world's literature on the durability of wood are vague references and tantalizing "leads" to cases where differences in resistance to wood destruction have been observed between timber species, or even in the same timber species from different sources. The present paper is based on the two premises that these isolated cases of natural durability might have a basis in (1) inherent genetic differences between resistant and nonresistant trees or (2) differences traceable to the soil types on which the resistant versus nonresistant trees were growing. The following remarks explore the possibilities of utilizing both the genetic variability and the specific ecological tolerances of tree species as they may relate to the natural durability of wood. The discussion will be restricted to the genetic and ecological aspects of the absorption of inorganic elements.

The subject can best be introduced by drawing upon a relevant phase of plant ecology, namely, plants as biological indicators of specific soil types. Long before botanical science began to record the remarkable association of certain plant species with particular soil types, men in various walks of life were observing and exploiting the vegetation indicators of soils. For example, there is the story of Columbus who, having to replace a mast on a ship of his first fleet, was counseled to choose a log of pine growing on red soil in nearby Cuba; the red limonitic soils of Cuba are known to be high in iron and to have furnished durable timbers. Inevitably, then, an impressive literature has developed out of observations on the striking association of certain plants with specific soils. In many instances, the particular selective feature of the soil has been chemical in nature. It may be the exceptional infertility of the soil in its mineral constituents and/or the presence of "toxic" inorganic elements in the soil that exclude most species but are absorbed by other, more tolerant species. These are the so-called edaphic endemics, that is, species (or subspecies, races, varieties, etc.) which are confined to a single natural area as defined by some exceptional feature of its soil. Among the tolerant species are those that are both able to take up adequate amounts of the essential elements and may as well accumulate excessive amounts of elements unique to the particular soil type. Plant accumulators of such elements as Zn, Mo, Mn, Pb, Se, Cu, Ni, Cr, Al, Au, Ag, and Si have been discovered in various parts of the world.

The Hypothesis

First, let us assume the existence of woody plants capable of taking up ions that act as inhibitors. Such inhibitors might be toxic or affect the taste or mechanical properties of the wood. If absorbed by certain tree species and deposited in wood in adequate amounts, certain of these elements might present to marine boring organisms a barrier to their activity. For example, as shown by Ray (1958), certain of the heavy metal ions might act as enzyme poisons in the metabolism of the wood-destroying animals. The suggested relationship between the accumulation of soil minerals by tree species and the preservation of the wood of these species against marine boring organisms is largely in the realm of speculation.

Examination of the Hypothesis

If the above speculation is worthy of further pursuit, then the problem has a number of facets that could be developed. The balance of this paper will attempt to review both the potential and the actual sides of the problem of wood durability produced by mineral ion uptake.

We can approach the general subject of natural durability of wood and mineralization under four headings: (1) soil types with unusual amounts and/or kinds of heavy metal or other elements; (2) amounts of mineral accumulation in wood; (3) instances of exceptional wood durability traceable to mineralization of woody tissues; (4) silvicultural practices and wood durability through mineralization.

Soil types with unusual amounts and/or kinds of heavy metal or other elements

Either the parent rock or the weathering processes peculiar to the area appear to be the major sources of exceptional soil mineralization. Certain sedimentary, igneous, and metamorphic rocks contain anomalous (*high* to the geologist) quantities of certain mineral elements. In arid regions, sedimentary rocks may contain high amounts of gypsum, selenium, etc. Under tropical conditions, sedimentary rocks may be high in iron or aluminum; characteristic soils derived from tropical sedimentary rocks are the limonitic and bauxitic soils of the Caribbean Islands and adjacent South America.

One of the most spectacular correlations between vegetation and soil minerals is associated with the ultrabasic metamorphic rock, serpentine, and its parent igneous rock, peridotite. In addition to high concentrations of iron and magnesium, soils derived from these rocks may contain high amounts of chromium and nickel. Serpentine soils often support a distinctive vegetation, which may be made up in part of species endemic to this soil type. Serpentine soils may support either herbaceous flora, or scrub-type vegetation, or, less often, tree species; these floristic differences appear to depend on differences in climate and degree of weathering of the parent rock. Serpentine soils are found widely scattered in tropical and temperate parts of the world. A summary of the distribution of serpentines and their ecological characteristics is to be found in a recent symposium (Whittaker, Walker, and Kruckeberg, 1954).

The aluminum of bauxite and alumina deposits has long been

known to be selectively absorbed by certain plant species. Hutchinson (1943) and Chenery (1948) have reviewed the literature of aluminum indicator species and provide long lists of these plants. Howard and Procter (1957), making a vegetation survey of the bauxite deposits on Jamaica, describe the succession of vegetation on the mined-out pits in which certain timber species are capable of normal growth.

In the course of developing new techniques for ore prospecting, geochemists in various parts of the world have resorted to biological indicators. The "biogeochemist" has rather successfully correlated the presence of ore deposits with anomalous (high) concentrations of minerals in the tissues of the surrounding vegetation. Thus the presence of anomalous amounts of Cu, Zn, Mo, Mn, etc. in plant tissue has been taken as an indication of ore deposits of these and associated minerals. The most extensive work of this sort has been done in Canada (Warren and Irish, 1952) and in northern Europe (Lounamaa, 1956). It seems reasonable that the data and even the techniques of the biogeochemist could be used to locate areas where tree species or other vegetation are growing on heavily mineralized areas.

Amounts of mineral accumulation in wood

Mineral content in trees is usually highest in leaves and twigs and lowest in older healthy wood (Delavault, 1957; Ellis, 1956). Moreover, the values are more frequently erratic in wood, these fluctuations probably being due to translocation to other living parts of the tree. Data on the accumulation of heavy metals in woody plants, sent to me by Helen L. Cannon, geologist of the geochemical exploration section of the USGS, are especially encouraging (table 1).

TABLE 1
ACCUMULATION OF HEAVY METALS IN WOODY PLANTS*

Copper	
Normal content, 2-15 ppm dry wt.	
Unusual concentrations	
Salix fluviatilis	leaves; 192 ppm dry wt. (Bateman, 1917)
	dead bark; 1426 ppm dry wt. (Bateman, 1917)

*Compiled by H. L. Cannon, U.S. Geological Survey, Denver, Colorado.

	dead wood; 644 ppm dry wt. (Bateman, 1917)
Quercus macrocarpa	500 ppm (MacDougal, 1899)
Juniperus	bark; 150 ppm (Lehmann, 1902)
	wood; 36 ppm

Zinc

Normal content, 5-40 ppm dry wt.
Unusual concentrations

Philadelphus	12, 000 ppm in ash (USGS)
White pine	320 ppm dry wt. (USGS)
Douglas fir	200 ppm in dry wt. (USGS)
Cottonwood	590 ppm in dry wt. (USGS)
Populus tremuloides	860 ppm in dry wt. (Cannon, 1955)
Salix sp.	2, 400 ppm in dry wt. (Cannon, 1955)
Betula populifolia	500 ppm in dry wt. (Brown, 1956)

Lead

Normal content, 1-4 ppm dry wt.
Unusual concentrations

Spruce wood	1, 300 ppm in ash (Thyssen, 1942)
White pine twigs	103 ppm dry wt. (USGS)
Western birch	100 ppm dry wt. (USGS)
Douglas fir	130 ppm dry wt. (USGS)
Vaccinium canadense	81 ppm dry wt. (USGS

Uranium

Normal content, 0.5 ppm in ash
Average content in mineralized ground, 1-2 ppm in ash
Unusual concentrations

Juniperus monosperma	tips; 10 ppm in ash (USGS)
Quercus gambelii	leaves; 49 ppm in ash (USGS)
	roots at depth; 1600 ppm in ash (USGS)
	surface roots; 190 ppm in ash (USGS)
Cowania stansburiana	tips; 51 ppm in ash (USGS)

Vanadium

Normal content, 2-20 ppm in ash

Unusual concentrations

Juniperus monosperma	tips; 20-30 in ash (USGS)
	roots; 1,680 ppm in ash (USGS)
Quercus gambelii	tips; 120 ppm in ash (USGS)
	roots; 952 ppm in ash (USGS)
Cowania stansburiana	tips; 220 ppm in ash (USGS)

Cobalt

Normal content, 0.03-0.6 ppm dry wt.

Unusual concentrations

Nyssa sylvatica	116 ppm dry wt. (Beeson, 1955)
Pinus sylvestris	45 ppm dry wt. (Maliuga, 1947)

Nickel

Normal content, 0.5-4 ppm dry wt.

Unusual concentrations

Pinus sylvestris	620 ppm dry wt. (Maliuga, 1947)
Birch	200 ppm in ash (Rankama, 1940)

H. V. Warren and R. Delavault, biogeochemists at the University of British Columbia, have kindly furnished me with some of their unpublished data. Their figures (table 2) indicate the order of magnitude of mineralization (mineral content) in woody tissue. The samples were taken from tree branches of 5 to 10 years in age; some samples were from areas of high mineralization in the soil and parent rock.

TABLE 2
COPPER AND ZINC CONTENTS OF WOOD SAMPLES FROM BRITISH COLUMBIA*

No.	Ppm in Dry Matter		Ppm in Ash		Ash as % of Dry Matter
	Copper	Zinc	Copper	Zinc	
106	2	7	1400	6100	0.12
108	2.5	26	1200	12000	0.22
109	2	19	380	3500	0.53
111	1	13	420	5200	0.25
133	1.5	18	350	4100	0.43
139	0.4	5	110	1600	0.24
94	1.8	8	350	1400	0.53

*Unpublished data, H. V. Warren, Univ. of British Columbia.

Instances of exceptional wood durability traceable to mineralization of woody tissues

Other than the case of silica in wood to be discussed below, I have found no recorded case of natural durability that has been specifically traced to high mineral content in woody tissues. Instances of variation in durability have led to speculation on probable habitat differences in sources of timbers, but these differences could not be specified (Clark, 1957). However, it may be useful to make certain predictions on the basis of what is known about plant distribution. (1) On areas of exceptional mineralization, some species may take up "luxury" amounts of essential elements (Cu, Mo, Mn, Fe, Si, etc.) or unusually high amounts of "nonessential" elements. (2) Certain genetic races of a species may have the capacity for exceptional ion accumulation. (3) Other species may accumulate exceptional amounts of minerals in areas of "normal" mineralization, while associated species are unable to do so.

An instance of a genetic race tolerant of high soil mineral content is given by Bradshaw (1952). He found a population of grass, *Agrostis tenuis*, growing on the barren tailings of a lead mine. Plants of nearby meadow populations of *A. tenuis* could not be grown on the soil of the tailings, whereas the tailings biotype thrived on the lead-saturated soil. Thus both interspecific and intraspecific variability with respect to mineral absorption may be an important consideration in the search for desirable tree species.

The presence of siliceous inclusions in timbers of the southwestern Pacific area and the relation of these silica-storing trees to resistance of wood to marine borers has led to an intensive survey of tropical tree species. G. L. Amos (1952) of the Commonwealth Scientific and Industrial Research Organization in Australia records that silica inclusions have been found in over 400 tropical species in 32 botanical families; the great preponderance of these were species with porous wood. Differences in deposition of silica were often found in different species of the same genus; in some instances an entire genus proved to have siliceous or nonsiliceous wood. These facts together with the evidence that silica in wood tissue is the essential property of certain timbers resistant to teredine borers gives much promise to the utilization and silviculture of species that store silica in wood.

A spectacular instance of the relation of silica content of wood to marine borer resistance was reported by Amos and

Dadswell (1948). *Syncarpia laurifolia* Tan. (Australian turpentine), which has a wide reputation for marine borer resistance, had been reported to have low resistance when grown in Hawaii. The authors traced the differences in resistance to the high silica content of Australian-grown timbers as compared with the silica content in Hawaiian-grown timbers of this species. These observations strongly suggest that for a single species, differences in edaphic (or soil) factors can lead to wide disparities in mineral accumulation, a fact long appreciated by agriculturally oriented soil scientists.

So far as I know, no survey of silica content in the woods of temperate zone timbers has been made. It is likely, however, that intra- and interspecific differences as well as locality differences with respect to silica would be found. Thus it may not be too rash to expect that a combination of the right species (or genetic race) and the right habitat in areas of the world could result in the production of timbers that are high enough in silica to enhance their durability.

Silvicultural practices and wood durability through mineralization

If trees growing on certain mineralized soils develop thereby exceptional durability, then the forester will undoubtedly be called upon to develop silvicultural practices that will ensure a continuing crop of such timbers. A number of possibilities present themselves. In some cases it might be feasible to harvest native tree species on a site of anomalous mineral content known to furnish the desirable product; the perpetuation of this native crop would then be a paramount objective. If the site did not support the requisite tree species, then steps could be taken to find a timber crop for the site. This would involve the introduction of suitable exotic species (or selected genetic races of native species) to the site. In instances where a species accumulates unusual amounts of some mineral on "normal" (nonanomalous) soils, the alternative of harvesting native or introduced tree species still exists. Finally, there remains the totally unexplored possibility of supplying the proper minerals to trees by foliar sprays or by soil amendments. The latter is, of course, an extension of the silvicultural practice of adding fertilizers to forest soils. In this connection, an indirect line of attack is suggested by the results of unbalanced fertilizer treatments with potassium, as described by Nemec (1940). He found that on soils high in SiO_2 and poor in K_2O, one-sided fertilization with K_2O considerably increased the SiO_2 absorption in needles of pine species. This is an example of how research in soil chemistry aimed at one constituent may indirectly lead to a desirable uptake of another element.

Conclusions and Summary

Wood preservation by natural means is a many-faceted problem that may draw upon the talents of a variety of specialists in biological and physical sciences. The field is largely unexplored and yet may merit serious consideration for the future. In this brief review, a largely untested hypothesis is offered: to wit, that the mineral content in timbers of certain species or from certain soil types may be both qualitatively and quantitatively significant as a deterrent to the activities of marine boring organisms. A qualitative and quantitative knowledge of mineral constituents of timbers and soils, as well as data on barrier thresholds of the animal pests to the mineral elements, must first be developed. Then, if a significant correlation can be made between the mineral content of woods and the inhibition of the activity of marine borers, it will remain for the silviculturalist to determine the economic feasibility of the necessary forest practices.

Should interest in this approach to wood preservation ever develop, likely exploratory phases of the research would be: (1) determination of toxicity or palatability thresholds of marine borers to specific inorganic elements; (2) collection of wood samples from timbers of known resistance to borers; (3) laboratory assay of these samples for specific mineral constituents of known or suspected toxicity to marine borers.

LITERATURE CITED

Amos, G.D. 1952. Silica in Timbers. Bull. No 267, Commonwealth Scientific and Industrial Research Organization, Melbourne, Australia.

-------, and H.E. Dadswell. 1948. Siliceous Inclusions in Wood in Relation to Marine Borer Resistance. J. Council for Sci. and Ind. Res., 21:190-98.

Bateman, W.G., and L.S. Wells. 1917. Cu in the Flora of a Copper Tailings Region (Anaconda Smelter). J. Am. Chem. Soc., 39:811-19.

Beeson, K.E., V.A. Lazar, and S.C. Boyce. 1955. Some

Plant Accumulators of the Micronutrient Elements. Ecology, 36:155-56.

Bradshaw, A.D. 1952. Populations of *Agrostis tenuis* Resistant to Lead and Zinc Poisoning. Nature, 169:1098.

Brown, J.S., and P.A. Meyer. (In press.) Geochemical Prospecting as Applied by the St. Joseph Lead Co., International Congress, Mexico City.

Buck, L.J. 1949. Association of Plants and Minerals. J. New York Bot. Gard., 50:265-69.

Cannon, H.L. 1955. Geochemical Relations of Zinc-bearing Peat to the Lockport Dolomite, Orleans County, N.Y., U.S. Geol. Sur. Bull., 1000-D:119-85.

Chenery, E.M. 1948. Aluminum in Plants in Its Relation to Plant Pigments. Ann. Bot., 12:121-36.

Clark, D.H. 1957. (Oral communication.) Director, Inst. For. Prod., Univ. of Washington, Seattle.

Delavault, R.E. 1957. (Correspondence.) Research Associate, Dept. of Geology and Geography, Univ. of British Columbia, Vancouver.

Ellis, E.L. 1956. The Spectrochemical Analysis of *Abies grandis* (Dougl.) Lindl. with Particular Reference to Decay by *Echinodontium tinctorium* (Ellis) E. and E. Ph.D. thesis, Univ. of Washington, Seattle.

Howard, R.A., and G.R. Proctor. 1957. The Vegetation on Bauxitic Soils in Jamaica. J. Arnold Arboretum, 38:1-41.

Hutchinson, G.E. 1943. The Biogeochemistry of Aluminum and of Certain Related Elements. Quart. Rev. Biol., 18:1-29, 128-53, 242-62, 331-63.

Lehmann, K.B. 1902. Action of Metallic Cu on Roots. Munch. Med. Woch. Schr., 49:340.

Lounamaa, J. 1956. Trace Elements in Plants Growing Wild on Different Rocks in Finland. A Semi-quantitative Spectrographic Survey. Ann. Bot. Soc. Zool. Bot. Fennicae "Vanamo," 29.

MacDougal, D.T. 1899. Cu in Plants. Bot. Gaz., 27:68-69.

Malluga, D.P. 1947. On Soils and Plants as Prospecting Indicators for Metals: 1 zv. Akad. Nauk. USSR Ser. Geol., 3:135-38.

Mohr, John L. 1959. On the Protozoan Associates of Limnoria. In this volume.

Nemec, A. 1940. The Effect of Unbalanced Fertilization with Potash Salt and Kainite on the Nutrition of the Pine in Forest Nurseries. III, The Effect of Fertilization on the Absorption of Mn, Fe, Al, and Si. Sbornik Ceskoslov. Akad. Zemidelske, 15:86-94 (from For. Abst., 5:12).

Rankania, H. K. 1940. On the Use of the Trace Elements in Some Problems of Practical Geology. Soc. Geol. Fenlande Comptes Rendus, 14:92-106.

Ray, D. L. 1959. Nutritional Physiology of Limnoria. In this volume.

Thyssen, S. W. 1942. Geochemical and Botanical Relationships in the Light of Applied Geophysics. Beitrage zur ang. Geophys., 10:35-84.

Warren, H. V., and R. Delavault. 1955. Some Biogeochemical Investigations in Eastern Canada. Canadian Mining J., July-August, pp. 1-12.

Warren, H. V., and Ruth I. Irish. 1952. Biogeochemical Investigations in the Pacific Northwest. Bull. Geol. Soc. Am., 63: 435-84.

Whittaker, R. H., R. B. Walker, and A. R. Kruckeberg. 1954. The Ecology of Serpentine Soils: A Symposium. Ecology, 35: 258-88.

(Top) Fernald; McBee; Barnes; Colley; Graham, Mann. *(Middle)* Ellis, Dreitzler; Reese; Galler; Henry, Bousfield; Whitaker. *(Bottom)* Turner, Lane; Tracey, Greenfield; Jermyn; Cloney; Hetzel

(Top) Stuntz; Galler, Hayes, Martin; van Niel, Henry, Edmondson; Illg, Bookhout, Schmidt. *(Middle)* Kadota; Galler, Schmitz; Mohr, Becker, Bousfield, Graham; Florey. *(Bottom)* Ray; Clogston; Gonor; Quayle; Bejuki

PARTICIPANTS IN THE SYMPOSIUM

Barnes, Harold
Marine Station
Millport, Isle of Cumbrae
Scotland

Becker, Günther
Bundesanstalt für Materialprüfung
Berlin-Dahlem
Germany

Bejuki, Walter M.
Prevention of Deterioration Center
National Academy of Sciences
National Research Council
Washington, D.C.

Bookhout, C. G.
Department of Zoology
Duke University
Durham, North Carolina

Bousfield, Edward L.
National Museum of Canada
Natural History Branch
Ottawa, Ontario
Canada

Bowman, Douglas
Graduate School
Department of Zoology
University of Washington
Seattle, Washington

Carl, G. Clifford
Director, Provincial Museum
Victoria, British Columbia
Canada

Chapman, D. D.
Department of Zoology
University of Washington
Seattle, Washington

Clogston, Frederick
Graduate School
Department of Zoology
University of Washington
Seattle, Washington

Cloney, Richard A.
Graduate School
Department of Zoology
University of Washington
Seattle, Washington

Culley, Reginald A.
Technical Director, Bernuth-
Lembcke Co., Inc.
New York, New York

Connell, Joseph H.
Department of Biological Sciences
University of California
Santa Barbara College
Goleta, California

Cornwall, I. E.
Victoria, British Columbia
Canada

Davies, Dan
Wood Preservation, Technical Division
Koppers Co., Inc.
Orrville, Ohio

Dreitzler, Ralph F.
West Coast Wood Preserving Co.
Seattle, Washington

Dudley, Patricia
Department of Zoology
University of Washington
Seattle, Washington

Duncan, R. S.
General Petroleum Corp.
Ferndale Refinery
Ferndale, Washington

Edmondson, C. H.
Bernice P. Bishop Museum
Honolulu, Hawaii

Edmondson, W. T.
Department of Zoology
University of Washington
Seattle, Washington

Ellis, Everett
Department of Wood Technology
School of Natural Sciences
University of Michigan
Ann Arbor, Michigan

Fahrenbach, Wolf H.
Graduate School
Department of Zoology
University of Washington
Seattle, Washington

Fernald. Robert L.
Department of Zoology
University of Washington
Seattle, Washington

Flammer, Larry
Graduate School
Department of Zoology
University of Washington
Seattle, Washington

Florey, Ernst
Department of Zoology
University of Washington
Seattle, Washington

Galler, S. R.
Biology Branch, Code 446
Office of Naval Research
Department of the Navy
Washington, D.C.

Gonor, Jefferson J.
Graduate School
Department of Zoology
University of Washington
Seattle, Washington

Gooding, Richard U.
Graduate School
Department of Zoology
University of Washington
Seattle, Washington

Graham, Robert D.
Wood Preservation, Forest Products Laboratory
State of Oregon
Corvallis, Oregon

Greenfield, Leonard J.
Graduate School
Stanford University
Hopkins Marine Station
Pacific Grove, California

Gurd, J. M.
Timber Preservers, Ltd.
New Westminster, British Columbia
Canada

Hayes, Helen L.
Biology Branch, Code 446
Office of Naval Research
Department of the Navy
Washington, D.C.

Henry, Dora
Department of Oceanography
University of Washington
Seattle, Washington

Hetzel, Howard
Graduate School
Department of Zoology
University of Washington
Seattle, Washington

Illg, Paul
Department of Zoology
University of Washington
Seattle, Washington

Jermyn, Michael A.
Wool Textile Research Laboratories
Commonwealth Scientific and Industrial Research Organization
Melbourne, Australia

Kadota, Hajime
Laboratory of Microbiology
Department of Fisheries
Kyoto University
Maizuru, Japan

Kruckeberg, Arthur R.
Department of Botany
University of Washington
Seattle, Washington

Lane, Charles E.
The Marine Laboratory
University of Miami
Coral Gables, Florida

Lasker, Reuben
Department of Biology
Compton College
Compton, California

Lauer, Gerald J.
Graduate School
Department of Zoology
University of Washington
Seattle, Washington

Mann, Ralph H.
American Wood Preservers Institute
New York, New York

Martin, Arthur W.
Department of Zoology
University of Washington
Seattle, Washington

McBee, Richard H.
Department of Botany and Bacteriology
Montana State College
Bozeman, Montana

Menzies, Robert J.
Columbia University Lamont Geological Observatory
Torrey Cliff
Palisades, New York

Mohr, John L.
Department of Zoology
University of Southern California
Los Angeles, California

Norkrans, Birgitta
Institute of Physiological Botany
University of Uppsala
Uppsala, Sweden

Ordal, Erling J.
Department of Microbiology
University of Washington
Seattle, Washington

Osterud, K. L.
Department of Zoology
Seattle, Washington

Quayle, Daniel B.
Pacific Biological Station
Nanaimo, British Columbia
Canada

Ray, Dixy Lee
Department of Zoology
University of Washington
Seattle, Washington

Reese, Elwyn T.
Pioneering Research Division
Quartermaster Research and Development Center, U.S. Army
Natick, Massachusetts

Richardson, Frank
Department of Zoology
University of Washington
Seattle, Washington

Schmitt, Waldo L.
Smithsonian Institution
U.S. National Museum
Washington, D.C.

Schmitz, Henry
President, University of Washington
Seattle, Washington

Stovner, R. H.
General Petroleum Corp.
Ferndale Refinery
Ferndale, Washington

Strunk, Stanley
School of Medicine
University of Washington
Seattle, Washington

Stunz, D. E.
Department of Botany
University of Washington
Seattle, Washington

Sweeney, T. R.
Organic and Biological Chemistry Branch
U.S. Naval Research Laboratory
Washington, D.C.

Tracey, Michael V.
Rothamsted Experimental Station
Harpenden, Herts
England

Trussel, Paul C.
British Columbia Research Council
University of British Columbia
Vancouver, British Columbia

Tsuchiya, Henry M.
Department of Chemical Engineering
Institute of Technology
University of Minnesota
Minneapolis, Minnesota

Turner, Ruth D.
Museum of Comparative Zoology
Harvard University
Cambridge, Massachusetts

Van Niel, C. B.
Stanford University
Hopkins Marine Station
Pacific Grove, California

Walker, Richard B.
Department of Botany
University of Washington
Seattle, Washington

Whitaker, D. R.
Division of Applied Biology
National Research Council
Ottawa, Ontario
Canada

Whiteley, Arthur H.
Department of Zoology
University of Washington
Seattle, Washington

Whiteley, Helen R.
Department of Microbiology
University of Washington
Seattle, Washington

Williams, H. B.
General Petroleum Corp.
Seattle, Washington